INTRODUCTORY DC/AC ELECTRONICS

Nigel P. Cook

Prentice Hall
Englewood Cliffs, NJ 07632

Library of Congress Cataloging-in-Publication Data

COOK, NIGEL P., 1961–
    Introductory DC/AC electronics.

    Includes index.
    1. Electronics. I. Title.
TK7816.C65 1989   621.381   88–32191
ISBN   0-13-500026-2

Editorial/production supervision: *Tom Aloisi*
Interior design and Cover design: *Lorraine Mullaney*
Cover Photograph: © *Don Carroll, The Image Bank*
Manufacturing buyer: *Bob Anderson*

If your diskette is defective or damaged in transit, return it directly to Prentice-Hall at the address below for a no-charge replacement within 90 days of the date of purchase. Mail the defective diskette together with your name and address.

Prentice-Hall, Inc.
Attention: Ryan Colby
College Operations
Englewood Cliffs, NJ 07632

LIMITS OF LIABILITY AND DISCLAIMER OF WARRANTY:
The author and publisher of this book have used their best efforts in preparing this book and software. These efforts include the development, research, and testing of the theories and programs to determine their effectiveness. The author and publisher make no warranty of any kind, expressed or implied, with regard to these programs or the documentation contained in this book. The author and publisher shall not be liable in any event for incidental or consequential damages in connection with, or arising out of, the furnishing, performance, or use of these programs.

Printed in the United States of America
10 9 8 7 6 5 4 3 2

ISBN: 0-13-500026-2

Prentice-Hall International (UK) Limited, *London*
Prentice-Hall of Australia Pty. Limited, *Sydney*
Prentice-Hall Canada Inc., *Toronto*
Prentice-Hall Hispanoamerican, S.A., *Mexico*
Prentice-Hall of India Private Limited, *New Delhi*
Prentice-Hall of Japan, Inc., *Tokyo*
Simon & Schuster Asia Pte. Ltd., *Singapore*
Editora Prentice-Hall do Brasil, Ltda., *Rio de Janeiro*

*To my loving wife Dawn
and beautiful children Candice
and Jonathan, whose love
inspires me.*

*And in loving memory of my
dear father John Cook.*

DEVELOPMENT OF AN ELECTRONIC PRODUCT

Following page 454, this 16-page insert combines color photographs, graphics, and a storyboard to describe the step-by-step development of an electronic product.

The advantages of this photographic essay are twofold in that it not only illustrates how a typical electronics company would develop a piece of equipment but also it shows the responsibilities of each person and how he or she fits into this process. This gives you a clearer picture of the career opportunities available in the field of electronics.

CONTENTS

3

RESISTANCE 64

4

RESISTORS 98

SECTION II
Direct-Current Electronics

7

SERIES DC CIRCUITS 208

8

PARALLEL DC CIRCUITS 242

9

SERIES–PARALLEL DC CIRCUITS 272

10

DC TROUBLESHOOTING AND MEASUREMENT EQUIPMENT 330

SECTION III

Alternating-current Electronics

11

ALTERNATING CURRENT (AC) 366

12

AC TROUBLESHOOTING AND MEASUREMENT EQUIPMENT 414

13

ELECTROMAGNETISM (AC) 436

14

CAPACITANCE AND CAPACITORS 458

15

CAPACITIVE REACTANCE AND CAPACITOR APPLICATIONS 510

16

INDUCTANCE AND INDUCTORS 548

17

TRANSFORMERS 596

OBJECTIVES 596
INTRODUCTION 597
WOOLEN MILL MAKES MINIS 598

18

RESISTIVE, INDUCTIVE, AND CAPACITIVE CIRCUITS (*RLC*) 628

APPENDIXES 676

INDEX 797

PREFACE

Having taught introductory DC/AC electronics to thousands of students, I have acquired a considerable amount of feedback regarding their needs and responses to the introduction to electronics. This text has been designed around the experiences I have gained and is geared to meet these needs.

In all introductory texts, readability and relativity should be prime concerns to provide a complete understanding of the topics presented. However, in the pursuit of simplicity to improve readability, boredom is often the price to be paid. Students are more involved in the learning process when they understand how the subject matter relates to their personal needs, career opportunities, or interests. Consequently, logical, current, and dynamic avenues of approach are needed to generate student enthusiasm, interest, and persistence.

This text has been specifically designed around the above philosophies, with extensive component applications, a strong emphasis on troubleshooting throughout, and a well-developed bank of questions to reinforce the areas of maximum concern. Although mathematics is essential to any technical field, it has been kept to a minimum so that you can concentrate on the concepts of each topic, and only use mathematics to reinforce—not explain—concepts.

UNIQUE MOTIVATIONAL FEATURES

The special learning tools and motivational enhancements included in this text are—

- A two-color format, highlighting important illustrations and concepts.
- A 16-page full color photo essay to acquaint your students with the development of an electronics product and the career opportunities available in electronics.
- Over 1,000 functionally highlighted illustrations and photographs.
- Troubleshooting procedures and applications integrated throughout.
- Over 250 step-by-step examples to help your students with various problem solving techniques.
- Calculator sequences in many examples to introduce new mathematical procedures.

- Objectives at the beginning of each chapter that specify the chapter's goals.
- Motivational vignettes detailing some of the electronic industry's entrepreneurs.
- Self-Test Review Questions after each chapter section.
- A detailed Summary at the end of each chapter that includes a list of Important Facts, New Terms, and Key Formulas.
- End-of-chapter question banks consisting of Multiple Choice, Essay, Practice Problems, and Troubleshooting Questions.

EXTENSIVE SUPPLEMENTS PACKAGE

A coordinated and complete set of supplements is available to accompany this text:

- A *Lab Manual*, authored by N. Cook and H. Scriven, contains 44 class tested experiments designed to translate this textbook's theory into practical action. Each experiment includes an overview, a list of material required, an introduction, and a step-by-step procedure that walks the student through the experiment. Interspersed throughout the experiments are discussions reviewing what has been achieved up to that point. Complete experiments dedicated solely to troubleshooting are incorporated where appropriate.

- An *Instructor's Resource Manual*, authored by N. Cook, begins with an introduction to the instructor and includes a complete course curriculum package and worked-out text solutions.

- A *Devices and Op Amps* text, authored by N. Cook. Designed to familiarize your students with all of the semiconductor devices (diodes, transistors, FET's, and thyristors). Basic and operational amplifiers are also discussed in detail in this text.

- An IBM compatible *Interactive Student Disk*, developed by S. Musser, contains an overview of each chapter and an additional 20 questions per chapter, which are graded automatically. *Free* with every textbook.

- An *Instructor's Test Item File*, authored by J. Fairbanks, contains over 400 additional test questions which can be used to generate weekly quizzes, and midterm and final examinations. Available in both manual or disk (IBM or Apple compatible) form.

- A *Student Study Guide*, authored by D. Bridgeman. Contains overviews for each chapter and additional review questions. Between the interactive student disk and this manual, the student can be assigned over 1,000 additional study questions and practice problems.

- A *VHS Soldering Inspection Video I free* upon adoption, developed by Video Training Resources, Inc. Introduces your students to high reliability soldering inspection procedures and standards. It covers circuit boards, tracks, and solder connections. This video is *free* to you upon adoption of this text (one copy per school).

The DC/AC circuit material covered in this text has been logically divided and sequenced to provide a gradual progression from the known to the unknown, and from the simple to the complex. The information is presented as follows:

1. Fundamentals of Electricity—Chapters 1 through 4.
2. DC Electronics—Chapters 5 through 10.
3. AC Electronics—Chapters 11 through 18.
4. Appendices.

Section I begins by discussing the world of electronics in which the past, present, and future of electronics are explored. This section outlines the reason and necessity for the fundamental DC/AC theory covered in this book. *A 16-page full-color insert* entitled ''The Development of an Electronic Product'' will acquaint your students with the people involved in the development of a product at an electronics company. This photographic tour shows the process from conception to shipping, and it details what function each person performs. The following chapters in this section introduces current voltage, power, resistance and also discusses the different types of resistors currently on the market.

Section II begins by discussing all the methods by which direct current can be generated. The following chapters in this section discuss in more detail DC electromagnetism, and then series, parallel, and series-parallel DC circuit analysis and troubleshooting. The last chapter is an in-depth discussion on DC troubleshooting and measurement equipment.

Section III starts by introducing the phenomena known as alternating current. The following chapters in this section introduces your students to the variety of AC troubleshooting and measurement equipment. The remaining six chapters cover in great detail the following topics: AC electromagnetism, capacitance and capacitors, inductance and inductors, transformers and RLC networks.

The *Appendices* are extensive and should be reviewed initially so you can discover what is available. This material includes—

- An *electronics dictionary* which can be used throughout your students' studies to obtain definitions of terms used. The *abbreviations* appendix can also be used at any time to obtain the full name(s) of either acronyms or abbreviations.

- The *safety* appendix discusses body resistance, precautions to be used when troubleshooting, and resuscitation. This section should be reviewed before lab experimentation.

- An appendix showing *electronic schematic symbols* and three others listing *conversion factors, constants, units, prefixes, formulas and codes* are available for reference.

- A six-step *equipment troubleshooting procedure* has been generated to outline the logical approach which should be followed to troubleshoot any piece of electronic equipment.

- A discussion on *protoboards* which should be reviewed before conducting any lab experiments.
- A guide to *selecting your scientific calculator* is found in Appendix J, and the *decibel* is discussed in detail in Appendix K.
- A complete *frequency spectrum chart* listing the sound, electromagnetic, and optical frequencies is available in Appendix L.
- Appendix M contains a detailed discussion on *soldering tools and techniques* and should be reviewed before attempting any soldering.
- The final two appendixes list all of the answers to the *Self-Test Review Questions* and the *Solutions to the Odd-Numbered Questions* at the end of each chapter.

DEVELOPMENT, CLASS TESTING AND REVIEWING

The first phase of development was conducted in the classroom with students and instructors as critics. Each and every topic was class-tested by videotaping each lesson; and the results were then evaluated and implemented. This feedback was invaluable and enabled me to fine-tune my presentation of topics and install understanding and confidence, rather than frustration, into the students.

The second phase of development was to forward a copy of the revised manuscript to several instructors at schools throughout the country. These technical and topical critiques helped to mold the text into a more accurate form.

The third and final phase was to class-test the final revised manuscript and then commission the last technical review of this text in the final stages of production.

ACKNOWLEDGMENTS

I would like to extend my thanks to the following people whose efforts helped me in the completion of this book. First, I would like to thank my good friend Hugh Scriven, who critiqued every last part of this textbook and was the audience for many of my ideas and philosophies. Secondly, a debt of thanks is extended to Dean Mackey, who computer-drafted every one of my illustrations and who conducted the technical review of the final manuscript, and to Celeste Pingtella, who had the arduous task of typing the manuscript. A special thank you goes to the professional, creative, and friendly people at Prentice Hall—namely, my editor, Alice Barr, Gregory Burnell, Tom Aloisi, Bill Thomas, Amy Ness, Joanne Jimenez, Lorraine Mullaney, Janet Schmid, Barbara Cassel, and Mariann Hutlak—all of whom worked tirelessly to bring this book to its fruition.

In addition, my thanks is extended to Dail Cooper, who illustrated the lab manual, and to Shari Berilla and Karl Kuessel for their ideas and help.

Also, a special thanks is extended to the following people and organizations who have supplied material that has been used in this book: Fred Bode, Naomi Dennis, Tom Kurtz, Jim Reeve and Bob Estus of Wavetek, and photographer

Dick Van Patten, who supplied all of the photographs used in the color insert, Mary Ann Kelly of the American Hospital Association, Fred Vaughn of San Diego Gas and Electric, and Stewart Tugman of Stackpole Electronics.

Finally, a special thank you goes to my sister Alexandra, and to my grandfather Jack and his wife, Betty, for their ever enthusiastic love and support.

TO THE STUDENT

The early pioneers in electronics were intrigued by the mystery and wonder of a newly discovered science, whereas people today are attracted by its ability to lend its hand to any application and accomplish almost anything imaginable.

If you analyze exactly how you feel at this stage, you will probably discover that you have mixed emotions about the journey ahead. On one hand, imagination, curiosity, and excitement are driving you on, while apprehension and reservations may be slowing you down. Your enthusiasm will overcome any indecision you have once you become actively involved in electronics and realize that it is as exciting as you ever expected it to be.

The style, format, approach, and content of this textbook was developed with you, the student, in mind. Introductory DC/AC electronics is the foundation on which your understanding of electronics will rest, and it is therefore critical that you get off to a good start. In the first chapter of this text, I will introduce you to the world of electronics; and in a 16-page full-color insert, the variety of exciting career opportunities available will be explored. At the beginning of each chapter, a vignette describes the rise to fame and fortune of many of the electronic industry's entrepreneurs. The inclusion of these and many other features—coupled with a concise text that presents the material in a gradual, understandable, and orderly manner—will, hopefully, provide you with an enjoyable and rewarding learning experience.

Establish good study habits and take advantage of the many supplements that accompany this textbook including the Interactive Study Disk and the Study Guide Manual. Good Luck!

PUBLISHER'S ACKNOWLEDGMENTS

Our appreciation and thanks are extended to the instructors who have reviewed and contributed greatly to the development of *Introductory DC/AC Electronics*. Also, a special thanks is extended to the instructors who participated in Prentice Hall's special summer/fall product development program:

Duane Bailey, *Southern Alberta Institute of Technology*

Richard Bridgeman, *DeVry Institute of Technology*

John Dunbar, *DeVry Institute of Technology*

Randall G. Epstein, *Total Technical Institute*

Jay N. Fairbanks, *Bryant & Stratton College*

John J. Hatch, *ITT Technical Institute*

Leonard E. Laabs, *Walla Walla College*

L. Leibensperger, *Lincoln Technical Institute*

Mike Merchant, *Microcomputer Technical Institute*

Scott Musser, *Lincoln Technical Institute*

Terry Nelson, *Lincoln Technical Institute*

Dan Nemanich, *ITT Technical Institute*

William L. Robertson, *ITT Technical Institute*

Hugh Scriven, *ITT Technical Institute*

David Slavin, *Capital City Business College*

Arlyn L. Smith, *Alfred State College*

David Tester, *ITT Technical Institute*

Tom West, *Durham Technical College*

1

THE WORLD
OF ELECTRONICS

AFTER COMPLETING THIS CHAPTER, YOU WILL BE ABLE TO:

1. State the four basic phenomena on which electronics is based.
2. List some of the components and circuits that exist within electronic equipment.
3. List the seven branches of electronics.
4. Describe the types of equipment that can be found in each of the different branches or fields of electronics.
5. List the persons and companies largely responsible for the progression of electronics.
6. Explain how the unit names of many electrical and electronic quantities are derived.
7. Explain the process and people involved in the development of a product.

INTRODUCTION

This chapter has been designed to introduce you to the past, present, and future, in short "the world of electronics." At present you have probably heard mention of current, voltage, components, circuits, and equipment. This chapter will aquaint you with all the different types and show you how they all relate to one another.

1.1
THE PRESENT (REALITY)

Figure 1-1 illustrates the tree of electronics. Working from the bottom up, you can see that the foundation of electronics rests on four basic roots or phenomena known as current, voltage, resistance, and power, all of which are related to one another, as can be seen by the formula circle in Figure 1-2. Electronic components were developed to manipulate or control voltage and current. Referring to Figure 1-3, you will see a listing of electronic components, some names of which you may recognize. When a group of components is used in conjunction with one another, it forms a circuit, examples of which can be seen in Figure 1-4. Just as components are the building blocks for circuits, circuits are in turn the building blocks for electronic equipment, which can be categorized into one of seven groups.

1.1.1 COMMUNICATIONS EQUIPMENT

Electronic communications allows the transmission and reception of information between two points. Radio and television are obvious communication devices, broadcasting entertainment between two points. Figure 1-5 illustrates and lists the different types of communication equipment that exist in what is probably the largest branch of electronics.

1.1.2 DATA PROCESSING EQUIPMENT

The computer is proving to be one of the most useful of all systems. Its ability to process, store, and manipulate large groups of information at an extremely fast rate seems to make it ideal for almost any and every application. Systems vary in complexity and capability, ranging from the Cray supercomputer to the home personal computer. No matter what their size, the applications of word processing, record keeping, inventory, analysis, and accounting are but a few reasons why data processing systems, illustrated and listed in Figure 1-6, are used extensively.

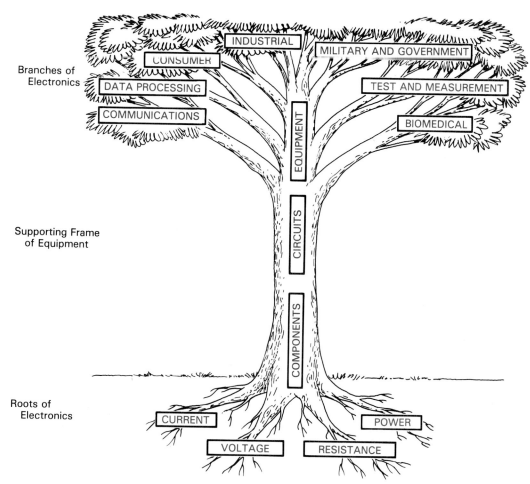

FIGURE 1-1 The Tree of Electronics.

FIGURE 1-2 The Roots of Electronics.

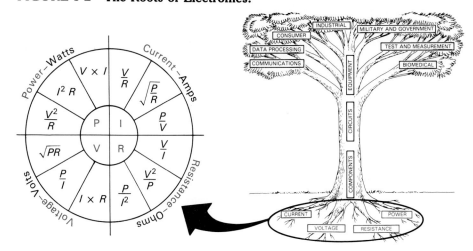

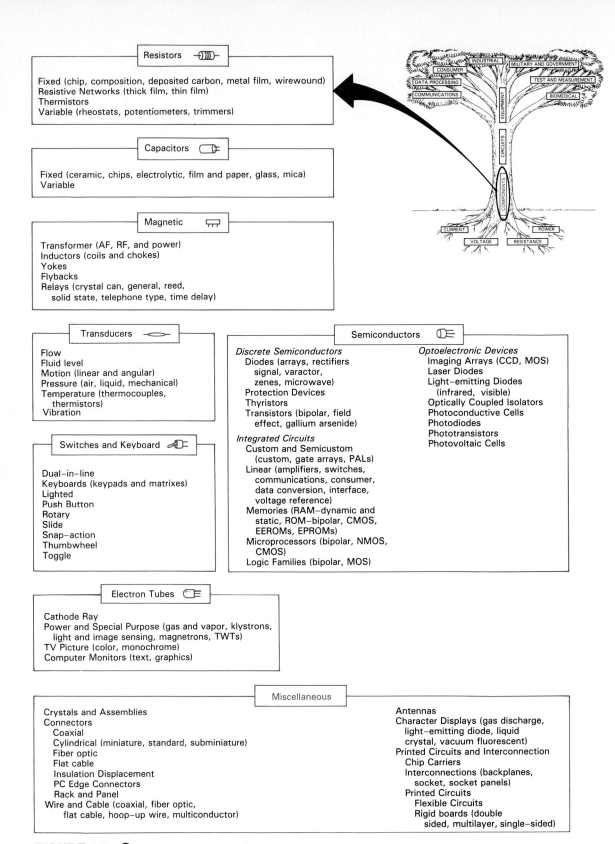

Resistors —◁▥▷—

Fixed (chip, composition, deposited carbon, metal film, wirewound)
Resistive Networks (thick film, thin film)
Thermistors
Variable (rheostats, potentiometers, trimmers)

Capacitors ◁▭◻

Fixed (ceramic, chips, electrolytic, film and paper, glass, mica)
Variable

Magnetic ▭

Transformer (AF, RF, and power)
Inductors (coils and chokes)
Yokes
Flybacks
Relays (crystal can, general, reed,
    solid state, telephone type, time delay)

Transducers ◁▱▷

Flow
Fluid level
Motion (linear and angular)
Pressure (air, liquid, mechanical)
Temperature (thermocouples,
    thermistors)
Vibration

Switches and Keyboard ◁▭◻

Dual-in-line
Keyboards (keypads and matrixes)
Lighted
Push Button
Rotary
Slide
Snap-action
Thumbwheel
Toggle

Semiconductors ◁▭◻

Discrete Semiconductors
    Diodes (arrays, rectifiers
        signal, varactor,
        zenes, microwave)
    Protection Devices
    Thyristors
    Transistors (bipolar, field
        effect, gallium arsenide)
Integrated Circuits
    Custom and Semicustom
        (custom, gate arrays, PALs)
    Linear (amplifiers, switches,
        communications, consumer,
        data conversion, interface,
        voltage reference)
    Memories (RAM–dynamic and
        static, ROM–bipolar, CMOS,
        EEROMs, EPROMs)
    Microprocessors (bipolar, NMOS,
        CMOS)
    Logic Families (bipolar, MOS)

Optoelectronic Devices
    Imaging Arrays (CCD, MOS)
    Laser Diodes
    Light–emitting Diodes
        (infrared, visible)
    Optically Coupled Isolators
    Photoconductive Cells
    Photodiodes
    Phototransistors
    Photovoltaic Cells

Electron Tubes ◁▭◻

Cathode Ray
Power and Special Purpose (gas and vapor, klystrons,
    light and image sensing, magnetrons, TWTs)
TV Picture (color, monochrome)
Computer Monitors (text, graphics)

Miscellaneous

Crystals and Assemblies
Connectors
    Coaxial
    Cylindrical (miniature, standard, subminiature)
    Fiber optic
    Flat cable
    Insulation Displacement
    PC Edge Connectors
    Rack and Panel
Wire and Cable (coaxial, fiber optic,
    flat cable, hoop-up wire, multiconductor)

Antennas
Character Displays (gas discharge,
    light–emitting diode, liquid
    crystal, vacuum fluorescent)
Printed Circuits and Interconnection
    Chip Carriers
    Interconnections (backplanes,
        socket, socket panels)
    Printed Circuits
        Flexible Circuits
        Rigid boards (double
            sided, multilayer, single–sided)

FIGURE 1-3 Components.

1.1.3 CONSUMER EQUIPMENT

Figure 1-7 illustrates and lists some of the current consumer electronic equipment available. From the smart computer controlled automobiles, which provide navigational information and monitor engine functions and braking, to the compact disk players, video camcorders, satellite TV receivers, and wide screen stereo TVs, this branch of electronics provides us with entertainment, information, safety, and, in the case of the pacemaker, life.

1.1.4 INDUSTRIAL EQUIPMENT

Almost any industrial company can be divided into basically three sections, all of which utilize electronic equipment to perform their functions. Figure 1-8 illustrates and lists the various types of equipment used with these three

FIGURE 1-4 Circuits.

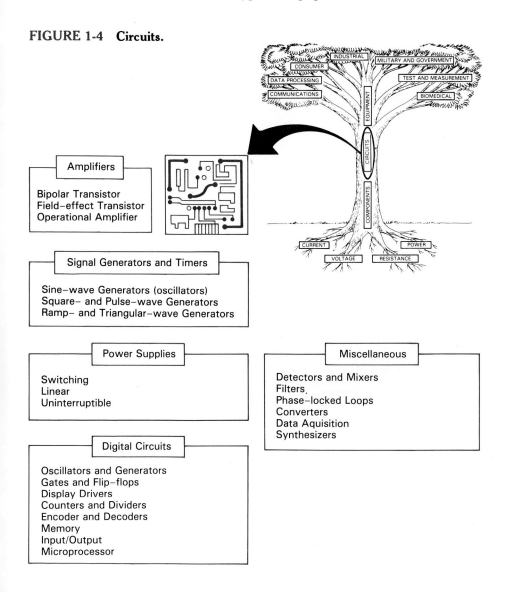

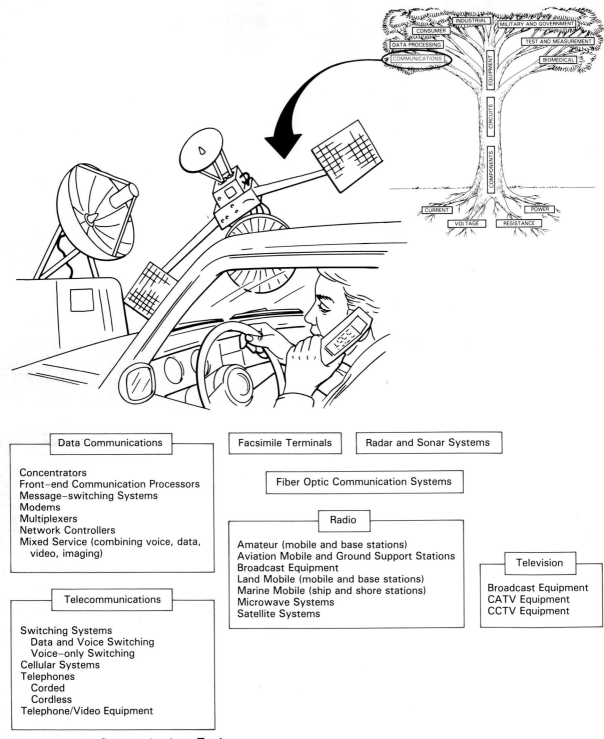

Tree diagram labels: CONSUMER, INDUSTRIAL, MILITARY AND GOVERNMENT, DATA PROCESSING, TEST AND MEASUREMENT, COMMUNICATIONS, BIOMEDICAL, EQUIPMENT, CIRCUITS, COMPONENTS, CURRENT, POWER, VOLTAGE, RESISTANCE

Data Communications

Concentrators
Front-end Communication Processors
Message-switching Systems
Modems
Multiplexers
Network Controllers
Mixed Service (combining voice, data,
    video, imaging)

Telecommunications

Switching Systems
    Data and Voice Switching
    Voice-only Switching
Cellular Systems
Telephones
    Corded
    Cordless
Telephone/Video Equipment

Facsimile Terminals

Radar and Sonar Systems

Fiber Optic Communication Systems

Radio

Amateur (mobile and base stations)
Aviation Mobile and Ground Support Stations
Broadcast Equipment
Land Mobile (mobile and base stations)
Marine Mobile (ship and shore stations)
Microwave Systems
Satellite Systems

Television

Broadcast Equipment
CATV Equipment
CCTV Equipment

FIGURE 1-5 **Communications Equipment.**

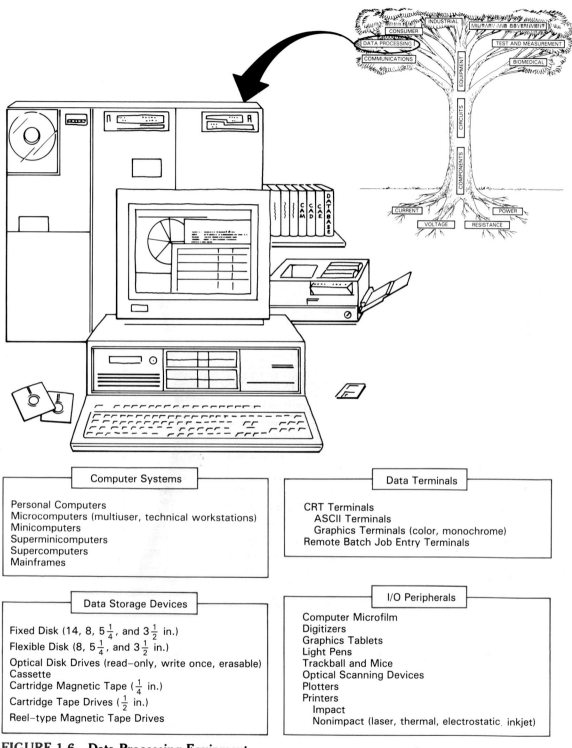

Computer Systems	Data Terminals

Personal Computers
Microcomputers (multiuser, technical workstations)
Minicomputers
Superminicomputers
Supercomputers
Mainframes
	CRT Terminals
	  ASCII Terminals
	    Graphics Terminals (color, monochrome)
	Remote Batch Job Entry Terminals

Data Storage Devices	I/O Peripherals

Fixed Disk (14, 8, $5\frac{1}{4}$, and $3\frac{1}{2}$ in.)
Flexible Disk (8, $5\frac{1}{4}$, and $3\frac{1}{2}$ in.)
Optical Disk Drives (read-only, write once, erasable)
Cassette
Cartridge Magnetic Tape ($\frac{1}{4}$ in.)
Cartridge Tape Drives ($\frac{1}{2}$ in.)
Reel-type Magnetic Tape Drives
	Computer Microfilm
	Digitizers
	Graphics Tablets
	Light Pens
	Trackball and Mice
	Optical Scanning Devices
	Plotters
	Printers
	  Impact
	  Nonimpact (laser, thermal, electrostatic, inkjet)

FIGURE 1-6 Data Processing Equipment.

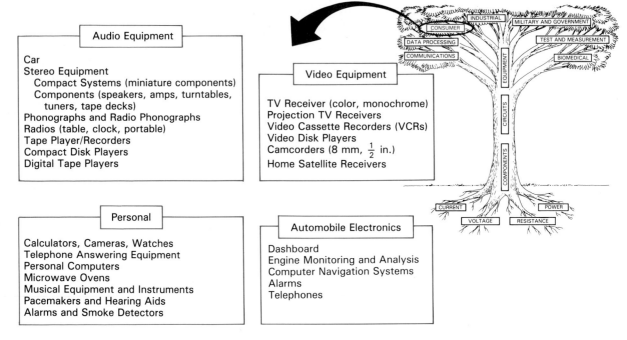

Audio Equipment

Car
Stereo Equipment
   Compact Systems (miniature components)
   Components (speakers, amps, turntables,
      tuners, tape decks)
Phonographs and Radio Phonographs
Radios (table, clock, portable)
Tape Player/Recorders
Compact Disk Players
Digital Tape Players

Video Equipment

TV Receiver (color, monochrome)
Projection TV Receivers
Video Cassette Recorders (VCRs)
Video Disk Players
Camcorders (8 mm, $\frac{1}{2}$ in.)
Home Satellite Receivers

Personal

Calculators, Cameras, Watches
Telephone Answering Equipment
Personal Computers
Microwave Ovens
Musical Equipment and Instruments
Pacemakers and Hearing Aids
Alarms and Smoke Detectors

Automobile Electronics

Dashboard
Engine Monitoring and Analysis
Computer Navigation Systems
Alarms
Telephones

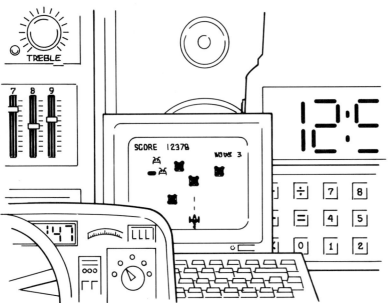

FIGURE 1-7 **Consumer Electronics Equipment.**

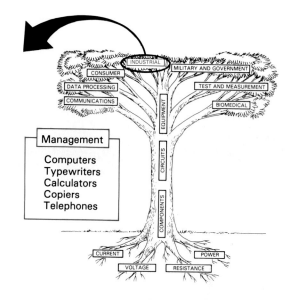

Manufacturing Equipment

Energy Management Equipment

Inspection Systems

Motor Controllers (speed, torque)

Numerical-control Systems

Process Control Equipment (data-acquisition systems, process instrumentation, programmable controllers)

Robot Systems

Vision Systems

Management

Computers
Typewriters
Calculators
Copiers
Telephones

Computer-aided Design and Engineering CAD/CAE

Hardware Equipment
    Design Work Stations (PC based, 32 microprocessor based platform, host based)
    Application Specific Hardware

Design Software
    Design Capture (schematic capture, logic fault and timing simulators, model libraries)
    IC Design (design rule checkers, logic synthesizers, floor planners-place and route, layout editors)
    Printed Circuit Board Design Software
    Project Management Software
    Test Equipment

FIGURE 1-8 Industrial Equipment.

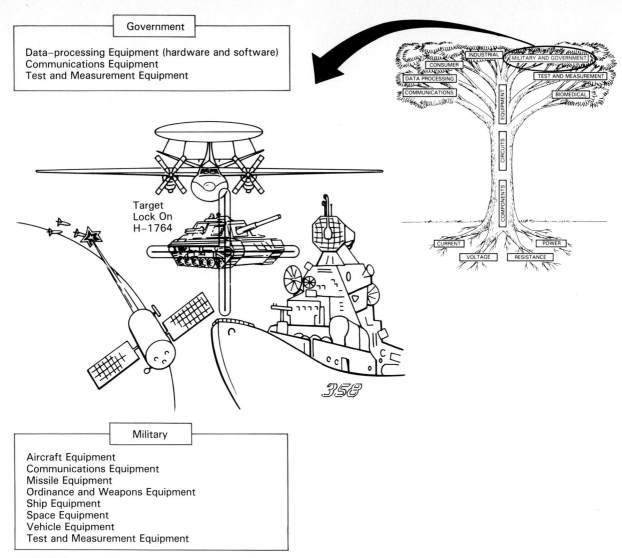

FIGURE 1-9 Military and Government Equipment

sections. The manufacturing section will typically use power, motor, and process control equipment, along with automatic insertion, inspection, and vision systems, for the fabrication of a product. The engineering section uses computers, test equipment, and the like for the design and testing of a product, while the management section uses electronic equipment such as typewriters, copiers, telephones, and more.

1.1.5 MILITARY AND GOVERNMENT EQUIPMENT

This branch of electronic equipment, produced by defense and government contractors, is illustrated and listed in Figure 1-9. Military equipment seems to advance in leaps and bounds, with the latest concentration of technology being focused on sensors, smart weapons, software programs, and space. Nonmilitary government gear mainly includes data processing and telecommunications equipment.

1.1.6 TEST AND MEASUREMENT EQUIPMENT

The rising complexity of electronic components, circuits, and equipment is causing a demand for sophisticated automatic test equipment for both the manufacturer and customer to test their products. Figure 1-10 illustrates and lists some of the test and measurement equipment, which can be classified as either stand-alone and/or computer-controlled test instruments.

FIGURE 1-10   Test and Measurement Equipment.

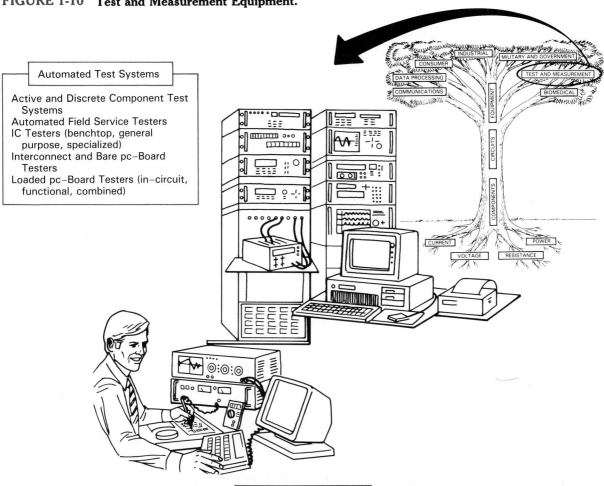

Automated Test Systems

Active and Discrete Component Test Systems
Automated Field Service Testers
IC Testers (benchtop, general purpose, specialized)
Interconnect and Bare pc-Board Testers
Loaded pc-Board Testers (in-circuit, functional, combined)

General Test Equipment

Amplifiers (lab)
Arbitrary Waveform Generators
Analog Voltmeters, Ammeters, and Multimeters
Audio Oscillators
Audio Waveform Analyzers and Distortion Meters
Calibrators and Standards
Dedicated IEEE-488 Bus Controllers
Digital Multimeters
Electronic Counters (RF, Microwave, Universal)
Frequency Synthesizers
Function Generators
Pulse/Timing Generators
Signal Generators (RF, Microwave)

Logic Analyzers
Microprocessor Development Systems
Modulation Analyzers
Noise-measuring Equipment
Oscilloscopes (Analog, Digital)
Panel Meters
Personal Computer (PC) Based Instruments
Recorders and Plotters
RF/Microwave Network Analyzers
RF/Microwave Power-measuring Equipment
Spectrum Analyzers
Stand-alone In-circuit Emulators
Temperature-measuring Instruments

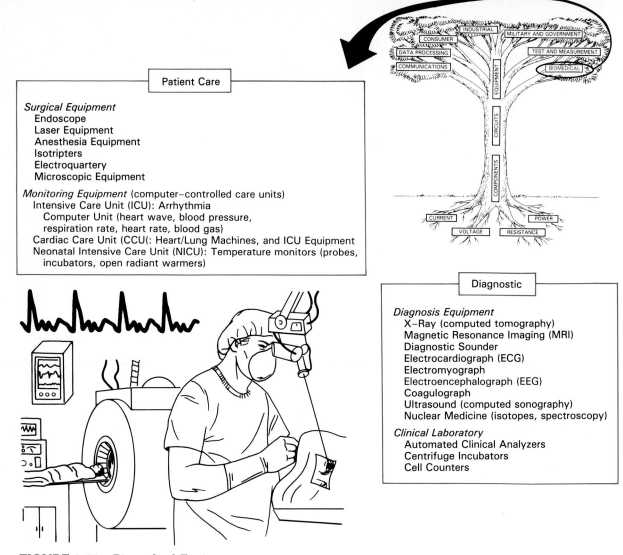

Patient Care

Surgical Equipment
  Endoscope
  Laser Equipment
  Anesthesia Equipment
  Isotripters
  Electroquartery
  Microscopic Equipment

Monitoring Equipment (computer–controlled care units)
  Intensive Care Unit (ICU): Arrhythmia
    Computer Unit (heart wave, blood pressure,
    respiration rate, heart rate, blood gas)
  Cardiac Care Unit (CCU(: Heart/Lung Machines, and ICU Equipment
  Neonatal Intensive Care Unit (NICU): Temperature monitors (probes,
    incubators, open radiant warmers)

Diagnostic

Diagnosis Equipment
  X–Ray (computed tomography)
  Magnetic Resonance Imaging (MRI)
  Diagnostic Sounder
  Electrocardiograph (ECG)
  Electromyograph
  Electroencephalograph (EEG)
  Coagulograph
  Ultrasound (computed sonography)
  Nuclear Medicine (isotopes, spectroscopy)

Clinical Laboratory
  Automated Clinical Analyzers
  Centrifuge Incubators
  Cell Counters

FIGURE 1-11 Biomedical Equipment.

1.1.7 BIOMEDICAL ELECTRONIC EQUIPMENT

Electronic equipment is used more and more within the biological and medical fields. Figure 1-11 illustrates and lists some of the medical equipment, which can be simply categorized as being either patient care or diagnostic equipment. In the operating room, the endoscope, which is an instrument used to examine the interior of a canal or hollow organ, and the laser, which is used to coagulate, cut, or vaporize tissue with extremely intense light, are both reducing the amount of invasive surgery. A large amount of monitoring equipment is used both in and out of operating rooms, and this equipment consists of generally large computer-controlled systems that can have a variety of modules inserted (based on the application) to monitor on a continuous basis body temperature, blood pressure, pulse rate, and so on. In the diagnostic group of equipment, the clinical laboratory test results are used as diagnostic tools, and with the advances in automation and computerized information systems, multiple tests can be carried out at increased speeds. Diagnostic imaging, in which a computer constructs an image of a cross-sectional plane of the body, is probably one of the most interesting equipment areas.

SELF-TEST REVIEW QUESTIONS (§ 1.1)

1. Name the four basic roots or phenomena on which the foundation of electronics rests.
2. List three examples of electronic components.
3. Name two circuits.
4. In which branch of electronics would a radar be catergorized?

1.2
THE PAST (HISTORY)

Our way of life has undergone enormous change since the unveiling of electricity and electronics. Each of us, with just a little thought, can easily envisage how difficult it would be to manage in a world without electronics. The people responsible for the discovery of this science are discussed in this section. Figure 1-12 illustrates how a section of the tree of electronics has been extracted, so the rings of time can be viewed. These rings represent the age of the tree, with one ring being equal to approximately one year, and in this illustration the rings are used to summarize the progression of electricity and electronics.

Our history begins in 1600 with an English physician, William Gilbert (1544–1603), who documented many years of research and experiments on magnets and magnetic bodies such as amber and loadstones. Probably his most important discovery was that amber would attract lightweight objects when rubbed with a cloth. In 1601 he was appointed physician to Queen Elizabeth I at a salary of $150 a year. He was the first to believe that the earth was nothing but a large magnet and that its magnetic field causes a needle to align itself between north and south.

Stephen Gray (1693–1736), who was also an Englishman, discovered that certain substances would conduct electricity. This lead was picked up by Charles du Fay a French experimenter, who believed that there were two types of electricity, which he called vitreous and resinous electricity.

One of the best known and most admired men in the latter half of the eighteenth century was the American Benjamin Franklin (1706–1790). He is reknowned for his kite in a storm experiment, which proved that lightning is electricity. Franklin also discovered that there was only one type of electricity, and the two previously believed types were simply two characteristics of electricity. Vitreous was renamed positive charge and resinous was called

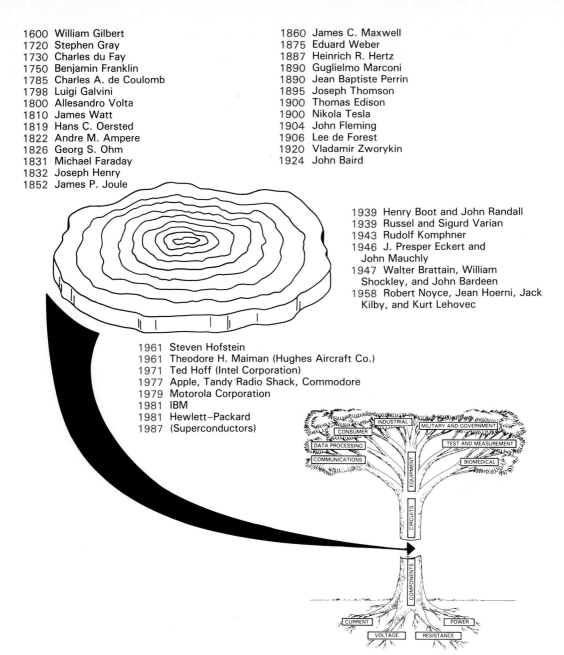

1600 William Gilbert	1860 James C. Maxwell
1720 Stephen Gray	1875 Eduard Weber
1730 Charles du Fay	1887 Heinrich R. Hertz
1750 Benjamin Franklin	1890 Guglielmo Marconi
1785 Charles A. de Coulomb	1890 Jean Baptiste Perrin
1798 Luigi Galvini	1895 Joseph Thomson
1800 Allesandro Volta	1900 Thomas Edison
1810 James Watt	1900 Nikola Tesla
1819 Hans C. Oersted	1904 John Fleming
1822 Andre M. Ampere	1906 Lee de Forest
1826 Georg S. Ohm	1920 Vladamir Zworykin
1831 Michael Faraday	1924 John Baird
1832 Joseph Henry	
1852 James P. Joule	

1939 Henry Boot and John Randall
1939 Russel and Sigurd Varian
1943 Rudolf Komphner
1946 J. Presper Eckert and
     John Mauchly
1947 Walter Brattain, William
     Shockley, and John Bardeen
1958 Robert Noyce, Jean Hoerni, Jack
     Kilby, and Kurt Lehovec

1961 Steven Hofstein
1961 Theodore H. Maiman (Hughes Aircraft Co.)
1971 Ted Hoff (Intel Corporation)
1977 Apple, Tandy Radio Shack, Commodore
1979 Motorola Corporation
1981 IBM
1981 Hewlett–Packard
1987 (Superconductors)

FIGURE 1-12 **History of Electricity and Electronics.**

negative charge, terms which were invented by Franklin, along with the terms battery and conductor, which are still used today.

The unit of electrical charge is the coulomb in honor of the French physicist Charles A. de Coulomb (1736–1806), who developed the laws of attraction and repulsion between charged bodies.

The galvanometer, which is used to measure electrical current, was named after Luigi Galvani (1737–1798), who conducted many experiments with electrical current, or as it was known at that time "galvanism."

The unit of voltage, the volt, was named in honor of Alessandro Volta

14   **The World of Electronics**

(1745–1827), an Italian physicist who is famous for his invention of the electric battery. In 1801, he was called to Paris by Napoleon to show his experiment on the generation of electric current.

The unit of power is the watt in honor of a Scottish engineer and inventor, James Watt (1736–1819), for his advances in the field of science.

In 1819 a Danish physicist, Hans C. Oersted (1777–1851), accidentally discovered an interesting phenomenon. Placing a compass near a current-carrying conductor, he noticed that the needle of the compass pointed to the conductor rather than north. He was quick to realize that electricity and magnetism were related, and in honor of his work, the unit oersted was adopted for the unit of magnetic field strength.

The unit of electrical current is the ampere or amp in honor of André M. Ampère (1775–1836), a French physicist who pioneered in the study of electromagnetism. After hearing of Oersted's discoveries, he conducted further experiments and discovered that two current-carrying conductors would attract and repel one another, just like two magnets.

Ohm's law is the most well known law in electrical circuits, and was formulated by Georg S. Ohm (1787–1854), a German physicist. His law was so coldly received that his feelings were hurt and he resigned his teaching post. When the law was finally recognized, he was reinstated. In honor of his accomplishments, the unit of resistance is called the ohm.

In 1831, Michael Faraday (1791–1867), an English physicist, explored further Oersted's discovery of electromagnetism and discovered that a magnetic field could be used to produce electric current. These findings are today referred to as Faraday's laws of electromagnetic induction. A German-born scientist working in Russia extended Faraday's findings and found that the current induced in a conductor is such that it opposes the change in the magnetic field producing it. This is known today as Lenz's law in honor of Heinrich F. E. Lenz. Michael Faraday also investigated static electricity and the lines of electric force, and it is in acknowledgment of his work in this area that the unit of capacitance is named the farad.

Joseph Henry (1797–1878), an American physicist, also conducted extensive studies into electromagnetism. Henry was the first to insulate the magnetic coil of wire and developed coils for telegraphy and motors. In recognition of his discovery of self-induction in 1832, the unit of inductance is called the henry.

James P. Joule (1818–1889), an English physicist and self-taught scientist, conducted extensive research into the relationships between electrical, chemical, and mechanical effects, which led him to the discovery that one energy form can be converted into another. For his achievements, his name was given to the unit of energy, the joule.

As a small boy, James C. Maxwell (1831–1879) was persistently inquisitive. He built many scientific toys before he was 8, at 14 he wrote a paper on how to construct oval curves, and at 18 two of his papers were published. The supreme achievement of this Scottish physicist was to translate Faraday's experiments into mathematical notation. This set of mathematical equations, known as Maxwell's equations, shows the relationship between electricity and magnetism.

Eduard W. Weber (1804–1891), a German physicist, made enduring contributions to the modern system of electrical units, and magnetic flux is measured in webers in honor of his work.

The Past (History)　　**15**

Heinrich R. Hertz (1857–1894), a German physicist, was the first to demonstrate the production and reception of electromagnetic (radio) waves. In honor of his work in this field, the unit of frequency is called the hertz.

Studying the experiments of Maxwell and Hertz, Guglielmo Marconi (1874–1937) invented a practical system of telegraphy communication. In an evolutionary process, Marconi extended his distance of communication from $1\frac{1}{2}$ miles in 1896 to 6000 miles in 1902. In September of 1899, Marconi equipped two U.S. ships with equipment and used them in the Atlantic Ocean to transmit to America the progress of the America cup yacht race.

The electron was first discovered by Jean B. Perrin (1870–1942), a French physicist who was awarded the Nobel prize for physics. Perrin discovered that cathode rays consisted of negatively charged particles and these particles, which later became known as electrons, were measured by an English physicist, Joseph Thomson (1846–1914).

Thomas Edison (1847–1931) was a self-educated inventor best known for his development of the phonograph and the incandescent lamp. The first version of the phonograph cost $18 and was turned with a hand crank. A decade later it was motor driven, with cylindrical wax and then disk-type records. In 1879, after $40,000 worth of fruitless experiments, he succeeded in developing an incandescent lamp that consisted of a loop of carbonized cotton thread that glowed in a vacuum for 40 hours. In 50 years, he took out 1033 patents. Nikola Tesla (1856–1943) was the inventor of the induction motor and worked to improve power transmission. Working for Edison for a short time, they developed a hatred for one another that prompted Tesla to begin his own business. Edison promoted dc power distribution, while Tesla believed in ac, and eventually Tesla's reasoning was adopted worldwide. In 1912, they were nominated together for the Nobel prize in physics, but Tesla would have nothing to do with Edison and so the prize went to a third party. A theory of Tesla's, which up to this time has not proven possible, is the wireless transmission of electrical power by high-energy electromagnetic or radiant beams.

John A. Fleming, a British scientist, saw the value of an effect that was discovered by Edison but for which he saw no practical purpose. "Edison effect" permitted Fleming to develop the "Fleming valve," which passes current in only one direction. Its operation made it the first device able to convert alternating current into direct current and to detect radio waves.

Although the Fleming valve was an advance, it could not amplify or boost a signal. The "audion" developed by U.S. inventor Lee de Forest (1873–1961) sparked an era known as "vacuum tube electronics" that brought about transcontinental telephony in 1915, radio broadcasting in 1920, radar in 1936, and television between 1927 to 1946, because of this triode vacuum tube's ability to amplify small signals.

The father of television, Vladamir C. Zworykin, developed the first television picture tube, called the kinescope, in 1920.

John L. Baird (1888–1946) was a British inventor and television pioneer. He was the first to transmit television over a distance. He reproduced objects in 1924, transmitted recognizable human faces in 1925, demonstrated the first true television in 1926, and in 1939 developed television, in natural color.

During World War II, there was a need for microwave frequency vacuum tubes. British inventor Henry Boot developed the magnetron in 1939, and

The World of Electronics

in the same year an American brother duo, Russel and Sigurd Varian, invented the klystron. In 1943, the traveling wave tube amplifier was invented by Rudolf Komphner, and up to this day these three microwave vacuum tubes are still extensively used.

In 1946, J. Presper Eckert and John Mauchly unveiled ENIAC, which used over 300,000 vacuum tubes. ENIAC, which is an acronym for Electronic Numerical Integrator and Computer, was the first large-scale electronic digital computer.

Walter Brattain, William Shockley, and John Bardeen sparked an even greater era, known a "solid-state electronics," with their invention of the transistor at Bell Laboratories. Transistorized equipment is smaller, cheaper, more reliable, and more robust and consumes less power than its vacuum tube counterparts.

In 1961, Steven Hofstein devised the field-effect transistor used in MOS (metal oxide semiconductor) integrated circuits, and in the same year Theodore H. Maiman, a scientist working at Hughes Aircraft Company, built the first operational laser using a synthetic ruby crystal.

In 1958, Robert Noyce, Jean Hoerni, Jack Kilby, and Kurt Lehovec all took part in the development of the integrated circuit, which incorporated many transistors and other components on a small chip of semiconductor material.

Ted Hoff of Intel Corporation designed a microprocessor, the 4004, that had all the basic parts of a central processor, in 1971. Intel improved on the 4-bit 4004 microprocessor and unveiled an 8-bit microprocessor in 1974 that could add two numbers in 2.5 millionths of a second.

Three mass-market personal computers emerged in 1977: the Apple II, Radio Shack's TRS-80, and the Commodore PET.

In 1979, Motorola Corporation continued to advance computers by creating a powerful and versatile 16-bit microprocessor that could multiply two numbers in 3.2 billionths of a second.

IBM, who had up to this time dominated the big computer market, entered the personal computer market with the IBM PC in 1981. Also in 1981, Hewlett-Packard unveiled its 32-bit microprocessor to further advance the speed and power of computers.

In scientific laboratories around the world, thousands of scientists began working furiously to develop a new technology that at the beginning of 1987 seemed little more than science fiction. Practical applications of superconductors cannot be fully realized at this time, just as no one could foresee all of the uses for the transistor when it was invented 40 years earlier.

SELF-TEST REVIEW QUESTIONS (§ 1.2)

Name the person who was honored by the use of his name for the unit of the following electrical properties:

1.	Resistance	**3.**	Voltage
2.	Current	**4.**	Power

1.3
THE FUTURE

So many things have happened in the past, so many things are happening at present, and so many things are under development to unfold in the future that it has become very difficult to keep tabs on the progression in all fields of electronics.

Electronics is an exciting and rewarding field to be involved in. The main purpose of this chapter is not only to give you an insight into the past and present, but also to give you the reasons why we begin our studies of electronics by discussing current, voltage, resistance, and power. These fundamentals or roots are necessary to understand components, as components need to be comprehended to understand circuits, and in turn circuits are necessary to understand the exciting, wide range of equipment that exists in the different branches of electronics.

Projections for the future of electronics are based mainly on our current knowledge, current technology, and the trends of the present. As we all well know, trends do not always help us to predict the future. Some trends carry on longer than ever expected, while those we think will carry on forever can suddenly end. But new developments in the uses of electronics are continually being realized, limited only by our initiative and ingenuity.

SELF-TEST REVIEW QUESTIONS (§ 1.3)

Which two of the following have been predicted by the government as having the largest use in the future?

1. Vacuum tubes
2. Lasers
3. Robots
4. Computers

1.4
PEOPLE IN ELECTRONICS

People are required in any industrial company to keep the industry moving, and an electronics company is not any different. The photo essay and associated storyboard following page 454 demonstrate how a typical electronics company would develop a product; but, more importantly, they show the people involved in this product development process and how each of them plays a very important role.

1. Electronics is built on four basic roots or phenomena: current, voltage, resistance, and power.

2. Electronic components control the four basic roots or phenomena of electronics.

3. Groups of components are interconnected to perform certain functions, and these arrangements are known as circuits.

4. Just as components are the building blocks for circuits, circuits are in turn the building blocks for electronic equipment.

5. All electronic equipment can be categorized into one of seven different branches or groups, which are communications, data processing, consumer, industrial, military and government, test and measurement, and biomedical.

6. The names for almost all the units of electricity and electronics are derived from early experimenters.

7. The four main quantities and units are:
   a. Current, which is measured in amperes or amps
   b. Voltage, which is measured in the unit of volts
   c. Resistance, which is measured in ohms
   d. Power, which is measured in watts

8. The engineer forms the concepts, designs, and modifications of all electronic circuits or systems.

9. Engineering technicians aid the engineer in the research and development of a product.

10. The technician is an expert in the diagnosis and repair of problems or malfunctions within electronic equipment.

11. The assembler is also an expert in his or her field, which is the fabrication of or manufacturing techniques used in the assembly of the final product.

Multiple Choice Questions

1. The unit of charge was named after:
   a. Alessandro Volta
   b. Heinrich Hertz
   c. Georg Ohm
   d. Charles Coulomb

2. _____ developed the basic laws relating to current flow.
   a. André Ampère
   b. Heinrich Hertz
   c. Michael Faraday
   d. Intel Corporation

3. The first 16-bit microprocessor was developed by:
   a. Intel Corporation
   b. Hewlett-Packard
   c. Texas Instruments
   d. Motorola Corporation
4. The first operational laser was built by:
   a. Walter Brattain
   b. Guglielmo Marconi
   c. Theodore Maiman
   d. James Maxwell
5. The unit of frequency was named in honor of:
   a. Heinrich Hertz
   b. Jean Perrin
   c. Georg Ohm
   d. Greek experimenters
6. The person largely responsible for the era termed "vacuum tube electronics" was:
   a. John Fleming
   b. William Gilbert
   c. Lee de Forest
   d. Benjamin Franklin
7. The carbon filament light bulb was invented by:
   a. Joseph Henry
   b. James Joule
   c. Thomas Edison
   d. Vladamir Zworykin
8. The persons responsible for the transistor and the subsequent "solid-state electronics" era were:
   a. Hertz and Marconi
   b. Eckert and Mauchly
   c. Noyce, Hoerni, Kilby, and Lehovec
   d. Brattain, Schockley, and Bardeen
9. The resistor, capacitor, and inductor are all examples of electronic circuits.
   a. True
   b. False
10. Amplifiers and oscillators are both examples of electronic _____.
    a. Circuits
    b. Components
    c. Both (a) and (b) are true
    d. None of the above are true
11. What new technology has just recently been unveiled?
    a. Microelectronics
    b. Solid-state electronics
    c. Superconductivity
    d. Two of the above could be true
12. Radio equipment would be classified in what branch of electronics?
    a. Data processing
    b. Industrial

c. Biomedical
d. Test and measurement
e. None of the above

13. In which branch of electronics would the oscilloscope be found?
   a. Data processing
   b. Test and measurement
   c. Military and government
   d. Biomedical
   e. None of the above

Essay Questions _____

14. What are the three sections into which an industrial company can be divided?   (1.1)

15. List the seven branches of electronic equipment   (1.1)

16. Give examples of five electronic components and three electronic circuits.   (1.1)

17. Give examples of four pieces of equipment in each branch of electronics.   (1.1)

18. List three attributes of the human body that medical electronic equipment can measure.   (1.1.7)

19. List the four roots of electronics.   (1.1)

20. Name the units for the following quantities: (1.1)
   a. Current, _____   **c.** Resistance, _____
   b. Voltage, _____   **d.** Power, _____

21. Who promoted the ac power distribution that is now used worldwide?   (1.2)

22. What was the ENIAC?   (1.2)

23. List two things for which Thomas Edison is best known?   (1.2)

24. State the responsibilities of the following people: (1.4)
   a. Engineer   **c.** Assembler
   b. Technician

25. What is the difference between a marketing person and a salesperson?   (1.4)

26. What function do the following people perform: (1.4)
   a. Test engineer
   b. Calibration technician
   c. Sustaining engineer

27. At which stage in the product development process would an assembler get involved?   (1.4)

28. What responsibilities do the following people have?   (1.4)
   a. Engineering/technical writer
   b. Quality assurance technician
   c. Customer service technician

2

VOLTAGE
AND CURRENT

AFTER COMPLETING THIS CHAPTER, YOU WILL BE ABLE TO:

1. Explain the atom's subatomic particles.
2. State the difference between an atomic number and atomic weight.
3. Understand the term natural element.
4. Describe what is meant by and state how many shells or bands exist around an atom.
5. Explain the difference between:
   a. An atom and a molecule.
   b. An element and a compound.
   c. A proton and an electron.
6. Describe the terms:
   a. Neutral atom.
   b. Negative ion.
   c. Positive ion.
7. Define electrical current.
8. Describe the ampere in relation to coulombs per second.
9. List the different current units and values.
10. Explain why the effect of current flow travels at the speed of light.
11. Describe the difference between conventional current flow and electron flow.
12. List the three rules to apply when measuring current.
13. Define electrical voltage.
14. List the different voltage units and values.
15. List the three rules to apply when measuring voltage.
16. Relate the electrical system to a fluid system.
17. Describe why current is directly proportional to voltage.

18. Explain the difference between:
    a. A conductor.
    b. An insulator.
    c. A semiconductor.
19. Describe conductance.
20. Explain what makes a good:
    a. Conductor.
    b. Insulator.
21. Describe what is meant by semiconductor material.
22. Explain the terms:
    a. Open circuit.
    b. Closed circuit.
    c. Short circuit.

INTRODUCTION

This chapter has been designed to introduce you to the building blocks of matter, which will help you to understand (1) the electrical quantities of voltage and current and (2) the difference between a conductor and an insulator.

2.1

THE STRUCTURE OF MATTER

All of the matter on the earth and in the air surrounding the earth can be classified as being either a solid, liquid, or gas. A total of approximately 92 different natural elements exist in, on, and around the earth. An element, by definition, is a substance consisting of only one type of atom; in other words, every element has its own distinctive atom, which makes it different from all of the other elements. This atom is the smallest particle into which an element can be divided without losing its identity, and a group of identical atoms is called an element, as seen in Figure 2-1.

For the sake of discussion, let us take a small amount of either a solid, a liquid, or a gas and divide it into two pieces. Then we divide a resulting piece into two pieces, and keep repeating the process until we finally end up with a tiny remaining part. Viewing the part under the microscope, as seen in Figure 2-2, the substance can still be identified as the original element as it is still made up of many of the original solid, liquid or gas atoms. A small amount of gold, for example, the size of a pinpoint, will still contain several billion atoms. If the element subdivision is continued, however, a point will be reached at which a single atom will remain. Let us now analyze the atom in more detail.

2.1.1 THE ATOM

The word *atom* is a Greek word meaning a particle that is too small to be subdivided. At present, we cannot clearly see the atom; however, physicists and researchers do have the ability to record a picture as small as 12 billionths of an inch (about the diameter of one atom), and this image displays the atom as a white fuzzy ball.

In 1913 a Danish physicist, Neils Bohr, put forward a theory about the atom, and his basic model outlining the subatomic particles that make up the atom is still in use today and is illustrated in Figure 2-3. Bohr actually combined the ideas of Lord Rutherford's (1871–1937) nuclear atom with

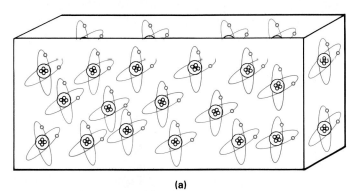 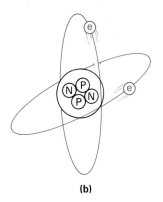

| (a) | (b) |

FIGURE 2-1 (a) Element: Many Similar Atoms. (b) Atom: Smallest Unit.

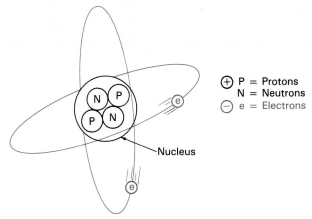

Small part of element under microscope–
original element still made up of many atoms.

Element

Microscope

FIGURE 2-2 Elements under the Microscope.

FIGURE 2-3 The Atom.

⊕ P = Protons
N = Neutrons
⊖ e = Electrons

Nucleus

TABLE 2-1
PERIODIC TABLE OF THE ELEMENTS

Atomic Number	Element Name	Symbol	Atomic Weight	Electrons/Shell							Discovered	Comments
				K	L	M	N	O	P	Q		
1	Hydrogen	H	1.007	1							1766	Active gas
2	Helium	He	4.002	2							1895	Inert gas
3	Lithium	Li	6.941	2	1						1817	Solid
4	Beryllium	Be	9.01218	2	2						1798	Solid
5	Boron	B	10.81	2	3						1808	Solid
6	Carbon	C	12.011	2	4						Ancient	Semiconductor
7	Nitrogen	N	14.0067	2	5						1772	Gas
8	Oxygen	O	15.9994	2	6						1774	Gas
9	Fluorine	F	18.998403	2	7						1771	Active gas
10	Neon	Ne	20.179	2	8						1898	Inert gas
11	Sodium	Na	22.98977	2	8	1					1807	Solid
12	Magnesium	Mg	24.305	2	8	2					1755	Solid
13	Aluminum	Al	26.98154	2	8	3					1825	Metal conductor
14	Silicon	Si	28.0855	2	8	4					1823	Semiconductor
15	Phosphorus	P	30.97376	2	8	5					1669	Solid
16	Sulfur	S	32.06	2	8	6					Ancient	Solid
17	Chlorine	Cl	35.453	2	8	7					1774	Active gas
18	Argon	Ar	39.948	2	8	8					1894	Inert gas
19	Potassium	K	39.0983	2	8	8	1				1807	Solid
20	Calcium	Ca	40.08	2	8	8	2				1808	Solid
21	Scandium	Sc	44.9559	2	8	9	2				1879	Solid
22	Titanium	Ti	47.90	2	8	10	2				1791	Solid
23	Vanadium	V	50.9415	2	8	11	2				1831	Solid
24	Chromium	Cr	51.996	2	8	13	1				1798	Solid
25	Manganese	Mn	54.9380	2	8	13	2				1774	Solid
26	Iron	Fe	55.847	2	8	14	2				Ancient	Solid (magnetic
27	Cobalt	Co	58.9332	2	8	15	2				1735	Solid
28	Nickel	Ni	58.70	2	8	16	2				1751	Solid
29	Copper	Cu	63.546	2	8	18	1				Ancient	Metal conductor
30	Zinc	Zn	65.38	2	8	18	2				1746	Solid
31	Gallium	Ga	69.72	2	8	18	3				1875	Liquid
32	Germanium	Ge	72.59	2	8	18	4				1886	Semiconductor
33	Arsenic	As	74.9216	2	8	18	5				1649	Solid
34	Selenium	Se	78.96	2	8	18	6				1818	Photosensitive
35	Bromine	Br	79.904	2	8	18	7				1826	Liquid
36	Krypton	Kr	83.80	2	8	18	8				1898	Inert gas
37	Rubidium	Rb	85.4678	2	8	18	8	1			1861	Solid
38	Strontium	Sr	87.62	2	8	18	8	2			1790	Solid

Max Planck's (1858–1947) and Albert Einstein's (1879–1955) quantum theory of radiation.

The three important particles of the atom are the proton, which has a positive charge, the neutron, which is neutral or has no charge, and the electron, which has a negative charge. Referring to Figure 2-3, you can see that the atom consists of a positively charged central mass called the nucleus, which is made up of protons and neutrons surrounded by a quantity of negatively charged orbiting electrons.

Atomic Number	Element Name	Symbol	Atomic Weight	Electrons/Shell							Discovered	Comments
				K	L	M	N	O	P	Q		
39	Yttrium	Y	88.9059	2	8	18	9	2			1843	Solid
40	Zirconium	Zr	91.22	2	8	18	10	2			1789	Solid
41	Niobium	Nb	92.9064	2	8	18	12	1			1801	Solid
42	Molybdenum	Mo	95.94	2	8	18	13	1			1781	Solid
43	Technetium	Tc	98.0	2	8	18	14	1			1937	Solid
44	Ruthenium	Ru	101.07	2	8	18	15	1			1844	Solid
45	Rhodium	Rh	102.9055	2	8	18	16	1			1803	Solid
46	Palladium	Pd	106.4	2	8	18	18	0			1803	Solid
47	Silver	Ag	107.868	2	8	18	18	1			Ancient	Metal conductor
48	Cadmium	Cd	112.41	2	8	18	18	2			1803	Solid
49	Indium	In	114.82	2	8	18	18	3			1863	Solid
50	Tin	Sn	118.69	2	8	18	18	4			Ancient	Solid
51	Antimony	Sb	121.75	2	8	18	18	5			Ancient	Solid
52	Tellurium	Te	127.60	2	8	18	18	6			1783	Solid
53	Iodine	I	126.9045	2	8	18	18	7			1811	Solid
54	Xenon	Xe	131.30	2	8	18	18	8			1898	Inert gas
55	Cesium	Cs	132.9054	2	8	18	18	8	1		1803	Liquid
56	Barium	Ba	137.33	2	8	18	18	8	2		1808	Solid
57	Lanthanum	La	138.9055	2	8	18	18	9	2		1839	Solid
72	Hafnium	Hf	178.49	2	8	18	32	10	2		1923	Solid
73	Tantalum	Ta	180.9479	2	8	18	32	11	2		1802	Solid
74	Tungsten	W	183.85	2	8	18	32	12	2		1783	Solid
75	Rhenium	Re	186.207	2	8	18	32	13	2		1925	Solid
76	Osmium	Os	190.2	2	8	18	32	14	2		1804	Solid
77	Iridium	Ir	192.22	2	8	18	32	15	2		1804	Solid
78	Platinum	Pt	195.09	2	8	18	32	16	2		1735	Solid
79	Gold	Au	196.9665	2	8	18	32	18	1		Ancient	Solid
80	Mercury	Hg	200.59	2	8	18	32	18	2		Ancient	Liquid
81	Thallium	Tl	204.37	2	8	18	32	18	3		1861	Solid
82	Lead	Pb	207.2	2	8	18	32	18	4		Ancient	Solid
83	Bismuth	Bi	208.9804	2	8	18	32	18	5		1753	Solid
84	Polonium	Po	209.0	2	8	18	32	18	6		1898	Solid
85	Astatine	At	210.0	2	8	18	32	18	7		1945	Solid
86	Radon	Rn	222.0	2	8	18	32	18	8		1900	Inert gas
87	Francium	Fr	223.0	2	8	18	32	18	8	1	1945	Liquid
88	Radium	Ra	226.0254	2	8	18	32	18	8	2	1898	Solid
89	Actinium	Ac	227.0278	2	8	18	32	18	9	2	1899	Solid

re earth series 58–71 and 90–107 have been omitted.

Table 2-1 lists the periodic table of the elements, in order of their atomic number. The atomic number of an atom describes the number of protons that exist within the nucleus.

The proton and the neutron are almost 2000 times heavier than the very small electron, and so if we ignore the weight of the electron, we can use the fourth column in Table 2-1 (weight of an atom) to give us a clearer picture of the protons and neutrons within the atom's nucleus. For example, a hydrogen atom, as seen in Figure 2-4(a), is the smallest of all atoms and

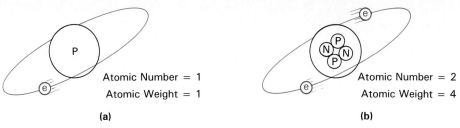

<table>
<tr><td>Atomic Number = 1
Atomic Weight = 1</td><td>Atomic Number = 2
Atomic Weight = 4</td></tr>
<tr><td>(a)</td><td>(b)</td></tr>
</table>

FIGURE 2-4 (a) Hydrogen Atom. (b) Helium Atom.

has an atomic number of 1, which means hydrogen has a one-proton nucleus. Helium, however [Figure 2-4(b)], is second on our table and has an atomic number of 2, indicating that two protons are within the nucleus. The atomic weight of helium, however, is 4, meaning that two protons and two neutrons make up the atom's nucleus.

The number of neutrons within an atom's nucleus can subsequently be calculated by subtracting the atomic number (protons) from the atomic weight (protons and neutrons).

For example, Figure 2-5 illustrates a beryllium atom:

Beryllium	
Atomic number: 4 (protons)	Atomic weight: 9 (protons and neutrons)

If the number of protons is 4, then the number of neutrons is 5 (9 − 4 = 5).

A neutral or balanced atom is one that has an equal number of protons and orbiting electrons, and so the net positive proton charge is equal but opposite to the net negative electron charge, resulting in a balanced or neutral state. For example, Figure 2-6 illustrates a copper atom, which is the most commonly used metal in the field of electronics. It has an atomic number

FIGURE 2-5 Beryllium Atom.

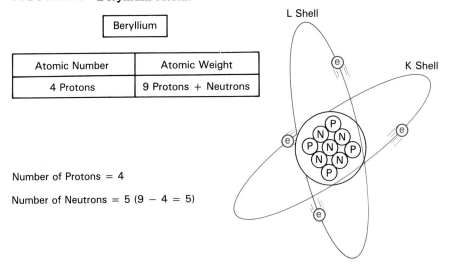

Beryllium	

Atomic Number	Atomic Weight
4 Protons	9 Protons + Neutrons

Number of Protons = 4

Number of Neutrons = 5 (9 − 4 = 5)

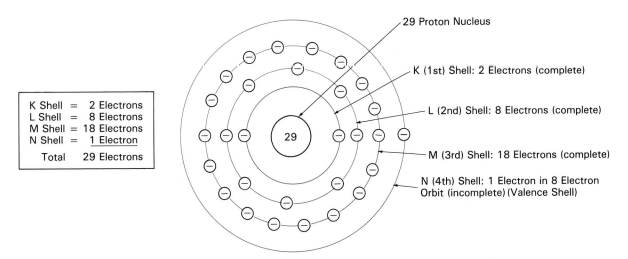

FIGURE 2-6 Copper Atom.

of 29, meaning that 29 protons and 29 electrons exist within the atom when it is in its netural state.

Orbiting electrons travel around the nucleus at varying distances from the nucleus, and these orbital paths are known as *shells* or *bands*. The orbital shell nearest the nucleus is referred to as the first or K shell. The second is known as the L, the third is M, the fourth is N, the fifth is O, the sixth is P, and the seventh is referred to as the Q shell. There are seven shells available for electrons (K, L, M, N, O, P, and Q) around the nucleus, and each of these seven shells can only hold a certain number of electrons, as seen in Figure 2-7. The outermost electron-occupied shell is referred to as the *valence shell* or ring, and these electrons are termed *valence electrons*. In the case of the copper atom, a single valence electron exists in the valence N shell.

All matter exists in one of three states: solids, liquids, and gases. The atoms of a solid are fixed in relation to one another but vibrate in a back and forth motion, unlike liquid atoms that can flow over each other. The atoms of a gas move rapidly in all directions and collide with one another. The far right column of Table 2-1 indicates whether the element is a gas, solid, or liquid.

FIGURE 2-7 Electrons and Shells.

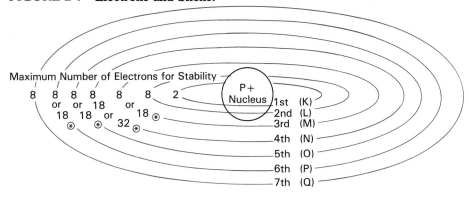

*The maximum number of electrons in these shells is dependent on the element's place in the periodic table.

The Structure of Matter   **29**

2.1.2 ATTRACTION AND REPULSION

For the sake of discussion and understanding, let us theoretically imagine that we are able to separate some positive and negative subatomic particles. Using these separated protons and electrons, let us carry out a few experiments, the results of which are illustrated in Figure 2-8. Studying Figure 2-8, you will notice that:

1. Like charges (positive and positive or negative and negative) repel one another.
2. Unlike charges (positive and negative or negative and positive) attract one another.

Orbiting negative electrons are therefore attracted toward the positive nucleus, which leads us to the question of why the electrons do not fly into the atom's nucleus. The answer is that the orbiting electrons remain in their stable orbit due to two equal but opposite forces. The centrifugal outward force exerted on the electrons due to the orbit counteracts the attractive inward force trying to pull the electrons toward the nucleus due to the unlike charges.

2.1.3 THE MOLECULE

An atom is the smallest unit of a natural element, or an element is a substance consisting of a large number of the same atom.

Combinations of elements are known as compounds, and the smallest unit of a compound is called a *molecule*, just as the smallest unit of an element is an atom. Figure 2-9 summarizes how elements are made up of atoms, and compounds are made up of molecules.

Water is an example of a liquid compound in which the molecule (H_2O) is a combination of an explosive gas (hydrogen) and a very vital gas (oxygen). Table salt is another example of a compound; here the molecule is made up of a highly poisonous gas atom (chlorine) and a potentially explosive solid atom (sodium). These examples of compounds each contain atoms that, when alone, are both poisonous and explosive, yet when combined the resulting substance is as ordinary and basic as water and salt.

FIGURE 2-8   Attraction and Repulsion. (a) Positive Repels Positive. (b) Negative Repels Negative. (c) Unlike Charges Attract.

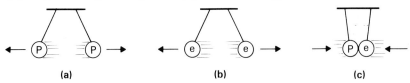

P = Proton (positive)
e = Electron (negative)

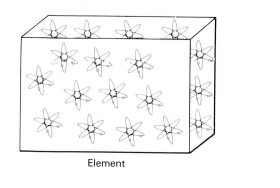

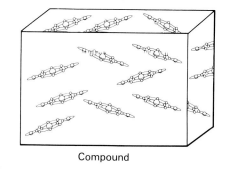

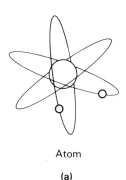

Element

Compound

Atom

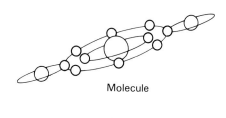

Molecule

(a)   (b)

FIGURE 2-9   (a) An Element Is Made up of Many Atoms. (b) A Compound Is Made up of Many Molecules.

SELF-TEST REVIEW QUESTIONS (§ 2.1)

1. Define the difference between an element and a compound.
2. Name the three subatomic particles that make up an atom.
3. What is the most commonly used metal in the field of electronics.
4. State the laws of attraction and repulsion.

2.2
CURRENT

The movement of electrons from one point to another is known as *electrical current*. Energy in the form of heat or light can cause an outer shell electron to be released from the valence shell of an atom. Once an electron is released, the atom is no longer electrically neutral and is called a *positive ion*, as it now has a net positive charge (more protons than electrons). The released electron tends to jump into a nearby atom, which will then have more electrons than protons and is referred to as a *negative ion*.

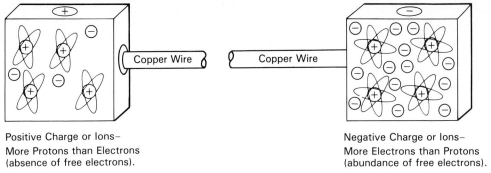

Positive Charge or Ions—
More Protons than Electrons
(absence of free electrons).

Negative Charge or Ions—
More Electrons than Protons
(abundance of free electrons).

FIGURE 2-10 **Positive and Negative Charges.**

Let us now take an example and see how electrons move from one point to another. Figure 2-10 illustrates a metal conductor connecting two charged objects. The metal conductor could be either gold, silver, or copper, but, whichever it is, one common trait can be noticed: the valence electrons in the outermost shell are very loosely bound and can easily be pulled from their parent atom.

The negative ions on the right of Figure 2-11 have more electrons than protons, while the positive ions on the left of Figure 2-11 have fewer electrons than protons and so display a positive charge. The strip of metal joining the two charges has its own atoms, which begin in the neutral condition. Let us now concentrate on one of the negative ions. In Figure 2-11(a), the extra electrons in the outer shells of the negative ions on the right side will feel the attraction of the positive ions on the left side and the repulsion of the outer negative ions, which will cause an electron in a negative ion to jump away from its parent atom's orbit and land in an adjacent atom to the left within the metal wire conductor, as seen in Figure 2-11(b). This adjacent atom now has an extra electron and is called a negative ion, while the initial parent negative ion becomes a neutral atom, which will now attract and receive an electron from one of the other negative ions, because their electrons are also feeling the attraction of the positive ions on the left side and the repulsion of the surrounding negative ions.

The electrons of the negative ion within the metal conductor feel the attraction of the positive ions, and eventually one of its electrons jumps to the left and into the adjacent atom, as shown in Figure 2-11(c). This continual movement to the left will produce a stream of electrons flowing from right to left.

Millions upon millions of atoms within the conductor pass a continuous movement of billions upon billions of electrons from right to left. This electron flow is known as electrical current.

To summarize, we could say that as long as a force or pressure, produced by the positive charge and negative charge, exists it will cause electrons to flow from the negative to the positive terminal. The positive side has a deficiency of electrons and the negative side has an abundance, and so a continuous flow or migration of electrons takes place between the negative and positive terminal through our metal conducting wire. This electric current or electron flow is a measurable quantity, as will now be explained.

2.2.1 COULOMBS PER SECOND

There are 6.24×10^{18} electrons in 1 coulomb, as illustrated in Figure 2-12. (If you are not familiar with the meaning of 10^{18}, refer to the discussion on exponents in the appendices). To calculate coulombs of charge (designated Q), we can use the formula

$$\text{charge, } Q = \frac{\text{Total number of electrons } (n)}{6.24 \times 10^{18}}$$

where Q = electric charge in coulombs

FIGURE 2-11 Electron Migration due to Forces of Positive Attraction and Negative Repulsion on Electrons.

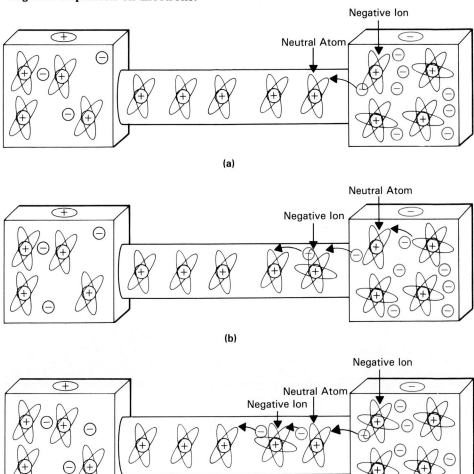

(a)

(b)

(c)

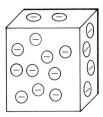

(a) 6.24×10^{18} Electrons
= 1 Coulomb of Charge

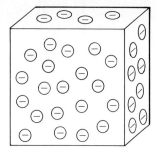

(b) 12.48×10^{18} Electrons
= 2 Coulombs of Charge

**FIGURE 2-12   (a) One Coulomb of Charge.
(b) Two Coulombs of Charge.**

EXAMPLE 2.1

If a total of 3.75×10^{19} free electrons exists within a piece of metal conductor, how many coulombs of charge would be within this conductor?

SOLUTION

$$Q = \frac{n}{6.24 \times 10^{18}}$$

$$= \frac{3.75 \times 10^{19}}{6.24 \times 10^{18}}$$

$$= 6 \text{ coulombs}$$

A total of 6 coulombs of charge exists within our conductor.

Calculator Sequence

Step	Keypad Entry	Display Response
1.	③ ⨀ ⑦ ⑤ ⒠ (Exponent) ① ⑨	3.75E19
2.	÷	
3.	⑥ ⨀ ② ④ ⒠ ① ⑧	6.24E18
4.	⊜	6.0096

EXAMPLE 2.2

A metal wire conductor has 14 coulombs of electrons passing through it. Calculate the number of electrons moving through the conductor.

SOLUTION

To calculate the number of electrons (n), we have to transpose our original formula. If

$$Q = \frac{n}{6.24 \times 10^{18}}$$

then

$$n = Q \times 6.24 \times 10^{18}$$

The total number of electrons (n) is

34   **Voltage and Current**

$$n = Q \times 6.24 \times 10^{18}$$
$$= 14 \text{ coulombs} \times 6.24 \times 10^{18}$$
$$= 8.7 \times 10^{19} \quad \text{electrons will}$$
pass through the conductor

2.2.2 THE AMPERE

A coulomb is a static amount of electric charge. In electronics, we are more interested in electrons in motion, and current is the flow of electrons, which involves time. Coulombs and time are therefore combined to describe the number of electrons and the rate at which they flow. This relationship is called *current* (I) flow and has the unit of amperes (A), abbreviated amps. If 6.24×10^{18} electrons (1 coulomb) were to drift past a specific point on a conductor in 1 second of time, then 1 ampere of current is said to be flowing:

$$\boxed{\text{current}, I = \frac{\text{coulombs } (Q)}{\text{time } (t)}}$$

$$1 \text{ ampere} = 1 \text{ coulomb per 1 second}$$
$$1 \text{ A} = \frac{1 \text{ C}}{1 \text{ s}}$$

To summarize, we could say that 1 ampere equals a flow rate of 1 coulomb per second, and current is measured in amps.

EXAMPLE 2.3

If 5×10^{19} electrons pass a point in a conductor in 4 seconds, what is the amount of current flow in amperes?

SOLUTION

Current (I) is equal to Q/t. We must first convert our electrons to coulombs.

$$Q = \frac{n}{6.24 \times 10^{18}}$$
$$= \frac{5 \times 10^{19}}{6.24 \times 10^{18}}$$
$$= 8 \text{ coulombs}$$

Now, to calculate the amount of current, we use the formula

$$I = \frac{Q}{t}$$

$$= \frac{8 \text{ coulombs}}{4 \text{ seconds}}$$
$$= 2 \text{ amperes}$$

This means that 2 amperes of 1.248×10^{19} electrons are passing a specific point in the conductor every second.

Calculator Sequence

Step	Keypad Entry	Display Response
1.	$\boxed{5}$ $\boxed{\text{E}}$ (exponent) $\boxed{1}$ $\boxed{9}$	5E19
2.	$\boxed{\div}$	
3.	$\boxed{6}$ $\boxed{.}$ $\boxed{2}$ $\boxed{4}$ $\boxed{\text{E}}$ $\boxed{1}$ $\boxed{8}$	6.24E18
4.	$\boxed{=}$	8.012
5.	$\boxed{\div}$	
6.	$\boxed{4}$	2.003
7.	$\boxed{=}$	2.003

2.2.3 UNITS OF CURRENT

Current within electronic equipment is normally a value in milliamps or microamps and very rarely exceeds 1 amp. Table 2-2 lists all the prefixes related to current. For example, one milliamp is one-thousandth of an amp, which means if 1 amp were divided into 1000 parts, one part of the 1000 would be flowing through the circuit.

TABLE 2-2
CURRENT UNITS

Name	Symbol	Value
Picoampere	pA	$10^{-12} = \dfrac{1}{1\,000\,000\,000\,000}$
Nanoampere	nA	$10^{-9} = \dfrac{1}{1\,000\,000\,000}$
Microampere	μA	$10^{-6} = \dfrac{1}{1\,000\,000}$
Milliampere	mA	$10^{-3} = \dfrac{1}{1\,000}$
Ampere	A	$10^{0} = 1$
Kiloampere	kA	$10^{3} = 1000$
Megaampere	MA	$10^{6} = 1\,000\,000$
Gigaampere	GA	$10^{9} = 1\,000\,000\,000$
Teraampere	TA	$10^{12} = 1\,000\,000\,000\,000$

EXAMPLE 2.4

Convert the following:
(a) 0.003 A = _____ mA (milliamps)
(b) 0.07 mA = _____ μA (microamps)
(c) 7333 mA = _____ A (amps)
(d) 1275 μA = _____ mA (milliamps)

SOLUTION

(a) In this example, 0.003 A has to be converted so it is represented in milliamps (10^{-3} or $\frac{1}{1000}$th of an amp). The basic algebraic rule to be remembered is that both equations on either side of the equals must be the same.

LEFT		RIGHT	
Number	Exponent	Number	Exponent
0.003×10^0		$=$ _____ $\times 10^{-3}$	

The exponent on the right in this example is going to be decreased a thousand times (10^0 to 10^{-3}) and so for the statement to balance the number on the right will have to be increased a thousand times; that is, the decimal point will have to be moved to the right three places (0.003 or 3). Therefore,

$$0.003 \times 10^0 = 3 \times 10^{-3}$$

or

$$0.003 \text{ Amp} = 3 \times 10^{-3} \text{ A or 3 mA}$$

(b) In this example the exponent is going from milliamperes to microamperes (10^{-3} to 10^{-6}) or 1000 times smaller, and so the number must be made 1000 times greater.

$$0.07 \text{ or } 70.0$$

Therefore, $0.07 \text{ mA} = 70 \text{ } \mu\text{A}$.

(c) The exponent is going from milliamperes to amperes, increasing 1000 times, so the number must decrease 1000 times.

$$7333. \text{ or } 7.333$$

Therefore, $7333 \text{ mA} = 7.333 \text{ A}$.

(d) The exponent is changing from microamperes to milliamperes, a 1000 times increase, so the number must decrease by the same factor.

$$1275.0 \text{ or } 1.275$$

Therefore, $1275 \text{ } \mu\text{A} = 1.275 \text{ mA}$.

2.2.4 THE SPEED OF CURRENT FLOW

Electrons will in fact move very slowly as they jump from parent to adjacent atom; however, the chain reaction occurs at the speed of light, which is 186,000 miles per second or 300,000,000 meters per second (3×10^8 m/s). This chain reaction is best understood by using an analogy where we relate free electrons within a conductor to a string of ping-pong balls within a tube.

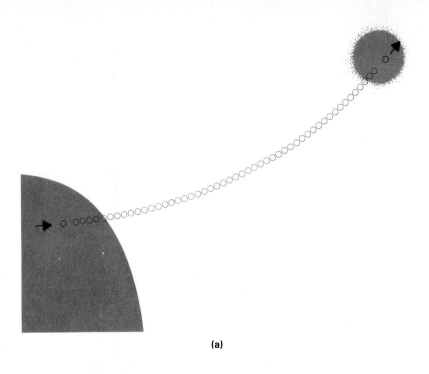

(a)

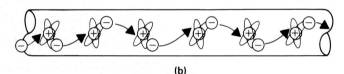

(b)

FIGURE 2-13   Chain Reaction. (a) Ping Pong Ball Analogy.
(b) Electrons.

If one extra ball is inserted in one end, a ball will appear out of the other end almost instantly. Although each ball within the tube has only moved a small distance, the effect has traveled toward the end of the tube at the speed of light, as seen in Figure 2-13(a).

As far as electrons go, the first electron jump from parent to adjacent atom causes the next to jump, and the next, and next, and so on, just as the first ball causes a chain reaction with the others; and even though the actual speed of the electrons is only a fraction of an inch per second, the effective velocity is at the speed of light, as seen in Figure 2-13(b).

2.2.5 CONVENTIONAL VERSUS ELECTRON FLOW

Electrons drift from a negative to a positive charge, as illustrated in Figure 2-14. This current, as already discussed, is known as *electron flow*.

In the eighteenth and nineteenth century, however, when very little was known about the atom, researchers believed wrongly that current was a flow of positive charges. Although this has now been proved incorrect, many texts still use conventional current flow, as shown in Figure 2-15.

Voltage and Current

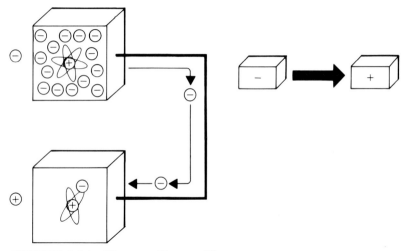

FIGURE 2-14   **Electron Current Flow.**

Whether conventional current flow or electron flow is used, the same answers to problems, measurements, or designs are obtained. The key point to remember is that direction is not important, but the amount of current flow is. On one hand, electron flow is utilized by some people who prefer working with the truth; on the other hand, some people involved with complex mathematical analysis tend to utilize conventional flow because it was used in the numerous volumes of formulas that were produced before the electron was discovered to be the moving subatomic particle.

Throughout this text, we will be using electron flow so that we can relate back to the atom when necessary; but if you desire to use conventional flow, then just reverse the direction of the arrows. But be consistent with your choice of flow to avoid confusion.

FIGURE 2-15   **Conventional Current Flow.**

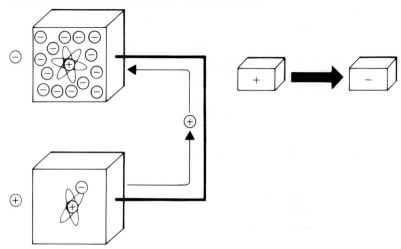

2.2.6 HOW IS CURRENT MEASURED?

Ammeters (ampere meters) are used to measure the current flow within a circuit, as seen in Figure 2-16. In electronics, where current is generally small, the milliamp (mA) or microamp (μA) scale will generally be used, whereas in electrical high-current circuits, the meter will generally be used on an amp scale.

Three rules must be remembered and applied when measuring current:

1. Always set the selector to the higher scale first (amps), and then reduce as needed to milli- or microamps, just in case a larger current than anticipated is within the circuit, in which case the meter could be damaged.

To explain this, let us take an example: If you were to weigh some potatoes on a set of scales and you were not aware of how much they weigh, the largest scale should initially be selected. For example, if you selected the 0 to 10 pound scale and the potatoes weighted 15 pounds, the meter needle would be forced violently to the right side and could damage the scales (this is known as pegging the meter). By always selecting the 0 to 50 pound scale and then stepping down to 0 to 40, 0 to 30, and 0 to 20 pounds, as needed, you would almost guarantee never pegging the meter. With the ammeter, we must apply the same philosophy and first select a higher scale and then work down if needed.

2. The current meter must be connected so that the positive lead of the ammeter (red) is connected to the positive charge or polarity,

FIGURE 2-16 **Measuring Current with the Ammeter.**

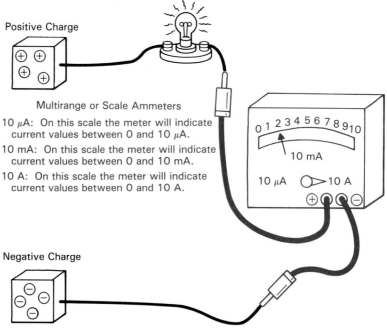

Positive Charge

Multirange or Scale Ammeters

10 μA: On this scale the meter will indicate current values between 0 and 10 μA.

10 mA: On this scale the meter will indicate current values between 0 and 10 mA.

10 A: On this scale the meter will indicate current values between 0 and 10 A.

Negative Charge

and the negative lead of the ammeter (black) is connected to the negative charge or polarity. Ammeters are sensitive to the polarity (positive or negative) of the charge.

3. If you wish to measure current flowing within a wire, the ammeter measuring the current must be placed in the path of current flow.

SELF-TEST REVIEW QUESTIONS (§ 2.2)

1. What is the unit of current?
2. Define current in relation to coulombs and time.
3. What is the difference between conventional and electron current flow?
4. List the three rules that should be applied when using the ammeter to measure current.

2.3
VOLTAGE

Voltage is the force or pressure exerted on electrons. Referring to Figure 2-17(a) and (b), you will notice two situations.

Figure 2-17(a) shows highly concentrated positive and negative charges or potentials connected to one another by a copper wire. In this situation, a large potential difference or voltage is being applied across the copper atom's electrons. This force or voltage causes a large amount of copper atom electrons to move from right to left.

Figure 2-17(b) illustrates a low concentration of positive and negative potentials, and so a small voltage or pressure is being applied across the conductor, causing a small amount of force and, therefore, current to move from right to left.

To summarize, then, we could say that a highly concentrated charge produces a high voltage, whereas a low concentrated charge produces a low voltage. Voltage is also appropriately known as the electron moving force or *electromotive force* (emf), and since two opposite potentials exist, one negative and one positive, the strength of the voltage can also be referred to as the amount of potential difference (pd) applied across the circuit. Referring back to Figure 2-17(a), we see that a large voltage, electromotive force, or potential difference exists across the copper conductor, while in Figure 2-17(b) a small voltage, potential difference, or electromotive force is exerted across the conductor.

Voltage is the force, pressure, potential difference (pd), or electromotive force (emf) that causes electron flow or current and is symbolized by italic uppercase V. The unit for voltage is the volt, symbolized by roman uppercase V. This can become a bit confusing when, for example, the voltage applied to a circuit equals 5 volts, the circuit notation would appear as

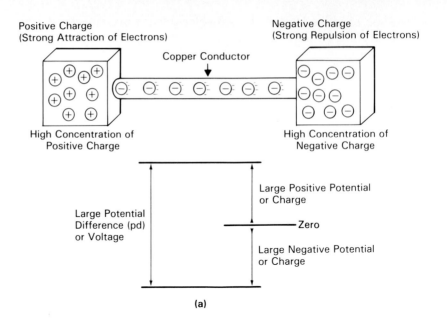

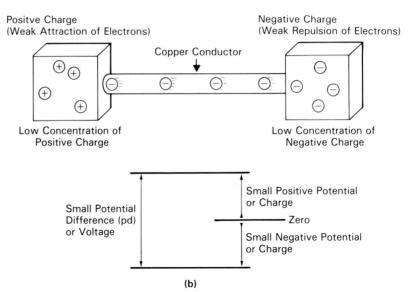

FIGURE 2-17 **(a) Large Potential Difference or Voltage. (b) Small Potential Difference or Voltage.**

$$V = 5 \text{ V}$$

This statement is algebraically incorrect because 1 volt cannot equal 5 volts. To avoid confusion, some texts and circuits use E, symbolizing electromotive force, to represent voltage; for example,

$$E = 5 \text{ V}$$

In this text, however, we will maintain the original designation for voltage (V).

2.3.1 SYMBOLS

Voltage is electrical energy and is produced by converting some other form of energy into electrical energy. A battery, which is illustrated in Figure 2-18, is such a device and converts chemical energy into electrical energy. At the positive terminal of the battery, positive charges or ions (atoms with more protons than electrons) are present, and at the negative terminal, negative charges or ions (atoms with more electrons than protons) are available to supply electrons for current flow within a circuit. A battery chemically generates negative and positive ions at its respective terminals. The symbol for the battery is illustrated in Figure 2-19.

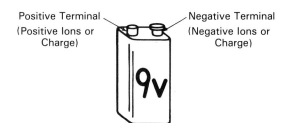

FIGURE 2-18 **Battery.**

In Figure 2-20, we have illustrated two other devices as they appear physically and in their electronic schematic symbols. The first new device is the light bulb, and the other is a piece of copper conductor wire with an alligator or crocodile clip on either end.

The 9-V battery seen in Figure 2-20(b) chemically generates positive and negative ions. The negative ions at the negative terminal force away the negative electrons, which are attracted by the positive charge or absence of electrons on the other terminal. Proceeding through the copper conductor wire, jumping from one atom to the next, they eventually reach the bulb. The electrons pass through the bulb, which glows due to this current passing through it. When emerging from the light bulb, the electrons travel through another connector cable and finally reach the positive terminal of the battery.

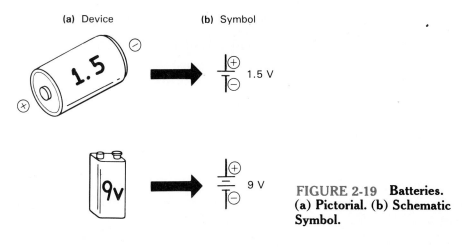

FIGURE 2-19 **Batteries. (a) Pictorial. (b) Schematic Symbol.**

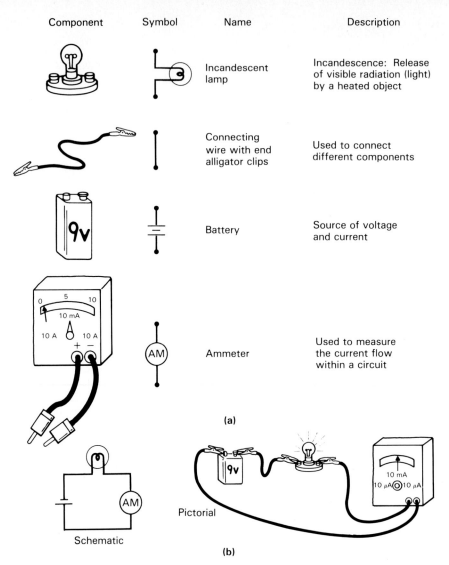

Component	Symbol	Name	Description
		Incandescent lamp	Incandescence: Release of visible radiation (light) by a heated object
		Connecting wire with end alligator clips	Used to connect different components
		Battery	Source of voltage and current
		Ammeter	Used to measure the current flow within a circuit

(a)

Schematic

Pictorial

(b)

FIGURE 2-20 (a) Components. (b) Example Circuit.

Studying Figure 2-20(b), you will notice two reasons why the circuit is drawn using symbols rather than illustrating the physical appearance:

1. A circuit with symbols is faster and more easily drawn.
2. A circuit with symbols has less detail and clutter and is consequently more easily comprehended because of the fewer distracting elements.

2.3.2 VOLTAGE UNITS

The unit for voltage is the volt (V). Voltage within electronic equipment is normally measured in volts, whereas heavy-duty industrial equipment normally requires high voltages that are generally measured in kilovolts (kV). Table 2-3 lists all the prefixes and values related to volts.

TABLE 2-3
VOLTAGE UNITS

Name	Symbol	Value
Picovolts	pV	$10^{-12} = \dfrac{1}{1\ 000\ 000\ 000\ 000}$
Nanovolts	nV	$10^{-9} = \dfrac{1}{1\ 000\ 000\ 000}$
Microvolts	μV	$10^{-6} = \dfrac{1}{1\ 000\ 000}$
millivolts	mV	$10^{-3} = \dfrac{1}{1000}$
volts	V	$10^{0} = 1$
kilovolts	kV	$10^{3} = 1000$
megavolts	MV	$10^{6} = 1\ 000\ 000$
gigavolts	GV	$10^{9} = 1\ 000\ 000\ 000$
teravolts	TV	$10^{12} = 1\ 000\ 000\ 000\ 000$

EXAMPLE 2.5

Convert the following:
a. 3000 V = _____ kV (kilovolts)
b. 0.14 V = _____ mV (millivolts)
c. 1500 kV = _____ MV (megavolts)

SOLUTION

a. 3000 V = 3 kV or 3×10^{3} volts (exponent ↑ 1000, number ↓ 1000)
b. 0.14 V = 140 mV or 140×10^{-3} volts (exponent ↓ 1000, number ↑ 1000)
c. 1500 kV = 1.5 MV or 1.5×10^{6} volts (exponent ↑ 1000, number ↓ 1000)

2.3.3 HOW IS VOLTAGE MEASURED?

Voltmeters (voltage meters) are used to measure electrical pressure (voltage) anywhere in an electronic circuit. Figure 2-21 illustrates a voltmeter being used to measure a battery's voltage.

A voltmeter is connected across a device, as illustrated in Figure 2-22(a) and (b), with the two leads of the voltmeter touching either side of the component, in this case a light bulb. Figure 2-22(a) indicates how to measure the voltage across light bulb 1 (L1), whereas Figure 2-22(b) indicates how to measure the voltage across light bulb 2 (L2).

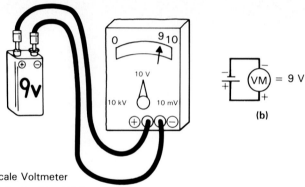

 = 9 V

(b)

Multirange or Scale Voltmeter

10 mV: This scale can be used to measure any voltage from 0 to 10 mV.

10 V: This scale can be used to measure any voltage from 0 to 10 V.

10 kV: This scale can be used to measure any voltage from 0 to 10 kV.

(a)

FIGURE 2-21   Using the Voltmeter to Measure Voltage. (a) Pictorial. (b) Schematic.

FIGURE 2-22   Measuring Components across (a) Lamp 1 and (b) Lamp 2.

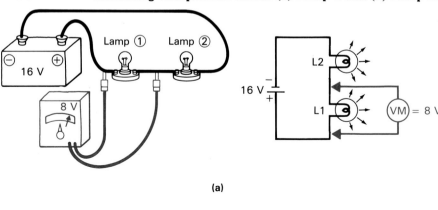

(a)

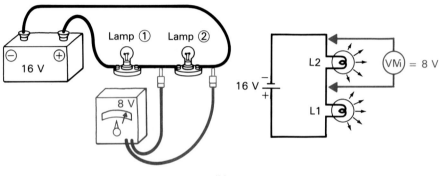

(b)

Three techniques must be remembered and applied when measuring voltage:

1. Always set to the higher scale first (kV) and then reduce as needed to the volt and millivolt range, just in case a larger voltage than anticipated is within the circuit, in which case the meter could be damaged.

2. Always connect the positive lead (red) of the meter to the component lead that is closer to the positive side of the battery, and the negative lead (black) of the meter to the component lead that is closer to the negative side of the battery.

3. If you are measuring the voltage across a component, then the voltmeter must be connected across the component and outside of the circuit.

In some applications, fixed-range voltmeters are used, for example, a voltage indicator within a car showing the condition of the battery (Figure 2-23). In this application, a fixed voltmeter of 0 to 15 V would be sufficient.

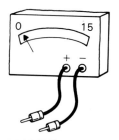

FIGURE 2-23   Fixed Range Voltmeter.

2.3.4 FLUID ANALOGY OF CURRENT AND VOLTAGE

In Figure 2-24(a), a system using a pump, pipes, and a water wheel is being used to convert electrical energy into mechanical energy. The electrical energy is in the form of voltage, which is applied to the pump to send it into operation and cause water to flow. The pump generates:

1. A high pressure at the outlet port, which forces the water molecules out and into the system.

2. A low pressure in the inlet port, which attracts the water molecules into the pump.

The water current flow is in the direction indicated, and the high pressure or potential within the piping will be used to drive the water wheel around, producing mechanical energy. The remaining water is attracted into the pump due to the suction or low pressure existing on the inlet port. In fact, the amount of water entering the inlet port is the same as the amount of water leaving the outlet port. It can therefore be said that the water flow rate is the same throughout the circuit. The only changing element is the pressure felt at different points throughout the system.

In Figure 2-24(b), our electric circuit containing a battery, conductors, and a bulb is being used to convert electrical energy into light energy. The battery generates a voltage just as the pump generates pressure. This voltage causes electrons to move through conductors, just as pressure causes water molecules to move through the piping. The amount of water flow is dependent on the pump's pressure, and the amount of current or electron flow is dependent on the battery's voltage. Water flow through the wheel can be compared to current flow through the bulb. The high pressure is lost in turning the wheel and producing mechanical energy, just as voltage is lost in producing light energy out of the bulb. We cannot say that pressure or

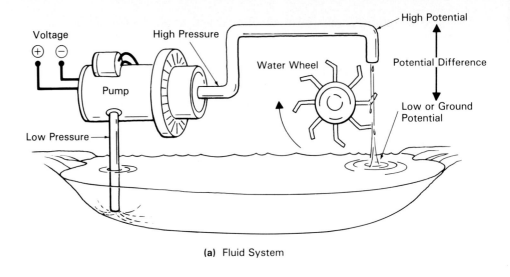

(a) Fluid System

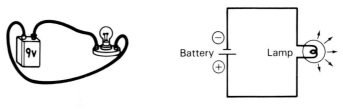

(b) Electrical System

(a) Fluid System		**(b)** Electrical System
Pump generates pressure, which is the water moving force.	Equivalent to (≅)	Battery generates voltage, which is the electron moving force.
Water current flow.		Electron current flow.
High pressure or potential.		High voltage or potential.
Low pressure or potential.		Low voltage or potential.

FIGURE 2-24 **Comparison between a Fluid System and an Electrical System.**

voltage flows: Pressure and voltage exist and cause water and current to flow, and it is this flow that is converted to mechanical energy in our fluid system and light energy in our electrical system. Voltage is the force of repulsion and attraction needed to cause current to flow through a circuit; without voltage, there cannot be current.

2.3.5 CURRENT IS DIRECTLY PROPORTIONAL TO VOLTAGE

Referring to either the fluid system or electrical circuit in Figure 2-24, you can easily see that the flow is proportional to the pressure, which means: If the pump were to generate a greater pressure, then a larger amount of water would flow through the system.

pressure ↑   water flow ↑

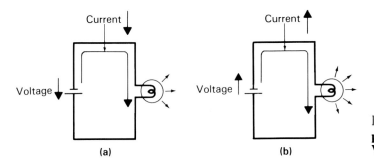

FIGURE 2-25 (a) Small Voltage produces a Small Current. (b) Large Voltage Produces a Large Current.

Similarly, if a larger voltage were applied to the electrical circuit, then this larger electron moving force (emf) would cause more electrons or current to flow through the circuit, as seen in Figure 2-25.

Current is therefore said to be directly proportional to voltage, as a voltage increase causes a current increase and, similarly, a voltage decrease results in a subsequent current decrease.

voltage ↑   current flow ↑

Current is directly proportional to voltage ($I \propto V$).

2.3.6 SUMMARY OF MEASURING CURRENT AND VOLTAGE

Current (Ammeter)	*Voltage (Voltmeter)*

1. Always set the meter to the highest scale first.

2. Ensure the positive lead of the meter is connected back to the positive side of the voltage supply (for example, battery), and the negative lead of the meter is connected back to the negative side of the supply voltage.

3. Always connect the ammeter in the path of current flow, so the electrons have to pass through the meter.	3. Always connect the voltmeter across the component so that the pressure or potential difference change across the component can be measured.

Figure 2-26 summarizes these measurements.

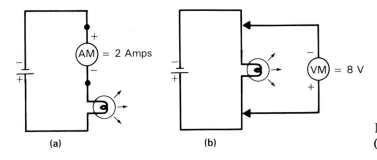

FIGURE 2-26 Measurement of (a) Current and (b) Voltage.

1. What is the unit of voltage?
2. Convert 3 MV to kilovolts.
3. List the three rules that should be applied when using the voltmeter to measure voltage.
4. What is the relationship between current and voltage?

2.4
CONDUCTORS

Materials that pass current easily, that is, offer little opposition (low resistance) to current, are called conductors.

As mentioned previously, the atom has a maximum of seven orbital paths known as shells, which are named K, L, M, N, O, P, and Q, stepping out toward the outermost or valence shell.

A lightning bolt that splits or sets fire to a tree and the operation of your calculator are both electrical results achieved by the flow of electrons. The only difference is that your calculators circuits control the flow of electrons, unlike the lightning bolt, which is the uncontrolled flow of electrons.

Conductors are materials or natural elements whose valence electrons can be easily removed from their parent atoms. They are therefore referred to as sources of free electrons, and these free electrons provide us with circuit current. The precious metals of silver and gold are the best conductors. More specifically, a better conductor has:

1. Electrons in shells the farthest away from the nucleus and these electrons feel very little nucleus attraction and can be broken away from their atom quite easily.
2. More electrons per unit volume amount
3. An incomplete valence shell. This means that the valence shell does not have the maximum amount of electrons possible in it. If the atom had its valence ring complete (full), there would be no holes (absence of an electron) in that shell, and so no encouragement for adjacent atom electrons to jump from their parent atom into the next atom and so produce the chain reaction known as current.

Economy must be considered when choosing a conductor. Large quantities of conductors using precious metals are obviously going to send the cost of equipment beyond reach.

The conductor must also satisfy some physical requirements in that we must be able to shape it into wires of different sizes and easily bend it to allow us to connect one circuit to the next.

Copper is the most commonly used conductor as it meets the following three requirements:

1. It is a good source of electrons.
2. It is inexpensive.
3. It is physically pliable.

Aluminum is also a very popular conductor and, although it does not possess as many free electrons as copper, it has the two advantages of being less expensive and lighter than copper.

2.4.1 CONDUCTANCE

Conductance is the measure of how good a conductor is at carrying current. Conductance (symbolized G) is equal to the reciprocal of resistance and is measured in the unit of siemens (S).

$$\text{conductance } (G) = \frac{1}{\text{resistance } (R)}$$

This means that conductance is inversely proportional to resistance. For example, if the opposition to current flow (resistance) is low, the conductance is high and the material is said to have a good conductance.

$$\text{high conductance } G \uparrow = \frac{1}{R \downarrow \text{ (low resistance)}}$$

Conductance (G) is in siemens (S) and resistance (R) is in ohms (Ω).

On the other hand, if the resistance of a conducting wire is high ($R \uparrow$), then its conductance value is low ($G \downarrow$) and it is therefore called a poor conductor. To summarize, we say that a good conductor has a good conductance value and a very small opposition or resistance to current flow.

EXAMPLE 2.6

A household electric blanket offers 25 ohms of resistance against current flow. Calculate the conductance of the electric blanket's heating element.

SOLUTION

$$\text{conductance} = \frac{1}{\text{resistance}}$$

$$G = \frac{1}{R}$$

$$= \frac{1}{25 \text{ ohms}}$$
$$= 40 \text{ millisiemens}$$
$$= 40 \text{ mS}$$

To further reinforce our understanding of a conductor, let's take an example and compare a good conductor (copper) to a poor conductor (carbon). Refer also to carbon and copper in the atomic periodic table (Table 2-1).

EXAMPLE 2.7

Carbon

1. The valence ring is the L or second shell, which feels a strong nucleus attractive force, discouraging the release of free electrons and so circuit current.
2. Only six electrons orbit the carbon atom, making it a bad source of electrons.
3. The valence shell (L, maximum eight electrons) is already half complete (four electrons) and so few holes (only four) exist, discouraging the jumping of electrons from parent atoms into adjacent atom holes.

Copper

1. The valence ring is the N or fourth shell, which feels only a weak nucleus attractive force, encouraging the release of free electrons and so circuit current.
2. A total of 29 electrons orbit the nucleus, making it a good source of electrons.
3. The valence shell (N, maximum 32 electrons) is incomplete as only one electron occupies it when neutral, and so 31 holes exist, encouraging the jumping of electrons from parent atoms into adjacent atom holes.

To express how good or poor a conductor is, we must specify a reference point. The reference point we use is the best conductor, silver, which has a conductivity value of 1.0. Table 2-4 lists other conductors and their relative conductivity values with respect to the best, silver.

TABLE 2-4
RELATIVE CONDUCTIVITY OF CONDUCTORS

Material	Conductivity (Relative)
Silver	1.000
Copper	0.945
Aluminum	0.575
Tungsten	0.297
Nichrome	0.015

EXAMPLE 2.8

What is the relative conductivity of tungsten if copper is used as the reference conductor?

SOLUTION

$$\text{tungsten} = 0.297$$
$$\text{copper} = 0.945$$
$$\text{relative conductivity} = \frac{\text{conductor}}{\text{reference}} = \frac{0.297}{0.945}$$
$$= 0.314$$

Calculator Sequence

Step	Keypad Entry	Display Response
1.	⓪ · ② ⑨ ⑦	0.297
2.	÷	
3.	⓪ · ⑨ ④ ⑤	0.945
4.	=	0.314

EXAMPLE 2.9

What is the relative conductivity of silver if copper is used as the reference?

SOLUTION

$$\text{relative conductivity} = \frac{\text{silver}}{\text{copper}}$$
$$= \frac{1.000}{0.945} = 1.058$$

SELF-TEST REVIEW QUESTIONS (§ 2.4)

1. A conductor is a material used to block the flow of current (true or false).
2. List the three atomic properties that make a better conductor.
3. What is the most commonly used conductor in the field of electronics?
4. Calculate the conductance of a 35-Ω heater element.

2.5
INSULATORS

Materials that are used to block current, that is, they offer high resistance or opposition to current flow, are called insulators. Just as certain materials permit the easy flow of current and so have good conductivity, certain materials allow small to almost no amount of free electrons to flow. These materials are known as insulators. Insulators can, with sufficient pressure or voltage applied across them, "break down" and conduct current; that is, the voltage must be great enough to dislodge the electrons from their close orbital shells (K, L shells) and send them off as free electrons.

A good insulator or dielectric should have the maximum possible resistance and conduct no current at all. To express how good or poor an insulator is, we list a voltage that, when applied across one-thousandth of an inch of this insulator material, will cause it to break down and conduct a large current. This measure of an insulator is known as its *dielectric strength*. Table 2-5 lists some of the more popular insulators and the value of kilovolts that will cause a centimeter of insulator to break down. From Table 2-5, we say that, if one centimeter of paper is connected to a variable voltage source, a voltage of 500 kilovolts is needed to break down the paper and cause current to flow.

TABLE 2-5
BREAKDOWN VOLTAGES OF CERTAIN INSULATORS

Material	Breakdown Strength (kV/cm)
Mica	2000
Glass	900
Teflon	600
Paper	500
Rubber	275
Bakelite	151
Oil	145
Porcelain	70
Air	30

EXAMPLE 2.10

What thickness of mica would be needed to withstand 16,000 V?

SOLUTION

$$\text{mica strength} = 2000 \text{ kv/cm}$$

$$\text{dielectric thickness} = \frac{16,000 \text{ V}}{2000 \text{ kV}} = 0.008 \text{ cm}$$

Calculator Sequence

Step	Keypad Entry	Display Response
1.	1 6 E (exponent) 3	16E3
2.	$\div$	
3.	2 E 6 (1000 + K = E6) $=$	2E6
4.		0.008

Voltage and Current

EXAMPLE 2.11

What maximum voltage could 1 mm of air withstand?

SOLUTION

There are 10 mm in 1 cm. If air can withstand 30,000 V/cm, then it can withstand 3000 V/mm.

SELF-TEST REVIEW QUESTIONS (§ 2.5)

1. An insulator is a material used to block the flow of current (true or false).
2. What is considered to be the best insulator material?
3. Define breakdown voltage.
4. Would the conductance figure of a good insulator be large or small?

2.6
SEMICONDUCTORS

Certain materials are neither insulators (high resistance to current) nor conductors (low resistance to current), but in fact fall between the two, as seen in Figure 2-27. These materials are known as *semiconductors* because they

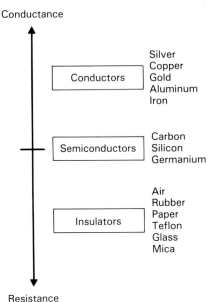

FIGURE 2-27 **Elements and Compounds Used in Electronics.**

conduct less than a metal conductor, but more than an insulator. Silicon is the most commonly used semiconductor for components such as transistors and integrated circuits.

2.7
THE OPEN, CLOSED, AND SHORT CIRCUIT

In Figure 2-28, a simple circuit using a battery, conductors, light bulb, and a switch is shown. A new component, the switch, is illustrated in Figure 2-28(b), and the schematic symbol is shown in Figure 2-28(a).

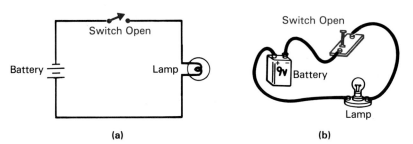

FIGURE 2-28   An Open Switch Causing an Open Circuit.
(a) Schematic. (b) Pictorial.

1. OPEN CIRCUIT (OPEN SWITCH)

An opened switch (Figure 2-28) produces an "open" in the circuit, which prevents current flow as the extra electrons in the negative terminal of the battery cannot feel the attraction of the positive terminal due to the break in the path. The opened switch has produced an open circuit.

2. OPEN CIRCUIT (OPEN COMPONENT)

If the switch is now closed so as to make a complete path in the circuit, an open could still exist due to the failure of one of the components. For example, in Figure 2-29, the light bulb filament has burnt out; in so doing, it creates an open in the circuit and again there will be no current flow. The open component also produced an open circuit.

FIGURE 2-29   An Open Lamp Filament Causing an Open Circuit. (a) Pictorial. (b) Schematic.

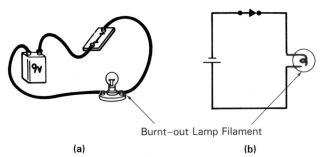

3. CLOSED CIRCUIT (CLOSED SWITCH)

A closed switch produces a closed circuit; current now has a complete path from the negative to positive terminal, as seen in Figure 2-30. The closed switch produces a closed circuit.

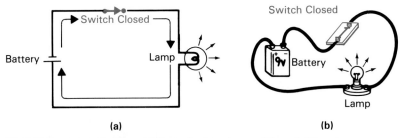

(a)   **(b)**

FIGURE 2-30   **A Closed Switch Causing a Closed Circuit. (a) Schematic. (b) Pictorial.**

4. SHORT CIRCUIT (SHORTED COMPONENT)

A short circuit normally occurs when one point is accidentally connected to another. Figure 2-31(a) illustrates the physical appearance of the circuit and how the accident occurred. A set of pliers was accidentally laid across the two contacts connecting the light bulb. Figure 2-31(b) shows the schematic illustration with the effect of the short across the light bulb drawn in. All the current will flow through the metal of the pliers, which offers no resistance or opposition to the current flow; since the light bulb does have some resistance or oppostition, almost no current will flow through it, and therefore no light will be produced.

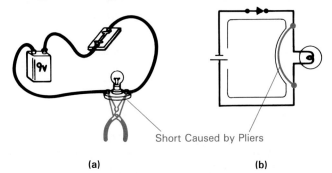

(a)   **(b)**

FIGURE 2-31   **Pliers Causing a Short Circuit across a Lamp. (a) Pictorial. (b) Schematic.**

SELF-TEST REVIEW QUESTIONS (§ 2.6 AND § 2.7)

1. What is a semiconductor?
2. What is the name of the most commonly used semiconductor material?
3. Does current flow through an open circuit?
4. Define the difference between a closed circuit and short circuit.

SUMMARY

1. All matter on, in, and around the earth can be classified as being either a solid, liquid, or gas.
2. An element is a material consisting of only one type of atom.
3. Protons, neutrons, and electrons are subatomic particles that make up the atom.
4. The atomic number of an atom describes the number of protons within the nucleus.
5. The atomic weight of an atom can be used to determine the number of protons and neutrons within an atom's nucleus.
6. Elliptically orbiting electrons travel in paths or shells that are labeled K, L, M, N, O, P, and Q and extend out from the nucleus.
7. Like charges repel one another, while unlike charges attract.
8. An atom is the smallest particle of an element. A molecule (which is the combination of two or more elements) is the smallest part of a compound.
9. Current is the movement of electrons from one point to another.
10. A neutral atom has an equal number of protons (positive charge) and electrons (negative charge).
11. A negative ion has more electrons than protons and subsequently possesses a net negative charge.
12. A positive ion has more protons than electrons and subsequently possesses a net positive charge.
13. There are 6.24×10^{18} electrons in 1 coulomb of charge.
14. One ampere of current is said to be flowing when 1 coulomb of charge passes a point in 1 second.
15. When measuring current with an ammeter, you should always:
    a. Set to a high scale initially.
    b. Ensure lead polarity is correct.
    c. Break current path and insert ammeter.
16. Voltage is the force or pressure that causes electrons to flow.
17. When measuring voltage with a voltmeter, you should always:
    a. Set to a high scale initially.
    b. Ensure lead polarity is correct.
    c. Connect the voltmeter across the component to be measured.
18. A material that has a low resistance to current flow is called a conductor.
19. Conductance is the measure of how good a conductor is and is measured in siemens.
20. A material that has a high resistance to current flow is called an insulator.
21. Semiconductor materials are neither insulators nor conductors.
22. An open in a circuit prevents the passage of current.
23. A closed circuit has a complete path for current to flow.
24. A short circuit occurs when an alternative path is made for current to flow.

Ammeter	Current	Open circuit
Ampere	Dielectric strength	Polarity
Atom	Electromotive force	Positive
Battery	Electron	Potential difference
Breakdown voltage	Electron current	Proton
Charge	flow	Relative conductivity
Closed circuit	Element	Resistance
Compound	Insulator	Semiconductor
Conductance	Ion	Short circuit
Conductor	Molecule	Switch
Conventional	Negative	Voltage
current flow	Neutron	Voltmeter
Coulomb	Nucleus	

NEW FORMULAS

$$Q = \frac{N}{6.24 \times 10^{18}}$$

Q = electric charge in coulombs
N = total number of electrons

$$I = \frac{Q}{t}$$

I = current in amps
Q = electric charge in coulombs
t = time in seconds

$$G = \frac{1}{R}$$

G = conductance in siemens
R = resistance in ohms

$$\text{relative conductivity} = \frac{\text{conductor's relative conductivity}}{\text{reference conductor's relative conductivity}}$$

$$\text{dielectric thickness} = \frac{\text{voltage to insulate}}{\text{insulator's breakdown voltage}}$$

REVIEW QUESTIONS

Multiple Choice Questions

1. Hydrogen is a:
   a. Gas
   b. Solid
   c. Liquid
   d. Other
2. Which of the following is a liquid?

    a. Calcium
    b. Magnesium
    c. Mercury
    d. Helium

3. The atomic number of an atom describes:
    a. The number of neutrons
    b. The number of electrons
    c. The number of neuclei
    d. The number of protons

4. The most commonly used metal in the field of electronics is:
    a. Silver
    b. Copper
    c. Mica
    d. Gold

5. The smallest unit of an element is:
    a. A compound
    b. An atom
    c. A molecule
    d. A proton

6. The smallest unit of a compound is:
    a. An element
    b. A neutron
    c. An electron
    d. A molecule

7. A water molecule is made up of:
    a. 2 parts hydrogen and 1 part oxygen
    b. Chlorine and sodium
    c. 1 part oxygen and 2 parts sodium
    d. 3 parts chlorine and 1 part hydrogen

8. A negative ion has:
    a. More protons than electrons
    b. More electrons than protons
    c. More neutrons than protons
    d. More neutrons than electrons

9. A positive ion has:
    a. Lost some of its electrons
    b. Gained extra protons
    c. Lost neutrons
    d. Gained more electrons.

10. One coulomb of charge is equal to:
    a. 6.24×10^{18} electrons
    b. 1018×10^{12} electrons
    c. 6.24×10^{8} electrons
    d. 6.24×10^{81} electrons

11. If 14 coulombs of charge passes by a point in 7 seconds, the current flow is said to be:
    a. 2 amps
    b. 98 amps

    c. 21 amps

    d. 7 amps

12. How many electrons are there within 16 coulombs of charge?
    a. 9.98×10^{19}
    b. 14
    c. 16
    d. 10.73×10^{19}

13. Current is measured in:
    a. Volts
    b. Coulombs/second
    c. Ohms
    d. Siemens

14. Voltage is measured in units of:
    a. Amperes
    b. Ohms
    c. Siemens
    d. Volts

15. Another word used to describe voltage is:
    a. Potential difference
    b. Pressure
    c. Electromotive force (emf)
    d. All the above

16. Conductors offer a _____resistance to current flow.
    a. High
    b. Low
    c. Medium
    d. Maximum

17. Conductors have:
    a. Electrons in shells farthest away from the nucleus
    b. Relatively more electrons per unit volume amount
    c. An incomplete valence shell
    d. All the above
    e. None of the above

18. Conductance is the measure of how good a conductor is at passing current, and is measured in:
    a. Siemens
    b. Volts
    c. Current
    d. Ohms

19. Insulators have:
    a. Electrons close to the nucleus
    b. Relatively few electrons per unit volume amount
    c. An almost complete valence shell
    d. All the above
    e. None of the above

20. An open circuit will cause:
    a. No current flow
    b. Maximum current flow

c. A break in the circuit
d. Both (a) and (c) are true

Essay Questions

21. Describe the three factors that make a good conductor. (2.4)
22. What three rules should be remembered and applied when measuring: (2.3.6)
    a. Current
    b. Voltage
23. Why is current directly proportional to voltage? (2.3.4)
24. List the four most commonly used current units and their values in terms of the basic unit. (2.2.3)
25. What is the speed of light in (a) miles and (b) meters per second? (2.2.4)
26. List the four most commonly used voltage units and their values in terms of the basic unit. (2.3.2)
27. Describe what is meant by: (2.7)
    a. An open circuit
    b. A closed circuit
    c. A short circuit
28. What is the difference between conventional and electron current flow? (2.2.5)
29. Give the unit and symbol for the following:
    a. Voltage (V) is measured in _____ (_____). (2.3)
    b. Current (I) is measured in _____ (_____). (2.2.2)
    c. Conductance (G) is measured in _____ (_____). (2.4.1)
    d. Resistance (R) is measured in _____ (_____). (2.4.1)
30. In relation to the structure of matter and the atom, describe: (2.1)
    a. The atom's subatomic particles
    b. An element (2.1)
    c. A compound (2.1.3)
    d. A molecule (2.1.3)
    e. A neutral atom (2.1.1)
    f. A positive ion (2.2)
    g. A negative ion (2.2)

Practice Problems

31. What is the value of conductance in siemens for a 100-ohm resistor?
32. Calculate the total number of electrons in 6.5 coulombs of charge.
33. Calculate the amount of current in amps passing through a conductor if 3 coulombs of charge passes by a point in 4 seconds.
34. Convert the following:
    a. 0.014 A to _____ mA
    b. 1374 A to _____ kA
    c. 0.776 μA to _____ nA
    d. 0.91 mA to _____ μA

35. Convert the following:
    a. 1473 mV to _____ V
    b. 7143 V to _____ kV
    c. 0.139 kV to _____ V
    d. 0.390 MV _____ kV
36. What is the relative conductivity of copper if silver is used as the reference conductor?
37. What minimum thickness of porcelain will withstand 24,000 volts?
38. To insulate a circuit from 10 V, what insulator thickness would be needed if the insulator is rated at 750 kV/cm?
39. Convert the following:
    a. 2000 kV/cm to _____ meters.
    b. 250 kV/cm to _____ mm.
40. What maximum voltage may be placed across 35 mm of mica without it breaking down?

3

RESISTANCE

AFTER COMPLETING THIS CHAPTER, YOU WILL BE ABLE TO:

1. Define resistance and the ohm.
2. Explain Ohm's law and its application.
3. Describe why:
   a. Current is proportional to voltage.
   b. Current is inversely proportional to resistance.
4. Describe and define the terms energy, work, and power.
5. Describe how the wattmeter can be used to measure power.
6. Explain what is meant by the kilowatt-hour.
7. Explain what factors affect the resistance of conductors.
8. Explain superconductivity and its advantages.
9. List some of the different types of conductors and their connectors.

INTRODUCTION

Voltage, current, resistance, and power are the four basic concepts of prime importance in our study of electronics. In Chapter 2, voltage and current were introduced, and in this chapter we will analyze resistance and power.

GENIUS OF CHIPPEWA FALLS

In 1960, Seymour R. Cray, a young vice-president of engineering for Control Data Corporation, informed president William Norris that in order to build the world's most powerful computer he would need a small research lab built near his home. Norris would have shown any other employee the door, but Cray was his greatest asset, and so in 1962 Cray moved into his lab, staffed by 34 and nestled in the woods near his home overlooking the Chippewa River in Minneapolis. Eighteen month later the press was invited to view the 14 by 6 foot 6600 supercomputer that could execute 3 million instructions per second and contained 80 miles of circuitry and 350,000 transistors, which were so densely packed that a refrigeration cooling unit was needed due to the lack of air flow.

Cray left Control Data in 1972 and founded his own company Cray Research. Four years later the 8.8 million Cray-1 scientific supercomputer outstripped the competition. It included some revolutionary design features, one of which is that since electronic signals cannot travel faster than the speed of light (one foot per billionth of a second) the wire length should be kept as short as possible, because the longer the wire the longer it takes for a message to travel from one end to the other. With this in mind, Cray made sure that none of the supercomputer's conducting wires exceeded 4 foot in length.

In the summer of 1985, the Cray-2, Seymour Cray's latest design, was installed at Lawrence Livermore Laboratory. The Cray-2 was 12 times faster than the Cray-1, and its densely packed circuits are encased in clear Plexiglas and submerged in a bath of liquid coolant. The 60-year-old genius has moved on from his latest triumph, nicknamed "Bubbles," and is working on another revolution in the supercomputer field, because for Seymour Cray a triumph is merely a point of departure.

3.1
WHAT IS RESISTANCE?

Resistance is the opposition to current flow accompanied by the dissipation of heat. In Figures 3-1 and 3-2, we have again used the fluid analogy to explain the concept of resistance.

In Figure 3-1(a), a valve has been opened almost completely, so a very small opposition to the water flow exists within the pipe. This small or low resistance within the pipe will not offer much opposition to water flow, and so a large amount of water will flow through the pipe and gush from the outlet.

In Figure 3-1(b), a small resistance has been placed in the circuit, providing very little resistance to the passage of current flow. This low resistance or small opposition will therefore allow a large amount of current to flow through the conductor, as illustrated by the heavy line.

In Figure 3-2(a), the valve is almost completely closed, resulting in a high resistance or opposition to water flow, and so only a trickle of water passes through the pipe and out from the outlet. In Figure 3-2(b), a large value of resistance has been placed in the circuit, causing a large opposition to the passage of current. This large resistance allows only a small amount of current to flow through the conductor, as illustrated by the light line.

In both examples of low and high resistance, you may have noticed that resistance and current are inversely proportional ($1/\propto$) to one another;

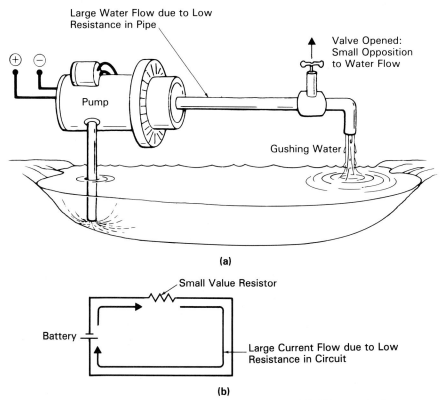

(a)

Large Water Flow due to Low
Resistance in Pipe

Valve Opened:
Small Opposition
to Water Flow

Pump

Gushing Water

Small Value Resistor

Battery

Large Current Flow due to Low
Resistance in Circuit

(b)

FIGURE 3-1   Low Resistance. (a) Fluid System. (b) Electrical Circuit.

FIGURE 3-2   High Resistance. (a) Fluid System. (b) Electrical Circuit.

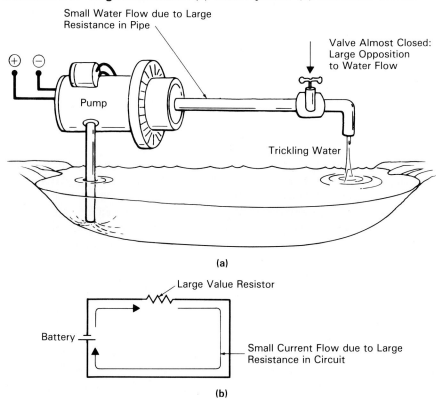

Small Water Flow due to Large
Resistance in Pipe

Valve Almost Closed:
Large Opposition
to Water Flow

Pump

Trickling Water

(a)

Large Value Resistor

Battery

Small Current Flow due to Large
Resistance in Circuit

(b)

What Is Resistance?   **67**

thus, if resistance is high, the current is low, and if resistance is low, the current is high.

$$\boxed{\text{resistance (ohms) is inversely proportional to current (amps)}}$$

If resistance is increased by some value, then current will be decreased by the same value. For example, if resistance is doubled, current is halved (assuming a constant voltage).

$$R\uparrow \; \frac{1}{\alpha} \; I\downarrow$$

In Figures 3-1 and 3-2, the fluid analogy has been used alongside the electrical circuit to help you understand the idea of low and high resistance. In between low and high resistance exist many different values of resistance, and we now need to analyze resistance further to be able to clearly define exactly how much resistance exists within a circuit, and not just say that resistance is generally low or high.

SELF-TEST REVIEW QUESTIONS (§ 3.1)

1. What is resistance?
2. What is the difference between a small and large resistance?
3. What is the relationship between current flow and resistance?
4. Would a very large resistance have a small or large conductance figure?

3.2
THE OHM

Current is measured in amps, voltage is measured in volts, and resistance is measured in ohms, in honor of Georg Simon Ohm and his work with current, voltage, and resistance. The larger the resistance, the larger the value of ohms and the more the resistor will oppose current flow. The ohm is given the symbol Ω, which is the Greek letter omega. By definition, 1 ohm is the value of resistance that will allow 1 ampere of current to flow through a circuit when a voltage of 1 volt is applied, as seen in Figure 3-3(b), where the resistor is drawn as a zizgag; however, in some schematics (circuit diagrams) it can be drawn as a rectangular block, as shown in Figure 3-3(a).

Figure 3-4 reinforces our understanding of the ohm by illustrating a 1-volt battery connected across a resistor whose resistance can be either increased or decreased. As the resistance in the circuit is increased, the current will decrease, and as the resistance of the resistor is decreased, the circuit current will increase. If the resistor is adjusted so that exactly 1 ampere of

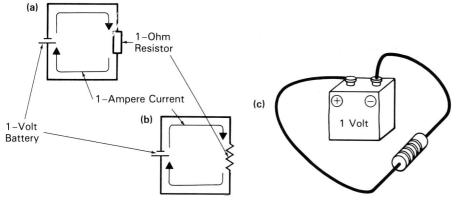

FIGURE 3-3 **One Ohm. (a) New and (b) Old Resistor Symbol. (c) Pictorial.**

current is flowing around the circuit, the value of resistance obtained is referred to as 1 ohm.

3.2.1 OHM'S LAW

Current flows in a circuit due to the force or voltage applied; however, the amount of current flow is limited by the amount of resistance in the circuit. In actual fact, the amount of current flow around a circuit is dependent on both voltage and resistance. This relationship between the three electrical properties of current, voltage, and resistance was first discovered by Georg Simon Ohm, a German physicist, in 1827. Published originally in 1826, Ohm's law states that: The current flow in a circuit is directly proportional ($\propto$) to the source voltage applied and inversely proportional ($1/\propto$) to the resistance of the circuit.

Stated in mathematical form, Ohm arrived at this formula:

$$\boxed{\text{Current } (I) = \frac{\text{voltage } (V)}{\text{resistance } (R)}}$$

$$\text{current } (I) \propto \text{voltage } (V)$$
$$\text{current } (I) \frac{1}{\propto} \text{resistance } (R)$$

FIGURE 3-4 **One Ohm Allows 1 Ampere to Flow with 1 Volt Applied.**

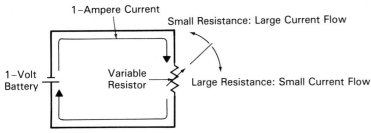

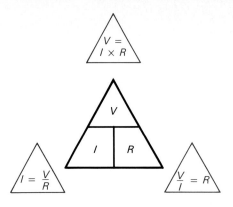

FIGURE 3-5 Ohm's Law Triangle.

By transposition, we can derive three forms of Ohm's law:

$$I = \frac{V}{R}, \qquad V = I \times R, \qquad R = \frac{V}{I}$$

3.2.2 THE OHM'S LAW TRIANGLE

The three forms of Ohm's law can be summarized by placing the three properties within a triangle for easy memory recall, as illustrated in Figure 3-5.

3.2.3 CURRENT IS PROPORTIONAL TO VOLTAGE ($I \propto V$)

Referring to Figure 3-6(a), you can see that an increase in the pump pressure will result in an increase in water flow. Similarly, an increase in voltage (electron moving force) will exert a greater pressure on the circuit electrons and cause an increase in current flow. This point is reinforced with Ohm's law:

$$I\uparrow = \frac{V\uparrow}{R}$$

EXAMPLE 3.1

Referring to Figure 3-7, if resistance remains constant at 1 Ω and the applied voltage equals 2 V, what is the value of current flowing through the circuit?

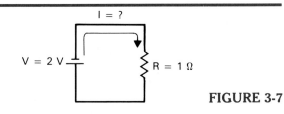

FIGURE 3-7

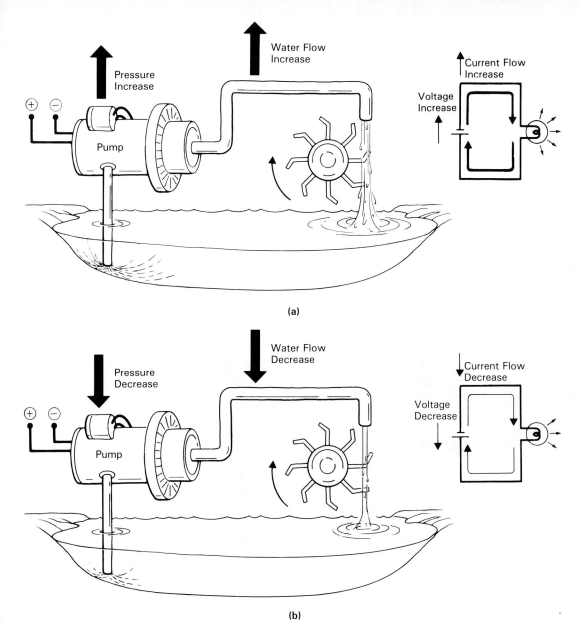

FIGURE 3-6 Current Flow Is Proportional to Voltage. (a) Pressure or Voltage Increase Causes a Water or Current Flow Increase ($V\uparrow$, $I\uparrow$). (b) Pressure or Voltage Decrease Results in a Water or Current Flow Decrease ($V\downarrow$, $I\downarrow$).

SOLUTION

$$\begin{aligned} \text{current } (I) &= \frac{\text{voltage } (V)}{\text{resistance } (R)} \\ &= \frac{2\ V}{1\ \Omega} \\ &= 2 \text{ amperes (A)} \end{aligned}$$

Calculator Sequence

Step	Keypad Entry	Display Response
1.	[2]	2
2.	[÷]	
3.	[1]	1
4.	[=]	2

EXAMPLE 3.2

Referring to Figure 3-8, if the voltage is now doubled to 4 V, what would be the change in current?

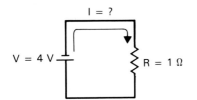

I = ?

V = 4 V

R = 1 Ω

FIGURE 3-8

SOLUTION

$$\text{current }(I) = \frac{\text{voltage }(V)}{\text{resistance }(R)}$$
$$= \frac{4\ V}{1\ \Omega}$$
$$= 4 \text{ amperes (A)}$$

On the other hand, a decrease in pump pressure or battery voltage will result in a small water or electron force, causing a decrease in current flow, as seen in Figure 3-6(b). This is again reinforced by Ohm's law.

$$I\downarrow = \frac{V\downarrow}{R}$$

Let's once again take an example to mathematically prove the point.

EXAMPLE 3.3

If the circuit resistance equals 2 Ω and the applied voltage equals 8 V, what would be the circuit current?

SOLUTION

$$\text{current }(I) = \frac{\text{voltage }(V)}{\text{resistance }(R)}$$
$$= \frac{8\ V}{2\ \Omega}$$
$$= 4 \text{ amperes (A)}$$

EXAMPLE 3.4

If the voltage is now halved to 4 V, what would be the change in circuit current?

SOLUTION

$$\text{current }(I) = \frac{\text{voltage }(V)}{\text{resistance }(R)}$$
$$= \frac{4\ V}{2\ \Omega} = 2 \text{ amperes (A)}$$

In conclusion, we can say that if resistance were to remain constant and the voltage were to double, the current within the circuit would also double. Similarly, if the voltage were halved, the current would also halve, proving that current and voltage increase or decrease by the same percentage for the same value of resistance, which makes them proportional to one another.

3.2.4 CURRENT IS INVERSELY PROPORTIONAL TO RESISTANCE $\left(I \frac{1}{\propto} R\right)$

With the valve opened, as seen in Figure 3-9, a small opposition to the water flow will result, and so a large water flow exists in a system with low

FIGURE 3-9 Current Is Inversely Proportional to Resistance. (a) Resistance Decrease Causes a Water or Current Flow Increase ($R\downarrow$, $I\uparrow$). (b) Resistance Increase Causes a Water or Current Flow Decrease ($R\uparrow$, $I\downarrow$).

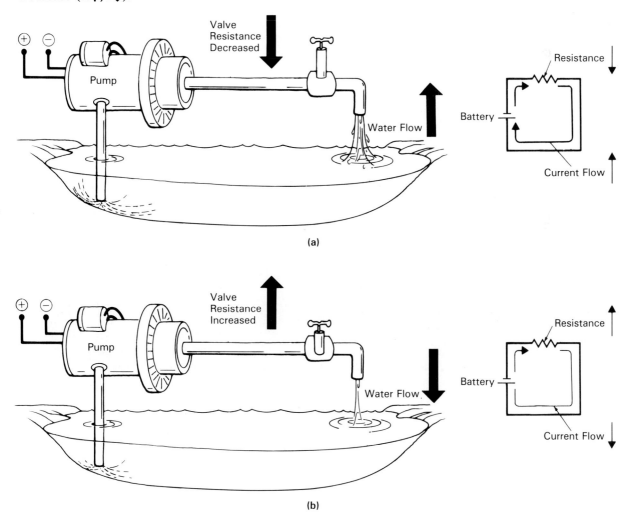

(a)

(b)

resistance. Similarly, with the electronic circuit a small resistance ($R\downarrow$) allows a large current ($I\uparrow$) flow around the circuit.

Consequently, in both the fluid system and electronic circuit, a decrease in resistance causes an increase in current flow. Ohm's law reinforces this point.

$$\text{(resistance)} \quad R\uparrow = \frac{V}{I\downarrow} \quad \text{(current)}$$

To prove this mathematically, let's take an example.

EXAMPLE 3.5

If the applied circuit voltage is 8 V and the circuit resistance equals 2 Ω, what is the total amount of current flow?

SOLUTION

$$\begin{aligned} \text{current } (I) &= \frac{\text{voltage } (V)}{\text{resistance } (R)} \\ &= \frac{8 \text{ V}}{2 \text{ }\Omega} \\ &= 4 \text{ amperes (A)} \end{aligned}$$

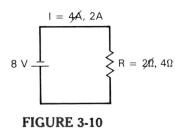

I = 4A, 2A

8 V   R = 2Ω, 4Ω

FIGURE 3-10

On the other hand, in Figure 3-9(b) the valve has been almost completely closed, offering a large opposition to the water flow; consequently, a small water flow results in a system with high resistance. With the electronic circuit in Figure 3-10, a large circuit resistance ($R\uparrow$) permits only a small current ($I\downarrow$) to flow around the circuit.

Ohm's law reinforces the illustration in Figure 3-10 by showing that a large resistance or opposition causes a small water or current flow.

$$\text{(resistance)} \quad R\uparrow = \frac{V}{I\downarrow} \quad \text{(current)}$$

To prove this mathematically, let's take an example:

EXAMPLE 3.6

If the applied voltage is still equal to 8 V and the circuit resistance is now 4 Ω, calculate the value of current flow through the circuit.

SOLUTION

$$\begin{aligned} \text{current } (I) &= \frac{\text{voltage } (V)}{\text{resistance } (R)} \\ &= \frac{8 \text{ V}}{4 \text{ }\Omega} \\ &= 2 \text{ amperes (A)} \end{aligned}$$

In conclusion, we can say that if voltage were to remain constant and the resistance were to double, the current within the circuit would be halved. On the other hand, if the circuit resistance were halved, the circuit current would double, confirming that current is inversely proportional to resistance.

SELF-TEST REVIEW QUESTIONS (§ 3.2)

1. Define one ohm in relation to current and voltage.
2. Calculate I if $V = 24$ V and $R = 6\ \Omega$.
3. What is the ohm's law triangle?
4. What is the relationship between current and voltage; between current and resistance?

3.3
ENERGY, WORK, AND POWER

The sun provides us with a consistent supply of energy in the form of light. Coal and oil are fossilized vegetation that grew, among other things, due to the sun, and are examples of energy that the earth has stored for millions of years. It can be said then that all energy begins from the sun. On the earth, energy is not created or destroyed; it is merely transformed from one form to another. The transforming of energy from one form to another is called *work*. The greater the energy transformed, the more work done.

The six basic forms of energy are light, heat, magnetic, chemical, electrical, and mechanical energy. The unit for energy is the *joule*. Potential (position) and kinetic (motion) are two terms used when describing energy. A cart on top of a hill has potential (position) energy, while a cart rolling down a hill has kinetic (motion) energy. Potential and kinetic energy are best described by looking at the example of a swinging pendulum, as seen in Figure 3-11. When the pendulum is in its upmost position [Figure 3-11(a)], it has potential energy, due to its position relative to the resting position, yet it has no motion and so no kinetic energy. When the pendulum is in the position shown in Figure 3-11(b), it has no potential energy, yet it has motion or kinetic energy.

FIGURE 3-11   Pendulum. (a) Potential Energy. (b) Kinetic Energy. (c) Potential and Kinetic Energy.

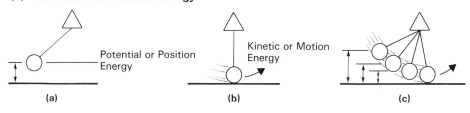

(a)　　　　　　　(b)　　　　　　　(c)

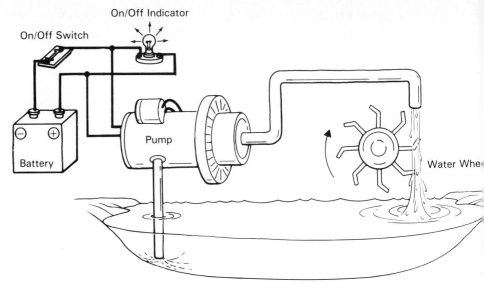

On/Off Indicator

On/Off Switch

Pump

Battery

Water Wheel

FIGURE 3-12 **Energy Transfer.**

In Figure 3-11(a), the pendulum had only potential energy and in 3-11(b) the pendulum had only kinetic energy. In between these two points, the pendulum possesses a combination of both potential and kinetic energy, as seen in Figure 3-11(c).

Looking at Figure 3-12, let us try to summarize our discussion so far on energy and work.

Chemical energy within the battery is converted to electrical energy when the electrons are attracted and repelled and set in motion. To use the two terms, we could say that the battery has the potential energy to set electrons in motion, and these moving electrons are said to possess kinetic energy. This electrical energy drives two devices:

1. The light bulb, which converts electrical energy into light and heat energy.
2. The pump, which uses the electrical energy to produce mechanical energy within the pump.

The pump has the potential energy to cause water flow or kinetic energy, just as the battery has the potential energy to cause kinetic energy in the form of electron flow. The kinetic energy within the water flow is finally used to turn the water wheel.

Work is being done everytime one form of energy is transformed to another. A device that converts one form of energy to another is called a *transducer*; in Figure 3-12, the battery, light bulb, and pump are all examples of transducers doing work at each stage of conversion.

The amount of work done is equal to the amount of energy transformed, and in both cases an equal amount of energy was transformed and so an equal amount of work was done. Energy and work have the same symbol (W), the same formula, and the same unit (the joule). Energy is merely the capacity, potential, or ability to do work, and work is done when a transformation of the potential, capacity, or ability takes place.

EXAMPLE 3.7

One person walks around a track and takes 50 minutes, while another person runs around the track and takes 50 seconds. Both were full of energy before they walked or ran around the track, and during their travels around the track they converted the chemical energy within their bodies into the mechanical energy of movement.

a. Who exerted the most energy?
b. Who did the most work?

SOLUTION

Both exerted the same amount of energy. The runner exerted all his energy (for example, 100 joules) in the short time of 50 seconds, while the walker spaced his energy (100 joules) over 50 minutes. They both did the same amount of work. So the only difference between the runner and the walker is time.

3.3.1 POWER

Power (P) is the rate at which work is performed and is given the unit of watt (W), which is joules per second. Thus power involves a time factor.

Returning to our two persons walking and running around the track, we could say that the number of joules of energy exerted in 1 second by the runner was far greater than the number of joules of energy exerted in 1 second by the walker, although the total energy exerted by both persons around the entire track was equal and, therefore, the same amount of work was done.

Now we have an understanding of power, work, and energy. Let's reinforce our knowledge by introducing the energy formula with some mathematical problems related to electronics.

3.3.2 CALCULATING ENERGY

EXAMPLE 3.8

If a 1-V battery can store 6.24×10^{18} electrons, how much energy is the battery said to have?

SOLUTION

$$\boxed{W = Q \times V}$$

$W = 1 \text{ coulomb} \times 1 \text{ volt}$

$\quad = 1 \text{ joule of energy}$

$W = $ Energy in joules
$Q = $ coulombs of charge (1 coulomb $= 6.24 \times 10^{18}$ electrons)
$V = $ voltage applied

EXAMPLE 3.9

How many coulombs of electrons would a 9-V battery have to store to have 63 joules of energy?

SOLUTION

If $W = Q \times V$, then

$$Q = \frac{W}{V}$$

$$\frac{\text{coulombs of}}{\text{electrons } (Q)} = \frac{\text{energy in joules } (W)}{\text{battery voltage } (V)}$$

$$= \frac{63 \text{ joules}}{9 \text{ volts}}$$

$$= 7 \text{ coulombs of electrons}$$

or

$$7 \times 6.24 \times 10^{18} = 4.36 \times 10^{19} \text{ electrons}$$

Calculator Sequence

Step	Keypad Entry	Display Response
1.	6 3	63
2.	÷	
3.	9	9
4.	=	7
5.	×	7
6.	6 . 2 4 E 1 8	6.24E18
7.	=	4.36E19

3.3.3 CALCULATING POWER

EXAMPLE 3.10

Power, in relation to electronics, is the rate at which electric energy is converted into some other form. In our example (Figure 3-13), it will be transformed from electric energy into light and heat energy by the light bulb. Power has the unit of watts, which is the number of joules of energy transformed per second (J/s). If 27 joules of electric energy is being transformed into light and heat per second, how many watts of power does the light bulb convert?

SOLUTION

$$\text{power} = \frac{\text{joules}}{\text{second}}$$

$$= \frac{27 \text{ joules}}{1 \text{ second}}$$

$$= 27 \text{ watts}$$

FIGURE 3-13 **Calculating Power.**

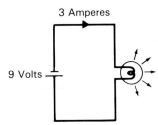

3 Amperes

9 Volts

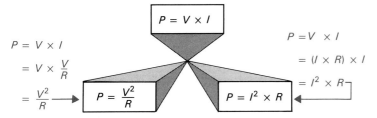

FIGURE 3-14 Power Formula.

The power could have easily been calculated by merely multiplying current by voltage to arrive at the same result.

$$\text{power} = I \times V = 3\,\text{A} \times 9\,\text{V}$$
$$= 27\,\text{W}$$

Therefore, we could say that the light bulb dissipates 27 watts, or 27 joules per second.

Looking at the power formula $P = I \times V$, we can say that 1 watt of power is expended when 1 amp of current flows through a circuit that has 1 volt applied or when 1 amp flows through 1 ohm.

On some occasions, the value of current or voltage may not be known; however, by substitution, we can arrive at alternative power formulas for wattage calculations, as seen in Figure 3-14.

EXAMPLE 3.11

In Figure 3-15 a 12-V battery is connected across a 36-Ω resistor. How much power does the resistor dissipate?

SOLUTION

$$\text{power} = \frac{\text{voltage}^2}{\text{resistance}}$$
$$= \frac{12\,\text{V}^2}{36\,\Omega} = \frac{144}{36}$$
$$= 4\,\text{W}$$

Four joules of energy per second is being dissipated.

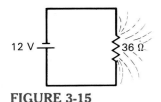

FIGURE 3-15

Calculator Sequence

Step	Keypad Entry	Display Response
1.	① ②	12
2.	x² (square key)	144
3.	÷	
4.	③ ⑥	36
5.	=	4

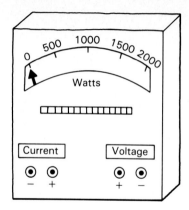

FIGURE 3-16 **Wattmeter.**

3.3.4 MEASURING POWER

Electrical power is measured with a wattmeter. The wattmeter seen in Figure 3-16 has four terminals, which must be connected correctly, as seen in Figure 3-17; in this example, we are measuring the amount of power being supplied to a radio.

The current terminals of the wattmeter are connected so that the current passing from the battery to the radio has to pass through the wattmeter's current terminals so that the wattmeter can sense current, while the voltage terminals of the wattmeter are connected across the radio so as to sense the radio's voltage supply. The wattmeter is therefore like an ammeter and voltmeter in one package, and through a multiplication process ($P = I \times$

FIGURE 3-17 **Wattmeter Connections or Hook-up.**
(a) Schematic Symbol. (b) Pictorial.

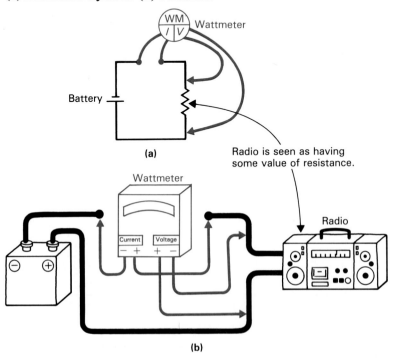

FIGURE 3-18   Kilowatt-hour Meter Being Tested for Accuracy by a Journeyman Meter Tester. (*Courtesy of Fred Vaughn, San Diego Gas and Electric*)

V), we can obtain an indication of the amount of power being supplied to the radio.

3.3.5 THE KILOWATT-HOUR

You and I pay for our electric energy in a unit called the kilowatt-hour (kWh). The kilowatt-hour meter, seen in Figure 3-18, measures how many kilowatt-

---

EXAMPLE 3.12

If a 100-watt light bulb is left on for 10 hours, how many kilowatt hours will we be charged for?

Calculator Sequence

SOLUTION

$$\begin{aligned}
\text{power} \\
\text{consumed} &= \text{power (kW)} \times \text{time (hours)} \\
\text{(kWh)} \\
&= 0.1 \text{ kW} \times 10 \text{ hours} \\
&\quad (100 \text{ watts} = 0.1 \text{ kW}) \\
&= 1 \text{ kilowatt-hour}
\end{aligned}$$

Step	Keypad Entry	Display Response
1.	0 . 1 E 3	0.1E3
2.	×	
3.	1 0	10
4.	=	1E3

Energy, Work, and Power　　**81**

EXAMPLE 3.13

Figure 3-19 illustrates a typical household heater and an equivalent electrical circuit. The heater has a resistance of 7 Ω and the electric company is charging 6 cents/kWh. Calculate
a. The power consumed by the heater.
b. The cost of running the heater for 7 hours.

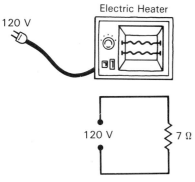

FIGURE 3-19

SOLUTION

a. Power $(P) = \dfrac{V^2}{R} = \dfrac{120^2}{7}$

$\qquad = \dfrac{14.4\,kV}{7}$

$\qquad = 2057$ watts (2 kW)

b. Power consumed $\quad =$ power (kW) × time (hour)
$\qquad\qquad\qquad\quad = 2.057 \times 7\,h$
$\qquad\qquad\qquad\quad = 14.399$ kWh
$\quad$ Cost $=$ kWh × rate $= 14.399 \times 6$ cents
$\qquad\qquad\qquad\qquad\quad = 86$ cents

hours are consumed, and the electric company then charges accordingly. One kilowatt-hour of power is consumed when 1000 watts (1 kW) is supplied in a 1-hour period.

$$\boxed{\text{power consumed (kWh)} = \text{power (kW)} \times \text{time (h)}}$$

SELF-TEST REVIEW QUESTIONS (§ 3.3)

1. List the six basic forms of energy.
2. What is the difference between energy and power?
3. List the formulas for calculating energy and power.
4. What is 1 kilowatt-hour of power?

3.4
CONDUCTORS AND THEIR RESISTANCE

Resistors are normally made out of materials that cause an opposition to current flow. However, conductors come in different shapes and sizes; some are good conductors and some are poor. Conductance is the measure of how

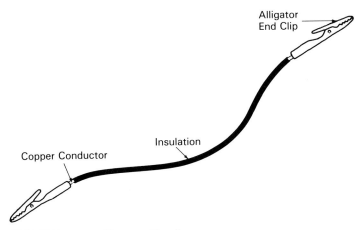

FIGURE 3-20 Copper Conductor.

good a conductor is, and even the best conductors have some value of resistance.

The total resistance of a typical conductor, as shown is Figure 3-20, is determined by four main factors:

1. The type of conducting material used for the flow of current.
2. The conductor's cross-sectional area.
3. The total length of the conductor.
4. The temperature of a conductor.

Let's now address each of these topics separately and discuss the reason why each will vary a circuit's resistance.

3.4.1 CONDUCTING MATERIAL

In our discussion on conductors, we said that copper, for instance, is a better conducting material than carbon as copper had more free electrons ready for current flow; and, as we have stated, a good conductor has a large conductance figure and a small resistance value. Carbon, however, has a smaller amount of free electrons than copper, resulting in a small conductance value and a high resistance figure.

3.4.2 CONDUCTOR'S CROSS-SECTIONAL AREA

The resistance of a conductor is inversely proportional to the conductor's cross-sectional area, which means: The greater the cross-sectional area, the lower the resistance [Figure 3-21(a)], and, similarly, the smaller the cross-sectional area, the higher the resistance [Figure 3-21(b)].

We can relate this to our fluid analogy, shown in Figure 3-22. A larger-sized pipe or conductor results in a larger water flow or current due to the lower resistance. On the other hand, the smaller the size of the pipe or

**FIGURE 3-21
Conduction. (a) Large
Cross-sectional Area.
(b) Small Cross-sectional
Area.**

Small Resistance R↓

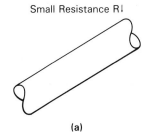

(a)

Large Resistance R↑

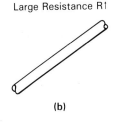

(b)

Conductors and Their Resistance   **83**

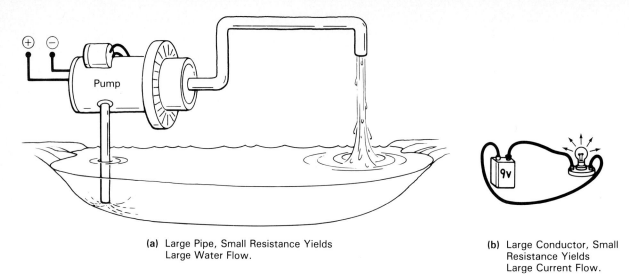

(a) Large Pipe, Small Resistance Yields
Large Water Flow.

(b) Large Conductor, Small
Resistance Yields
Large Current Flow.

FIGURE 3-22   Large Cross-sectional Area Causes Low Resistance. (a) Large Pipe (Small Resistance) Yields Large Water Flow. (b) Large Conductor (Small Resistance) Yields Large Current Flow.

FIGURE 3-23   Small Cross-sectional Area Causes High Resistance. (a) Small Pipe (Large Resistance) Yields Small Water Flow. (b) Small Conductor (Large Resistance) Yields Small Current Flow.

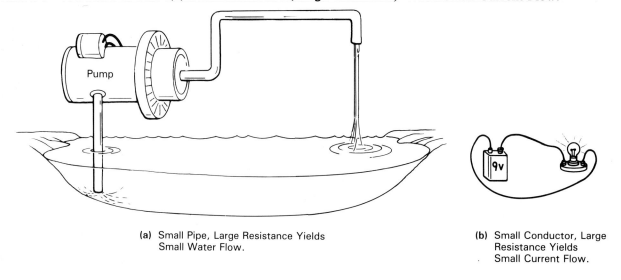

(a) Small Pipe, Large Resistance Yields
Small Water Flow.

(b) Small Conductor, Large
Resistance Yields
Small Current Flow.

conductor, the less the amount of water or electron flow due to the larger resistance (Figure 3-23).

We need a unit of measure to compare one thickness of wire conductor to another. Figure 3-24 illustrates a conductor with a diameter of 0.001 inch (1/1000th of an inch). This is termed 1 mil, and the diameters of all conductors are measured in mils.

As nearly all conductors are circular in cross section, the area of a

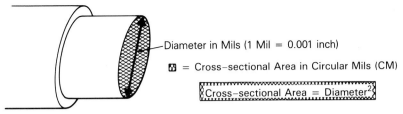

Diameter in Mils (1 Mil = 0.001 inch)

⊠ = Cross-sectional Area in Circular Mils (CM)

Cross-sectional Area = Diameter²

FIGURE 3-24   Measuring a Conductor's Cross-sectional Area.

conductor is measured in circular mils (cmil), which can be calculated by squaring the diameter.

$$\text{circular mils (cmil)} = \text{diameter}^2$$

EXAMPLE 3.14

If a conductor has a diameter of 71.96 mils, what is the circular mil area?

SOLUTION

$$
\begin{aligned}
\text{cmil} &= \text{diameter}^2 \\
&= 71.96^2 \\
&= 5178 \text{ cmil}
\end{aligned}
$$

3.4.3   CONDUCTOR LENGTH

Increasing the length of the conductor used increases the amount of resistance within the circuit. For example, if a conductor has a resistance of 1 ohm for every 10 feet of conductor, then 30 feet of conductor is going to triple the conductor's resistance within the circuit to 3 ohms.

The resistance of any conductor is calculated by use of the following formula:

$$R = \frac{\rho \times l}{a}$$

where R = resistance of conductor in ohms
  l = length of conductor in feet
  a = area of conductor in circular mils
  ρ = resistivity of conducting material

Resistivity, by definition, is the resistance (in ohms) that a certain length

TABLE 3-1
MATERIAL RESISTIVITY

Material	Resistivity (cmil/ft) in Ohms
Silver	9.9
Copper	10.7
Gold	16.7
Aluminum	17.0
Tungsten	33.2
Zinc	37.4
Brass	42.0
Nickel	47.0
Platinum	60.2
Iron	70.0

of material (in centimeters) will offer to the flow of current. Table 3-1 lists the resistivity of the more commonly used conductors.

EXAMPLE 3.15

Calculate the resistance of 333 feet of copper conductor with a conductor area of 3257 circular mils.

Calculator Sequence

Step	Keypad Entry	Display Response
1.	$\boxed{1}$ $\boxed{0}$ $\boxed{.}$ $\boxed{7}$	10.7
2.	$\boxed{\times}$	
3.	$\boxed{3}$ $\boxed{3}$ $\boxed{3}$	333
4.	$\boxed{\div}$	3563.1
5.	$\boxed{3}$ $\boxed{2}$ $\boxed{5}$ $\boxed{7}$	3257
6.	$\boxed{=}$	1.09

SOLUTION

$$R = \frac{\rho \times l}{a}$$
$$= \frac{10.7 \times 333}{3257}$$
$$= 1.09 \ \Omega$$

EXAMPLE 3.16

Calculate the resistance of 1274 feet of aluminum conductor with a diameter of 86.3 mils.

SOLUTION

$$R = \frac{\rho \times l}{a}$$

If the diameter is equal to 86.3 mils, then the circular mil area equals d^2:
$$86.3^2 = 7447.7 \text{ cmil}$$

Therefore,

$$R = \frac{17 \times 1274}{7447.7}$$
$$= 2.9\ \Omega$$

3.4.4 TEMPERATURE EFFECTS ON CONDUCTORS

When heat is applied to a conductor, the atoms within the conductor convert this thermal energy into another form of energy, in this case mechanical energy or movement. These random moving atoms cause collisions between the directed electrons (current flow) and the adjacent atoms, resulting in an opposition to the current flow (resistance).

Metallic conductors are said to have a positive temperature coefficient (+ temperature coefficient), because the greater the heat applied to the conductor, the greater the atom movement causing more collisions of atoms to occur, and consequently the greater the conductor's resistance.

| heat ↑ | resistance ↑ |

3.4.5 MAXIMUM CONDUCTOR CURRENT

Anytime current flows through any conductor, a certain resistance or opposition is inherent in that conductor. This resistance will convert current to heat, and the heat further increases the conductor's resistance, causing more heat to be generated due to the opposition.

A conductor, consequently, must be chosen carefully for each application so that it can carry the current without developing excessive heat. This is achieved by selecting conductors with a greater cross-sectional area to decrease its resistance.

Conducting wires are normally covered with a plastic or rubber type of material, known as insulation, to protect the users and technicians from electrical shock and also to keep the conductor from physically contacting other conductors within the equipment, as seen in Figure 3-25. If the current through the conductor is too high, this insulation will burn due to the heat and may cause a fire hazard. The National Fire Protection Association has

FIGURE 3-25   Conductor with Insulator.

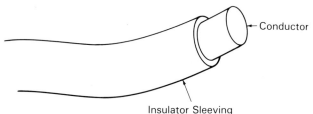

TABLE 3-2
THE AMERICAN WIRE GAUGE (AWG) FOR COPPER CONDUCTOR

AWG #	Diameter (Mils)	Maximum Current (A)	Ω/1000 feet
0000	460.0	230	0.0490
000	409.6	200	0.0618
00	364.8	175	0.0780
0	324.9	150	0.0983
1	289.3	130	0.1240
2	257.6	115	0.1563
3	229.4	100	0.1970
4	204.3	85	0.2485
5	181.9	75	0.3133
6	162.0	65	0.3951
7	144.3	55	0.4982
8	128.5	45	0.6282
9	114.4	40	0.7921
10	101.9	30	0.9981
11	90.74	25	1.260
12	80.81	20	1.588
13	71.96	17	2.003
14	64.08	15	2.525
15	57.07		3.184
16	50.82	6	4.016
17	45.26	Wires	5.064
18	40.30	of this 3	6.385
19	35.89	size	8.051
20	31.96	have	10.15
22	25.35	current	16.14
26	15.94	measured	40.81
30	10.03	in mA.	103.21
40	3.145		1049.0

Note: the larger the AWG number the smaller the size of the conductor.

developed a set of standards known as the American Wire Gauge for all copper conductors, which lists their diameter, area, resistance, and maximum safe current in amps. This table is given in Table 3-2. A rough guide for measuring wire size is shown in Figure 3-26.

3.4.6 SUPERCONDUCTIVITY

Conductors have a positive temperature coefficient, which means if temperature increases then so does resistance; but what happens if the temperature is decreased? In 1911, a Dutch physicist, Heike Onnes, discovered that mercury (a liquid conductor) lost its resistance to electrical current when the temperature was decreased to -459.7 Fahrenheit (0 on the Kelvin temperature scale). Mercury actually became a superconductor, allowing a supercurrent to flow and not encounter any resistance and therefore not generate any heat. (Heat dissipated by any resistance can be calculated by the power formula $P = I^2 \times R$. If $R = 0$, then the power dissipated by the conductor is zero watts.)

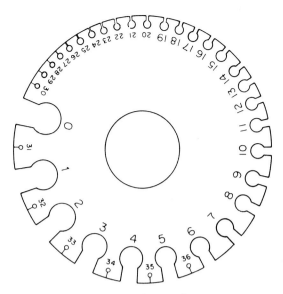

FIGURE 3-26 Wire Gauge Size.

In the spring of 1986, two IBM scientists discovered that a conductor compound made up of barium, lanthanum, copper, and oxygen would superconduct (have no resistance to current flow) at $-406°F$. Since then other scientists have increased the temperature to a point that now conductor compounds can be made to superconduct at $-300°F$. Some scientific projects that need to make use of the advantages of superconductivity immerse these conductor compounds in a bath of liquid nitrogen so they can achieve superconductivity. Liquid nitrogen, however, is difficult to handle, and so the real scientific achievement will be to find a conductor compound that will superconduct at room temperature.

When discussing the AWG table (Table 3-2), a maximum value of current was stated for a given thickness of wire. This is because a large current will generate a large amount of heat if the resistance is too large, and so to decrease the resistance and therefore heat generated ($P\uparrow = I^2R\uparrow$) a thicker conductor with a lower resistance is used. Superconductivity allows a standard 1000-A conductor, which would be approximately 2 in. thick, to be replaced by a superconductor about the thickness of a human hair. Not only will conductors be dramatically reduced in size, but the increased efficiency (as no power is being wasted in the form of heat) will result in high energy savings.

3.4.7 CONDUCTOR TYPES

A cable is made up of two or more wires. Figure 3-27 illustrates the different types of wires and cables. In Figure 3-27(a), (b), and (c), only one conductor exists within the insulation, and so they are classified as wires, whereas the

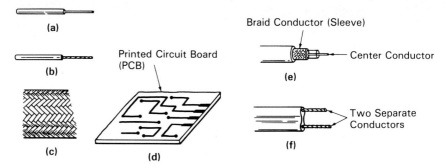

FIGURE 3-27 **Wires and Cables. (a) Solid Wire. (b) Stranded Wire. (c) Braided Wire. (d) Printed Wire. (e) Coaxial Cable. (f) Twin Lead.**

FIGURE 3-28 **Connectors. (a) Temporary Connectors. (b) Plugs. (c) Lug and Binding Post. (d) Sockets.**

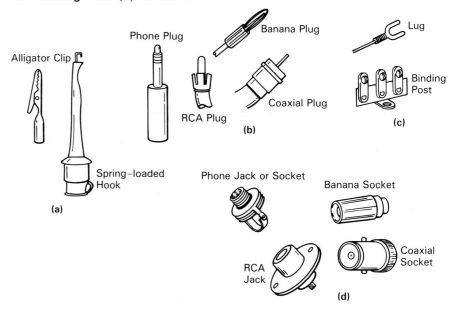

cables seen in Figure 3-27(e) and (f) have two conductors. The coaxial and twin-lead cables are most commonly used to connect TV signals into television sets. Figure 3-27(d) illustrates how printed conducting copper, silver, or gold paths exist on a plastic insulating board and are used to connect components such as resistors that are mounted on the other side.

Wire and cables have to connect from one point to another. Some are soldered directly, while others are attached to plugs that plug into sockets. A sample of connectors is shown in Figure 3-28.

SELF-TEST REVIEW QUESTIONS (§ 3.4)

1. List the four factors that determine the total resistance of a con-

2. What is the relationship between the resistance of a conductor and its cross-sectional area, length, and resistivity?

3. Conductors are said to have a positive temperature coefficient (true or false).

4. The smaller the AWG number the larger the size of the conductors (true or false).

SUMMARY

1. Resistance is opposition to current flow, with the dissipation of energy.

2. A low resistance results in a large current flow, while a high resistance allows only a small current to flow.

3. Resistance, which is measured in ohms, is inversely proportional to current, which is measured in amperes.

4. One ohm is the value of resistance that will allow 1 ampere of current to flow through a circuit when a voltage of 1 volt is being applied.

5. The relationship between the three electrical properties of current, voltage, and resistance was first discovered by Georg Simon Ohm.

6. Ohm's law states: The current flow in a circuit is proportional to the source voltage applied and inversely proportional to the resistance in the circuit.

7. The six basic forms of energy are light, heat, magnetic, chemical, electrical, and mechanical energy.

8. Work has been achieved everytime one form of energy is transformed to another.

9. A device that converts one form of energy to another is called a transducer.

10. Power is the rate at which energy is transformed and is measured in joules per second or watts.

11. Electrical power is measured with a wattmeter.

12. One kilowatt-hour of power is consumed when 1000 watts (1 kW) is consumed in a 1-hour period.

13. The resistance of a conductor is determined by:
    a. The material used
    b. The conductor's cross-sectional area
    c. The length of the conductor
    d. The temperature of the conductor

14. The larger the American Wire Gauge (AWG) number, the smaller the size of the conductor.

15. A superconductor is a compound that when cooled will conduct a current with no electrical resistance.

16. A wire has only one conductor within the insulating sheath unlike a cable that can have two or more insulated conductors within the outer insulating sheath.

NEW TERMS

Alligator clip	**Kelvin**	**Resistivity**
American Wire Gauge (AWG)	**Kilowatt-hour meter**	**Socket**
	Lug	**Superconductor**
Binding post	**Mil**	**Temperature coefficient of resistance**
Cable	**Ohm**	
Circular Mil	**Phone**	**Transducer**
Connector	**Plug**	**Watt**
Energy	**Power**	**Wattmeter**
Joule	**RCA plug and jack**	**Wire**

NEW FORMULAS

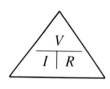

$$\boxed{V = I \times R}$$

$$I = \frac{V}{R}$$

$$R = \frac{V}{I}$$

V = voltage in volts
I = current in amps
R = resistance in ohms

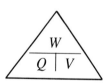

$$\boxed{W = Q \times V}$$

$$Q = \frac{W}{V}$$

$$V = \frac{W}{Q}$$

W = energy in joules
Q = electric charge in coulombs
V = voltage in volts

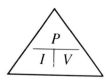

$$\boxed{P = I \times V}$$

$$V = \frac{P}{I}$$

$$I = \frac{P}{V}$$

P = power in watts
I = current in amps
V = voltage in volts

$$P = I \times V \qquad \boxed{P = I^2 \times R} \qquad \boxed{P = \frac{V^2}{R}}$$

$$\boxed{\text{power consumed} = \text{power} \times \text{time}}$$

Power consumed is in kilowatt hours (kWh).
Power is in kilowatts (kW).
Time is in hours (h).

$$\boxed{R = \frac{\rho \times l}{a}}$$

R = resistance of conductor in ohms
ρ = resistivity of conducting material
l = length of conductor in feet
a = area of conductor in circular mils.

REVIEW QUESTIONS

Multiple Choice Questions

1. Resistance is measured in:
   a. Ohms
   b. Volts
   c. Amps
   d. Siemens

2. Current is proportional to:
   a. Resistance
   b. Voltage
   c. Both (a) and (b)
   d. None of the above

3. Current is inversely proportional to:
   a. Resistance
   b. Voltage
   c. Both (a) and (b)
   d. None of the above

4. If the applied voltage is equal to 15 V and the circuit resistance equals 5 Ω, the total circuit current would be equal to:
   a. 4 amps
   b. 5 amps
   c. 3 amps
   d. 75 amps

5. Calculate the applied voltage if 3 mA flows through a circuit resistance of 25 kΩ.
   a. 63 mV
   b. 25 V
   c. 77 μV
   d. 75 V

6. Energy is measured in:
   a. Volts
   b. Joules
   c. Amps
   d. Watts

7. Chemical energy within a battery is converted into:
   a. Electrical energy
   b. Mechanical energy
   c. Magnetic energy
   d. Heat energy

8. The water pump has the potential to cause water flow, just as the battery has the potential to cause:
   a. Voltage
   b. Electron flow
   c. Current
   d. Both (b) and (c) are true

9. Work is measured in:
   a. Joules
   b. Volts
   c. Amps
   d. Watts

10. The device that converts one energy form to another is called a:
    a. Transformer
    b. Transducer
    c. Transit
    d. Transistor

11. Power is the rate at which energy is transformed and is measured in:
    a. Joules
    b. Watts
    c. Volts
    d. Amps

12. Power is measured by using a (an):
    a. Ammeter
    b. Voltmeter
    c. Ohmmeter
    d. Wattmeter

13. A good conductor has a:
    a. Large conductance figure
    b. Small resistance figure
    c. Both (a) and (b) are true
    d. None of the above are true

14. The resistance of a conductor is:
    a. Proportional to the length of the conductor
    b. Inversely proportional to the area of the conductor
    c. Both (a) and (b)
    d. None of the above are true

15. AWG is an abbreviation for:
    a. Alternate Wire Gauge
    b. Alternating Wire Gauge

Resistance

c. American Wave Guide
d. American Wire Gauge

Essay Questions ─────────────────────

16. What is resistance?   (3.1)
17. Briefly describe why:
    a. Current is proportional to voltages.   (3.2.3)
    b. Current is inversely proportional to resistance.   (3.2.4)
18. State Ohm's law.   (3.2–1)
19. List the three forms of Ohm's law.   (3.2.2)
20. List the six basic forms of energy.   (3.3)
21. Briefly describe the following terms:
    a. Potential energy.   (3.3)
    b. Kinetic energy.   (3.3)
22. What is a transducer?   (3.3)
23. Define power.   (3.3–1)
24. Give three formulas for electric power.   (3.3.3)
25. Give the units for each of the following:
    a. Energy **b.** Power   **c.** Voltage   **d.** Resistance **e.** Work
    f. Current **g.** Charge **h.** Conductance
26. What is the difference between work and power?   (3.3)
27. What instrument is used to measure electrical power?   (3.3.4)
28. State the formula used by electric companies to determine the amount of power consumed?   (3.3.5)
29. What is 1 kilowatt hour?   (3.3.5)
30. List the four factors that determine a conductor's resistance.   (3.4)
31. What is a circular mil?   (3.4.2)
32. Define the resistivity of a conducting material?   (3.4.3)
33. Describe why conductors have a positive temperature coefficient of resistance.   (3.4.4)
34. What is the purpose(s) of placing an insulating sheath over conducting wires?   (3.4.5)
35. What is the American Wire Gauge?   (3.4.5)
36. What is a superconductor?   (3.4.6)
37. List some of the advantages of superconductivity.   (3.4.6)
38. What is the difference between a wire and cable?   (3.4.7)
39. Give some examples of different wires and cables.   (3.4.7)
40. List examples of different conductor connectors.   (3.4.7)

Practice Problems ─────────────────────

41. An electric heater with a resistance of 6 Ω is connected across a 120-V wall outlet.
    a. Calculate the current flow.   (3.2)
    b. Draw the schematic diagram.

42. What source voltage would be needed to produce a current flow of 8 mA through a 16-kΩ resistor?   (3.2)

43. If an electric toaster draws 10 amps when connected to a power outlet of 120 V, what is its resistance?   (3.2)

44. Calculate the power used in questions 41, 42, and 43.   (3.3.3)

45. Calculate the current flowing through the following light bulbs when they are connected across 120 V: (3.3.3)
   a. 300 watt
   b. 100 watt
   c. 60 watt
   d. 25 watt

46. If an electric company charges 9 cents/kWh, calculate the cost for each light bulb in question 45 if on for 10 hours.   (3.3.5)

47. Indicate which of the following unit pairs is the larger: (3.3)
   a. Millivolts or volts
   b. Microamps or milliamps
   c. Kilowatts or watts
   d. Kilohm or megohm

48. Calculate the resistance of 200 feet of copper having a diameter of 80 mils?   (3.4.3)

49. What AWG size wire should be used to safely carry just over 15 amps?   (3.4.5)

50. Calculate the voltage dropped across 1000 feet of No. 4 copper conductor when a current of 7.5 amps is flowing through it.   (3.4.5)

51. Calculate the unknown resistance in a circuit when an ammeter indicates that a current of 12 mA is flowing and a voltmeter indicates 12 V.   (3.3)

52. What battery voltage would use 1000 joules of energy to move 40 coulombs of charge through a circuit?   (3.3.2)

53. Calculate the resistance of a light bulb that passes 500 mA of current when 120 V is applied. What is the bulb's wattage?   (3.3)

54. Which of the following circuits has the largest resistance and which has the smallest?   (3.3)
   a. $V = 120$ V, $I = 20$ mA
   b. $V = 12$ V, $I = 2$ A
   c. $V = 9$ V, $I = 100$ μA
   d. $V = 1.5$ V, $I = 4$ mA

55. Calculate the power dissipated in each circuit in question 54.   (3.3.3)

56. How many watts are dissipated if 5000 joules of energy are consumed in 25 seconds?   (3.3.3)

57. Convert the following:
   a. 1000 W = _____ kw
   b. 0.345 W = _____ mW
   c. 1250×10^3 W = _____ MW
   d. 0.00125 W = _____ μW

58. What is the value of the resistor when a current of 4 amps is causing 100 watts to be dissipated?   (3.3.3)

59. Convert the following to kilowatt-hours: (3.3.5)
  a. 7500 wattseconds
  b. 542,300 wattminutes

60. What is the output of a 12-V power supply of 300 mA? (3.3.3)

4

RESISTORS

AFTER COMPLETING THIS CHAPTER, YOU WILL BE ABLE TO:

1. Describe the difference between a fixed and variable resistor.
2. Explain the differences between the six basic types of fixed valve resistors: carbon composition, carbon film, metal film, wirewound, metal oxide, and thick film.
3. Identify the different resistor wattage ratings.
4. Describe the SIP, DIP, and chip thick film resistor packages.
5. Explain the following types of variable resistors.
   a. Mechanically adjustable: rheostat and potentiometer.
   b. Thermally adjustable: RTD, TFD, and thermistor.
   c. Optically adjustable: photoresistor.
6. List the three rules to remember when measuring resistance with an ohmmeter.
7. Explain how a resistor's value and tolerance are printed on the body of the component by either:
   a. Color coding, or
   b. Typographically
8. Describe the difference between a general-purpose and precision resistor.
9. State the purpose of the filament and ballast resistor.
10. Describe some of the more common resistor problems.

INTRODUCTION

Resistance would seem to be an undesirable effect as it reduces current flow and wastes energy as it dissipates heat. Resistors, however, are probably used more than any other component in electronics, and this chapter will discuss all the different types.

4.1

RESISTOR TYPES

Conductors offer a certain small amount of resistance; however, in electronics this resistance is not normally enough, and so additional resistance is needed to control the amount of current flow. The component used to supply this additional resistance is called a *resistor*.

There are two basic types of resistors: fixed and variable. The fixed resistor, examples of which can be seen in Figure 4-1, has a value of resistance

FIGURE 4-1 **Fixed-value Resistors.**

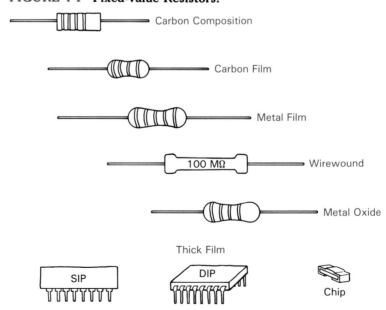

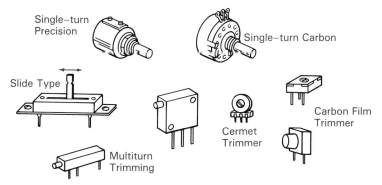

Single-turn
Precision

Single-turn Carbon

Slide Type

Cermet
Trimmer

Carbon Film
Trimmer

Multiturn
Trimming

FIGURE 4-2 Variable Resistors.

that cannot be changed and is the more common of the two. The variable resistor, on the other hand, has a range of values that can be selected generally by mechanically adjusting a control, as illustrated in Figure 4-2.

Let's now discuss the different types of fixed and then variable resistors.

4.1.1 FIXED-TYPE RESISTORS

Fixed-type resistors can be divided into six basic categories:

1. Carbon composition
2. Carbon film
3. Metal film
4. Wirewound
5. Metal oxide
6. Thick film

(1) CARBON COMPOSITION RESISTORS

This is the most common and least expensive type of fixed resistor, the appearance of which can be seen in Figure 4-3(a). It is constructed by placing a piece of resistive material, with embedded conductors at each end, within an insulating cylindrical molded case, as illustrated in Figure 4-3(b).

This resistor type is called carbon composition because powdered car-

FIGURE 4-3 Carbon Composition Resistors. (a) Appearance. (b) Construction. (*Courtesy of Stackpole Electronics, Inc.*)

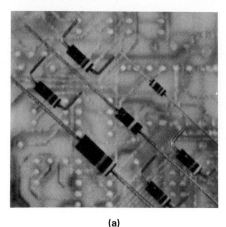

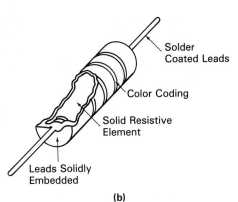

Solder
Coated Leads

Color Coding

Solid Resistive
Element

Leads Solidly
Embedded

(a)

(b)

Resistor Types **101**

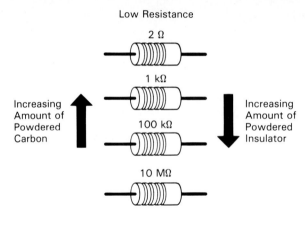

FIGURE 4-4 Carbon Composition Resistor Ratios.

bon and powdered insulator are bonded into a compound and used as the resistive material. By changing the ratio of powdered insulator to carbon, the value of resistance can be changed within the same area. For example, Figure 4-4 illustrates four fixed-value resistors ranging from 2 Ω to 10 MΩ.

The 2-Ω resistor is exactly the same size as the 10-MΩ (10 million Ω) resistor. This is achieved by having more powdered carbon and less powdered insulator in the 2 Ω and less powdered carbon and more powdered insulator in the 10 MΩ.

The color-coded rings or bands on the resistors in Figure 4-3 are a means of determining the value of resistance. This and other coding systems will be discussed later.

The physical size of the resistors lets the user know how much power in the form of heat can be dissipated, as seen in Figure 4-5. As you already know, resistance is the opposition to current flow or electrons, and this opposition causes heat to be generated when current flows. The amount of heat dissipated per unit of time is measured in watts. The bigger the resistor, the more heat that can be dissipated, and so a large-sized resistor could

FIGURE 4-5 Resistor Wattage Rating Guide.

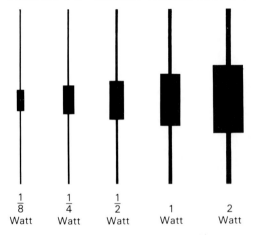

dissipate heat at a rate of 2 watts, while a small-sized resistor could dissipate heat at a rate of only $\frac{1}{4}$ watt. The key point is that heat should be dissipated faster than it is generated; if not, a larger-sized resistor with a larger surface area to dissipate away the additional heat should be used so that the resistor will not burn up.

Another factor to consider when discussing resistors is their tolerance. Tolerance is the amount of deviation or error from the specified value. For example, a 1000-Ω (1 kΩ) resistor with a $\pm 10\%$ (plus and minus 10%) tolerance when manufactured could be anywhere between 900 to 1100 Ω.

$$\pm 10\% \text{ of } 1000 = 100$$

$$10\% - \boxed{1000} + 10\%$$
$$\downarrow \qquad\qquad\qquad \downarrow$$
$$900 \qquad\qquad\qquad 1100$$

This means two identically marked resistors when measured could be from 900 to 1100 Ω, a difference of 200 Ω. In some applications, this may be acceptable, although in others where high precision is required this deviation could be too large, and so more expensive, smaller tolerance resistors are used.

(2) CARBON FILM RESISTORS

Figure 4-6(a) illustrates the physical appearance of carbon film resistors. They are constructed, as seen in Figure 4-6(b), by first depositing a thin layer or film of resistive material (a blend of carbon and insulator) on a ceramic (insulating) substrate. The film is then cut to form a helix or spiral. A greater ratio of carbon to insulator will achieve a low-resistance helix; on the other hand, a greater ratio of insulator to carbon will create a higher-resistance helix. Carbon film resistors have smaller tolerance figures ($\pm 5\%$ to $\pm 2\%$), are more stable (maintain same resistance value over a wide range

FIGURE 4-6 **Carbon Film Resistors. (a) Appearance. (b) Construction.**
(Courtesy of Stackpole Electronics, Inc.)

(a)

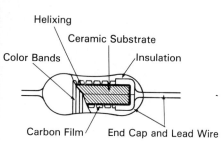

Helixing
Ceramic Substrate
Color Bands
Insulation
Carbon Film
End Cap and Lead Wire

(b)

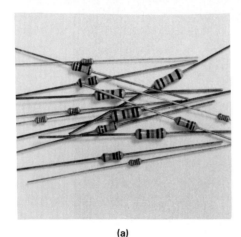

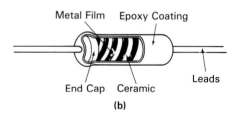

Metal Film Epoxy Coating

End Cap Ceramic

Leads

(a)

(b)

FIGURE 4-7   Metal Film Resistors. (a) Appearance. (b) Construction.
(Courtesy of Stackpole Electronics, Inc.)

of temperatures), and have less internally generated noise (random small bursts of voltage) than carbon composition resistors.

(3) METAL FILM RESISTORS

Figure 4-7(a) illustrates the physical appearance of some typical metal film resistors. They are constructed by spraying a thin film of metal on a ceramic cylinder (substrate) and then cutting the film to form a substrate, as seen in Figure 4-7(b). Metal film resistors have possibly the best tolerances commercially available of $\pm 1\%$ to $\pm 0.1\%$. They also maintain a very stable resistance over a wide range of temperatures (good stability) and generate very little internal noise compared to any carbon resistor.

(4) WIREWOUND RESISTORS

Figure 4-8(a) and (c) illustrates the appearance of different types of wirewound resistors. They are constructed, as can be seen in Figure 4-8(b), by wrapping a length of wire uniformly around a ceramic insulating core, with terminals making the connections at each end. The length and thickness of the wire are varied to change the resistance, which ranges from 1 Ω to 150 kΩ.

Since current flow is opposed by resistors and this opposition generates heat, the larger the physical size of the resistor, the greater the amount of heat that can be dissipated away and so the greater the current that can be passed through the resistor. Wirewound resistors are generally used in applications requiring low resistance values, which means the current and therefore power dissipated are high ($R\downarrow = V/I\uparrow$, $P\uparrow = I^2\uparrow R$). Consequently, these resistors are designed to have large surface areas so they can safely dissipate away the heat. The amount of heat that can be dissipated is measured in watts and is indicated on the resistor. A 10-watt resistor can be used to dissipate 20 watts if air is blown across its surface by a cooling fan or if it is immersed in a coolant.

Wirewound resistors typically have good tolerances of $\pm 1\%$; however, their large physical size and difficult manufacturing process make them very expensive.

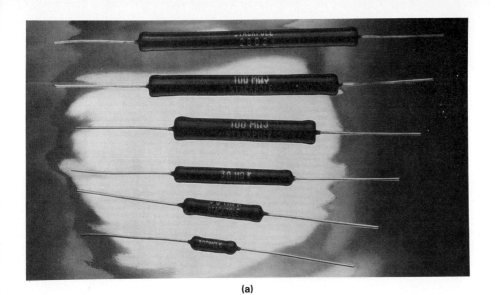

(a)

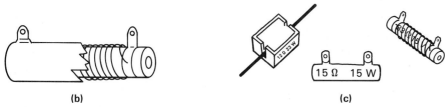

(b)   (c)

FIGURE 4-8   Wirewound Resistors. (a) Appearance. (b) Construction.
(c) Other Types. *(Courtesy of Stackpole Electronics, Inc.)*

(5) METAL OXIDE RESISTORS

Figure 4-9(a) illustrates the physical appearance of some metal oxide resistors. They are constructed, as can be seen in Figure 4-9(b), by depositing an oxide of a metal such as tin onto an insulating substrate. The ratio of

FIGURE 4-9   Metal Oxide Resistors. (a) Appearance. (b) Construction.
(Courtesy of Stackpole Electronics, Inc.)

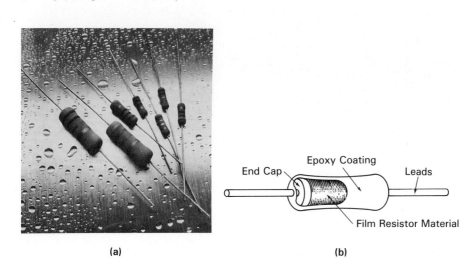

(a)   (b)

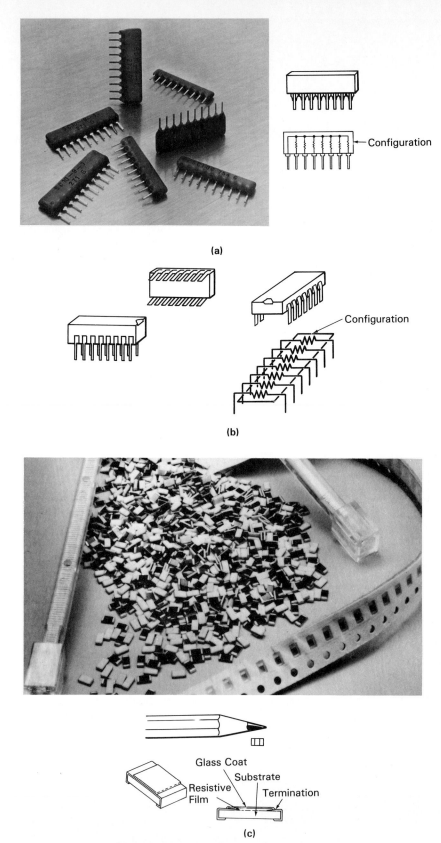

FIGURE 4-10 Thick Film Resistors. (a) SIPs: Single In-line Packages. (b) DIPs: Dual In-line Packages. (c) Chips. (*Courtesy of Stackpole Electronics, Inc.*)

oxide (insulator) to tin (conductor) will determine the resistor's resistance.

Metal oxide resistors have excellent temperature stability, which is the ability of the resistor to maintain its value of resistance without change even when temperature is changed. There is a tendency when heat is increased for atoms to move, and in so doing they cause collisions, resulting in an opposition to current; so a temperature increase will cause a resistor's resistance to increase. This increase from the desired resistance is very small with metal oxide resistors.

(6) THICK FILM RESISTORS

Figure 4-10 illustrates examples of thick film resistors. Figure 4-10(a) and (b) shows the two different types of resistor networks called SIPs and DIPs. The single in-line package is so called because all its lead connections are in a single line, whereas the dual in-line package has two lines of connecting pins. The chip resistors seen in Figure 4-10(c) are small thick film resistors that are approximately the size of a pencil lead.

The SIP and DIP resistor networks are constructed by first screening on the internal conducting strip (silver) that connects the external pins to the resistive material, and then screening on the thick film of resistive paste

---

EXAMPLE 4.1

Calculate the amount of deviation of the following resistors:

a. $2.2\text{k}\Omega \pm 10\%$
b. $5\text{ M}\Omega \pm 2\%$
c. $3\text{ }\Omega \pm 1\%$

SOLUTION

a. 10% of 2.2 kΩ = 220 Ω. For +10%, the value is

$$2200 + 220 \text{ }\Omega = 2420 \text{ }\Omega$$
$$= 2.42 \text{ k}\Omega$$

For −10%, the value is

$$2200 - 220 \text{ }\Omega = 1080 \text{ }\Omega$$
$$= 1.98 \text{ k}\Omega$$

Resistor will measure anywhere between 1.98 and 2.42 kΩ.

Calculator Sequence

Step	Keypad Entry	Display Response
1.	① ⓪	10
2.	⑧	10
3.	② · ② E ③	2.2E3
4.	=	220

b. 2% of 5 MΩ = 100 kΩ

$$5 \text{ M}\Omega + 100 \text{ k}\Omega = 5.1 \text{ M}\Omega$$
$$5 \text{ M}\Omega - 100 \text{ k}\Omega = 4.9 \text{ M}\Omega \qquad \text{Deviation} = 4.9 \text{ to } 5.1 \text{ M}\Omega$$

c. 1% of 3 Ω = 0.03 Ω or 30 milliohms (MΩ).

$$3 \text{ }\Omega + 0.03 \text{ }\Omega = 3.03 \text{ }\Omega$$
$$3 \text{ }\Omega - 0.03 \text{ }\Omega = 2.97 \text{ }\Omega \qquad \text{Deviation} = 2.97 \text{ to } 3.03 \text{ }\Omega$$

(bismuth/ruthenate), the blend of which will determine resistance. The chip resistor uses the same resistive film paste material, which is deposited onto an insulating substrate with two end terminations and protected by a glass coat.

The SIP and DIP resistor networks, once constructed, are trimmed by lasers to obtain close tolerances of typically $\pm 2\%$. Resistance values ranging from 22 Ω to 2.2 MΩ are available with a power rating of $\frac{1}{2}$ watt.

The chip resistor is commercially available with resistance values from 10 Ω to 3.3 MΩ, with a $\pm 2\%$ tolerance and a $\frac{1}{8}$ watt heat dissipation capability. They are ideally suited for applications requiring physically small sized resistors.

4.1.2 VARIABLE RESISTORS

In certain applications, we may require a variation in resistance while in circuit, for example, in the volume adjustment on a radio or television. The component that achieves this for us is the variable resistor.

The resistance of variable resistors can basically be varied in one of three ways:

1. Mechanically (user) adjustable:
   a. Rheostat
   b. Potentiometer
2. Thermally (heat) adjustable:
   a. Thermistor
   b. Resistive temperature detector (RTD)
   c. Thin film detector (TFD)
3. Optically (light) adjustable:
   a. Photoresistor

(1) MECHANICALLY (USER) ADJUSTABLE VARIABLE RESISTORS

This category includes two types, both of which will cause a change in resistance when a shaft is rotated.

(a) Rheostat (Two Terminals: *A* and *B*) Figure 4-11(a) shows the physical appearance of different rheostats, while Figure 4-11(b) illustrates schematic symbols. As can be seen in the construction of a circular rheostat in Figure 4-11(c), one terminal is connected to one side of the track and the other terminal of this two-terminal device is connected to a movable wiper. As the wiper is moved away from the end of the track with the terminal, the resistance between the stationary end terminal and the mobile wiper terminal increases. This is summarized in Figure 4-11(d), where the wiper has been moved down by a clockwise rotation of the shaft. Current would have to flow through a large resistance as it travels from one terminal to the other. On the other hand, as the wiper is moved closer to the end of the track connected to the terminal, the resistance decreases. This is summarized in Figure 4-11(e), which shows that as the wiper is moved up, as a result

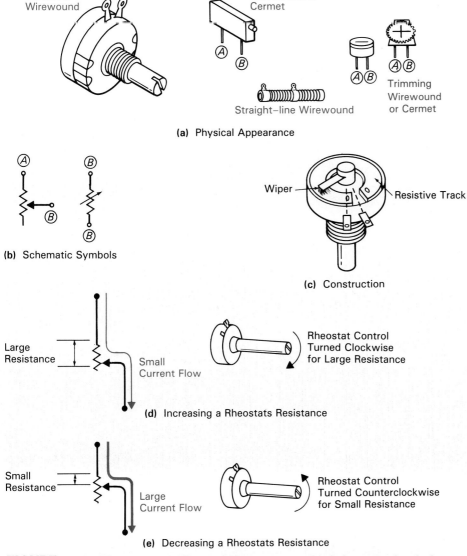

(a) Physical Appearance

(b) Schematic Symbols

(c) Construction

(d) Increasing a Rheostats Resistance

(e) Decreasing a Rheostats Resistance

FIGURE 4-11 Rheostat. (a) Physical Appearance. (b) Schematic Symbols. (c) Construction. (d) Increasing a Rheostat's Resistance. (e) Decreasing a Rheostat's Resistance.

of turning the shaft counterclockwise, current will only see a small resistance between the two terminals.

Rheostats come in many shapes and sizes, as can be seen in Figure 4-11(a). Some employ a straight-line motion to vary resistance, while others are classified as circular-motion rheostats. The resistive elements also vary; wirewound and carbon tracks are very popular, and cermet, which is a ceramic (insulator)–metal (conductor) mix is also used, the ratio of which can be used to produce different values of resistive tracks. A trimming rheostat is a miniature device used to change resistance by a small amount. Other circular-motion rheostats are available that require between two to ten turns to cover the full resistance range.

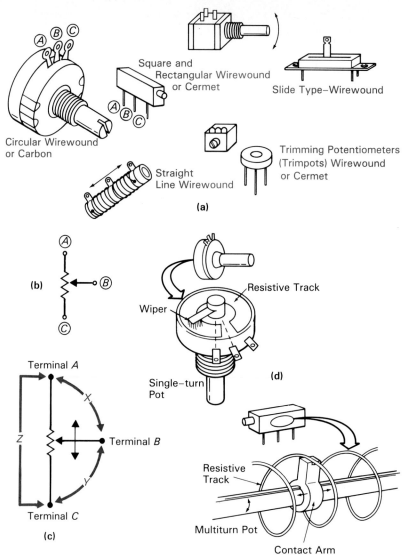

FIGURE 4-12 Potentiometer. (a) Physical Appearance. (b) Schematic Symbol. (c) Operation. (d) Construction.

(b) Potentiometer (Three Terminals: *A*, *B*, and *C*) Figure 4-12(a) illustrates the physical appearance of a variety of potentiometers, also called pots (slang), while Figure 4-12(b) shows its schematic symbol. You will probably notice that the difference between a rheostat and potentiometer is the number of terminals; the rheostat has two terminals and the potentiometer has three. With the rheostat, there were only two terminals and the resistance between the wiper and terminal varied as the wiper was adjusted. For the potentiometer illustrated in Figure 4-12(c), you can see that resistance can actually be measured across three separate combinations: between *A* and *B* (*X*), between *B* and *C* (*Y*), and between *C* and *A* (*Z*).

The only difference between the rheostat and the potentiometer in construction is the connection of a third terminal to the other end of the resistive track, as seen in Figure 4-12(d), showing the single turn potentiometer. Also

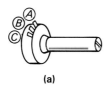

(a)

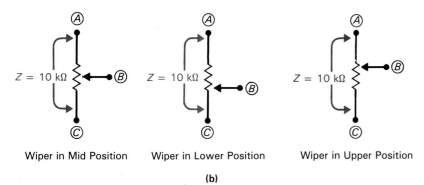

Wiper in Mid Position   Wiper in Lower Position   Wiper in Upper Position

(b)

FIGURE 4-13 A 10-Kilohm Potentiometer. (a) Physical Appearance. (b) End to End Resistance of a Potentiometer Remains Constant.

illustrated in this section of the figure is the construction of a multiturn potentiometer in which a contact arm slides along a shaft and the resistive track is formed into a helix of two to ten coils.

Figure 4-13(a) illustrates a 10-kΩ potentiometer. The resistance measured between terminal A and C will always be the same (10 kΩ) no matter where we put the wiper, because current still has to travel through the complete resistance between A and C, as illustrated in Figure 4-13(b) with a 10 kΩ potentiometer. The resistance between A and B (X) and B and C' (Y) will, however, vary as the wiper's position is moved, as illustrated in Figure 4-14.

If the user mechanically turns the shaft in the clockwise direction, the

FIGURE 4-14 Varying a Potentiometer's Resistance.

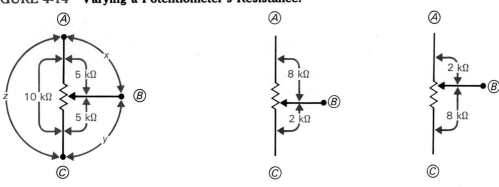

(a) Wiper in Mid Position ($x = y$)

$x = 5$ kΩ
$y = 5$ kΩ
$z = 10$ kΩ

(b) Wiper Moved Down ($x > y$)

$x = 8$ kΩ
$y = 2$ kΩ

(c) Wiper Moved Up ($x < y$)

$x = 2$ kΩ
$y = 8$ kΩ

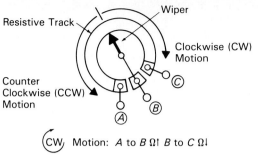

CW Motion: A to B $\Omega\uparrow$ B to C $\Omega\downarrow$

CCW Motion: A to B $\Omega\downarrow$ B to C $\Omega\uparrow$

FIGURE 4-15
Potentiometer's Operation.

resistance between A and B increases, while the resistance between B and C decreases. Similarly, if the user mechanically turns the shaft counterclockwise, a decrease occurs between A and B and there is a resulting increase in resistance between B and C. This point is summarized in Figure 4-15.

In some applications, you may see the symbol illustrated in Figure 4-16, where B is connected to C and only two terminals are hooked up by the user. In this situation, the potentiometer is being used as a rheostat.

Like the rheostat, the potentiometer comes in many different shapes and sizes. Wirewound, carbon, and cermet resistive tracks, circular or straight-line motion, two to ten multiturn, and other variations are available for different applications.

Whether rheostat or potentiometer, the resistive track can be classified as having either a linear or tapered (nonlinear) resistance. In Figure 4-17, we have taken a 1 kΩ rheostat and illustrated the resistance value changes between A and B for a linear and tapered one-turn rheostat.

The definition of linear is having an output that varies in direct proportion to the input. The input, in this case, is the user turning the shaft, and the output, as can be seen, is the linearly increasing resistance between A and B.

With a tapered rheostat or potentiometer, the resistance varies nonuniformly along its resistor element, being sometimes greater or less for equal shaft movement at various points along the resistance element, as seen in Figure 4-17.

In Figure 4-18, you can see an application of a potentiometer as a volume control in a television set. By increasing or decreasing the poten-

FIGURE 4-16   The Potentiometer as a Rheostat.
(a) Schematic Symbol. (b) Physical Appearance.

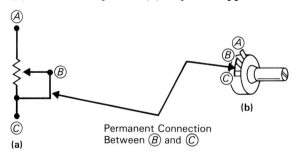

Resistors

	Fully CCW	$\frac{1}{4}$ CW	$\frac{1}{2}$ CW	$\frac{3}{4}$ CW	Fully CW
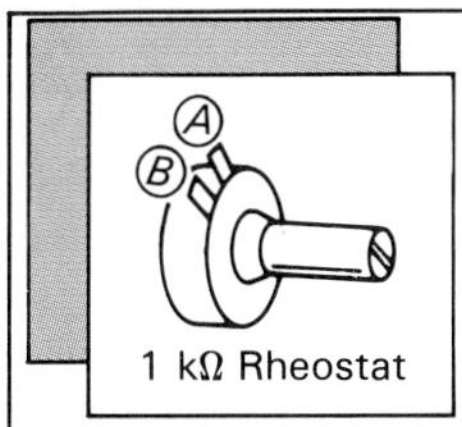 1 kΩ Rheostat	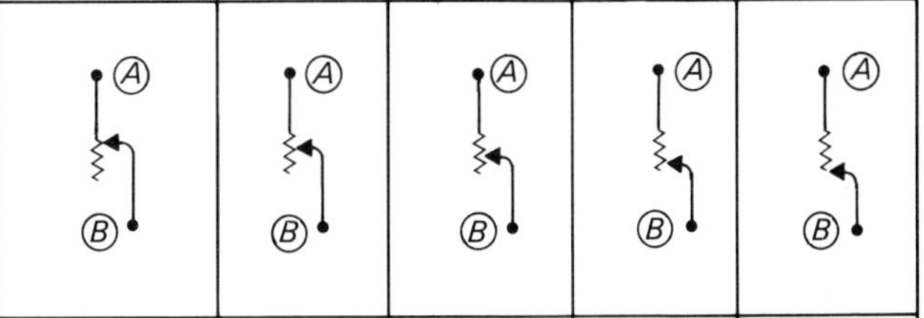				
Linear	0 Ω	250 Ω $\left(\frac{1}{4} \text{ of } 1000 \, \Omega\right)$	500 Ω $\left(\frac{1}{2} \text{ of } 1000 \, \Omega\right)$	750 Ω $\left(\frac{3}{4} \text{ of } 1000 \, \Omega\right)$	1000 Ω
Tapered	0 Ω	350 Ω	625 Ω	900 Ω	1000 Ω

These values have been arbitrarily chosen to illustrate a nonlinear change.

FIGURE 4-17 **Linear versus Tapered Resistive Track.**

tiometer's resistance, the amount of current passing to the loudspeaker is varied and so is volume.

(2) THERMALLY (HEAT) ADJUSTABLE VARIABLE RESISTOR

When first discussing variable resistors, we talked about the rheostat and potentiometer, both of which have a mechanical input (the user turning the shaft) to produce a change in resistance.

A bolometer is a device that, instead of changing its resistance when mechanical energy is applied, changes its resistance when heat energy is applied. There are basically three different types of temperature detectors, all of which are illustrated in Figure 4-19.

The measurement of temperature (thermometry) is probably the most

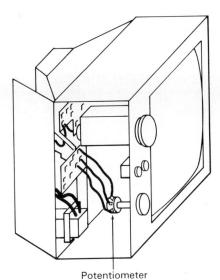

Potentiometer

FIGURE 4-18 **Potentiometer as a Volume Control.**

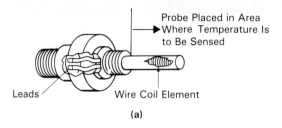

Probe Placed in Area Where Temperature Is to Be Sensed

Leads

Wire Coil Element

(a)

Platinum Winding			
°C	Ohms	°C	Ohms
−200	18.53	+200	175.84
−150	39.65	+250	194.08
−100	60.20	+300	212.03
−50	80.25	+350	229.69
±0	100.0	+400	247.06
+50	119.40	+450	264.14
+100	138.50	+500	280.93
+150	157.32	+550	297.16

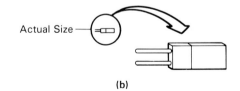

Actual Size

(b)

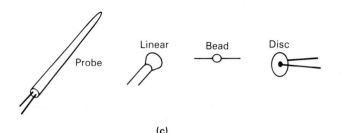

Probe   Linear   Bead   Disc

°C	Ohms
−50	100,000
0	7,500
+50	7,400
+100	100
+150	50
+200	27
+250	10
+300	7.5

(c)

FIGURE 4-19   Temperature Sensors. (a) Resistive Temperature Detector (RTD). (b) Thin Film Detector (TFD). (c) Thermistor.

common type of measurement used in industry today. Before discussing our three temperature sensors, let's first consider the four units of temperature measurement detailed in Table 4-1.

Commercially, temperature is normally expressed in degrees Celsius (°C) or degrees Fahrenheit (°F); however, although not commonly known, kelvins (k) and degrees Rankine (°R) are often used in industry, with the kelvin being the international unit of temperature.

TABLE 4-1
FOUR TEMPERATURE SCALES AND CONVERSION FORMULAS

$F = (\frac{9}{5} \times C) + 32$
$C = \frac{5}{9} \times (F - 32)$
$R = F + 460$
$F = R - 460$
$K = C + 273$
$C = K - 273$

	Fahrenheit	*Celsius*	*Kelvin*	*Rankine*
Absolute zero	−459.69°F	−273.16°C	0°K	0°R
Melting point of ice (x)	32°F	0°C	273.16°K	491.69°R
(Division between x and y)	(180°F)	(100°C)	(100°K)	(180°R)
Boiling point of water (y)	212°F	100°C	373.16°K	671.69°R

114   Resistors

EXAMPLE 4.2

Convert the following:
a. $74°F = $ _____ $°C$
b. $45°C = $ _____ $°F$
c. $25°C = $ _____ K
d. $10°F = $ _____ $°R$

Calculator Sequence

Step	Keypad Entry	Display Response
1.	$\boxed{7}$ $\boxed{4}$	74
2.	$\boxed{-}$	
3.	$\boxed{3}$ $\boxed{2}$	32
4.	$\boxed{=}$	42
5.	$\boxed{STO}$ (store result in memory)	
6.	$\boxed{C/CE}$ (cancel display)	0
7.	$\boxed{5}$	5
8.	$\boxed{÷}$	
9.	$\boxed{9}$	9
10.	$\boxed{=}$	0.55555
11.	$\boxed{×}$	
12.	$\boxed{RCL}$ (Recall value from memory)	42
13.	$\boxed{=}$	23.3

SOLUTION

a. $C = \frac{5}{9} \times (F - 32)$
   $= \frac{5}{9} \times (74 - 32)$
   $0.555 \times 42 = 23.3°C$

b. $F = (\frac{9}{5} \times C) + 32$
   $= (\frac{9}{5} \times 45) + 32$
   $= (1.8 \times 45) + 32 = 113°F$

c. $K = C + 273$
   $= 25 + 273 = 298 \ K$

d. $R = F + 460$
   $= 10 + 460 = 470°R$

The resistive temperature detector (RTD) and thin film detector (TFD) seen in Figure 4-19(a) and (b) are both temperature sensors that contain a conducting material such as copper, nickel, or platinum and consequently have a positive temperature coefficient: reistance increases as temperature increases. The thermistor, on the other hand, contains a semiconductor material that has a negative temperature coefficient: resistance decreases as temperature increases.

Referring to the construction of the RTD in Figure 4-19(a), you can see that the sensing element consists of a coil of fine wire generally made of platinum, which gives a relatively linear increase in resistance as temperature increases, as indicated in the table in Figure 4-19(a).

The thin film detector (TFD) seen in Figure 4-19(d) is constructed by placing a thin layer of platinum, for very precise temperature readings, on a ceramic substrate. Because of its small size, the TFD responds rapidly to temperature change and is ideally suited for surface temperature sensing.

The thermistor, a variety of which can be seen in Figure 4-19(c), is the most common type of temperature sensor and produces rapid and extremely large changes in resistance for very small changes in temperature, as seen in the associated table. The termistor can be used as a temperature probe inside an oven. As the oven heats up, the thermistor's resistance decreases, and at a certain decreased resistance the thermistor turns off the oven. Once the temperature starts to drop, the thermistor's resistance increases, and this increase of resistance turns on the oven.

temperature↑	thermistor's R↓	oven off
temperature↓	thermistor's R↑	oven on

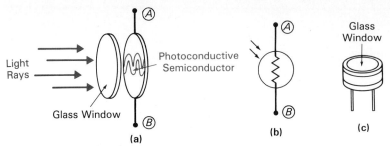

FIGURE 4-20 **Photoresistor. (a) Construction. (b) Schematic Symbol. (c) Physical Appearance.**

(3) OPTICALLY (LIGHT) ADJUSTABLE VARIABLE RESISTOR

Photo means illumination, and the photoresistor is a resistor that is photoconductive. This means that, as the material is exposed to light, it will become more conductive and less resistive. Figure 4-20 illustrates the photoresistor, which is also called a light-dependent resistor (LDR), and its schematic symbol.

The photoresistor is a thin slice of photoconductive material whose resistance decreases as light is applied. The light energy is absorbed by the atoms within the photoconductive material, causing these atoms to release their valence electrons. This results in an increase in electrons, and therefore current passing through the photoresistor, and so its resistance will have decreased. To summarize, we can say:

light energy↑   conduction↑   resistance↓

Photoresistors can be used to turn on and off outdoor home security lights. During the day, the natural sunlight decreases the resistance of the photoresistor and this low resistance keeps the lights off. At dusk, the sunlight is almost gone, and the photoresistor's resistance increases. This increased resistance is used to turn on the security lights.

sun up↑   photoresistance↓   security lights OFF
sun down↓   photoresistance↑   security lights ON

SELF-TEST REVIEW QUESTIONS (§ 4.1)

1. List the six types of fixed-value resistors.
2. What is the difference between SIP and DIPs?
3. Name the two types of mechanically adjustable variable resistors and state the difference between the two.
4. What is meant by a linear and tapered potentiometer?
5. A thermistor has a negative temperature coefficient of resistance (true or false).
6. What name is given to the optically adjustable resistor?

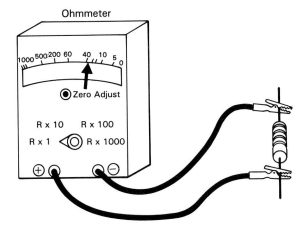

FIGURE 4-21 **Measuring Resistance.**

4.2
HOW IS RESISTANCE MEASURED?

Resistance is measured with an ohmmeter. Figure 4-21 illustrates a multirange ohmmeter with a resistor's resistance being measured. The meter has four ranges. The simplest range has been selected ($R \times 1$), and on this scale the resistance value indicated can be read directly from the meter; in this example the value is 36Ω. If we had selected the $R \times 10$ range, then 36 would have to be multiplied by 10 ($36 \times 10 = 360$) to obtain the value of 360 Ω. If we were on the $R \times 100$ range, then $R \times 100 = 36 \times 100$, which equals 3.6 kΩ. The $R \times 1000$ scale, if selected, means that you would interpret 36 as $R \times 1000 = 36 \times 1000$, which equals 36 kΩ.

Three rules must be remembered and applied when measuring resistance.

1. Short the meter leads together and adjust the zero ohms control so the pointer is at 0 Ω. This calibrates the meter scale.
2. Turn off the power and remove the component (if practicable) to be measured from circuit to protect the ohmmeter.
3. Connect the leads across the component, adjust the range scale until the pointer is approximately in the middle of the scale (for accuracy), and then multiply for the range scale selected.

SELF-TEST REVIEW QUESTIONS (§ 4.2)

1. Name the instrument used to measure resistance.
2. On the $R \times 1000$ range, how would a reading of 22 be interpreted?
3. On the $R \times 1$ range, how would a reading of 470 be interpreted?
4. List the three rules that should be applied when using the ohmmeter.

4.3
RESISTOR CODING

Using an ohmmeter, you can find the value of any resistor. However, manufacturers indicate the value and tolerance of resistors on the body of the component in one of two ways:

1. Color code: colored rings or bands
2. Typographically: printed alphanumerics (alphabet and numerals)

Examples of both of these can be seen in Figure 4-22.

4.3.1 GENERAL-PURPOSE AND PRECISION COLOR CODE

There are basically two different types of fixed resistors: general purpose and precision. Resistors with tolerances of $\pm 2\%$ or less are precision resistors and have five bands, while resistors with tolerances of $\pm 5\%$ or greater have four bands and are referred to as general purpose. The color code and differences between precision and general-purpose resistors are illustrated in Figure 4-23.

When you pick up a resistor, notice that the bands are nearer to one end; this end should be held in your left hand. If there are four bands on the resistor, follow the general-purpose resistor code; if five bands are present, adopt the precision resistor code.

(1) GENERAL-PURPOSE RESISTOR CODE

1. The first band on either a general-purpose or precision resistor can never be black, and it is the first digit of our numbers.
2. The second band indicates the second digit of the number.
3. The third band specifies the multiplier to be applied to the number, which ranges from $\times 1/100$ to $100 \text{ k}\Omega$.
4. The fourth band describes the tolerance or deviation from the specified resistance, which is $\pm 5\%$ or greater.

FIGURE 4-22 **Resistor Value and Tolerance Indication.**
(a) Typographical. (b) Color Code.

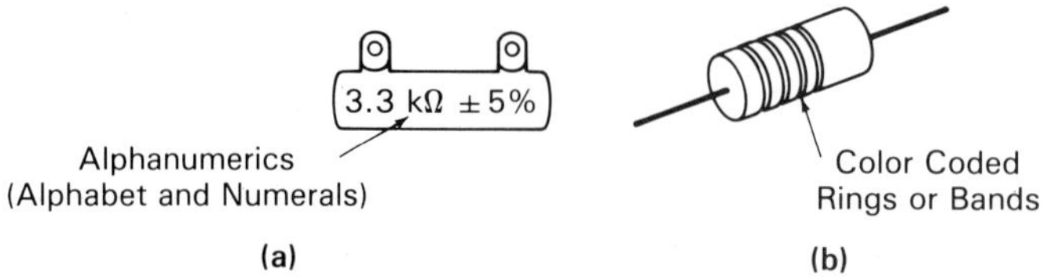

3.3 kΩ ±5%

Alphanumerics
(Alphabet and Numerals)

Color Coded
Rings or Bands

(a)

(b)

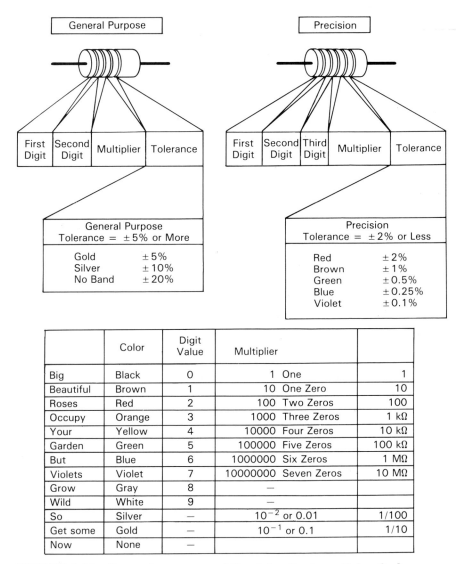

	Color	Digit Value	Multiplier	
Big	Black	0	1 One	1
Beautiful	Brown	1	10 One Zero	10
Roses	Red	2	100 Two Zeros	100
Occupy	Orange	3	1000 Three Zeros	1 kΩ
Your	Yellow	4	10000 Four Zeros	10 kΩ
Garden	Green	5	100000 Five Zeros	100 kΩ
But	Blue	6	1000000 Six Zeros	1 MΩ
Violets	Violet	7	10000000 Seven Zeros	10 MΩ
Grow	Gray	8	—	
Wild	White	9	—	
So	Silver	—	10^{-2} or 0.01	1/100
Get some	Gold	—	10^{-1} or 0.1	1/10
Now	None	—		

FIGURE 4-23   General-purpose and Precision Resistor Color Code.

(2) PRECISION RESISTOR CODE

1. This band, like the general-purpose resistor, is never black and is the first digit of our three-digit number.
2. The band provides the second digit.
3. The third band indicates the third and final digit of the number.
4. The fourth band specifies the multiplier to be applied to the number.
5. The fifth and final band indicates the tolerance figure of the precision resistor, which is always less than ±2%, which accounts for why precision resistors are more expensive than general purpose.

Band 1 = Green
Band 2 = Blue
Band 3 = Brown
Band 4 = Gold

FIGURE 4-24

EXAMPLE 4.3

Figure 4-24 illustrates a ½-watt resistor.
a. Is this a general-purpose or precision resistor?
b. What is this resistor's value of resistance?
c. What tolerance does this resistor have, and what deviation plus and minus could occur to this value?

SOLUTION

a. General purpose (4 bands)
b. green   blue × brown
     5   6 × 10   = 560 Ω
c. Tolerance band is gold, which is ±5%.

$$\text{deviation} = 5\% \text{ of } 560 = 28$$
$$560 + 28 = 588 \ \Omega$$
$$560 - 28 = 532 \ \Omega$$

The resistor could, when measured, be anywhere from 532 to 588 Ω.

EXAMPLE 4.4

State the resistor's value and tolerance and whether it is general purpose or precision for the examples shown in Figure 4-25.

Orange/Green/Black/Silver   Green/Blue/Red/None   Red/Red/Green/Gold/Blue

(a)   (b)   (c)

FIGURE 4-25

SOLUTION

a. orange green   black silver (4 bands = general purpose)
     3   5 × 1   10%   = 35 Ω ± 10%

b. green blue   red   none (4 bands = general purpose)
     5   6 × 100 20%   = 5.6 kΩ ± 2%

c. red red green   gold   blue   (5 bands = precision)
     2   2   5 × 0.1   0.25%   = 22.5 Ω ± 0.25%

Resistors

4.3.2 ZERO-OHM RESISTOR

Figure 4-26(a) illustrates the only resistor color code that you will probably have trouble with, the zero ohm resistor, which has one black band. Zero ohms is equivalent to a straight piece of wire. You may ask yourself—Why do they manufacture a resistor of zero ohms since it really is not a resistor at all?

Printed circuit boards, like the one in Figure 4-26(b) no longer have all their resistor and other components inserted by hand. Automatic insertion equipment replaces the human assembler and places the correct component in the correct position in a matter of seconds rather than hours. In some applications, a direct contact between two points is needed, in which case a piece of wire needs to be inserted, as seen in Figure 4-26(b). However, the resistor automatic insertion equipment will only handle resistors, not wire (called jumpers). In the past, manufacturers had to do this insertion manually after all the components were inserted, which caused enormous delays; but now the zero ohm resistor can be inserted automatically during the resistor insertion process and have exactly the same effect, as seen in Figure 4-26(c).

FIGURE 4-26 **Zero-ohm Resistor. (a) Appearance. (b) Printed Circuit Board. (c) Zero-ohm Resistor as Jumper.**

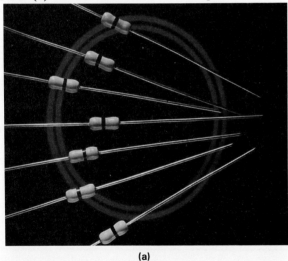

(a)

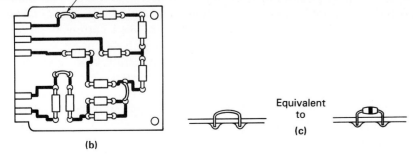

Direct Contact between Two Pads

(b)

Equivalent to

(c)

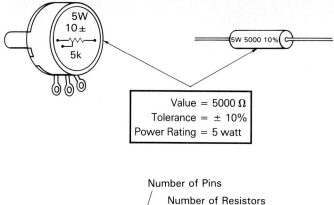

Value = 5000 Ω
Tolerance = ± 10%
Power Rating = 5 watt

(a)

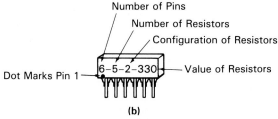

Number of Pins
Number of Resistors
Configuration of Resistors
Dot Marks Pin 1
Value of Resistors
6–5–2–330

(b)

FIGURE 4-27 **Typographical Value Indication.**
(a) Standard Variable and Fixed. (b) SIP and DIP.

4.3.3 OTHER RESISTOR IDENTIFICATION METHODS

If a resistor color code is not found, then some other form of marking will always be made on the resistor. The larger, fixed, wirewound resistors and nearly all variable resistors normally have the resistance value, tolerance, and wattage printed on the resistor, as seen in Figure 4-27(a). SIP and DIP packages are normally coded with both letters and numerals, all of which have a particular meaning, as seen in Figure 4-27(b). Chip resistors tend to be too small to have any intelligible form of marking and so will have to be measured with an ohmmeter. They are packaged, as seen in Figure 4-10(c), in either polythene bags, paper tape, or plastic magazines, and in all these cases the packaging will be used to specify values, tolerance, and wattage.

4.4
FILAMENT RESISTOR

Figure 4-28 illustrates the filament resistor within a glass bulb. This component is more commonly known as the household light bulb. The filament resistor is just a coil of wire that glows white hot when current is passed through it and, in so doing, dissipates both heat and light energy.

A coil of wire similar to the one found in the light bulb is called a ballast resistor and is used to maintain a constant current despite variation in voltage. Since the coil of wire is a conductor and conductors have a positive temperature coefficient, an increase in voltage will cause a corresponding increase in current and therefore in the heat generated, which will result in an increase in the wire's resistance. This increase in resistance will

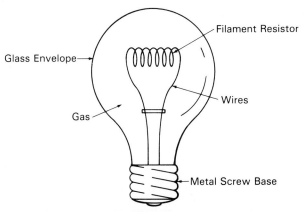

FIGURE 4-28 **Filament Resistor within the Incandescent Light Bulb.**

decrease the initial current rise. Similarly, a decrease in voltage and therefore current ($I \propto V$) will result in a decrease in heat and wire resistance. This decrease in resistance will permit an increase in current to counteract the original decrease. Current is therefore regulated or maintained constant by the ballast resistor, despite variations in voltage.

SELF-TEST REVIEW QUESTIONS (§ 4.3 AND § 4.4)

1. What tolerance differences occur between a general-purpose and precision resistor?
2. A red/red/red/gold/blue resistor has what value and tolerance figure?
3. What color code would appear on 4.7 kΩ, ±10% tolerance resistor?
4. The filament resistor is like all other resistors in that it converts all the electrical energy into heat (true or false).

4.5
RESISTOR TROUBLESHOOTING

It is virtually impossible for resistors to internally short-circuit (no resistance or zero ohms); however, they may be shorted out by some other device providing a low resistance path around the resistor. Generally, the resistor's internal resistive elements will begin to develop a higher resistance than its specified value (due to a partial internal open) or in some cases go completely open circuit (maximum resistance or infinite ohms).

Fixed resistors tend to go either partially or fully open as the resistive track begins to break down. Variable resistors have problems with the wiper making contact with the resistive track at all times. Faulty variable resistors in audio systems can normally be detected because they generate a scratchy noise whenever, for example, you adjust the volume or tone control.

The ohmmeter is the ideal instrument for verifying whether a resistor is functioning correctly or not. When checking a resistor's resistance, here are some points that should be remembered.

1. Always calibrate the meter first. If the meter indicates 100 Ω when the two meter leads are shorted together (0 Ω), then every reading you take will be 100 Ω out.

2. The ohmmeter has its own internal power source (a battery), so always turn off the power and remove the resistor to be measured from the circuit. If this is not carried out, you will not only obtain inaccurate readings, but you can also damage the ohmmeter.

3. The range scales on a digital multimeter often confuse people; for example, if a range is chosen so that the highest reading is 100 kΩ and a 1 MΩ resistor is measured, the meter will indicate an infinite ohms reading and you may be misled into believing that you have found the problem. In the case of an analog ohmmeter, the range scale multiplier is not applied and a misleading result is obtained. For example, if the $R \times 100$ range is selected, the resistance reading (R) must be multiplied by 100 to obtain the resistance.

4. Another point to keep in mind is tolerance. For example, if a suspected 1-kΩ (1000-Ω) resistor is measured with an ohmmeter and reads 1.2 kΩ (1200 Ω), it could be within tolerance if no tolerance band is present on the resistor's body ($\pm 20\%$). A 1-kΩ resistor with a $\pm 20\%$ tolerance could measure anywhere between 800 and 1200 Ω.

5. The ohmmeter's internal battery voltage is really too small to deliver an electrical shock; however, you should avoid touching the bare metal parts of the probes or resistor leads as your body resistance of approximately 50 kΩ will affect your meter reading.

SELF-TEST REVIEW QUESTIONS (§ 4.5)

1. What problem normally occurs with resistors? (opens/shorts)
2. What instrument is used to verify a resistor's resistance?

SUMMARY

1. Resistors are classified as being either fixed or variable.
2. There are six different types of fixed resistors:

    a. Carbon composition
    b. Carbon film
    c. Metal film
    d. Wirewound
    e. Metal oxide
    f. Thick film

3. The physical size of a resistor lets you know how much power can be dissipated in the form of heat.

4. The tolerance of a resistor describes the amount the resistor deviates from its specified value.

5. Variable resistors can have their resistance varied in one of three ways:
   a. Mechanically (rheostat, potentiometer)
   b. Thermally (RTD, TFD, thermistor)
   c. Optically (photoresistor)

6. Three rules should be remembered when using the ohmmeter to measure resistance:
   a. Short the meter leads, and adjust the zero ohms control to align the pointer to 0 Ω.
   b. Turn off the power and remove the component to be measured.
   c. Connect the leads across the component, adjust the range scale until the pointer is in mid scale position, and then multiply for the range scale selected.

7. Fixed resistors with tolerances of ±2% or less are called precision resistors and have a five band color code, while resistors with tolerances of ±5% or greater have four bands and are referred to as general-purpose resistors.

8. The household light bulb uses a coil of wire referred to as a filament resistor, which glows white hot when current is passed through it, releasing heat and light.

9. Resistors will generally, when developing a problem, either partially or completely open circuit.

10. Variable resistors develop problems with the wiper making contact with the resistive track for the entire resistance range.

NEW TERMS

Alphanumeric
Ballast resistor
Carbon composition resistor
Carbon film resistor
Celsius (C)
Cermet resistor
Chip resistor
Color code
Dual in-line package (DIP)
Fahrenheit (F)
Filament resistor
Fixed resistor
General-purpose resistor
Helix
Infinite (∞)

Jumper	Rankine (R)	Thermistor
Kelvin (K)	Resistive tempera- ture detector (RTD)	Thermometry
Linear		Thick film resistor
Metal film resistor	Rheostat	Thin film detector
Metal oxide resistor	Single in-line package (SIP)	(TFD)
Ohmmeter		Tolerance
Photoresistor	Substrate	Variable resistor
Potentiometer	Tapered	Wattage rating
Precision resistor	Temperature stability	Wirewound resistor

NEW FORMULAS

$$F = (\tfrac{9}{5} \times C) + 32$$
$$C = \tfrac{5}{9} \times (F - 32)$$
$$R = F + 460$$
$$F = R - 460$$
$$K = C + 273$$
$$C = K - 273$$

F = Fahrenheit
C = Celsius
R = Rankine
K = Kelvin

REVIEW QUESTIONS

Multiple Choice Questions

1. The most common type of fixed resistor is the:
   a. Wirewound
   b. Carbon composition
   c. Film
   d. Metal oxide

2. A SIP package has:
   a. A single line of connectors
   b. A double line of connectors
   c. No connectors
   d. None of the above are true

3. The most commonly used mechanically adjustable variable resistor is the:
   a. Rheostat

b. Thermistor

c. Potentiometer

d. Photoresistor

4. A thermistor has a negative temperature coefficient, which means that:
   a. As temperature increases, resistance increases
   b. As temperature increases, conductance decreases
   c. As temperature increases, resistance decreases
   d. All of the above are true

5. A _____ color band resistor is called a general-purpose resistor, while a _____ band is known as a precision resistor.
   a. four, five
   b. three, four
   c. one, four
   d. two, seven

6. What would be the power dissipated by a 2-kΩ carbon composition resistor when a current of 20 mA is flowing through it?
   a. 40 watts
   b. 0.8 watt
   c. 1.25 watts
   d. None of the above

7. Which of the following is the most common type of temperature detector?
   a. RTD
   b. TFD
   c. Thermistor
   d. Barretter

8. A rheostat is a _____ terminal device, while the potentiometer is a _____ terminal device.
   a. 2, 3
   b. 1, 2
   c. 2, 4
   d. 3, 2

9. Which of the following is the unit of temperature in the International System of Units?
   a. Celsius
   b. Fahrenheit
   c. Kelvin
   d. Rankine

10. Which of the following temperature detectors has a positive temperature coefficient?
    a. RTD
    b. Thermistor
    c. TFD
    d. Both a and b
    e. Both a and c

11. Photoresistors are also called:
    a. TFDs

b. LDRs

c. RTDs

d. TGFs

12. Photoresistors have:
    a. A positive temperature coefficient
    b. A negative temperature coefficient
    c. Neither a nor b is true

13. Resistance is measured with a (an):
    a. Wattmeter
    b. Milliampmeter
    c. Ohmmeter
    d. A 100 meter

14. Which of the following is *not* true:
    a. Precision resistors have a tolerance of >5%.
    b. General-purpose resistors can be recognized because they have either three or four bands.
    c. The fifth band indicates the tolerance of a precision resistor.
    d. The third band of a general-purpose resistor specifies the multiplier.

15. The first band on either a general-purpose or precision resistor can never be:
    a. Brown
    b. Red or black stripe
    c. Black
    d. Red

16. What is the name of the resistor used to maintain a constant current despite variations in voltage by changing its resistance?
    a. RTD
    b. Thermistor
    c. Ballast resistor
    d. LDR

17. The term infinite ohms describes:
    a. A small finite resistance
    b. A resistance so large that a value cannot be placed on it
    c. A resistance between maximum and minimum
    d. None of the above

18. To calibrate an ohmmeter means:
    a. To make sure it indicates 100 Ω when 0 Ω is being measured
    b. To zero the meter when the meter leads are touching
    c. To make sure it indicates 0 Ω when the meter leads are apart
    d. Both a and b are true
    e. Both a and c are true

19. The typical problem with resistors is that they will:
    a. Develop a partial internal open
    b. Completely open circuit
    c. Develop a higher resistance than specified
    d. All of the above are true
    e. Both a and c are true

20. A 10% tolerance, 2.7 MΩ carbon composition resistor measures 2.99 MΩ when checked with an ohmmeter. Is it:
  a. Within tolerance
  b. Outside tolerance
  c. Faulty
  d. Both a and c are true
  e. Both a and b are true

Essay Questions

21. Describe the difference between a fixed and variable resistor.   (4.1)
22. List the six basic categories of fixed value resistors.   (4.1.1)
23. Explain why a resistor's size determines its wattage rating.   (4.1.1)
24. Why are resistors given a tolerance figure, and which is best, a small or large tolerance?   (4.1.1)
25. Briefly describe the construction of:
  a. Carbon composition fixed resistors   (4.1.1)
  b. Rheostats   (4.1.2)
  c. Potentiometers   (4.1.2)
  d. Multiturn precision potentiometers   (4.1.2)
26. Draw the schematic symbol for a rheostat and potentiometer and describe the difference. Also show how a potentiometer can be used as a rheostat.   (4.1.2)
27. Define linear and tapered resistive tracks.   (4.1.2)
28. List the three rules that should be applied when using an ohmmeter to measure resistance.   (4.2)
29. Describe the construction, operation, and temperature coefficient of the following: (4.1.2)
  a. RTD
  b. Photoresistor
  c. Thermistor
  d. TFD
30. List the four temperature scales in use today.   (4.1.2)
31. Give the color code for the following resistor values: (4.3)
  a. 1.2 MΩ, ± 10%
  b. 10 Ω, ± 5%
  c. 27.3 kΩ, ± 20%
  d. 273 kΩ, ± 0.5%
32. Give the resistor values for the following color codes: (4.3)
  a. Orange, orange, black
  b. Red, red, green, red, red
  c. Brown, black, orange, gold
  d. White, brown, brown, silver
33. For what application would the zero-ohm resistor be used.   (4.3.2)
34. Describe the difference between a SIP and DIP resistor package.   (4.1.1)

35. What are the common problems encountered with fixed and variable resistors. (4.5)

Practice Problems

36. If a 5.6 kΩ resistor has a tolerance of ±10%. What would be the allowable deviation in resistance above and below 5.6 kΩ? (4.3)

37. If a current of 50 mA is flowing through a 10-kΩ, 25-W resistor, (a) how much power is the resistor dissipating? (b) can the current be increased and if so by how much? (4.3)

38. What minimum wattage size could be used if a 12-Ω wirewound resistor were connected across a 12-V supply? (4.3)

39. Calculate the resistance deviation for the tolerances of the resistors listed in questions 31 and 32. (4.1.1)

40. If a 1-kΩ rheostat has 50 V across it, calculate the different values of current if the rheostat is varied in 100-Ω steps. Fill in the graph and table below.

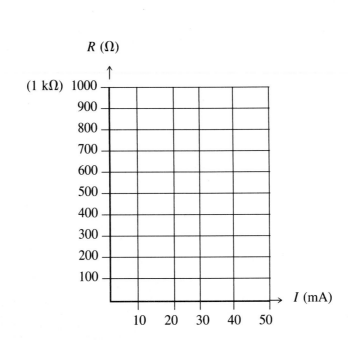

R (Ω)	I (mA)
100	
200	
300	
400	
500	
600	
700	
800	
900	
1000	

V = 50 V

Troubleshooting Questions

41. Briefly describe some of the problems that can occur with fixed and variable resistors. (4.5)

42. Describe why the tolerance of a resistor can make you think it has a problem when in fact it doesn't. (4.5)

43. If a 20-kΩ, 1-W, ±10% tolerance resistor measures 20.39 kΩ, is it in or out of tolerance? (4.1.1)

44. If 33 μA is flowing through a carbon film 4.7-kΩ, ¼-W, ±5% resistor, will it burn up or not? (4.3)

45. What points should be remembered when using the ohmmeter to verify a resistor's value? (4.5)

5

DIRECT CURRENT (DC)

AFTER COMPLETING THIS CHAPTER, YOU WILL BE ABLE TO:

1. Describe how direct current can be generated:
   a. Mechanically with friction and pressure
   b. Thermally with a thermocouple
   c. Optically with a photocell
   d. Magnetically with a generator
   e. Chemically with a battery
   f. Electrically with a power supply
2. Understand the nature of magnetism and why certain materials will magnetize, and certain materials will not magnetize.
3. Describe the magnetic field and the rules behind magnetic attraction and repulsion.
4. Explain how magnetic energy can be used to generate direct current.
5. State the differences between the three artificial magnets.
6. Describe the operation of both a primary and secondary battery.
7. State the difference between a primary and secondary cell.
8. Describe the features, applications, and construction of all the popular primary and secondary cells.
9. Explain the advantages of connecting batteries in series and parallel.
10. Define what is meant by a battery's:
    a. Internal resistance
    b. Maximum power transfer
    c. Optimum power transfer
11. Describe the operation and various types of fuses, circuit breakers, and switches.

INTRODUCTION

Direct current, abbreviated dc, is a flow of continuous current in only one direction and it is what we have been dealing with up to this point. A dc current is produced when a dc voltage source, such as a battery, is connected across a closed circuit. The requirement for a dc voltage source is that its output voltage polarity remain constant, and the fixed polarity of the voltage source will produce a unidirectional (one direction) current through a dc circuit.

THE FIRST COMPUTER BUG

Mathematician Grace Murray Hopper, an extremely feisty and independent U.S. naval officer, was assigned to the Bureau of Ordnance Computation Project at Harvard during World War II. As Hopper recalled, "We were not programmers in those days, the word had not yet come over from England. We were 'coders,'" and with her colleagues she was assigned to compute ballistic firing tables on the Harvard Mark 1 computer. In carrying out this task, Hopper developed programming method fundamentals that are still used today.

Hopper is also credited, on a less important note, with creating a term frequently used today falling under the category of computer jargon. During the hot summer of 1945, the computer developed a mysterious problem. Upon investigation, Hopper discovered that a moth had somehow strayed into the computer and prevented the operation of one of the thousands of electromechanical relay switches. In her usual meticulous manner, Hopper removed the remains and then taped and entered it into the logbook. In her own words, "From then on, when an officer came in to ask if we were accomplishing anything, we told him we were 'debugging' the computer," a term which is still used today to describe the process of finding problems in a computer program.

5.1
SOURCE AND LOAD

Figure 5-1(a) illustrates a dc circuit with a battery connected across a light bulb. In this circuit, the potential energy of the battery produces current flow or kinetic energy that is used to produce light energy from the bulb. The battery is therefore the *source* in this circuit and the bulb is called a *load*, which by definition is a device that absorbs the energy being supplied and converts it into the desired form. The *load resistance* will determine how hard the voltage source has to work. For example, the bulb filament has a

FIGURE 5-1 **Load on a Voltage Source. (a) Pictorial. (b) Schematic.**

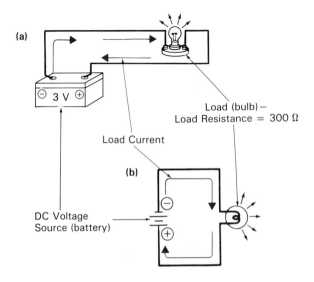

(a)

$\ominus$ 3 V $\oplus$

Load (bulb) —
Load Resistance = 300 Ω

Load Current

(b)

DC Voltage
Source (battery)

resistance of 300 Ω. The connecting wires have a combined resistance of 0.02 Ω (20 mΩ), and since this value is so small compared to the filaments resistance, it will be ignored as it will have no noticeable effect. Consequently, a 300-Ω load resistance will permit 10 mA of *load current* to flow when a 3-V source is connected across the dc circuit ($I = V/R = 10$ mA). A change in filament resistance will result in a change in current; so if the load resistance is increased, the load current decreases ($I\downarrow = V/R\uparrow$), and, conversely, if the load resistance is decreased, the load current increases ($I\uparrow = V/R\downarrow$) for a constant voltage source. Figure 5-2 illustrates a heater, light bulb, motor, computer, robot, television, and microwave oven. All these and all other devices or pieces of equipment are connected to some form of supply (for example, a battery). The supply does not see all the circuitry and internal workings of the equipment; it simply sees the whole device or piece of equipment as some value of resistance. The resistance of the equipment and the supply voltage determine the value of current flow from the supply to the device or equipment. The battery or any other current source sees these

FIGURE 5-2 **(a) Load Resistance of Equipment. (b) Schematic Symbols and Terms.**

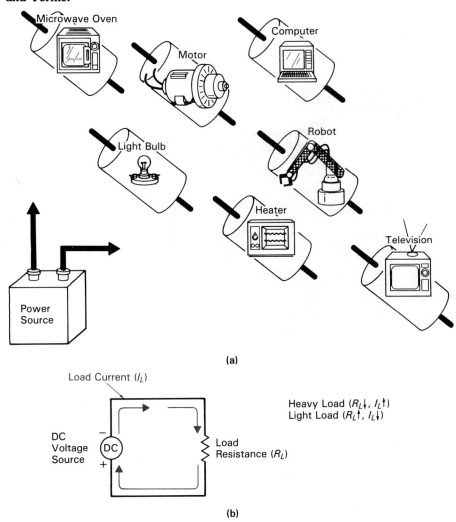

(a)

Load Current (I_L)

DC Voltage Source

Load Resistance (R_L)

Heavy Load ($R_L\downarrow$, $I_L\uparrow$)
Light Load ($R_L\uparrow$, $I_L\downarrow$)

(b)

devices or pieces of equipment as having some value of resistance, which is referred to as the device's *load resistance*.

EXAMPLE 5.1

A component or piece of equipment offers a 390-Ω load resistance in a 9-V circuit. What current will flow in this circuit?

SOLUTION

$$\text{current } (I) = \frac{\text{voltage (V)}}{\text{resistance (R)}}$$
$$= \frac{9 \text{ V}}{390 \text{ }\Omega}$$
$$= 23 \text{ mA}$$

In summary, the phrase load resistance describes the device or equipment's circuit resistance, whereas the phrase load current describes the amount of current drawn by the device or equipment. A device that causes a large load current to flow, due to its small load resistance, is called a large or heavy load because it is heavily loading down or working the supply or battery. On the other hand, a device that causes a small load current to flow, due to its large load resistance, is referred to as a small or light load because it is only lightly loading down or working the supply or battery.

Load current and load resistance are therefore inversely proportional to one another in that a large load resistance ($R_L\uparrow$) causes a small load current ($I_L\downarrow$), referred to as a light load, and vice versa.

SELF-TEST REVIEW QUESTIONS (§ 5.1)

1. Define load resistance, load current, dc voltage source and dc current.
2. A large load current will flow when a large load resistance is connected across the supply (true/false).
3. What is the relationship between load resistance and load current?
4. What is meant by a heavy load?

5.2
DIRECT-CURRENT SOURCES

The six basic forms of energy are mechanical, heat, light, magnetic, chemical, and electrical. To produce direct current (electrical energy), one of the other forms of energy must be used as a source and then converted or

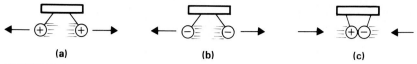

FIGURE 5-3   **Attraction and Repulsion. (a) Positive Repels Positive.**
(b) Negative Repels Negative. (c) Unlike Charges Attract.

transformed into electrical energy in the form of a dc voltage that can be used to produce direct current.

5.2.1   MECHANICALLY GENERATED DC: MECHANICAL $\longrightarrow$ ELECTRICAL

When initially discussing current, we said that a positive charge was a material that has a deficiency or lack of electrons, while a negative charge has an excess of electrons. These positive and negative charges are known as static (stationary) charges or static electricity. Attraction and repulsion occur with positive and negative charges, as reviewed in Figure 5-3.

(1) FRICTION

Friction is a form of mechanical energy, and Figure 5-4 illustrates how the electrons in a glass rod (insulator) can be forced out of their orbits and captured by another material, such as wool, by the work of rubbing. A static negative charge now exists in the wool, while a static positive charge exists in the glass. These mechanically generated static charges can be discharged in one of three ways, producing electrical energy as illustrated in Figure 5-5(a), (b), and (c).

Figure 5-5(a) illustrates the discharge of a static charge by placing a

FIGURE 5-4   **Using Friction to Generate Static Electricity: Mechanical Energy to Electrical Energy.**

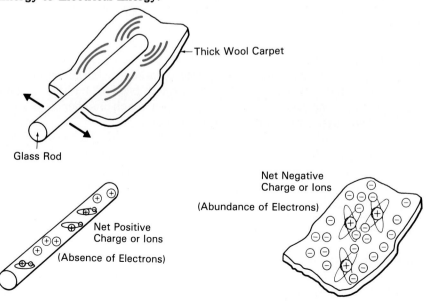

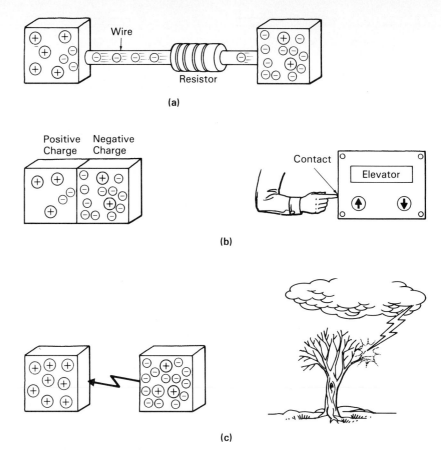

FIGURE 5-5 Static Discharge. (a) Connection Static Discharge. (b) Contact Static Discharge. (c) Arc Static Discharge.

conductor between the two charges and allowing electrons to flow from one material to the other. Once a state of equilibrium is achieved, the two static charges will have neutralized, and no further current will flow. If the connection has a small resistance, the discharge current will be brief and intense; on the other hand, if the connection has a large resistance, the discharge current will be prolonged and less intense.

Figure 5-5(b) illustrates the discharge of a static charge by contact. You may have noticed that your body can, through friction, build up a static charge that is discharged by contact once you touch another oppositely charged or neutralized material, giving you a shock.

Figure 5-5(c) illustrates how large static charges will discharge through the air, causing an arc between the two static charges. Moving clouds in a storm cause friction between the molecules in the air. The buildup of a natural static charge in a cloud, if large enough, can cause a discharge or release of the electrical energy to the ground.

(2) PRESSURE

Mechanical energy in the form of pressure can also be used to generate electrical energy. Quartz and Rochelle-salts are solid compounds that will produce electricity when a pressure is applied, as seen in Figure 5-6. Quartz is a natural or artificially grown crystal composed of silicon dioxide, from which thin slabs or plates are carefully cut and ground. Rochelle salt is a crystal of sodium potassium tartrate. They are both piezoelectric crystals. *Piezoelectric effect* is the generation of a voltage between the opposite faces of a crystal as a result of pressure being applied. Piezoelectric effect also describes the reverse of what has just been discussed, as it was discovered that crystals would not only generate a voltage when mechanical force was applied, but would generate mechanical force when a voltage was applied.

Figure 5-7(a) illustrates the physical appearance of a piezoresistive diaphragm pressure sensor. Piezoresistance of a semiconductor is described as a change in resistance due to a change in the applied strain, and therefore semiconductor resistors can be used as pressure sensors. Four piezoresistors [Figure 5-7(b)] are buried in the surface of a thin, circular (silicon) diaphragm that has two pads connected to it making the connection. When pressure is applied, a stress or strain is applied to the diaphragm and therefore to the buried piezoresistors. The resistance between the pads will change depending on the amount of pressure applied to the diaphragm, and consequently a change in pressure (mechanical input) will cause a corresponding change in resistance (electrical output). Unlike quartz, which generates a voltage, the piezoresistive sensor needs an excitation voltage applied, and its value will determine the output voltage swing. Figure 5-7(c) illustrates the operation of this pressure transducer, which could be used for example to measure either the cooling system, hydraulic transmission, or fuel injection pressure in an automobile, which all vary in the 0 to 100 psi (pounds per square inch)

FIGURE 5-6 **Direct Current from Pressure Using Crystals. (a) Quartz Crystal. (b) Physical Appearance. (c) Construction. (d) Operation.**

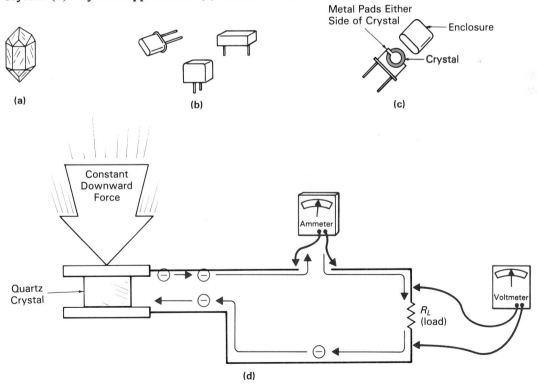

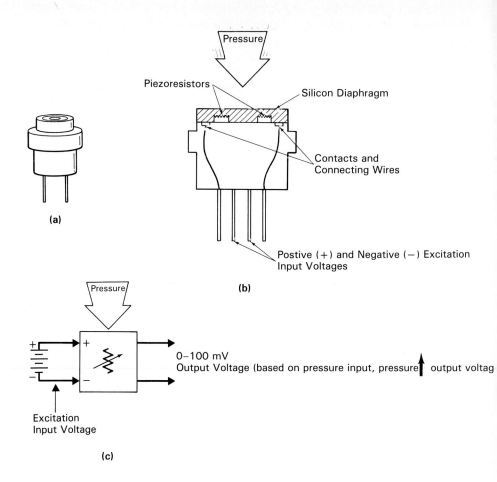

(a)

Pressure

Piezoresistors

Silicon Diaphragm

Contacts and Connecting Wires

Postive (+) and Negative (−) Excitation Input Voltages

(b)

Pressure

+
−

Excitation Input Voltage

0–100 mV Output Voltage (based on pressure input, pressure output voltag

(c)

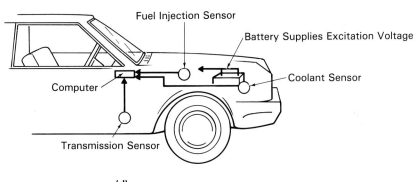

Fuel Injection Sensor

Battery Supplies Excitation Voltage

Coolant Sensor

Computer

Transmission Sensor

(d)

FIGURE 5-7   Direct Current from Pressure Using Piezoresistive Sensors. (a) Physical Appearance. (b) Construction. (c) Operation. (d) Application.

range. The piezoresistive transducer would receive its excitation voltage from the car's battery (12 V) and produce a 0 to 100-mV output voltage, which would be sent to the car's computer, where the data would be analyzed and, if necessary, acted upon, as seen in Figure 5-7(d).

SELF-TEST REVIEW QUESTIONS (§ 5.2.1)

1. How is friction used to generate dc?
2. How is pressure used to generate dc?

5.2.2 THERMALLY GENERATED DC: HEAT ⟶ ELECTRICAL

Heat can be used to generate a charge, as illustrated in Figure 5-8(a). When two dissimilar metals, such as copper and iron, are welded together and heated, an electrical charge is produced. This junction is called a *thermo-couple*, and the heat causes the release of electrons from their parent atoms, resulting in a meter deflection indicating the generation of a charge. The size of the charge is proportional to the temperature difference between the two metals. Figure 5-8(b) lists some typical thermocouples and their voltages generated when heat is applied, and also shows the physical appearance of two types.

Thermocouples, like the one in Figure 5-8(c), are normally used to indicate temperature on a scale calibrated in degrees. The thermocouple seen in Figure 5-8(d) is being heated by the gas pilot light, and the electrical energy generated will allow a valve to open and let gas through to the water heater. If, by accident, the pilot is extinguished by a gust of wind, the thermocouple will not receive any more heat and therefore will not generate electrical energy, which will consequently close the electrically operated valve and prevent a large explosion.

SELF-TEST REVIEW QUESTIONS (§ 5.2.2)

1. What is the name given to the device used to convert heat to dc?
2. Name an application in which the transducer in question 1 could be used.

5.2.3 OPTICALLY GENERATED DC: LIGHT ⟶ ELECTRICAL

The photocell, seen in Figure 5-9(a), also makes use of two dissimilar metals or semiconductor materials in its transformation of light to electrical energy. Figure 5-9(b) illustrates the construction and operation of a photoelectric cell, which is also called a solar cell.

A light-sensitive metal or semiconductor is placed behind a transparent piece of dissimilar metal. When the light illuminates the light-sensitive material, a charge is generated, causing current to flow and the meter to

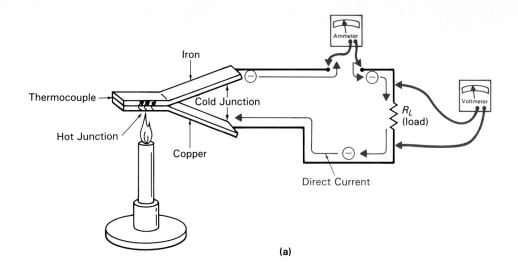

(a)

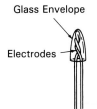

Two Metals	Temperature Range (°F)	Voltage Generated for Temperature Range (mV)
Iron vs. Copper–Nickel	32 to 1382	0 to 42.3
Nickel–Chromium vs. Nickel–Aluminum	−328 to 2282	−5.9 to 50.6
Platinum–6% Rhodium vs. Platinum–30% Rhodium	32 to 3092	0 to 12.4

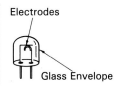

(b)

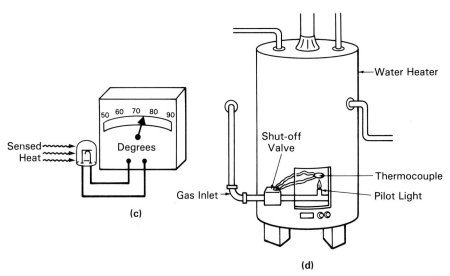

FIGURE 5-8 Direct Current from Heat Using the Thermocouple. (a) Operation. (b) Typical Thermocouples. (c) Indicating Temperature. (d) Water Heater Application.

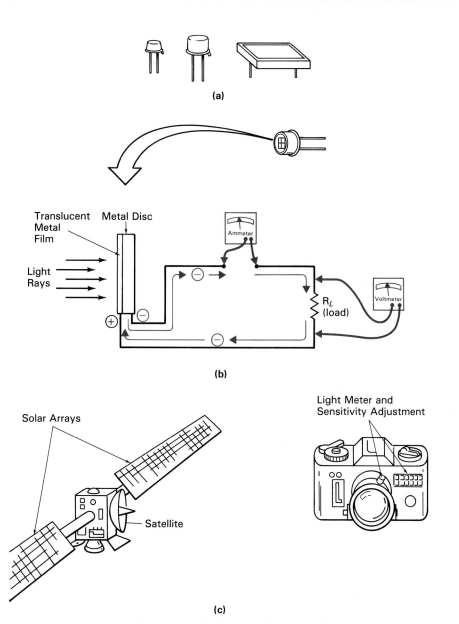

FIGURE 5-9 Direct Current from Light Using the Solar Cell. (a) Physical Appearance. (b) Construction and Operation. (c) Applications.

indicate current flow. This phenomenon is known as *photovoltaic* action, and it finds application as a light meter for photographic purposes and as an electrical power source, for use by the satellite seen in Figure 5-9(c).

Electric power companies make use of solar cells, as can be seen in Figure 5-10(a), which shows a generation engineer testing a variety of solar panels being used to convert sunlight into electrical energy for the consumer. The electrical energy generated by the solar panels on the roof of the homes in Figure 5-10(b) will be used to provide hot water for the occupants.

Out in space the solar cell may reach 75% efficiency, whereas on the earth they are only about 20% efficient as the sun's rays are inhibited by the earth's atmosphere. The power obtained is on average about 100 milliwatts

FIGURE 5-10   **Solar Panels.**

per square centimeter, and their biggest application would seem to be the generation of power in remote areas.

SELF-TEST REVIEW QUESTIONS (§ 5.2.3)

1. What is the name of the component used to convert light energy to dc?
2. Name an application in which the transducer in question 1 could be used.

5.2.4 MAGNETICALLY GENERATED DC: MAGNETIC ⟶ ELECTRICAL

Before discussing how magnetism can be used to generate electrical energy, let us first begin with a discussion of the principles of magnetism.

(1) THE NATURE OF MAGNETISM

In the ancient city of Magnesia in Asia Minor, stones composed of iron oxide were found to have the strange ability to attract and repel one another. It was also noticed that these stones, when suspended from a string, would always point in only one direction. Historians are not sure whether the Chinese or Europeans first began using these "leading stones," or *lodestones* as they are now known, for long-distance land and sea travel, but the original compass was probably a lodestone suspended from a string. Later it was found that a metal needle could be magnetized by stroking it across a lodestone, and so the next form of compass consisted of a magnetized needle attached to a piece of wood in a bowl of water; hence the origin of the term needle of a compass.

The lodestone, seen in Figure 5-11(a), is an example of a natural magnet, unlike the needle, which is called an artificial magnet because it must be artificially magnetized in order to possess a magnetic field but in both cases they have the ability to point in only one direction by interacting with the earth's natural magnetic field, seen in Figure 5-11(b).

Magnetism was recorded by the Greeks as early as 800 B.C., but it was not until 1269 that Petrus Peregrinus de Maricout found that the invisible magnetic lines produced by lodestones intersected at two points, which he called poles. In 1600, William Gilbert discovered that the magnetic field of a lodestone was lost when it was heated, but reappeared when the lodestone cooled. In 1819, Hans Christian Oersted discovered that any current-carrying conductor generated a magnetic field, and this interrelationship between electricity and magnetism became known as *electromagnetism*.

In 1838, a German mathematician and physicist, Johann Gauss, published a paper on the earth's natural magnetic field, and in honor of his contribution the strength of a magnetic field is measured in gauss. A magnetic field strength of 1 gauss (G) is produced 1 centimeter away from a wire

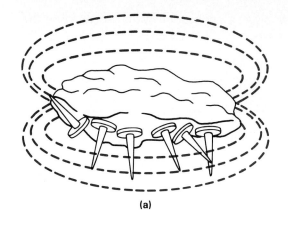

(a)

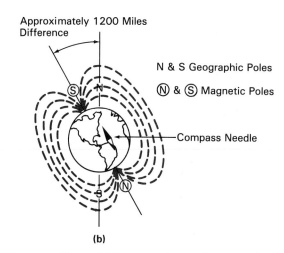

Approximately 1200 Miles Difference

N & S Geographic Poles

Ⓝ & Ⓢ Magnetic Poles

Compass Needle

(b)

FIGURE 5-11   Natural Magnets. (a) Lodestone. (b) Earth.

FIGURE 5-12   (a) Magnetizing a Screwdriver. (b) Magnetized Screwdriver.

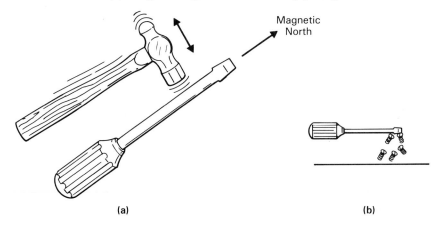

Magnetic North

(a)   (b)

carrying a current of 5 amperes. To give some examples, the field strength of a magnetized needle may only be a fraction of a gauss, whereas a lodestone 1 inch long may have a strength of 100 G. The earth's magnetic field has a magnetic field strength of 0.3 G at the equator and 0.7 G at either pole, and, although this sounds small, a simple experiment can be conducted to prove how influential this invisible magnetic field is. If a screwdriver is positioned so that it points north and the screwdriver is tapped with a hammer several times, the screwdriver will actually become magnetized by the earth's magnetic field, as illustrated in Figure 5-12. The screwdriver is in fact an artificial magnet that was magnetized by a natural magnet (the earth). The atoms within the steel were originally misaligned. However, after being tapped and influenced by the earth's magnetic field, the atoms became aligned to reinforce one another.

(2) MAGNETISM AND THE ATOM

As you are already aware, an atom consists of electrons spinning around a nucleus; some atoms have few, while others have many, electrons orbiting the atom's nucleus. A question arises when we find out that we can magnetize a screwdriver made of iron or nickel but cannot magnetize a screwdriver if it were made of copper or tungsten. Why will certain materials magnetize while others will not? The answer is found in our smallest subatomic particle, the electron. Each electron is in fact a miniature magnet spinning around the nucleus. An electron at rest will not generate a magnetic field; however, orbiting electrons produce a magnetic field because of their high orbital velocity. Referring to your atomic periodic table of the elements (Table 2-1), you will notice that tungsten, for example, has 74 electrons (will not magnetize) and iron has 26 electrons (will magnetize). With tungsten, 37 electrons rotate around the nucleus in the clockwise direction, while 37 electrons orbit in the counterclockwise direction. The clockwise magnetic field will cancel with the counterclockwise magnetic field, and so the net magnetic field is zero, which is why tungsten cannot be magnetized.

Iron has less electrons (26); however, 10 of the iron electrons spin clockwise and 16 spin counterclockwise producing a net counterclockwise magnetic field of 6 electrons, which is why iron can be magnetized. If approximately 100 trillion atoms of the same magnetic polarity are all arranged in one direction, they are called a *domain*. Domains within unmagnetized materials are all misaligned, as seen in Figure 5-13(a), while aligned domains produce magnetized materials, as seen in Figure 5-13(b).

(3) THE MAGNETIC FIELD

The magnetic field produced by a magnet is an invisible force concentrated at two opposite points called the *poles*, as seen in Figure 5-14. The north pole is the pole that would point north toward the earth's south magnetic pole if the magnet were free to rotate, and the other is termed the south pole because it points to the south. As you can see by again referring to Figure 5-14, the magnetic field is made up of invisible *flux lines* that leave the north pole and return at the south.

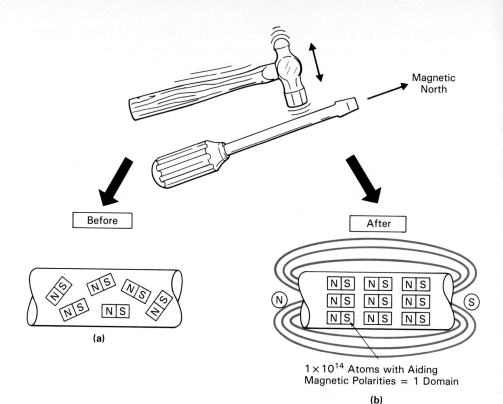

FIGURE 5-13   (a) Misaligned Domains. (b) Aligned Domains.

FIGURE 5-14   Magnetic Field (Made Visible for Demonstration).

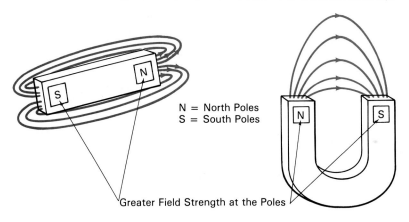

(4) ATTRACTION AND REPULSION

When analyzing magnetic lines of flux within the magnetic field, it is noticed that the law of attraction and repulsion could be applied to magnetic poles, just as it can be to positive and negative charges. This is summarized in Figure 5-15. A north against a north pole and south against a south pole will cause a force of repulsion between the two like poles, while a south

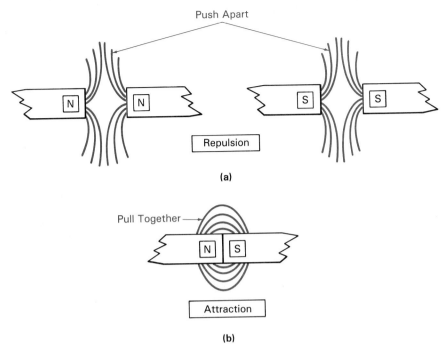

FIGURE 5-15   Attraction and Repulsion. (a) Like Poles Repel Each Other. (b) Unlike Poles Attract Each Other.

pole against a north pole will result in a force of attraction between the two unlike poles.

(5) ARTIFICIAL MAGNETS

There are three types of artificial magnets: (1) permanent magnets, (2) temporary magnets, and (3) electromagnets.

(a) Permanent Magnets A permanent magnet is a piece of hardened steel or other special alloy that, when magnetized by the presence of a magnetic field, will retain its magnetism indefinitely (high retentivity), even when the material is removed from the magnetizing force or the magnetizing force is no longer present, as seen in Figure 5-16.

(b) Temporary Magnets A temporary magnet is a piece of soft iron that will be magnetized while in the presence of a magnetizing field, but will be demagnetized the moment the soft iron is taken away from the magnetizing force or the magnetizing force is no longer present, as seen in Figure 5-17. Soft iron possesses low retentivity and will not retain a magnetic field.

(c) Electromagnets An electromagnet is a temporary magnet that consists of an insulated wire wrapped around a soft iron cylinder (core) to form a coil, as shown in Figure 5-18(a). A magnetic field is produced only while current flows through the solenoid or coil, as shown in Figure 5-18(b). If a

Magnetic Field Present

Magnetic Field Removed

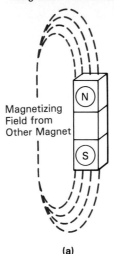

Magnetizing
Field from
Other Magnet

(a)

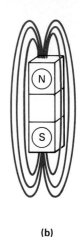

(b)

FIGURE 5-16   Permanent Magnets. (a) A permanent magnet is magnetized when a magnetizing field is present (domains aligned). (b) Permanent magnet retains the magnetic energy even when the magnetizing field is no longer present (domains remain aligned).

FIGURE 5-17   Temporary Magnets. (a) Temporary magnet is magnetized when the magnetizing field is present (domains aligned). (b) Temporary magnet loses magnetic energy when the magnetizing field is removed (domains no longer aligned).

Magnetic Field Present

Magnetic Field Removed

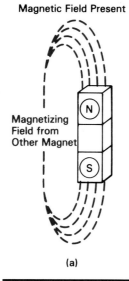

Magnetizing
Field from
Other Magnet

(b)

(a)

FIGURE 5-18   Electromagnets. (a) No Current, No Magnetic Field. (b) Small Current, Magnetic Field. (c) Large Current, Magnetic Field.

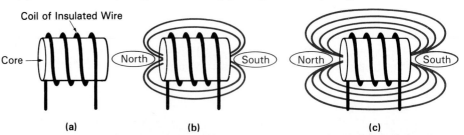

(a)   (b)   (c)

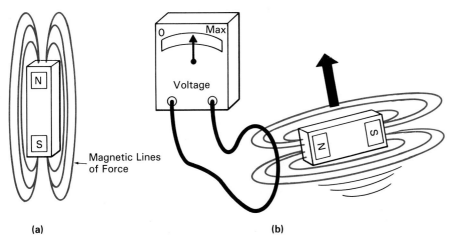

(a)   **(b)**

**FIGURE 5-19 Direct Current from Magnetism. (a) Stationary Magnet.
(b) Moving Magnet.**

larger current is passed through the coil, a corresponding stronger magnetic
field is generated, as seen in Figure 5-18(c).

(6) CONVERTING MAGNETIC ENERGY INTO ELECTRICAL ENERGY

A magnet like the one in Figure 5-19(a) can be used to generate electrical
current by movement of the magnet past a conductor, as seen in Figure 5-
19(b). When the magnet is stationary, no current flows through the circuit,
and so the current meter indicates this zero current. If the magnet is moved
so that the magnetic lines of force cross the wire conductor, current is induced
in the conductor as indicated by the meter. To achieve an energy conversion
from magnetic to electric, you can either move a magnet past a wire or move
the wire past the magnet.

   To produce a continuous supply of electric current, the magnet or wire
must be constantly in motion. The device that achieves this is the dc electric
generator illustrated in Figure 5-20.

FIGURE 5-20 Generator.

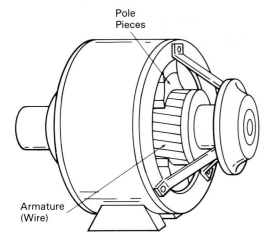

1. What is the name given to the natural magnetic rock?
2. How do the laws of attraction and repulsion apply to magnetism?
3. List the three types of artificial magnets.
4. What device is used to convert magnetic energy into electrical?

5.2.5 CHEMICALLY GENERATED DC: CHEMICAL ⟶ ELECTRICAL

The chemical voltage source (the battery) is called the voltaic cell, and the principle of operation was first discovered by Alessandro Volta, an Italian physicist, in 1800.

A car uses a lead–acid battery, a portable radio has a carbon–zinc or alkaline battery, and, in some instances, you may be using the rechargeable nickel–cadmium batteries. No matter what type of battery is used, they all produce voltage by the same basic principle and all have three basic components, as seen in Figure 5-21: (1) a negative plate, (2) a positive plate, and (3) an electrolyte.

(1) OPERATION

Two dissimilar, separated metal plates such as copper and zinc are placed within a container that is filled with a liquid known as the *electrolyte*, usually an acid. A chemical reaction causes electrons to be repelled from one plate and attracted to the other, passing through the electrolyte. A large amount of electrons collect on one plate (negative plate), while an absence or deficiency of electrons exists on the opposite, positive plate. The electrolyte acts on the two plates and transforms chemical energy into electrical energy,

FIGURE 5-21 **Basic Battery.**

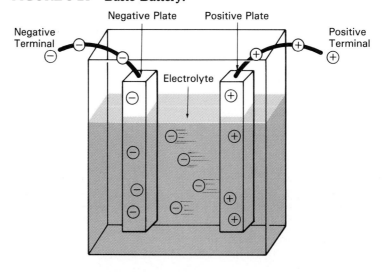

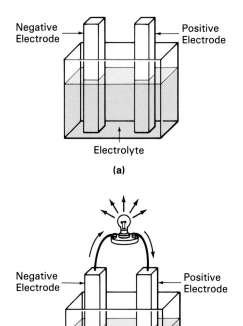

Negative Electrode

Positive Electrode

Electrolyte

(a)

Negative Electrode

Positive Electrode

(b)

FIGURE 5-22 **Battery Operation.**

which can be taken from the cell at its two output terminals as an electrical current flow.

If nothing is connected across the battery, as seen in Figure 5-22(a), a chemical reaction between the electrolyte and the negative electrode produces free electrons that travel from atom to atom and are then held in the negative electrode.

In Figure 5-22(b), a light bulb has been connected between the negative and positive electrodes. The mutual repulsion of the free electrons at the negative electrode combined with the attraction of the positive electrode results in a migration of free electrons (current flow) through the light bulb, causing its filament to produce light.

(2) PRIMARY CELLS

The chemical reaction within the cell will actually dissolve the negative plate until eventually it will be completely eaten away. This discharging process also results in hydrogen gas bubbles forming around the positive plate, causing a resistance between the two plates known as the batteries internal resistance to increase (called *polarization*). To counteract this problem, all dry cells have a chemical within them known as a depolarizer agent, which reduces the buildup of these gas bubbles. With time, however, the depolarizer's effectiveness will be reduced, and the battery's internal resistance will increase as the battery reaches a completely discharged condition. These types of batteries are known as primary (first time, last time) cells because they discharge once and must then be discarded. Almost all primary cells

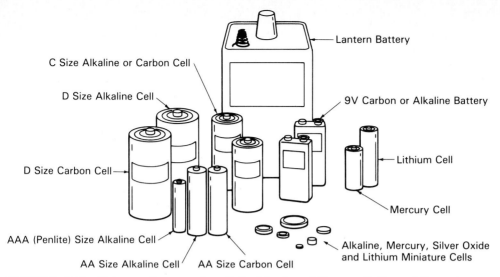

FIGURE 5-23   Primary Rectangular, Cylindrical, and Button Cells and Batteries.

have their electrolyte in paste form and are therefore referred to as *dry cells*, as opposed to some wet cells whose electrolyte is in liquid form.

(a) Primary Cell Types Figure 5-23 illustrates some of the different types of primary cells on the market today. The shelf life of a battery is the length of time a battery can remain on the shelf (in storage) and still retain its usability. Most primary cells will deteriorate in a three-year period to approximately 80% of their original capacity.

Of all the types of primary cells, five dominate the market as far as applications and sales are concerned. These are the carbon–zinc, alkaline–manganese, mercury, silver oxide and lithium, all of which are described and illustrated in Table 5-1.

The cell voltage in Table 5-1 describes the voltage produced by one cell (one set of plates). If two or more sets of plates are installed in one package, the component is called a battery.

(3) SECONDARY CELLS

Secondary cells operate on the same basic principle as the previously discussed primary cells. However, in this case the plates are not eaten away or dissolved; they only undergo a chemical change during discharge. Once discharged, the secondary cell can have the chemical change that occurred during discharge reversed by recharging, resulting in a fully charged cell once more. Restoring a secondary cell to the charged condition is achieved by sending a current through the cell in the direction opposite to that of the discharge current, as illustrated in Figure 5-24.

In Figure 5-24(a), we have first connected a light bulb across the battery, and free electrons are supplied by the negative electrode moving through the light bulb or load and back to the positive electrode.

After a certain amount of time, any secondary cell will run down or will have discharged to such an extent that no usable value of current can

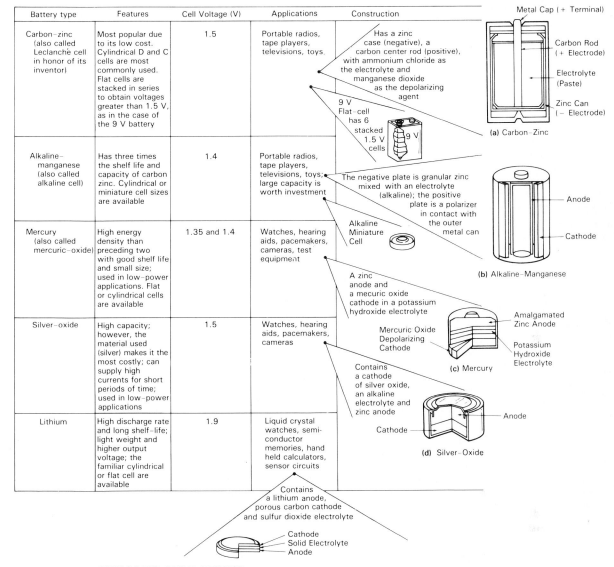

Battery type	Features	Cell Voltage (V)	Applications	Construction
Carbon–zinc (also called Leclanché cell in honor of its inventor)	Most popular due to its low cost. Cylindrical D and C cells are most commonly used. Flat cells are stacked in series to obtain voltages greater than 1.5 V, as in the case of the 9 V battery	1.5	Portable radios, tape players, televisions, toys.	Has a zinc case (negative), a carbon center rod (positive), with ammonium chloride as the electrolyte and manganese dioxide as the depolarizing agent
Alkaline–manganese (also called alkaline cell)	Has three times the shelf life and capacity of carbon zinc. Cylindrical or miniature cell sizes are available	1.4	Portable radios, tape players, televisions, toys; large capacity is worth investment	The negative plate is granular zinc mixed with an electrolyte (alkaline); the positive plate is a polarizer in contact with the outer metal can
Mercury (also called mercuric–oxide)	High energy density than preceding two with good shelf life and small size; used in low–power applications. Flat or cylindrical cells are available	1.35 and 1.4	Watches, hearing aids, pacemakers, cameras, test equipment	A zinc anode and a mercuric oxide cathode in a potassium hydroxide electrolyte
Silver–oxide	High capacity; however, the material used (silver) makes it the most costly; can supply high currents for short periods of time; used in low–power applications	1.5	Watches, hearing aids, pacemakers, cameras	Contains a cathode of silver oxide, an alkaline electrolyte and zinc anode
Lithium	High discharge rate and long shelf–life; light weight and higher output voltage; the familiar cylindrical or flat cell are available	1.9	Liquid crystal watches, semi-conductor memories, hand held calculators, sensor circuits	Contains a lithium anode, porous carbon cathode and sulfur dioxide electrolyte

Labels in the Construction illustrations:

9 V Flat–cell has 6 stacked 1.5 V cells — 9 V

Metal Cap (+ Terminal); Carbon Rod (+ Electrode); Electrolyte (Paste); Zinc Can (– Electrode)

(a) Carbon–Zinc

Anode; Cathode

(b) Alkaline–Manganese

Alkaline Miniature Cell

Amalgamated Zinc Anode; Mercuric Oxide Depolarizing Cathode; Potassium Hydroxide Electrolyte

(c) Mercury

Anode; Cathode

(d) Silver–Oxide

Cathode; Solid Electrolyte; Anode

TABLE 5-1 PRIMARY CELL TYPES

be produced. The surfaces of the plates are changed, and the electrolyte lacks the necessary chemicals.

In Figure 5-24(b), during the recharging process, we use a battery charger, which reverses the chemical process of discharge by forcing electrons back into the cell and restoring the battery to its charged condition. The battery charger voltage is normally set to about 115% of the battery voltage. If this voltage is set too high, then an excessive current can result, causing the battery to overheat.

Secondary cells are often referred to as wet cells because the electrolyte is not normally in paste form (dry) as in the primary cell, but is in liquid (wet) form and is free to move and flow within the container.

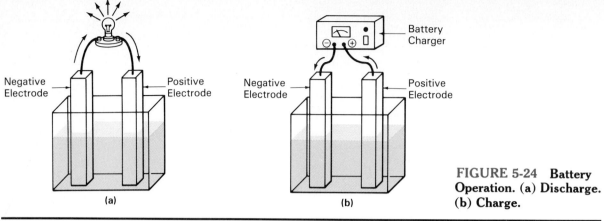

Negative
Electrode

Positive
Electrode

(a)

Battery
Charger

Negative
Electrode

Positive
Electrode

(b)

**FIGURE 5-24 Battery
Operation. (a) Discharge.
(b) Charge.**

(a) Discharging

FIGURE 5-25

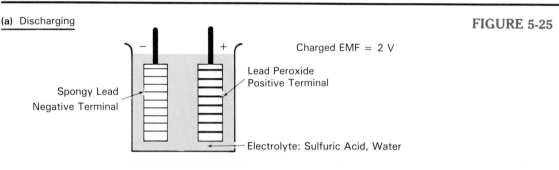

Charged EMF = 2 V

Spongy Lead
Negative Terminal

Lead Peroxide
Positive Terminal

Electrolyte: Sulfuric Acid, Water

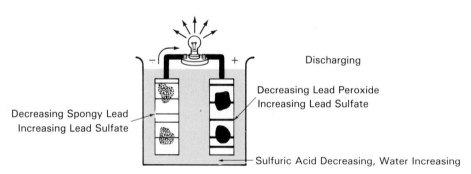

Discharging

Decreasing Spongy Lead
Increasing Lead Sulfate

Decreasing Lead Peroxide
Increasing Lead Sulfate

Sulfuric Acid Decreasing, Water Increasing

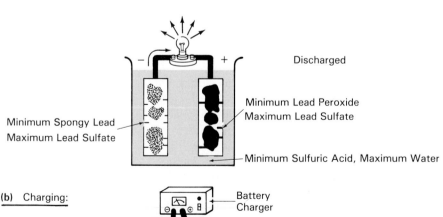

Discharged

Minimum Spongy Lead
Maximum Lead Sulfate

Minimum Lead Peroxide
Maximum Lead Sulfate

Minimum Sulfuric Acid, Maximum Water

(b) Charging:

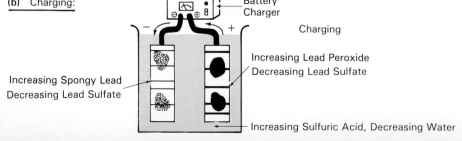

Battery
Charger

Charging

Increasing Spongy Lead
Decreasing Lead Sulfate

Increasing Lead Peroxide
Decreasing Lead Sulfate

Increasing Sulfuric Acid, Decreasing Water

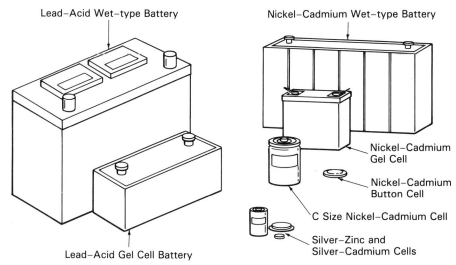

Lead–Acid Wet-type Battery

Nickel–Cadmium Wet-type Battery

Nickel–Cadmium
Gel Cell

Nickel–Cadmium
Button Cell

C Size Nickel–Cadmium Cell

Silver–Zinc and
Silver–Cadmium Cells

Lead–Acid Gel Cell Battery

FIGURE 5-26 Secondary Rectangular, Cylindrical, and Button Cells and Batteries.

Figure 5-25(a) gives more information on the discharge and Figure 5-25(b) on the charge process that occurs to the plates and electrolyte of a lead–acid secondary cell.

(a) Secondary Cell Types Figure 5-26 illustrates the physical appearance of some of the different types of secondary cells on the market today. The lead–acid and nickel–cadmium are the two most popular and are illustrated and described in Table 5-2.

The two other types of secondary cells are the silver–zinc and silver–cadmium. Both have a large silver content and are therefore quite costly; however, they have the highest energy of all secondary cells and are used in specialized applications like video-tape recorders, portable televisions, and missiles.

(4) SECONDARY CELL CAPACITY

Capacity (C) is measured by the amount of ampere-hours (Ah) a battery can supply during discharge. If, for example, a battery has a discharge capacity of 10 ampere-hours, the battery could supply 1 amp for 10 hours, 10 amps for 1 hour, 5 amps for 2 hours, and so on, during discharge. Automobile or batteries (12 V) will typically have an ampere-hour rating of between 100 and 300 Ah.

Ampere-hour units are actually specifying the coulombs of charge in the battery. If, for example, a lead–acid car battery was rated at 150 Ah, this value would have to be converted to ampere-seconds to determine coulombs of charge, as 1 ampere-second (As) is equal to 1 coulomb. Ampere-hours are quite easily converted to ampere-seconds by simply multiplying ampere-hours by 3600, which is the number of seconds in 1 hour. In our example, a 150-Ah battery will have a charge of 54×10^4 coulombs.

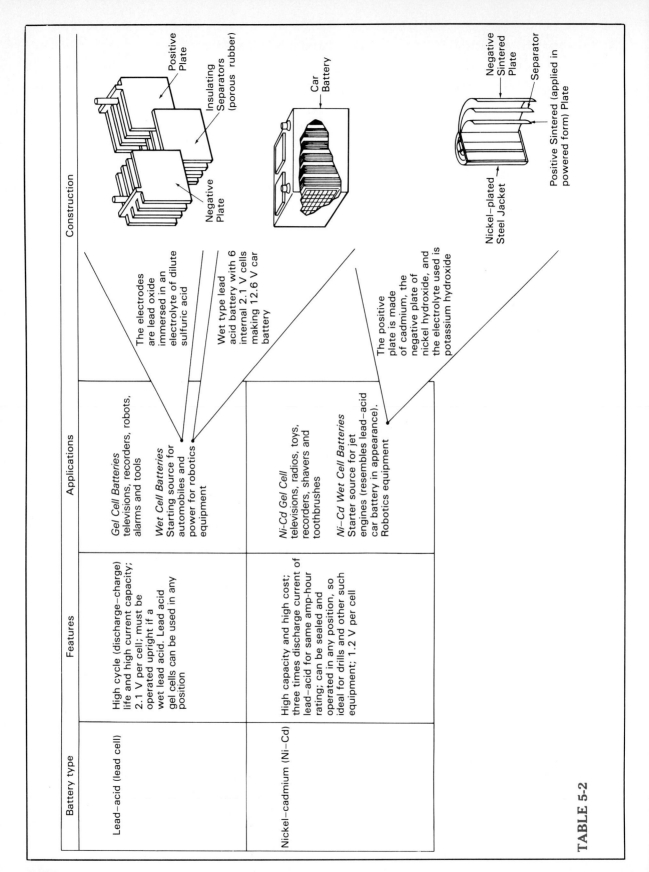

Battery type	Features	Applications	Construction
Lead–acid (lead cell)	High cycle (discharge–charge) life and high current capacity; 2.1 V per cell; must be operated upright if a wet lead acid. Lead acid gel cells can be used in any position	*Gel Cell Batteries* televisions, recorders, robots, alarms and tools *Wet Cell Batteries* Starting source for automobiles and power for robotics equipment	The electrodes are lead oxide immersed in an electrolyte of dilute sulfuric acid Wet type lead acid battery with 6 internal 2.1 V cells making 12.6 V car battery
Nickel–cadmium (Ni–Cd)	High capacity and high cost; three times discharge current of lead–acid for same amp-hour rating; can be sealed and operated in any position, so ideal for drills and other such equipment; 1.2 V per cell	*Ni-Cd Gel Cell* televisions, radios, toys, recorders, shavers and toothbrushes *Ni-Cd Wet Cell Batteries* Starter source for jet engines (resembles lead–acid car battery in appearance). Robotics equipment	The positive plate is made of cadmium, the negative plate of nickel hydroxide, and the electrolyte used is potassium hydroxide

Construction labels: Positive Plate, Insulating Separators (porous rubber), Negative Plate, Car Battery, Negative Sintered Plate, Separator, Nickel-plated Steel Jacket, Positive Sintered (applied in powered form) Plate

TABLE 5-2

EXAMPLE 5.2

How many coulombs of charge does a 3.8-Ah lead–acid gel battery store?

SOLUTION

A 3.8-Ah battery can supply 3.8 amps in 1 hour or 13,680 amps in a second (3.8 × 3600). Since current is measured in coulombs/second, this battery when fully charged can store 13.68×10^3 coulombs.

EXAMPLE 5.3

A Ni–Cd battery is rated at 300 Ah and is fully discharged.
(a) How many coulombs of charge must the battery charger put into the battery to restore it to full charge?
(b) If the charger is supplying a charging current of 3 amps, how long will it take the battery to fully charge?
(c) Once fully charged, the battery is connected across a load that is pulling a current of 30 A. How long will it take until the battery is fully discharged?

SOLUTION

(a) The same amount that was taken out: 300 Ah × 3600 = 108×10^4 C.

(b) $\dfrac{300 \text{ Ah}}{3 \text{ A}} = 100$ hours until charged

(c) $\dfrac{300 \text{ Ah}}{30 \text{ A}} = 10$ hours until discharged

(5) BATTERIES IN SERIES AND PARALLEL

Batteries are often connected together to gain a higher voltage or current than can be obtained from one cell. Let's first analyze what can be obtained by connecting two cells in series, as seen in Figure 5-27.

When batteries are connected in series, the negative terminal of *A* is connected to the positive terminal of *B*. The total voltage across the bulb will be the sum of the two cell voltages, 18 V.

When batteries are connected in parallel, the negative terminal connects to the other negative terminal and the positive terminal connects to the positive. The total current produced will be the sum of the two cell currents, while the voltage remains the same as for one cell, as seen in Figure 5-28. Each cell provides half the total current flow through the load.

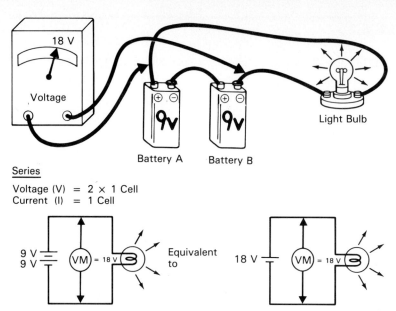

Series

Voltage (V) = 2 × 1 Cell
Current (I) = 1 Cell

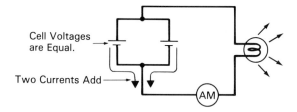

FIGURE 5-27 **Batteries in Series.**

FIGURE 5-28 **Batteries in Parallel.**

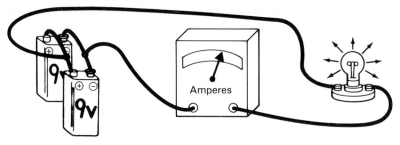

Parallel

Voltage (V) = 1 Cell
Current (I) = 2 × 1 Cell

(6) INTERNAL RESISTANCE

Figure 5-29 illustrates a typical lead–acid secondary cell on the left and its symbol on the right. When a circuit or piece of equipment is connected across the battery, as seen in Figure 5-29(a), the circuit or piece of equipment can be represented by its equivalent value of resistance, called the load resistance (R_L). In Figure 5-29(a), the switch is open and so the battery is

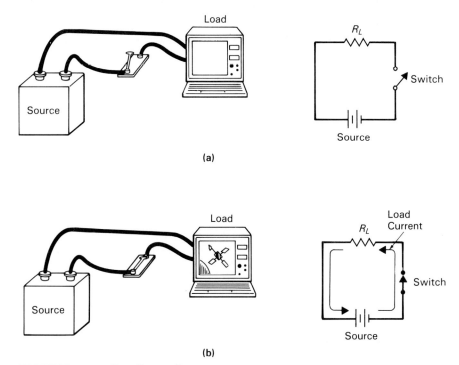

(a)

(b)

FIGURE 5-29   Loading a Battery.

not loaded and no load current flows; in Figure 5-29(b), the battery has a completed current path, as the switch is now closed, and so a load current will flow the value of which depends on the load resistance and battery voltage.

The battery just discussed is known as an ideal voltage source; however, in reality, there is no such thing as an ideal voltage source. Batteries or any other type of voltage source are not 100% efficient; they all possess some form of internal resistance, as seen in Figure 5-30(a), represented by a resistor connected in series with the battery symbol. As you can see in the schematic circuit illustrated in Figure 5-30(b), R_L and R_S form a voltage divider.

Very little voltage will be dropped across R_S as it is normally small with respect to R_L. Internal inefficiencies of batteries and all voltage sources must always be kept small because a large R_S would drop a greater amount of the voltage, resulting in less output voltage to the load, and therefore a waste of electrical power. For example, a lead–acid cell will typically have an internal resistance of 0.01 Ω (10 mΩ).

FIGURE 5-30   Source and Load Resistance Divide the Voltage.

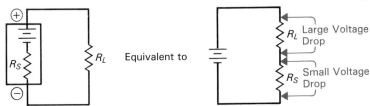

(7) MAXIMUM POWER TRANSFER

Maximum power will be delivered to the load when the resistance of the load (R_L) is equal to the resistance of the source (R_s). The best way to see if this theorem is correct is to apply it to a series of examples and then make a comparison.

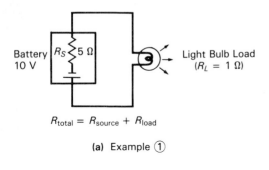

$R_{total} = R_{source} + R_{load}$

(a) Example ①

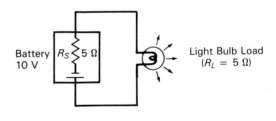

(b) Example ②

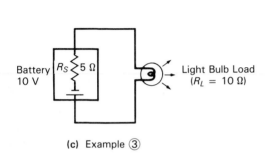

(c) Example ③

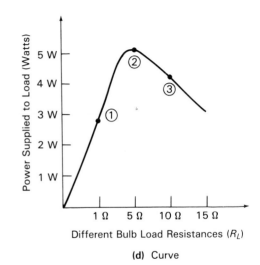

Different Bulb Load Resistances (R_L)

(d) Curve

FIGURE 5-31   Maximum Power Transfer.

EXAMPLE 5.4

Figure 5-31(a) illustrates a 10-V battery with a 5-Ω internal resistance connected across a load. The load in this case is a light bulb, which has a load resistance of 1 Ω.

$$\text{circuit current, } I = \frac{V}{R}$$

$$= \frac{10 \text{ V}}{R_S + R_L}$$

$$= \frac{10 \text{ V}}{6 \text{ Ω}}$$

$$= 1.66 \text{ amperes}$$

The power supplied to the load is $P = I^2 \times R = 1.66^2 \times 1 \text{ Ω} = 2.8$ watts.

EXAMPLE 5.5

Figure 5-31(b) illustrates the same battery and R_S connected across a 5-Ω light bulb.

$$
\begin{aligned}
I &= \frac{V}{R} \\
&= \frac{10 \text{ V}}{R_S + R_L} \\
&= \frac{10 \text{ V}}{10 \text{ }\Omega} \\
&= 1 \text{ ampere}
\end{aligned}
$$

Power supplied is $P = I^2 \times R = 1^2 \times 5 \text{ }\Omega = 5$ watts.

EXAMPLE 5.6

Figure 5-31(c) illustrates the same battery once again and in this case a 10-Ω light bulb is connected in the circuit.

$$
\begin{aligned}
I &= \frac{V}{R} \\
&= \frac{10 \text{ V}}{R_S + R_L} \\
&= \frac{10 \text{ V}}{15 \text{ }\Omega} \\
&= 0.67 \text{ amperes}
\end{aligned}
$$

Thus $P = I^2 \times R = 0.67^2 \times 10 \text{ }\Omega = 4.4$ watts.

As can be seen by the graph in Figure 5-31(d), which plots the power supplied to the load against load resistance, maximum power is delivered to the load (5 watts) when the load resistance is equal to the source resistance.

The maximum power transfer condition is only used in special cases such as the automobile starter, where the load resistance remains constant. In most other cases, where load resistance can vary over a range of values, circuits are designed with a load resistance that will cause the best amount of power to be delivered. This is known as optimum power transfer.

SELF-TEST REVIEW QUESTIONS (§ 5.2.5)

1. List the three basic components in a battery.
2. List all the different primary cell types.
3. List the different types of secondary cells available.

Direct-Current Sources

4. Two 1.5-V batteries in series would produce a total voltage of 1.5 V (true or false).
5. How does a battery's internal resistance relate to the maximum power transfer theorem?
6. Why do batteries have an internal resistance?

5.2.6 ELECTRICALLY GENERATED DC

Up till now we have only discussed electrical energy in the form of direct current (dc); however, as we will discover in Section III of this book, electrical energy is more readily available in the form of alternating current (ac); in fact, that is what arrives at the wall receptacle at your home or work place. We can use ac electrical energy to generate dc electrical energy by use of a piece of equipment called a dc power supply, which is illustrated in Figure 5-32.

Power in the laboratory or repair shop can be obtained from a battery, but since ac power is so accessible, a dc power supply is used. The power supply has a large advantage over the battery as a source of voltage in that it can quickly provide an accurate voltage that can easily be varied by a control on the front panel, and it also never runs down.

During your electronic studies, you will use a power supply in the lab to provide different voltages for various experiments. In Figure 5-32, you can see that the power supply is being used to supply 12 volts to a 12-kΩ resistor.

The on/off switch is used to turn on the power supply, while the volts/amperes switch is used to switch the meter between a voltmeter or ammeter. In the volts position, the meter indicates whatever voltage has been set by the voltage adjustment control, in this example, 12 volts. If the volt/amperes switch is placed in the amperes position, the meter will now monitor the amount of current flowing out of the power supply, in this example 1 milliampere.

FIGURE 5-32 **Direct Current from Alternating Current.**

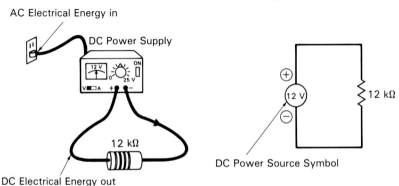

1. An electric company supplies what form of electrical energy?
2. A dc power supply converts dc electrical energy into ac (true or false).
3. A volts/amperes switch on a typical power supply will determine whether the output is voltage or current (true or false).
4. What advantage does the dc power supply have over a battery?

5.3
EQUIPMENT PROTECTION

Current must be monitored and not be allowed to exceed a safe level so as to protect users from shock, to protect the equipment from damage, and to prevent fire hazards. There are three basic types of protective devices: (1) fuses, (2) circuit breakers, and (3) switches. Let's now discuss each of these in more detail.

5.3.1 FUSES

A fuse is a type of metal resistor that consists of a wire link or element of low-melting point inside a casing, as illustrated in Figure 5-33. Fuses have a current rating that is the maximum amount of current they will allow to pass through the fuse to the equipment. Figure 5-34, for example, illustrates a battery supplying power to a radio with a 2-A fuse connected between the two. When current passes through the fuse, some of the electrical energy is transformed into heat. If the current through the thin metal element exceeds the current rating of this fuse, in our example 2 amperes, this excessive current will create enough heat to melt the element and "open" or "blow" the fuse, thus disconnecting the radio and protecting it from damage.

One important point to remember is that if the current increases to a damaging level, then the fuse will open and protect the radio. This implies that something went wrong with the battery and it started to supply too much

FIGURE 5-33 Fuse. (a) Physical Appearance. (b) Schematic Symbol.

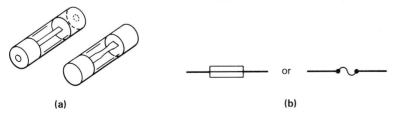

(a)   (b)

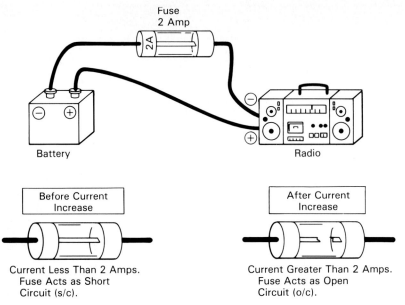

FIGURE 5-34 **Application of a Fuse.**

current. This is almost never the case. What actually happens is that, as with all equipment, eventually something internally breaks down. This can cause the overall load resistance of the piece of equipment to increase or decrease. An increase in load resistance (R_L↑) means that the battery sees a higher resistance in the current path and, therefore, a small current would flow from the battery (I↓), and the user would be aware of the problem due to the nonoperation of the equipment. If an internal equipment breakdown

FIGURE 5-35 **Fuse Types, Sizings, and Casings.**

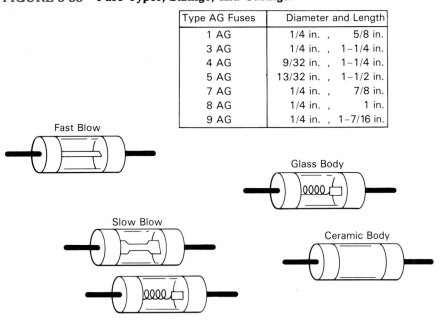

Type AG Fuses	Diameter and Length
1 AG	1/4 in. , 5/8 in.
3 AG	1/4 in. , 1–1/4 in.
4 AG	9/32 in. , 1–1/4 in.
5 AG	13/32 in. , 1–1/2 in.
7 AG	1/4 in. , 7/8 in.
8 AG	1/4 in. , 1 in.
9 AG	1/4 in. , 1–7/16 in.

causes the equipment's load resistance to decrease ($R_L\downarrow$), then the battery will see a smaller circuit resistance and will supply a heavier circuit current, ($I\uparrow$), which could severely damage the equipment before the user had time to turn it off. Protection is needed to automatically disconnect current in this situation to protect our equipment from damage.

Fuse elements come in various shapes and sizes so as to either produce quick heating and then melting (fast blow) or delayed heating and then melting (slow blow) as illustrated in Figure 5-35. The reason for the variety is in the application differences. Some pieces of equipment, when turned on, will be of so low a resistance that a short momentary current surge, sometimes in the region of four times the fuse's current rating, will result in the first couple of seconds. If a fast blow fuse were placed in the circuit, it would blow at the instant the equipment was turned on, even though the equipment did not need to be protected; it merely needs a large amount of current initially to start up. A slow blow fuse would be ideal in this application, as it would permit the initial heavy current and would begin to heat up and yet not blow due to the delay; then the surge would end, the current would reduce, and the fuse would still be intact (some slow blow fuses will allow a 400% overload current to flow for a few seconds).

Some equipment, when on cannot take any increase in current; otherwise, damage will occur. The slow blow would not be good in this application as it would allow an increase of current to pass to the equipment for too long a period of time, whereas the fast blow would disconnect the current instantly if the current rating of the fuse were exceeded, consequently preventing equipment damage. The automobile was the first application for a fuse in a glass holder, and a size standard named AG (automobile glass) was established (see Figure 5-35).

Fuses also have a voltage rating that indicates the maximum circuit voltage that can be applied across the fuse by the circuit in which the fuse resides. This rating, which is important after the fuse has blown, prevents arcing across the blown fuse contacts, as illustrated in Figure 5-36. Once the fuse has blown, the circuit's positive and negative voltages are now connected across the fuse contacts; and if the voltage is too great, then an arc can jump across the gap, causing a sudden surge of current, damaging the equipment connected.

FIGURE 5-36 **Voltage Rating of Fuses.**

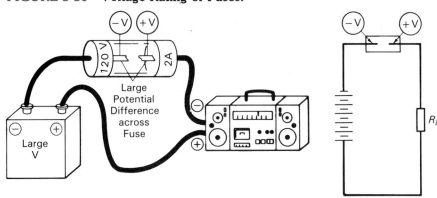

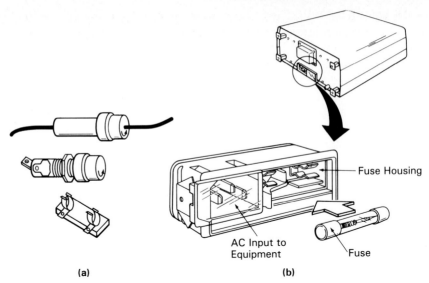

Fuse Housing

AC Input to
Equipment

Fuse

(a)   (b)

**FIGURE 5-37   Fuse Holders. (a) Holders. (b) Fuse and Holder in
Equipment.**

      Fuses are mounted within fuse holders and placed normally at the back
of the equipment for easy access. Figure 5-37 illustrates fuse holders and
how they are installed in a piece of test equipment. When replacing fuses,
you must be sure that the power is off because a potential is across it and
that the fuse with the correct current and voltage rating is used. A good
fuse, since it is merely a piece of wire, should have a resistance of 0 ohms;
that is, no voltage drop can be measured across it. A blown or burned out
fuse will when removed read infinite ohms and when in its holder and power
on have the full applied voltage across its two terminals.

5.3.2 CIRCUIT BREAKERS

      In many appliances and also in your home electrical system, circuit breakers
are used in place of fuses to prevent damaging current. A circuit breaker
can open a current-carrying circuit without damaging itself, then be manually
reset, and used repeatedly, unlike a fuse, which must be replaced when it

**FIGURE 5-38   Circuit Breaker. (a) Symbol. (b) Appearance. (c) Typical
Application.**

Reset Button

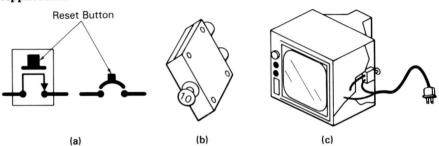

(a)   (b)   (c)

blows. To summarize, we can say that a circuit breaker is a reusable fuse. Figure 5-38 illustrates a circuit breaker's symbol, appearance, and typical application.

There are three types of circuit breakers: (1) thermal type, (2) magnetic type, and (3) thermomagnetic type. Let's now discuss each of these in more detail.

(1) THERMAL-TYPE CIRCUIT BREAKERS

The operation of this type of circuit breaker depends on temperature expansion due to electrical heating. Figure 5-39 illustrates the construction of a thermal circuit breaker. A U-shaped bimetal (two metal) strip is attached to the housing of the circuit breaker; it is composed of a layer of brass on one side and a layer of steel on the other. The current arrives at terminal A, enters the right side of the bimetallic U-shaped strip, leaves the left side, and travels to the upper contact, which is engaging the bottom contact, where it leaves and exits the circuit breaker from terminal B.

If current were to begin to increase beyond the circuit breaker's current rating, heating of the bimetallic strip would occur. As with all metals, heat causes the strip to expand. Some metals expand more than others, and this measurement is called the thermal expansion coefficient. In our situation, brass will expand more than steel, resulting in the lower end of the shaped bimetal strip bending to the right. This allows a catch attached to a pivoted arm to be pulled down by the tension of the spring, which lifts the left side of the arm and an attached top contact. The result is to separate the top contact surface from the bottom contact surface, thereby opening the current path and protecting the circuit from the excessive current. This is also known as tripping the breaker. The reset button must now be engaged to close the contacts; however, if the problem still exists, the breaker will just trip once more, due to the excessive current.

FIGURE 5–39 **Thermal-type Circuit Breaker.**

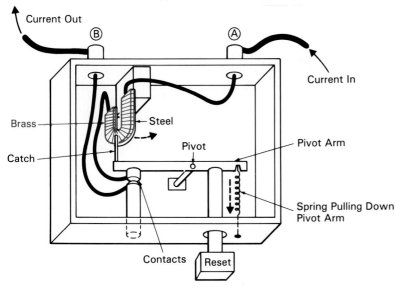

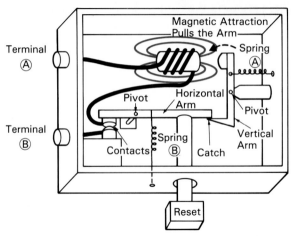

FIGURE 5-40 **Magnetic-type Circuit Breaker.**

(2) MAGNETIC-TYPE CIRCUIT BREAKERS

This type of circuit breaker depends on the response of an electromagnet to break the circuit for protection. Figure 5-40 illustrates the construction of a magnetic-type circuit breaker.

A small level of current flows through the coil of the electromagnet, providing a small amount of magnetic pull on the iron arm, pulling it to the left; however, this magnetic force cannot overcome the pull to the right being generated by the spring A. This safe value of current will therefore pass from the A terminal, through the coil to the top contact, out of the bottom contact, and then exit the breaker at terminal B.

If the current exceeds the current rating of the circuit breaker, an increase in current causes an increase in the current through the coil of the electromagnet, which generates a greater magnetic force on the vertical arm, which pulls the top half of the vertical arm to the left and the lower half below the pivot to the right. This releases the catch holding the horizontal arm, allowing spring B to pull the right side of the lower arm down and consequently open the contact and disconnect or trip the breaker. The reset button must now be engaged to close the contacts; however, if the problem still exists, the breaker will trip once again, as the excessive circuit current still exists.

(3) THERMOMAGNETIC-TYPE CIRCUIT BREAKERS

Figure 5-41 illustrates the typical home- or residential-type thermomagnetic circuit breaker.

The thermal circuit breaker is similar to a slow blow fuse in that it is ideal for passing momentary surges without tripping because of the delay caused by the heating of the bimetallic strip. The magnetic circuit breaker, however, is most similar to a fast flow fuse in that it is immediately tripped when an increase of current occurs.

The thermomagnetic circuit breaker combines the advantages of both previously mentioned circuit breakers by incorporating both a bimetallic strip

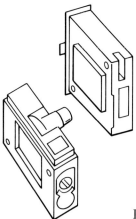

FIGURE 5-41 **Thermomagnetic-type Circuit Breaker.**

and an electromagnet as an actuating mechanism, and it operates as follows: For currents at and below the current rating, the circuit breaker is a short circuit and connects current source to load. For momentary overcurrent surges, the electromagnet is activated, but does not have enough force to trip the breaker. However, if overcurrent continues, then the bimetallic strip will have heated and the combined forces of the bimetallic strip and electromagnet will trip the breaker and disconnect the current source from the load.

For a very large surge of current, the electromagnet receives enough current to trip the breaker independently of the bimetallic strip and disconnect the current source from the load.

5.3.3 SWITCHES

A switch is a device that completes (short circuits) or breaks (open circuits) the path of current, as seen in Figure 5-42. All mechanical switches can be classified into one of eight categories, which are illustrated in Figure 5-43. Many different variations of these eight different classifications exist. Figure 5-44 illustrates some of the different types on the market; but remember that they are all just basically a switch that merely opens or closes connections.

FIGURE 5-42 **Open and Closed Switch.**

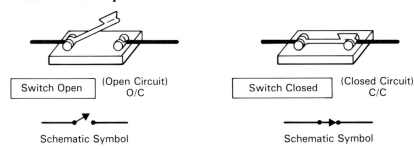

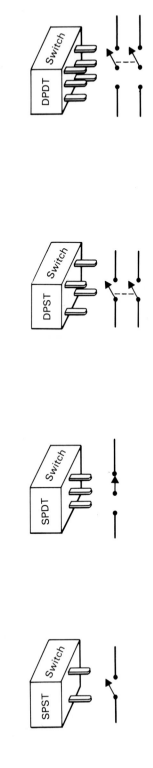
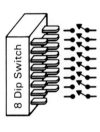

(a) Single–pole, single–throw (SPST). A two–terminal switch with only one pole or moving contact, which can be cast in one direction only (single throw).

(b) Single–pole, double–throw (SPDT). A three–terminal switch for connecting one terminal to either of two other terminals.

(c) Double–pole, single–throw (DPST). A four–terminal switch that is used to connect or disconnect two pairs of terminals simultaneously.

(d) Double–pole, double–throw (DPDT). A switch that has six terminals and is used to connect one pair of terminals to either of the other two pairs.

(e) Normally open push button (NOPB). A switch that will make contact and pass current when it is pressed.

(f) Normally closed push button (NCPB). A switch that normally makes contact and passes current, but disconnects or opens when pressed.

(g) Rotary. An electromechanical device that is capable of making (closing contacts) or breaking (opening contacts) in a circuit.

(h) Dual-in-line package (DIP) switch. A dip switch is a group of separate miniature switches within two (dual) rows of external connecting pins or terminals.

Single pole: one moving contact.
Double pole: two moving contacts.
Single throw: pole can be thrown or cast in only one direction.
Double throw: pole can be thrown or cast in two directions.

FIGURE 5-43 Eight Basic Types of Switches.

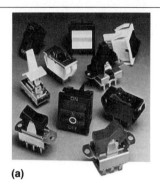

(a)

(g)

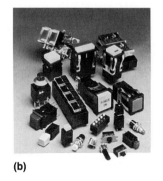

(b)

(f)

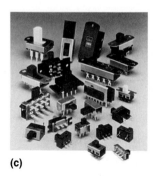

(c)

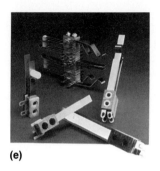

(e)

(d)

FIGURE 5-44 **Switches.** (*Courtesy of ITW Switches*)

1. What do the current and voltage ratings of a fuse indicate?
2. What is the difference between a slow blow and fast blow fuse?
3. List the three types of circuit breakers.
4. List the eight basic types of switches.

SUMMARY

1. Direct current (dc) is a constant flow of current in only one DIRECTion.
2. Direct current can be generated:
   a. Mechanically, by rubbing two materials together (friction) or by applying force to a crystal (pressure).
   b. Thermally, with a thermocouple, which consists of a junction of two dissimilar metals that when heated will generate dc.
   c. Optically, with a solar or photocell, which utilizes a light-sensitive material to generate dc.
   d. Magnetically, by moving a magnet past a conductor, which is the principle behind a dc generator.
   e. Chemically, by a battery that produces dc electrical energy from a chemical reaction that occurs between its positive plate, negative plate, and electrolyte.
   f. Electrically, with a dc power supply that generates a direct current (dc) out from an alternating current (ac) in.
3. The earth is a natural magnet that possesses a magnetic field between its magnetic poles.
4. Every electron in motion is in fact a miniature magnet, with some electrons rotating around the nucleus in a clockwise direction and some in a counterclockwise direction. Some metals:
   a. Will not magnetize because their cumulative electron counterclockwise magnetic field cancels with the cumulative electron clockwise magnetic field.
   b. Will magnetize because a net clockwise or counterclockwise magnetic flux results.
5. If 1×10^{14} atoms of the same magnetic polarity are arranged in one direction, they are called a domain.
6. A magnetic field is an invisible force that exists between and is concentrated at the poles.
7. A permanent magnet retains its magnetism indefinitely, even after the magnetizing field has been removed, and so it is said to have a high retentivity.
8. A temporary magnet will lose its magnetic force after the magnetizing field is removed, and it is said to have a low retentivity.
9. An electromagnet consists of a coil of conductor wrapped around a core that generates a magnetic field when current is passed through it.

174   Direct Current (DC)

10. A primary cell can only be used once, whereas a secondary cell can be recharged and used over and over again.
11. Batteries connected in series combine to produce increased voltage, whereas batteries connected in parallel combine to develop increased current.
12. Maximum power is transferred from source to load when the source resistance is equal to the load resistance.
13. A fuse consists of a wire link inside a casing that will melt and disconnect power from the connected equipment if the current exceeds a safe level.
14. A circuit breaker is a reusable fuse.
15. A switch is a device that completes or breaks the path of current and is generally used to turn circuits on and off.

NEW TERMS

Alkaline–manganese cell
Alternating current (ac)
Ampere-hour
Ampere-second
Artificial magnet
Battery
Battery charger
Blown fuse
Capacity
Carbon–zinc cell
Cell
Coil
Core
DC generator
Depolarizer
Direct current (dc)
Discharge
Domain
Double-pole, double-throw (DPDT) switch
Double-pole, single-throw (DPST) switch
Dry cell

Dual in-line package (DIP) switch
Electrolyte
Electromagnet
Electromagnetism
Fast blow fuse
Fuse
Flux line
Friction
Gauss
Generator
Lead-acid cell
Lithium cell
Load
Load current
Load resistance
Lodestone
Magnetic-type circuit breaker
Maximum power transfer
Mercuric oxide cell
Natural magnet
Nickel–cadmium cell
Normally closed push button (NCPB) switch

Normally open push button (NOPB) switch
Permanent magnet
Photocell
Photovoltaic
Piezoelectric effect
Piezoresistance
Pole
Power supply
Pressure
Primary cell
Quartz
Rating
Retentivity
Rochelle salts
Rotary switch
Secondary cell
Shelf life
Silver–cadmium cell
Silver–oxide cell
Silver–zinc cell
Single-pole, double-throw (SPDT) switch

Single-pole, single-throw (SPST) switch	Source current	Thermal-type circuit breaker
Slow blow fuse	Source resistance	Thermocouple
Solar cell	Source voltage	Thermomagnetic-type circuit breaker
Solar panels	Static charge	
Solenoid	Supply	
Source	Supply voltage	Wet cell
	Switch	
	Temporary magnet	

NEW FORMULAS

Identical cells

(a) In series: $V_t = V_1 + V_2 + V_3 \ldots$
(Total voltage is equal to the sum of all the cell voltages.)

(b) In parallel: $V_t = V_1$ or V_2 or $V_3 \ldots$
(Total voltage is the same as for one cell voltage.)

REVIEW QUESTIONS

Multiple Choice Questions ———————————————

1. Direct current is:
   a. A reversing of current continually in a circuit
   b. A flow of current in only one direction.
   c. Produced by an ac voltage
   d. None of the above

2. If load resistance were doubled, load current will ———— .
   a. Halve
   b. Double
   c. Triple
   d. Remain the same

3. Quartz is a solid compound that will produce electricity when ———— is applied.
   a. Friction
   b. An electrolyte
   c. A magnetic field
   d. Pressure

4. Direct current is generated thermally by use of a:
   a. Crystal
   b. Thermocouple
   c. Thermistor
   d. None of the above

5. An application of a photovoltaic cell would be:
   a. A satellite power source
   b. To turn on and off security lights
   c. Both (a) and (b) are true
   d. None of the above

6. The atomic theory of magnetism states that every _____ in motion is a miniature magnet.
   a. Electron
   b. Proton
   c. Neutron
   d. Photon

7. Magnetic flux lines _____ the north pole and _____ at the south pole.
   a. Arrive, leave
   b. Leave, arrive
   c. Are never present, always present
   d. None of the above are true

8. Like magnetic poles _____, while unlike magnetic poles _____.
   a. Attract, repel
   b. Repel, attract
   c. Repel, repel
   d. Attract, attract

9. Of the three artificial magnets, the _____ holds its magnetic force even after the magnetizing force has been removed.
   a. Permanent magnet
   b. Temporary magnet
   c. Electromagnet
   d. Magnetite

10. Which type of battery can be used for only one discharge (and cannot be rejuvenated)?
    a. A lead–acid cell
    b. A secondary cell
    c. A nickel–cadmium cell
    d. A primary cell

11. A battery with a capacity of 12 amperes per hour could supply:
    a. 12 amps for 1 hour
    b. 24 amps for ½ hour
    c. 6 amps for 2 hours
    d. All the above

12. Batteries are normally connected in series to obtain a higher total:
    a. Voltage
    b. Current
    c. Resistance
    d. None of the above

13. The dc power supply converts:
    a. ac to dc
    b. dc to ac
    c. High dc to a low dc
    d. Low ac to a high ac

14. A fuse's current rating states the:
    a. Maximum amount of current allowed to pass
    b. Minimum amount of current allowed to pass
    c. Maximum permissible circuit voltage
    d. None of the above

15. A thermal-type circuit breaker is equivalent to a _____ fuse, whereas a magnetic-type circuit breaker is equivalent to a _____ fuse.
    a. Slow blow, slow blow
    b. Fast blow, slow blow
    c. Slow blow, fast blow
    d. Glass case, ceramic case

16. The type of circuit breaker found in the home is the:
    a. Thermal type
    b. Thermomagnetic type
    c. Magnetic type
    d. Carbon–zinc type

17. A single-pole, double-throw switch would have _____terminals.
    a. Two
    b. Three
    c. Four
    d. Five

18. Maximum power will be transferred from the source to the load when the source resistance is equal to:
    a. 25 Ω
    b. Load current
    c. Load resistance
    d. Source voltage

19. A switch will very simply _____ (open) or _____ (close) a path for current to flow.
    a. Break, make
    b. make, break
    c. s/c, o/c
    d. o/c, s/c

20. With switches, the word pole describes a:
    a. Stationary contact
    b. Moving contact
    c. Path of current
    d. None of the above

Essay Questions _____

21. Describe briefly how direct current can be generated:
    a. Mechanically with pressure and friction   (5.2.1)
    b. Thermally   (5.2.2)
    c. Optically   (5.2.3)
    d. Magnetically   (5.2.4)
    e. Chemically   (5.2.5)
    f. Electrically   (5.2.6)

22. List the three artificial magnets: (5.2.4)

23. Briefly describe the operation of a battery.   (5.2.5)

24. Simply state the difference between a primary and secondary cell.   (5.2.5)

25. Describe what can be gained by connecting batteries in series and parallel.   (5.2.5)

26. Describe a fuse's: (5.3.1)
    a. Current rating
    b. Voltage rating
27. Briefly describe the operation of the following types of circuit breakers: (5.3.2)
    a. Thermal type
    b. Magnetic type
28. List the eight different switch classifications: (5.3.3)
29. List the four popular primary cells. (5.2.5)
30. List the two popular secondary cells. (5.2.5)

Practice Problems _____

31. If a 50-W light bulb is connected across a 120-V source and the following fuses are available, which should be used to protect the circuit? (5.3.1)
    a. 0.5 A/120 V (fast blow)
    b. ¾ A/120 V (fast blow)
    c. 1 A/12 V (slow blow)
    d. 0.5 A/120 V (slow blow)
32. Draw a diagram and indicate the total source voltage of: (5.2.5)
    a. Six 1.5-V cells connected in series with one another.
    b. Six 1.5-V cells connected in parallel with one another.
33. Draw a diagram and indicate the polarities of a 12-V lead–acid battery (5.2.5) being charged by a 15-V battery charger. If this battery is rated at 150 Ah, (a) how many coulombs of charge will be stored in the fully charged condition? (b) How long will it take the battery to charge if a charging current of 5 A is flowing?
34. Show how to connect two 12-V batteries to double the voltage and current rating of a single battery. (5.2.5)
35. If a 6-V nickel–cadmium battery discharges at a rate of 3.5 amps in 4 hours, calculate (a) its ampere-hour and ampere-second rating, and (b) how long it would take the battery to discharge if a current of 2 A is flowing? (5.2.5)

6

ELECTROMAGNETISM (DC)

AFTER COMPLETING THIS CHAPTER, YOU WILL BE ABLE TO:

1. Describe what is meant by the word electromagnetism.
2. Explain the atomic theory of electromagnetism.
3. Understand the left-hand rule of electromagnetism.
4. Describe how a magnetic field is generated by current flow through a:
   a. Conductor
   b. Coil
5. Understand the following magnetic terms:
   a. Magnetic flux
   b. Flux density
   c. Magnetizing force
   d. Magnetomotive force
   e. Reluctance
   f. Permeability (relative and absolute)
6. Understand the following applications of electromagnetism:
   a. Magnetic-type circuit breaker
   b. Electric bell
   c. Relay and its applications
   d. Solenoid-type electromagnet and its application

INTRODUCTION

Electromagnetism can be broken down into two words, electro or electrical and magnetism. Electromagnetism is therefore, by definition, the magnetism resulting from the flow of an electrical current. There are actually two areas of interest that need to be covered when discussing electromagnetism; these are conductors (wire), and coils (electromagnets), both of which will generate a magnetic field when current is present.

6.1

CONDUCTORS

In both instances in Figure 6-1, the magnetic field produced by the current flow encircles the wire carrying the direct current. This phenomena was originally discovered by Hans Christian Oersted, a Danish physicist, in 1820.

6.1.1 ATOMIC THEORY OF ELECTROMAGNETISM

Every electron generates a magnetic field. When electrons are forced to leave their parent atom by voltage and flow toward the positive polarity, they are all moving in the same direction and each electron's magnetic field will add with the next. The accumulation of all these electron fields will create the magnetic field around the conductor, as seen in Figure 6-2.

A simple experiment can be performed to prove that this invisible

FIGURE 6-1   Conductor and Coil Electromagnetism.

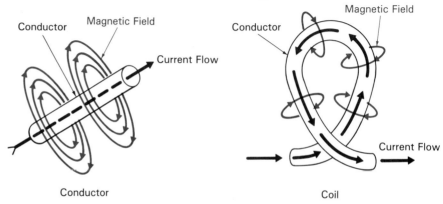

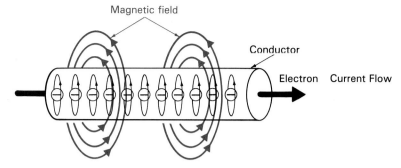

FIGURE 6-2 Electron Magnetic Fields.

magnetic field does in fact exist around a conductor, the setup of which is illustrated in Figure 6-3. With the switch open, as seen in Figure 6-3(a) no current will flow and, therefore, no magnetic field is generated around the conductor and the iron filings on the cardboard will be disorganized.

With the switch closed, a current of 3 A ($I = V/R = 9$ V$/3$ $\Omega = 3$ A) will flow through the circuit and a magnetic field will be set up around the conductor. The iron filings become organized in circles due to the influence of the magnetic field, as seen in Figure 6-3(b).

FIGURE 6-3 Magnetic Field Experiment. (a) Switch Open, No Current Flow. (b) Switch Closed, Current Flow, Magnetic Field.

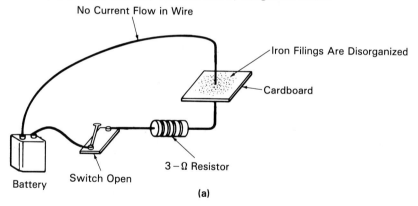

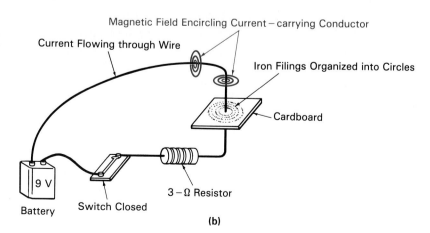

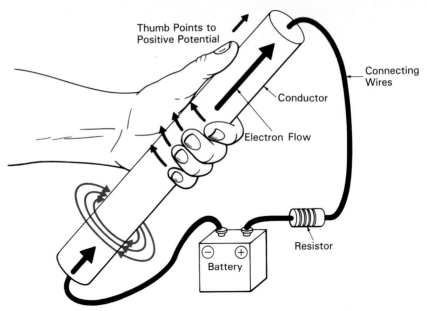

FIGURE 6-4   The Left-hand Rule of Magnetism.

6.1.2 ELECTROMAGNETISM'S LEFT-HAND RULE

As we have seen so far, there is a relationship between the current in a conductor and the magnetic field around the conductor. The left-hand rule states that, if a conductor is held with your left hand so that your thumb is pointing in the direction of electron flow (to the positive potential), your fingers will be pointing in the direction of the magnetic force, as illustrated in Figure 6-4.

SELF-TEST REVIEW QUESTIONS (§ 6.1)

1. An electric field always encircles a current-carrying conductor (true or false).
2. What subatomic particle is said to have its own magnetic field?
3. For what purpose could the left-hand rule be used?
4. From which pole do the magnetic lines of force originate?

6.2
ELECTROMAGNET (SPIRAL COIL)

If our conductor is wound to form a coil, as illustrated in Figure 6-5(a), the conductor will generate its own magnetic field, which will combine additively within the coil, and the net effect of all these conductor magnetic fields can be used to generate a strong coil magnetic field, as seen in Figure 6-5(a).

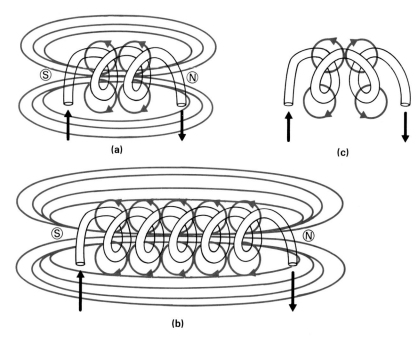

FIGURE 6-5   Electromagnet. (a) Small Number of Turns. (b) Large Number of Turns. (c) Opposing Turns.

If a greater number of loops are added, a greater or stronger magnetic field is generated, as seen in Figure 6-5(b). If the coils were wound in an opposing opposite direction, the opposing magnetic fields would cancel and so no magnetic field would be generated, as seen in Figure 6-5(c).

The left-hand rule can also be applied to electromagnets to determine which of the poles will be the north end of the electromagnet. This is illustrated in Figure 6-6, which shows that, if you wrap the fingers of your left hand around the coil so that your fingers are pointing in the direction of current flow, your thumb will be pointing to the north end of the electromagnet.

FIGURE 6-6   The Left-hand Rule for Coils.

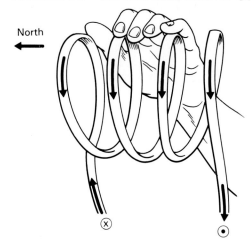

North

1. Which other name could be used instead of electromagnet?
2. The greater the number of loops in an electromagnet, the greater or stronger the magnetic field (true or false).

6.3

MAGNETIC MEASUREMENT AND PROPERTIES

6.3.1 MAGNETIC FLUX (φ) AND FLUX DENSITY (β)

One magnetic line of flux (symbolized φ, or phi) or force is called a maxwell in honor of James Maxwell's work in this field. However, this unit is too small and impractical, and because of this the weber (wb) was introduced. One weber is equal to 10^8 magnetic lines of force or maxwells, and

magnetic flux (φ) = number of lines of force (or maxwells) in webers
1 weber = 10^8 lines of force (or maxwells)

If 10^8 lines of force exist in 1 square meter, and 10^8 lines of force exist in 1 square centimeter, which is the stronger magnetic field? As far as magnetic flux (φ) is concerned, 1 weber exists in both. So some other way is needed to specify how many lines of force exist in a given area, and this will tell us if it is a strong or weak magnetic field.

Flux density (β) is equal to the number of magnetic lines of flux (φ) per square meter, and it is given in the unit of teslas (T).

$$\text{flux density } (\beta) = \frac{\text{magnetic flux } (\phi)}{\text{area}}$$

Here flux density is in teslas, magnetic flux is in webers, and area is in square meters.

EXAMPLE 6.1

If magnet A produces 10^8 (100,000,000) lines of flux (1 weber) in 1 square centimeter, and magnet B produces 10^8 lines of flux in 1 square meter, which magnet is producing the more concentrated or intense magnetic field?

SOLUTION

$$\text{Magnet } A \quad \text{flux density} = \frac{1 \text{ weber}}{0.01 \text{ (m}^2)}$$
$$= 100 \text{ teslas}$$
$$\text{Magnet } B \quad \text{flux density} = \frac{1 \text{ weber}}{1 \text{ (m}^2)}$$
$$= 1 \text{ tesla}$$

Flux density determined that magnet A produced the more concentrated magnetic force.

6.3.2 MAGNETOMOTIVE FORCE (MMF)

The magnetic flux produced by an electromagnet is produced by current flowing through a coil of wire. As previously discussed, electromotive force (emf) is the pressure or voltage that forces electrons to move. Magnetomotive force (mmf), is the magnetic pressure that produces the magnetic field. The formula for mmf is

$$\boxed{\text{mmf} = I \times N \text{ (ampere-turns)}}$$

where mmf = magnetomotive force in ampere-turns $(A \cdot t)$
I = current in amperes
N = number of turns in the coil

The formula basically says that, if you increase the current through the coil or increase the number of turns in the coil, you will increase the magnetic pressure (magnetomotive force), and by increasing magnetic pressure, you will increase the magnetic field produced, as seen in Figure 6-7.

FIGURE 6-7 **Magnetomotive Force.**

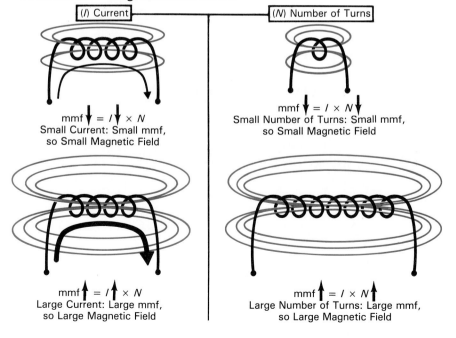

6.3.3 MAGNETIZING FORCE (H)

Magnetomotive force (mmf) is equal to the product of current and the number of turns; however, if a 20-turn coil or solenoid is stretched out to twice its length, the magnetic field or force will be half as strong, because there would be less of a reinforcing effect between the coils due to the greater distance between the coils. The length of the coil is consequently a factor that also determines the field intensity, and this new term, magnetizing force (*H*), is the reference we commonly use to describe the magnetic field intensity.

$$\text{magnetizing force } (H) = \frac{I \times N}{l}$$

$$\text{or } H = \frac{\text{mmf}}{l}$$

where H = magnetizing force in ampere-turns/meter (A · t/m)
I = current in amperes
l = length of coil in meters
N = number of coil turns

6.3.4 RELUCTANCE ($\mathcal{R}$)

Reluctance, in reference to magnetic energy, is equivalent to resistance in electrical energy. Reluctance is the opposition or resistance to the establishment of a magnetic field in an electromagnet. The formula for calculating reluctance is the magnetic energy equivalent of the electrical Ohm's law as shown:

$$\textit{Magnetic}$$
$$\text{reluctance } (\mathcal{R}) = \frac{\text{mmf (magnetomotive force)}}{\phi \text{ (magnetic flux)}}$$
$$\text{(measured in ampere-turns/weber)}$$

$$\textit{Electrical}$$
$$\text{resistance } (R) = \frac{V \text{ (electromotive force)}}{I \text{ (current)}}$$
$$\text{(measured in ohms)}$$

6.3.5 PERMEABILITY (μ)

As magnetic reluctance is equivalent to electrical resistance, magnetic permeability is equivalent to electrical conductance. Permeability is a measure of how easily a material will allow a magnetic field to be set up within it.

Electromagnetism (DC)

Permeability is symbolized by the Greek letter μ and is measured with respect to air, which has a permeability of 1. A high permeability figure (μ↑) indicates that a magnetic field can easily be established within a material and, therefore, this material's reluctance must be low (ℛ↓). On the other hand, a low permeability figure (μ↓) indicates that there will be a large reluctance (ℛ↑) to establishing a magnetic field. This is mathematically stated as:

$$
\text{Magnetic permeability } (\mu) = \frac{1}{\mathscr{R} \quad \text{(reluctance)}}
$$

(Permeability is inversely proportional to reluctance and is measured in henrys per meter.)

$$
\text{Electric conductance } (G) = \frac{1}{R \quad \text{(resistance)}}
$$

(Conductance is inversely proportional to resistance and is measured in siemens.)

Relative (with respect to) is a word which means that a comparison has to be made. Relative permeability (μ_r) is the measure of how well another given material will conduct magnetic lines of force with respect to, or relative to, our reference material, air, which has a relative permeability value of 1.

Table 6-1 lists the relative permeability of various materials. The relative permeability (μ_r) of air, which is equal to 1, should not be confused with

TABLE 6-1
RELATIVE AND ABSOLUTE PERMEABILITIES OF VARIOUS MATERIALS

Material	Relative Permeability (μ_r)	Absolute Permeability (μ_0)
Air or vacuum	1	1.26×10^{-6}
Nickel	50	6.28×10^{-5}
Cobalt	60	7.56×10^{-5}
Cast iron	90	1.1×10^{-4}
Machine steel	450	5.65×10^{-4}
Transformer iron	5500	6.9×10^{-3}
Silicon iron	7000	8.8×10^{-3}
Permaloy	100,000	0.126
Supermalloy	1,000,000	1.26

Absolute permeability of air (μ_0) = $4\pi \times 10^{-7}$ or 1.26×10^{-6}.

the absolute permeability (μ_0) of free space of air, which is equal to $4\pi \times 10^{-7}$ or 1.26×10^{-6}.

$$\text{permeability } (\mu) = \text{relative permeability } (\mu_r) \times \text{absolute}$$
$$\text{permeability of air } (\mu_0)$$

Relative permeability is always greater than 1.

EXAMPLE 6.2

What is the magnetomotive force produced when 3 amperes of current flows through 5 turns in a coil?

SOLUTION

$$\begin{aligned}
\text{mmf} &= I \text{ (current)} \times N \text{ (number of turns)} \\
&= 3 \text{ amps} \times 5 \text{ turns} \\
&= 15 \text{ A} \cdot \text{t (ampere-turns)}
\end{aligned}$$

EXAMPLE 6.3

If the magnetic flux produced by a material is equal to 335 μWb and the mmf equals 15 At:
a. What is the reluctance of the material?
b. What is the material's permeability?

SOLUTION

$$\begin{aligned}
\text{reluctance } (\mathcal{R}) &= \frac{\text{mmf (magnetomotive force)}}{\phi \text{ (magnetic flux)}} \\
&= \frac{15 \text{ A} \cdot \text{t (ampere-turns)}}{335 \text{ }\mu\text{Wb}} \\
&= 44.8 \times 10^3 \text{ A} \cdot \text{t/Wb} \\
\text{permeability } (\mu) &= \frac{1}{44.8 \times 10^3} \quad \text{(reluctance)} \\
&= 22.3 \times 10^{-6} \text{ henrys/meter} \\
&\quad \text{(H/M)}
\end{aligned}$$

Calculator Sequence

Step	Keypad Entry	Display Response
1.	① ⑤	15
2.	÷	
3.	③ ③ ⑤ E ⑥ +/−	335E-6
4.	=	44776.1
5.	1/x	22.3E-6

6.3.6 SUMMARY

The magnetic field strength of an electromagnet can be increased by increasing the number of turns in its coil, increasing the current through the electromagnet, or decreasing the length of the coil ($H = A \times N/l$). The field strength can be further increased by placing an iron core within the

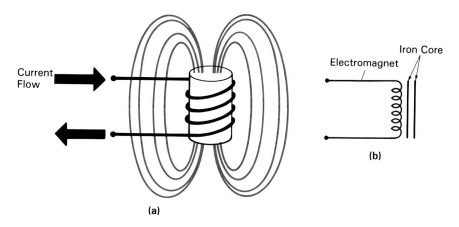

FIGURE 6-8   **Electromagnet. (a) Operation. (b) Symbol.**

electromagnet, as illustrated in Figure 6-8(a). An iron core has less reluctance or opposition to the magnetic lines of force than air, and so the flux density (β) is increased. Another way of saying this is that the permeability or conductance of magnetic lines of force is greater than that of air, and if the permeability of iron is large, then its reluctance must be small. The symbol for an iron core electromagnet is seen in Figure 6-8(b).

SELF-TEST REVIEW QUESTIONS (§ 6.3)

Define the following:

1.   Magnetic flux
2.   Flux density
3.   Magnetomotive force
4.   Magnetizing force
5.   Reluctance
6.   Permeability

6.4
APPLICATIONS OF ELECTROMAGNETISM

6.4.1   MAGNETIC-TYPE CIRCUIT BREAKER

The magnetic-type circuit breaker (Figure 6-9) was first discussed in Chapter 5 and is one application of an electromagnet. This and all other circuit breakers are used for current protection. If the rated current value or below

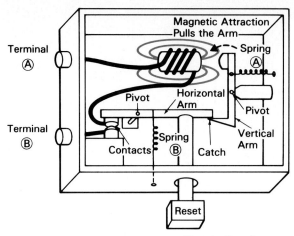

FIGURE 6-9   **Magnetic-type Circuit Breaker.**

is passing through the circuit breaker, the electromagnet will not generate a strong enough magnetic field to pull the iron arm to the left and open the contacts. However, if the current rating of the circuit breaker is exceeded, the increase in current will cause a corresponding increase in magnetic flux, attracting the iron arm, opening the contacts, and protecting the equipment from the dangerous level of current.

6.4.2   ELECTRIC BELL

The electric bell in Figure 6-10 utilizes a soft iron core electromagnet, a striker, and a gong. When the bell push is pressed, a complete electrical circuit is made from the positive terminal to the electromagnet, the contact,

FIGURE 6-10   **Electric Bell.**

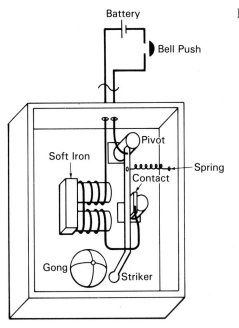

the striker, and then back to the negative terminal. The electromagnet energizes (produces a magnetic field) and attracts the soft iron striker, which strikes the gong and at the same time breaks the circuit, de-energizing the electromagnet. The striker is pulled back by the spring, and the circuit is reestablished, resulting in a continuous ringing of the bell.

6.4.3 RELAY

A relay is an electromechanical device that either makes (closes) or breaks (opens) a circuit by moving contacts together or apart. Figure 6-11 illustrates the relay; Figure 6-11(a) shows the normally open (NO) relay and Figure 6-11(b) shows the normally closed (NC) relay.

(1) OPERATION

In both cases, the relay consists basically of an electromagnet connected to lines x and y, a movable iron arm known as the armature, and some contacts. When current passes through x to y, the electromagnet generates a magnetic field, or is said to be energized, which attracts the armature toward the electromagnet. When this occurs, it closes or makes the normally open relay's contacts and opens or breaks the normally closed relay contacts.

FIGURE 6-11   Relays. (a) Single-pole, Single-throw (SPST), Normally Open (NO) Relay (Contacts Are Open Until Activated). (b) Single-pole, Single-throw (SPST), Normally Closed (NC) Relay (Contacts Are Closed Until Activated).

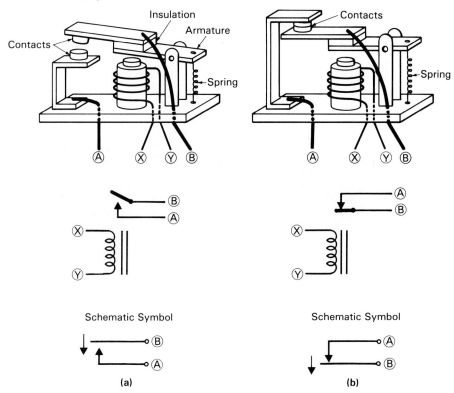

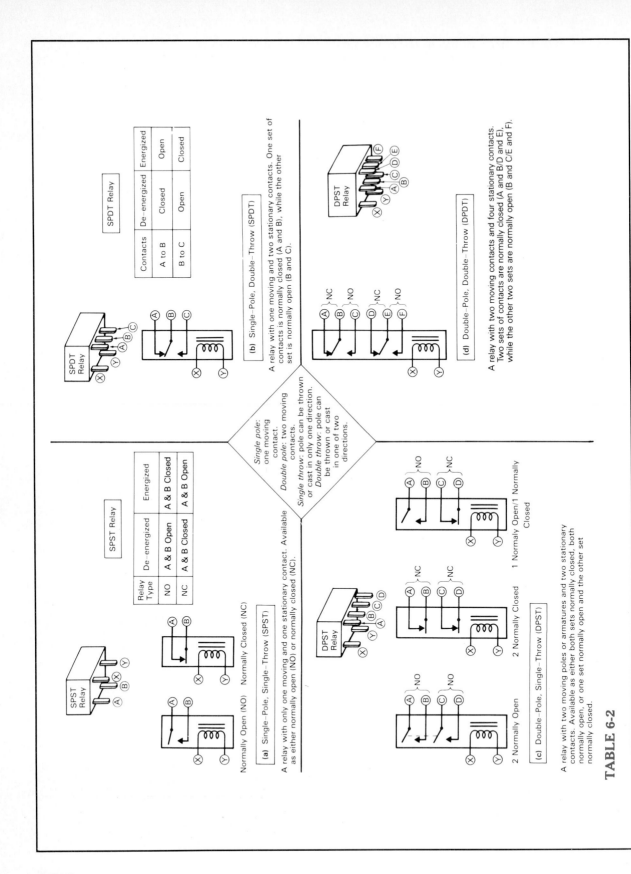

SPDT Relay

Contacts	De-energized	Energized
A to B	Closed	Open
B to C	Open	Closed

SPST Relay

Relay Type	De-energized	Energized
NO	A & B Open	A & B Closed
NC	A & B Closed	A & B Open

Normally Open (NO)   Normally Closed (NC)

(a) Single–Pole, Single–Throw (SPST)

A relay with only one moving and one stationary contact. Available as either normally open (NO) or normally closed (NC).

Single pole: one moving contact.
Double pole: two moving contacts.
Single throw: pole can be thrown or cast in only one direction.
Double throw: pole can be thrown or cast in one of two directions.

(b) Single–Pole, Double–Throw (SPDT)

A relay with one moving and two stationary contacts. One set of contacts is normally closed (A and B), while the other set is normally open (B and C).

(d) Double–Pole, Double–Throw (DPDT)

A relay with two moving contacts and four stationary contacts. Two sets of contacts are normally closed (A and B/D and E), while the other two sets are normally open (B and C/E and F).

1 Normaly Open/1 Normally Closed

2 Normally Closed

2 Normally Open

(c) Double–Pole, Single–Throw (DPST)

A relay with two moving poles or armatures and two stationary contacts. Available as either both sets normally closed, both normally open, or one set normally open and the other set normally closed.

TABLE 6-2

194

If the electromagnet is de-energized by stopping the current through the coil, the spring will pull back the armature to open the NO relay contacts or close the NC relay contacts between *A* and *B*.

To summarize then, you can see that the "normal" condition for the contacts between *A* and *B* is when the electromagnet is de-energized. In the de-energized condition, the normally open relay contacts are open and the normally closed relay contacts are closed.

The two relays discussed so far are actually single-pole, single-throw relays as they have one movable contact (single pole) and one stationary contact that the pole can be thrown to, as seen in Figure 6-11. There are actually four basic configurations for relays and all are illustrated in Table 6-2. Variations on these basic four can come in all shapes and sizes, with one relay controlling sometimes several sets of contacts. Figure 6-12 illustrates several different styles and packages of relays that are available.

(2) APPLICATIONS OF A RELAY

The relay is generally used in two basic applications:

(1) To enable one master switch to operate several remote or difficultly placed contact switches, as illustrated in Figure 6-13. When the master switch is closed, the relay is energized, closing all its contacts and turning on all the lights. The advantage of this is twofold in that, first, the master switch can turn on three lights at one time, which saves time for the operator,

FIGURE 6-12   Relay Styles and Packages.

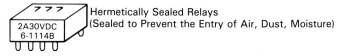
Hermetically Sealed Relays
(Sealed to Prevent the Entry of Air, Dust, Moisture)

2A30VDC
6-1114B

Magnetically Shielded Relays
R56 24 V

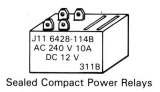
J11 6428-114B
AC 240 V 10A
DC 12 V
311B
Sealed Compact Power Relays

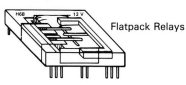
Flatpack Relays
H6B   12 V

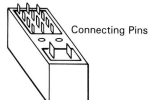
Connecting Pins

Power Relays
12 V 99

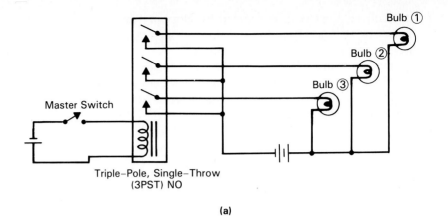

Master Switch

Triple-Pole, Single-Throw
(3PST) NO

Bulb ①
Bulb ②
Bulb ③

(a)

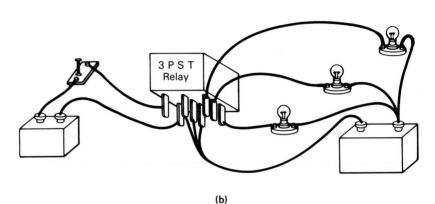

3 P S T
Relay

(b)

FIGURE 6-13 One Master Switch Operating Several Remotes.
(a) Schematic. (b) Pictorial.

FIGURE 6-14 Low-current Switch Enabling a High-current Circuit.
(a) Schematic. (b) Pictorial.

Single-Pole, Single-Throw (SPST) NO   Heavy Gauge Cable

Operator's
Switch

Low
V

Light Gauge Cable

Motor

High
V

(a)

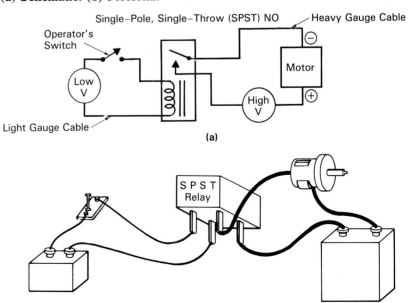

S P S T
Relay

(b)

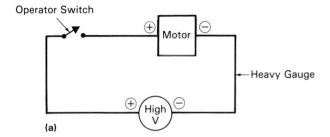

(a)

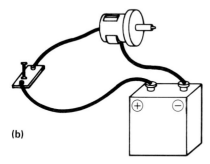

(b)

FIGURE 6-15 **Without Relay (More Expensive).**

and, second, only one set of wires need be taken from the master switch to the lights, rather than three sets for all three lights.

(2) The second basic application of the relay is to enable a switch in a low-voltage circuit to operate relay contacts in a high-voltage circuit, as seen in Figure 6-14. The operator activates the switch in the low-voltage circuit, which will energize the relay, closing its contacts and connecting the high voltage to the motor. Figure 6-15 illustrates the circuit arrangement that would have to be used without a relay.

In a high-voltage circuit, a large current will be flowing and so a large or heavy-gauge wire will be needed to handle this current. With the relay circuit in Figure 6-14, large-gauge wire is only needed for the short distance between the relay and the motor, while in the circuit without the relay (Figure 6-15), heavy-gauge wire, which is more expensive than smaller-gauge wire, is needed for the long distance between the operator's switch and the motor.

(a) **Application of a Relay: Automobile Starting Circuit.** A starter relay is illustrated in Figure 6-16. In this application of a relay, we will see how a relay can be used to supply the large dc current needed to activate a starter motor of an automobile.

When the ignition switch is engaged in the passenger compartment by the driver, current flows through a light-gauge wire from the negative side of the battery, through the ignition switch, through the relay's electromagnet, and back to the positive side of the battery. This current flow through the electromagnet of the relay energizes the relay and closes the relay's contacts. Closing the relay's contacts makes a path for the current to flow through the heavy-gauge cable from the negative side of the battery, through the relay contacts and starter motor, and back to the positive side of the battery. The starter motor's output shaft spins the engine causing it to start.

This application is a perfect example of how a relay can be used to

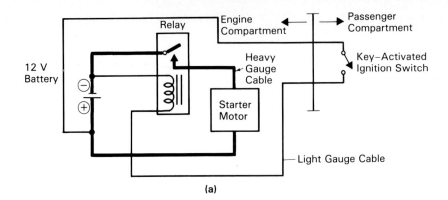

(a)

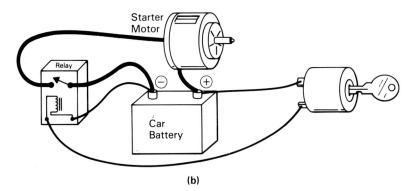

(b)

FIGURE 6-16 **Automobile Starting Circuit.**

close contacts in a heavy-current (heavy-gauge cable) circuit, while the driver has only to close contacts in a small-current (light-gauge cable) circuit.

If the relay were omitted, the driver's ignition switch would have to be used to connect the 12 V and large current to the starter motor. This would require that:

1. Heavy-gauge expensive cable be connected in a longer path between starter motor and passenger compartment.

2. The ignition switch would need to be larger to handle the heavier current.

(3) REED RELAYS AND SWITCHES

The reed relay or switch consists of two flat magnetic strips mounted inside a capsule, which is normally made of glass. The reed relay and reed switch are illustrated in Figure 6-17(a) and (b).

The reed relay differs from the reed switch in that a reed relay has its own energizing coil, while the reed switch needs an external magnetic force to operate.

The reed relay and reed switch are both operated by a magnetic force that is provided by a coil for the reed relay and by a separate permanent magnet for the reed switch. When the magnetic force is present, opposite magnetic polarities are induced in the overlapping, high-permeability reed blades, causing them to attract one another by induced magnetism and snap

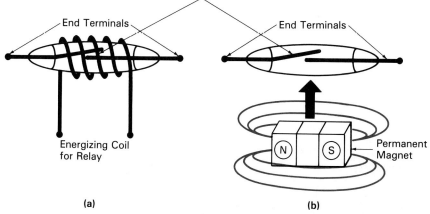

FIGURE 6-17   Reed Relay and Switch. (a) Reed Relay: Contacts Close When Electromagnetic Coil Is Energized. (b) Reed Switch: Contacts Close When External Magnetic Field Comes in Close Proximity, Due to Induced Magnetism in the Contacts.

together, thus closing the circuit. When the magnetic force is removed, the blades spring apart due to spring tension and open the circuit.

(a) Application of a Reed Switch: Home Security Circuit (Figure 6-18) When the window is closed in the normal condition, the permanent magnet is directly adjacent to the reed switch and, therefore, its contacts are closed, allowing current to flow from the negative side of the small 3-V battery, through the closed reed switch contacts, the relay's electromagnet, and back to the positive side of the 3-V battery. The relay is a normally closed (NC) SPST; but as current is flowing through the coil of the relay, the contacts are open, preventing the large positive and negative voltage from the 12-V battery from reaching the siren.

If the window is forced and opened by an intruder, the permanent magnet will no longer be in close proximity to the reed switch and the reed switch's contacts will open. When the reed switch's contacts open, the relay's coil will no longer have current flowing through it and, therefore, the relay will

FIGURE 6-18   Reed Switch as Part of a Home Security System.

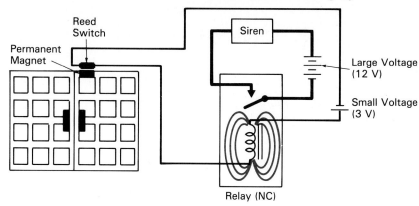

de-energize and its contacts will return to their normal closed condition. A large current is now permitted to flow from the negative side of the large 12-V battery, through the relay's contacts, the siren, and back to the positive side of the battery. The siren will sound, as it now has 12-V applied to it, and alert the occupant of the home.

6.4.4   SOLENOID-TYPE ELECTROMAGNET

Up to now, electromagnets have been used to close or open a set, or sets, of contacts to either make or break a current path. These electromagnets have used a stationary soft iron core; however, some electromagnets are constructed with movable iron cores, as seen in Figure 6-19, which can be used to open or block the passage of a gas or liquid through a valve. These are known as solenoid-type electromagnets.

When no current is flowing through the solenoid coil, no magnetic field is generated and so no magnetic force is exerted on the movable iron core, as seen in Figure 6-19(a), and therefore the compression spring maintains it in the up position, with the valve plug on the end of the core preventing the passage of either a liquid or gas through the valve (valve closed).

When a current flows through the electromagnet, the solenoid coil is energized, creating a magnetic field, as seen in Figure 6-19(b). Due to the

FIGURE 6-19   Solenoid-type Electromagnet. (a) De-energized. (b) Energized.

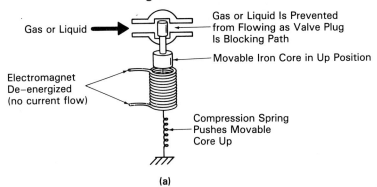

(a)

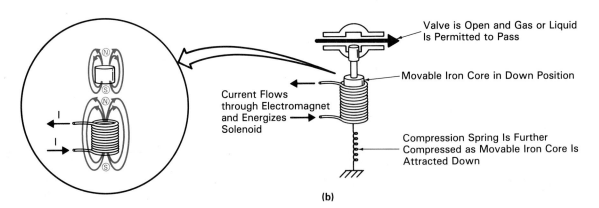

(b)

influence of the coil's magnetic field, the movable soft iron core will itself generate a magnetic field, as seen in the insert in Figure 6-19(b). This condition will create a north pole at the top of the solenoid coil and a south pole at the bottom of the movable core, and the resulting attraction will pull down the core (which is free to slide up and down), pulling with it the valve plug and opening the valve.

Solenoid-type electromagnets are actually constructed with the core partially in the coil and are used in washing machines to control water and in furnaces to control gas.

SELF-TEST REVIEW QUESTIONS (§ 6.4)

1. List two applications of the electromagnet.
2. What is the difference between a NO and NC relay?
3. What is the difference between a reed relay and a reed switch?
4. Give an application for the conventional relay and the reed switch.

SUMMARY

1. Electromagnetism (dc) is the *magnetism* resulting from the flow of direct *electrical* current.
2. Every current-carrying conductor generates a magnetic field, and a conductor wound into a coil is called an electromagnet and produces a large concentrated magnetic field.
3. The atomic theory of electromagnetism states that each moving electron has its own magnetic field encircling it.
4. When electrons are forced into motion in a common direction (current flow) by voltage, each electron's magnetic field will add together, producing a joint magnetic field that encircles the current-carrying conductor.
5. The left-hand rule of electromagnetism states that:
   a. If a conductor is held with your left hand so that your thumb is pointing in the direction of current flow, your fingers will point in the direction of the magnetic force.
   b. If you wrap your left hand around a coil so that your fingers are pointing in the direction of current flow, your thumb will point to the north pole of the electromagnet.
6. One magnetic line of flux (ϕ) is called a maxwell.
7. One weber (wb) is equal to 10^8 magnetic lines of force or maxwells.
8. Flux density (β) is equal to the number of magnetic lines of flux (ϕ) per square meter and is measured in teslas.
9. Magnetomotive force (mmf) is the magnetic pressure that produces a magnetic field and is equal to the product of current and the number of coil turns.

10. The magnetizing force (H) describes the magnetic field intensity and takes into account the length of the coil, which determines the reinforcing effect between the coils.

11. Reluctance is the opposition to the establishment of a magnetic field in an electromagnet.

12. Permeability is a measure of how easily a material will allow a magnetic field to be set up within it.

13. Relative permeability is the measure of how well another given material will conduct magnetic lines of force relative to our reference material air, which has a relative permeability of 1.

14. The absolute permeability (magnetic conductance) of air or free space is equal to $4\pi \times 10^{-7}$.

15. Electromagnetism is used in such applications as:
    a. Magnetic-type circuit breaker
    b. Electric bell
    c. A relay, which is used to:
       i. Allow a master switch to operate several remote or difficultly placed switches.
       ii. To allow a switch in a low-current circuit to control the current flow in a high-current circuit, such as an automobile starter.

16. Reed relays have a coil surrounding a capsule that encases two flat magnetic strips.

17. The reed switch is similar to the reed relay except that the magnetic force needed to close the two reeds or blades is provided by a separate magnetic field source; it finds its major application in home security systems.

18. A solenoid-type electromagnet incorporates a movable iron core that can be used to open and close valves by energizing or de-energizing the solenoid coil.

NEW TERMS

Absolute permeability (μ_0)
Ampere turns per meter ($A \cdot t/m$)
Ampere turns per weber ($A \cdot t/Wb$)
Armature
Coil
De-energized
Electromagnet
Electromagnetism
Energized

Flux density (β)
Henrys per meter (H/m)
Left-hand rule
Magnetic field
Magnetic flux (ϕ)
Magnetizing force (H)
Magnetomotive force (mmf)
Maxwell (Mx)
Permeability (μ)

Reed relay
Reed switch
Relative permeability (μ_r)
Relay
Reluctance
Solenoid
Tesla
Weber

1 magnetic line of force = 1 maxwell
10^8 lines of force or maxwells = 1 weber

$$\beta = \frac{\phi}{A}$$

β = flux density in teslas (T)
ϕ = magnetic flux in webers (Wb)
A = area in square meters (m²)

$$mmf = I \times N$$

mmf = magnetomotive force in ampere turns (A · t)

$$H = \frac{I \times N}{l}$$

I = current in amperes
N = number of turns in coil
H = magnetizing force in ampere turns per meter (A · t/m)
l = length of coil in meters

$$\mathcal{R} = \frac{mmf}{Q}$$

$\mathcal{R}$ = reluctance in ampere turns per weber (A · t/Wb)

$$\mu = \frac{1}{\mathcal{R}}$$

μ = permeability in henrys per meter (H/m)

$$\mu = \mu_r \times \mu_0$$

μ_r = relative permeability
μ_0 = absolute permeability = $4\pi \times 10^{-7}$

REVIEW QUESTIONS

Multiple Choice Questions

1. Electromagnetism was first discovered by:
   a. Hans Christian Oersted
   b. Heinrich Hertz
   c. James Watt
   d. James Clark Maxwell

2. Every current-carrying conductor generates a (an):
   a. Electric field
   b. Magnetic field
   c. Both (a) and (b)
   d. None of the above

3. If a greater number of loops of a conductor in a coil are added, a _____ magnetic field is generated.
   a. Weaker
   b. Stronger

4. If a smaller current is passed through a coil, a _____ magnetic field is generated.
   a. Weaker
   b. Stronger

5. One magnetic line of flux is known as a _____.
   a. Weber
   b. Tesla
   c. Maxwell
   d. Oersted

6. 10^8 magnetic lines of flux are referred to as one _____.
   a. Weber
   b. Tesla
   c. Maxwell
   d. Oersted

7. Flux density is equal to the magnetic flux divided by area and is measured in _____.
   a. Webers
   b. Teslas
   c. Maxwells
   d. Oersteds

8. By increasing either current flow through or the number of turns in a coil, the mmf, which is an abbreviation for _____, will increase.
   a. Multiple magnetic formulas
   b. Magnetomotive force
   c. Electromotive force
   d. None of the above

9. _____ is a term used to describe the magnetic field intensity and is equal to the mmf divided by the length of the coil.
   a. Flux density
   b. Magnetomotive force
   c. Magnetizing force
   d. Reluctance

10. Reluctance is equivalent to electrical:
    a. Current
    b. Voltage
    c. Resistance
    d. Power

11. Permeability is equivalent to electrical:
    a. Current
    b. Conductance
    c. Resistance
    d. Voltage

12. The absolute permeability of air is equal to:
    a. 1
    b. 1.26×10^{-6}
    c. 4π
    d. None of the above

13. An electromagnet can be used in the application of:
    a. A relay
    b. An electric bell
    c. A circuit breaker
    d. All of the above are true

14. Relays can be used to:
    a. Allow one master switch to enable several others.
    b. Allow several switches to enable one master.
    c. Allow a switch in a low-current circuit to close contacts in a high-current circuit.
    d. Both (a) and (b) are true.
    e. Both (a) and (c) are true.

15. The starter relay in an automobile is used to:
    a. Allow one master switch to enable several others.
    b. Allow several switches to enable one master.
    c. Allow a switch in a low-current circuit to close contacts in a high-current circuit.
    d. Both (a) and (b) are true.

16. The _____ uses an electromagnet around two flat magnetic strips mounted inside a glass capsule.
    a. Reed switch
    b. Magnetic circuit breaker
    c. Reed relay
    d. Starter relay

17. The reed switch could be used in a (an):
    a. Home security system
    b. Automobile starter
    c. Magneic-type circuit breaker
    d. All of the above

18. The relative permeability of air or a vacuum is equal to:
    a. 4π
    b. 6.26×10^{-6}
    c. 1
    d. None of the above

19. An electromagnet is also known as a:
    a. Coil
    b. Solenoid
    c. Resistor
    d. Both (a) and (c) are true
    e. Both (a) and (b) are true

20. A normally open relay (NO) will have:
    a. Contacts closed until activated
    b. Contacts open until activated
    c. All contacts permanently open
    d. All contacts permanently closed

21. Describe what is meant by the word electromagnetism. (6.1)
22. Briefly describe how the left-hand rule of electromagnetism is applied to: (6.2)
    a. Conductors
    b. Coils
23. How many maxwells make up 1 weber? (6.3.1)
24. Give the formulas for the following:
    a. Flux density (6.3.1)
    b. Magnetomotive force (6.3.2)
    c. Magnetizing force (6.3.3)
    d. Reluctance (6.3.4)
    e. Permeability (6.3.5)
25. Describe the operation of:
    a. The magnetic-type circuit breaker (6.4.1)
    b. The electric bell (6.4.2)
    c. The NO and NC relay (6.4.3)
26. Describe the difference and operation of the: (6.4.3)
    a. Reed relay
    b. Reed switch
27. Explain the operation and application of the solenoid-type electro-magnet. (6.4.4)
28. Why is a wire looped many times to form an electromagnet more useful than a straight piece of wire? (6.2)
29. From which pole does the magnetic flux emerge and into which pole end does the flux return? (6.1)
30. What are the differences between magnetic flux and flux density? (6.3.1)

Practice Problems

31. If a magnet has a pole area of 6.4×10^{-3} m² and a 1200-μWb total flux, what flux density would the pole produce? (6.3.1)
32. Calculate the magnetomotive force produced when 760 mA flow through 25 turns. (6.3.2)
33. Calculate the magnetizing force (H) or field intensity if a 15 cm, 40-turn coil has a current of 1.2 A flowing through it. (6.3.3)
34. Calculate the relative permeability (μ_r) of the following materials ($\mu_0 = 4\pi \times 10^{-7}$): (6.3.5)
    a. Cast iron
    b. Nickel
    c. Machine steel
35. Calculate the reluctance of a magnetic circuit if the magnetic flux produced is equal to 2.3×10^{-4} Wb, and is produced by 3 A flowing through a solenoid of 36 turns. (6.3.4)
36. If a coil of 50 turns is passing a current of 4 A, calculate the mmf. (6.3.2)

37. Calculate the reluctance of an iron core when mmf = 150 A · t and ϕ = 360 µWb.   (6.3.4)
38. Calculate mmf when a 9-V battery is connected across a 50-turn, 23-Ω coil.   (6.3.2)
39. Calculate the permeability of a permalloy core.   (6.3.5)
40. Calculate the magnetizing force (H) of the coil in question 38 if it were 0.7 m long.   (6.3.3)

7

SERIES DC CIRCUITS

AFTER COMPLETING THIS CHAPTER, YOU WILL BE ABLE TO:

1. Define what is meant by a series circuit.
2. Identify series circuits.
3. Connect components so that they are in series with one another.
4. Describe why current remains the same throughout a series circuit.
5. Explain how to calculate total resistance in a series circuit.
6. Explain how Ohm's Law can be applied to calculate current, voltage, and resistance.
7. Describe why the series circuit is known as a voltage divider.
8. Explain the fixed and variable voltage divider.
9. Explain how to calculate power in a series circuit.
10. Describe how to troubleshoot and recognize;
    a. An open component
    b. A component value variation
    c. A short circuit in a series circuit.

INTRODUCTION

A series circuit, by definition, is the connecting of components end to end in a circuit to provide a single path for the current. This is true not only for resistors, but also for other components that can be connected in series; in all cases the components are connected in succession or strung together one after another so that only one path for current exists between the negative ($-$) and positive ($+$) terminals of the supply.

HITLER'S COMPUTER MISTAKE

In the early days of World War II, German scientist Konrad Zuse, who designed and built the first general-purpose computer, proposed constructing a computer that would operate 1000 times faster than anything else at that time. He was going to redesign his Z3 computer, which was being used at that time to solve engineering problems in aircraft and missile design, and include vacuum tubes instead of electromechanical relay switches. This proposal was rejected by Hitler, who was not interested in this long-term, 2-year project, as he was sure that the war was going to be, for him, a certain, quick victory. Due to Hitler's shortsightedness, this powerful computer, which could have been used to break British communication codes, was never developed. However, unknown to both Hitler and Zuse, the British code-breaking computer project, called Ultra, had highest priority and was moving rapidly toward completion.

EXAMPLE 7.1

Figure 7-1 illustrates five examples of series resistive circuits. In Figure 7-2(a), there are seven resistors laid out on a table top. With connecting wire, string all the resistors in series, starting at $R1$, and proceeding in numerical order through the resistors until reaching $R7$.

FIGURE 7-1   Five Series Resistive Circuits.

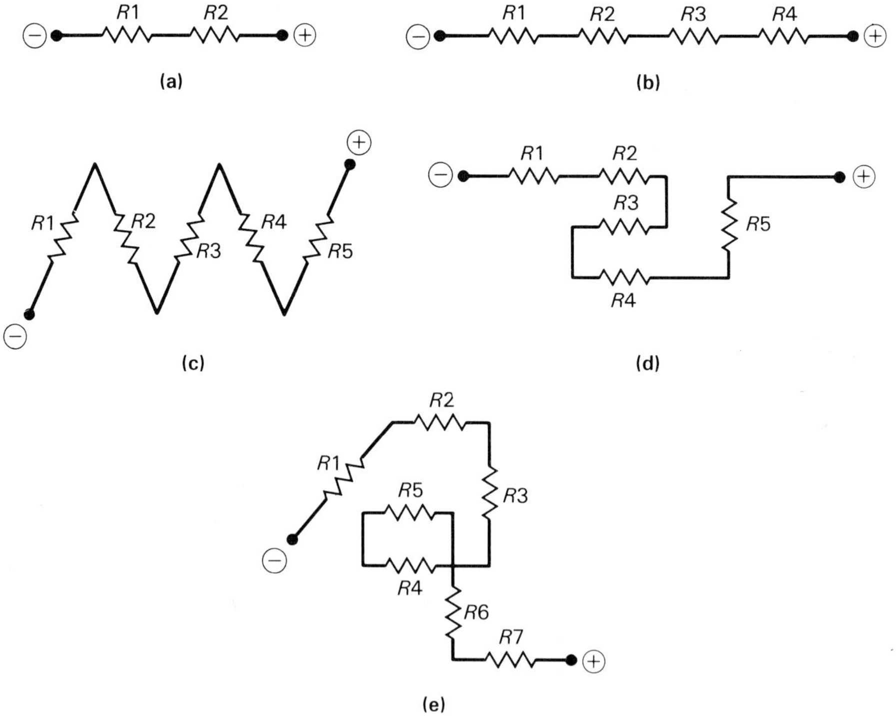

SOLUTION

In Figure 7-2(b), you can see that all the resistors are now connected in series, and the current has only one path to follow from negative to positive.

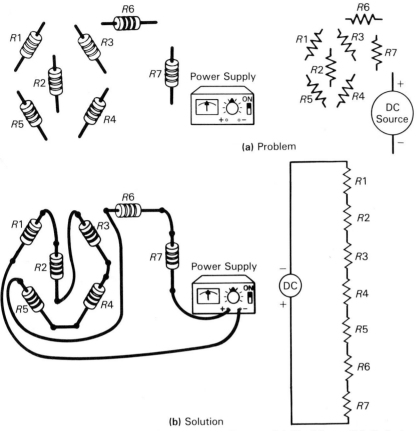

(a) Problem

(b) Solution

FIGURE 7-2   Connecting Resistors in Series. (a) Problem. (b) Solution.

7.1
CURRENT IN A SERIES CIRCUIT

The current in a series circuit has only one path to follow and cannot divert in any other direction; consequently, the current through a series circuit is the same throughout that circuit.

Returning once again to the water analogy, you can see in Figure 3(a) that, if two gallons of water per second is being supplied by the pump, then two gallons per second must be pulled into the pump; and if the rate at which water is leaving and arriving at the pump is the same, then two gallons of water per second must be flowing throughout the circuit. The same value of water flow exists throughout a series-connected fluid system. If the values were adjusted to double the opposition to flow, then half, or

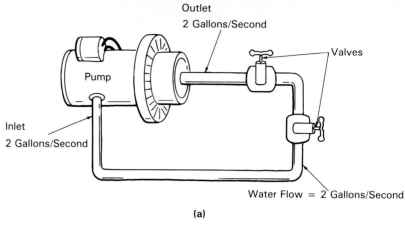

Outlet
2 Gallons/Second

Valves

Pump

Inlet
2 Gallons/Second

Water Flow = 2 Gallons/Second

(a)

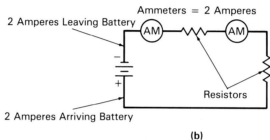

Ammeters = 2 Amperes

2 Amperes Leaving Battery

AM — AM

Resistors

2 Amperes Arriving Battery

(b)

FIGURE 7-3   **Series Circuit Current. (a) Fluid System. (b) Electric System.**

one gallon per second, would be leaving the pump; but that same value of 1 gallon/second would be flowing throughout the system.

Similarly, with the electronic series circuit, seen in Figure 7-3(b), there is a total of 2 amperes leaving and 2 amperes arriving at the battery, so the same value of current exists throughout the series-connected electronic circuit. If the resistance of the circuit is doubled, then half, or 1 ampere of current, will leave the battery, but that same value of 1 ampere will flow throughout the entire circuit. Current is, therefore, exactly the same value at every point in a series circuit. This can be mathematically stated as

$$I_T = I_1 = I_2 = I_3 = \cdots$$

(Total current = current through R_1 = current through R_2 = current through R_3, and so on).

EXAMPLE 7.2

In Figure 7-4, a total current (I_T) of 1 ampere is flowing out of a battery and through two resistors R_1 and R_2. Calculate:
a. The current through R_1 (I_1).
b. The current through R_2 (I_2).

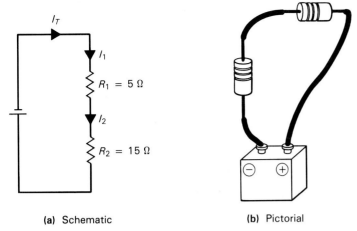

(a) Schematic   (b) Pictorial

FIGURE 7-4   **(a) Schematic. (b) Pictorial.**

SOLUTION

Since R_1 and R_2 are connected in series, the current through both will be the same as the circuit current, which is equal to 1 ampere.

SELF-TEST REVIEW QUESTIONS (§ 7.1)

1. What is a series circuit?
2. What is the current flow through each of eight 8-Ω resistors if 8A total current is flowing out of a battery?

7.2
RESISTANCE IN A SERIES CIRCUIT

Resistance is the opposition to current flow, and in a series circuit every resistor in series offers opposition to the current flow. In the water analogy, in Figure 7-3, the total resistance or opposition to water flow is the sum of the two individual valve opposition values. The pump, like the battery, senses the total opposition in the circuit offered by all the valves or resistors, and the amount of current that flows is dependent on this resistance or opposition.

The total resistance, therefore, in a series-connected electronic resistive circuit is equal to the sum of all the individual resistances, as seen in Figure 7-5(a) through (d). An equivalent circuit could therefore be drawn for each

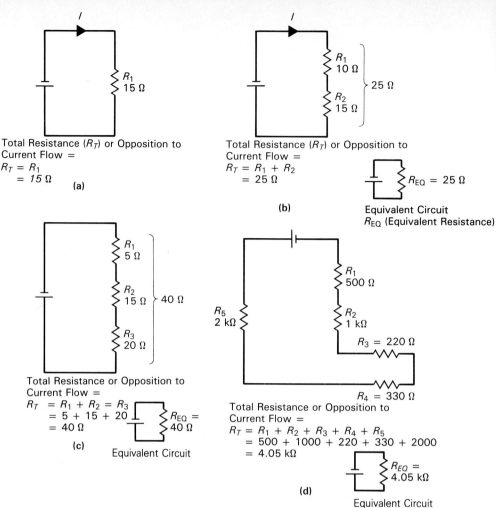

Total Resistance (R_T) or Opposition to
Current Flow =
$R_T = R_1$
    $= 15\ \Omega$

(a)

Total Resistance (R_T) or Opposition to
Current Flow =
$R_T = R_1 + R_2$
    $= 25\ \Omega$

(b)

$R_{EQ} = 25\ \Omega$

Equivalent Circuit
R_{EQ} (Equivalent Resistance)

Total Resistance or Opposition to
Current Flow =
$R_T = R_1 + R_2 = R_3$
    $= 5 + 15 + 20$
    $= 40\ \Omega$

(c)

$R_{EQ} = 40\ \Omega$

Equivalent Circuit

Total Resistance or Opposition to
Current Flow =
$R_T = R_1 + R_2 + R_3 + R_4 + R_5$
    $= 500 + 1000 + 220 + 330 + 2000$
    $= 4.05\ k\Omega$

(d)

$R_{EQ} = 4.05\ k\Omega$

Equivalent Circuit

FIGURE 7-5 Total Resistance.

of the circuits in Figures 7-5(b), (c), and (d) with one resistor, of a value
equal to the sum of all the series resistance values.

No matter how many resistors are connected in series, the total resistance
or opposition to current flow is always equal to the sum of all the resistor
values. This formula can be mathematically stated as:

$$R_T = R_1 + R_2 + R_3 + \cdots$$

(Total resistance = value of R_1 + value of R_2 + value of R_3, and so on.)

Total resistance (R_T) is the only opposition the battery can sense; it does
not see the individual separate resistors, but one equivalent resistance. Based
on its voltage and this total resistance, a value of current will be produced
to flow through the circuit (Ohm's law, $I = V/R$).

SELF-TEST REVIEW QUESTIONS (§ 7.2)

1. State the total resistance formula for a series circuit.
2. Calculate R_T if $R_1 = 2\ k\Omega$, $R_2 = 3\ k\Omega$, and $R_3 = 4700\ \Omega$.

EXAMPLE 7.3

Referring to Figure 7-6, calculate:
a. The circuit's total resistance.
b. The current flowing through R_2.

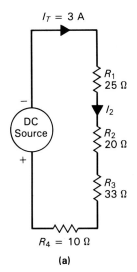

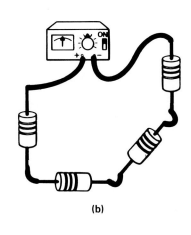

(b)

(a)

FIGURE 7-6 (a) Schematic. (b) Pictorial.

SOLUTION

a. $R_T = R_1 + R_2 + R_3 + R_4$
$= 25\ \Omega + 20\ \Omega + 33\ \Omega + 10\ \Omega$
$= 88\ \Omega$

b. $I_T = I_1 = I_2 = I_3 = I_4$. Therefore, $I_2 = I_T = 3$ amperes.

7.3
VOLTAGE IN A SERIES CIRCUIT

A potential difference or voltage drop will occur across each resistor in a series circuit when current is flowing. The amount of voltage drop is dependent on the value of the resistor and the amount of current flow.

Let's examine this idea of potential difference or voltage drop by returning, once more, to the water analogy. In Figure 7-7(a), you can see that the high pressure from the pump's outlet is present on the left side of the valve; however, on the right side of the valve the high pressure is no longer present. The high potential that exists on the left of the value is not present

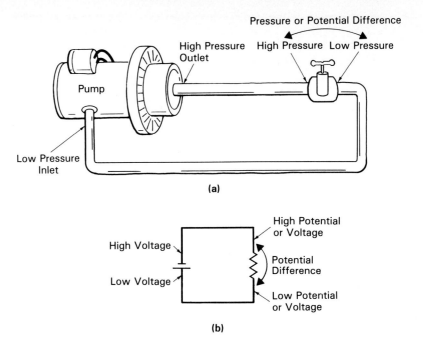

(a)

(b)

FIGURE 7-7 **Series Circuit Voltage. (a) Fluid Analogy of Potentional Difference. (b) Electric Potential Difference.**

on the right, so a potential or pressure difference is said to exist across the valve.

Similarly, with the electronic circuit in Figure 7-7(b), the battery produces a high voltage or potential that is present at the top of the resistor; however, the high voltage that exists at the top of the resistor is not present at the bottom. Therefore, a potential or voltage drop is said to occur across the resistor.

The voltage drop across resistors can be found by utilizing Ohm's law: $V = I \times R$.

EXAMPLE 7.4

Referring to Figure 7-8, calculate:
a. Total resistance (R_T)
b. Amount of series current flowing throughout the circuit (I_T)

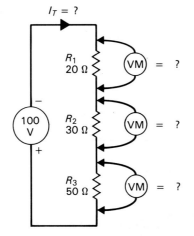

216

FIGURE 7-8

c. Voltage drop across R_1
d. Voltage drop across R_2
e. Voltage drop across R_3

SOLUTION

a. Total resistance $(R_T) = R_1 + R_2 + R_3$
$$= 20\ \Omega + 30\ \Omega + 50\ \Omega$$
$$= 100\ \Omega$$

b. Total current $(I_T) = \dfrac{V_S}{R_T}$
$$= \dfrac{100\ \text{V}}{100\ \Omega}$$
$$= 1\ \text{A}$$

The same current will flow through the complete series circuit, so the current through R_1 will equal 1 A, the current through R_2 will equal 1 A, and the current through R_3 will equal 1 A.

c. Voltage across R_1 $(V_{R1}) = I_1 \times R_1$
$$= 1\ \text{A} \times 20\ \Omega$$
$$= 20\ \text{V}$$

d. Voltage across R_2 $(V_{R2}) = I_2 \times R_2$
$$= 1\ \text{A} \times 30\ \Omega$$
$$= 30\ \text{V}$$

e. Voltage across R_3 $(V_{R3}) = I_3 \times R_3$
$$= 1\ \text{A} \times 50\ \Omega$$
$$= 50\ \text{V}$$

Figure 7-9 shows the result of this example and, as you can see, the 20-Ω resistor drops 20 V, the 30-Ω resistor has 30 V across it, and the 50-Ω resistor has dropped 50 V.

From this example, you will notice that the larger the resistor value, the larger the voltage drop. Resistance and voltage drops are consequently proportional to one another.

Another interesting point you may have noticed from Figure 7-9 is that, if you were to add up all the voltage drops around a series circuit, they would equal the source (V_S) applied. In the example, this is true, since

$$\text{Total voltage applied } (V_S \text{ or } V_T) = V_{R1} + V_{R2} + V_{R3}$$
$$100\ \text{V} = 20\ \text{V} + 30\ \text{V} + 50\ \text{V}$$
$$100\ \text{V} = 100\ \text{V}$$

The series circuit has in fact divided up the applied voltage, and it appears proportionally across all the individual resistors. This characteristic

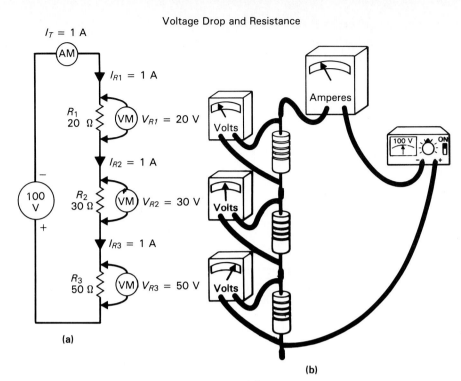

FIGURE 7-9 (a) Schematic. (b) Pictorial.

was first observed by Gustav Kirchhoff in 1847; in honor of his discovery, this effect is known as Kirchhoff's voltage law, which states: The sum of the voltage drops in a series circuit is equal to the total voltage applied.

To summarize the effects of current, resistance, and voltage in a series circuit so far, we can say that:

1. The current in a series circuit has only one path to follow.
2. The value of current in a series circuit is the same throughout the entire circuit.
3. The total resistance in a series circuit is equal to the sum of all the resistances.
4. Resistance and voltage drops in a series circuit are proportional to one another, so a large resistance will have a large voltage drop and a small resistor will have a small voltage drop.
5. The sum of the voltage drops in a series circuit is equal to the total voltage applied.

EXAMPLE 7.5

Calculate the voltage drop across the resistor R_1 in the circuit in Figure 7-10(a) for a resistance of 4 Ω and 2 Ω.

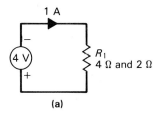

(a)

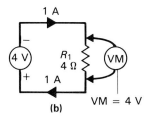

VM = 4 V
(b)

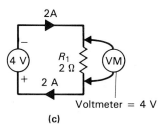

Voltmeter = 4 V
(c)

FIGURE 7-10

SOLUTION

The voltage across R_1 can be calculated by using Ohm's law [Figure 7-10(b)].

$$V_{R1} = I_1 \times R_1$$
$$= 1 \text{ A} \times 4 \text{ }\Omega$$
$$= 4 \text{ V}$$

If the resistance is changed to 2 Ω, the current flow within the circuit would be equal to [Figure 7-10(c)].

$$I = \frac{V_S}{R} - \frac{4 \text{ V}}{2 \text{ }\Omega} = 2 \text{ A}$$

The voltage dropped across the resistor, however, would still be equal to

$$V_{R1} = I_1 \times R_1$$
$$= 2 \text{ A} \times 2 \text{ }\Omega$$
$$= 4 \text{ V}$$

   If only one resistor is connected in a series circuit, then the entire applied voltage appears across this resistor. The amount of current flow is determined by the value of this single resistor and remains the same throughout the circuit.

EXAMPLE 7.6

Referring to Figure 7-11(a), calculate:
a. Total circuit resistance
b. Value of current (I_T)
c. Voltage drop across each resistor
   Then draw the circuit with respective voltages, resistances, and current inserted.

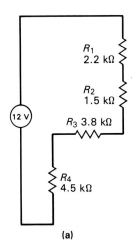
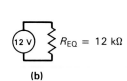

(a)   (b)   **FIGURE 7-11**

SOLUTION

a. $R_T = R_1 + R_2 + R_3 + R_4$
$$= 2.2\ k\Omega + 1.5\ k\Omega + 3.8\ k\Omega + 4.5\ k\Omega$$
$$= (2.2 \times 10^3) + (1.5 \times 10^3) + (3.8 \times 10^3) + (4.5 \times 10^3)$$
$$= 12\ k\Omega \quad \text{[Figure 7-11(b)]}$$

b. $I_T = \dfrac{V_S}{R_T} = \dfrac{12\ V}{12\ k\Omega} = 1\ mA$

c. Voltage drop across each resistor:
$$V_{R1} = I_T \times R_1$$
$$= 1\ mA \times 2.2\ k\Omega$$
$$= 2.2\ V$$
$$V_{R2} = I_T \times R_2$$
$$= 1\ mA \times 1.5\ k\Omega$$
$$= 1.5\ V$$
$$V_{R3} = I_T \times R_3$$
$$= 1\ mA \times 3.8\ k\Omega$$
$$= 3.8\ V$$
$$V_{R4} = I_T \times R_4$$
$$= 1\ mA \times 4.5\ k\Omega$$
$$= 4.5\ V$$

d. See Figure 7-12.

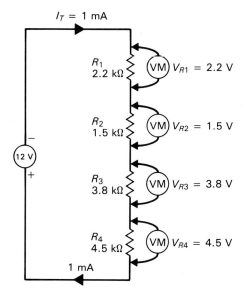

$I_T = 1\ \text{mA}$

R_1
2.2 kΩ
(VM) $V_{R1} = 2.2\ \text{V}$

R_2
1.5 kΩ
(VM) $V_{R2} = 1.5\ \text{V}$

12 V

R_3
3.8 kΩ
(VM) $V_{R3} = 3.8\ \text{V}$

R_4
4.5 kΩ
(VM) $V_{R4} = 4.5\ \text{V}$

1 mA

FIGURE 7-12

7.3.1 FIXED VOLTAGE DIVIDER

A series-connected circuit is often referred to as a voltage-divider circuit. The total voltage applied (V_S) is divided and dropped across the resistors in the series circuit. The amount of voltage dropped across a resistor is proportional to the value of resistance, and so a larger resistance causes a larger voltage drop across that resistor, while a smaller resistance causes a smaller voltage drop.

The voltage dropped across a resistor is normally a factor that needs to be calculated. The voltage-divider formula allows you to calculate the voltage drop across any resistor without having to work out current. This formula is stated as

$$V_X = \left(\frac{R_X}{R_T}\right) \times V_S$$

where V_X = voltage dropped across selected resistor
R_T = total series circuit resistance
R_x = selected resistor's value
V_S = source or applied voltage

Figure 7-13 illustrates a circuit from a previous problem. Normally, it would be necessary to:

1. Calculate the total resistance by adding up all the resistance values.

2. Once we have the total resistance and source voltage (V_S), we could then calculate current.

3. Having calculated the current flowing through each resistor, then we use the current value to calculate the voltage dropped across any one of the four resistors by merely multiplying current by the individual resistance value.

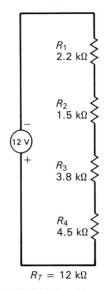

$R_T = 12 \text{ k}\Omega$

FIGURE 7-13

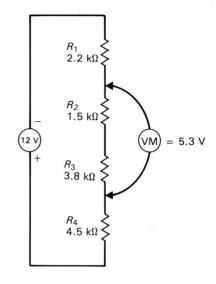

FIGURE 7-14

The voltage-divider formula allows us to bypass the last two steps. If we know total resistance, supply voltage, and the resistance value, we can calculate the voltage drop across the resistor without having to calculate steps 2 and 3. For example, what would be the voltage dropped across R_2 and R_4 in Figure 7-13? The voltage dropped across R_2 is

$$V_{R2} = \frac{R_2}{R_T} \times V_s$$

$$= \frac{1.5 \text{ k}\Omega}{12 \text{ k}\Omega} \times 12 \text{ V}$$

$$= 1.5 \text{ V}$$

The voltage dropped across R_4 is

$$V_{R4} = \frac{R_4}{R_T} \times V_s$$

$$= \frac{4.5 \text{ k}\Omega}{12 \text{ k}\Omega} \times 12 \text{ V}$$

$$= 4.5 \text{ V}$$

Calculator Sequence

Step	Keypad Entry	Display Response
1.	1 . 5 E 3	1.5E3
2.	÷	
3.	1 2 E 3	12E3
4.	×	0.125
5.	1 2	12
6.	=	1.5

The voltage-divider formula could also be used to find the voltage drop across multiple resistors. For example, what would be the voltage dropped across R_2 and R_3? (This is illustrated in Figure 7-14.) The voltage across $R_2 + R_3$ is

$$V_{R2} \text{ and } V_{R3} = \frac{R_2 + R_3}{R_T} \times V_s$$

$$= \frac{5.3 \text{ k}\Omega}{12 \text{ k}\Omega} \times 12 \text{ V}$$

$$= 5.3 \text{ V}$$

222   *Series DC Circuits*

EXAMPLE 7.7

Calculate the voltage drop across (see Figure 7-15):
a. R_1, R_2, and R_3 separately
b. R_2 and R_3 together
c. R_1, R_2, and R_3 together

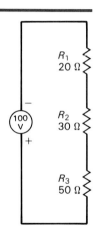

FIGURE 7-15

SOLUTION

a. The voltage drop across a resistor is proportional to the resistance value. The total resistance (R_T) in this circuit is 100 Ω. R_1 is 20% of the total resistance, and so 20% of the source voltage will appear across R_1. R_2 is 30% of the total resistance, and so 30% of the source voltage will appear across R_2. R_3 is 50% of the total resistance, and so 50% of the source voltage will appear across R_3. This was a very simple problem in which the figures worked out very neatly. The voltage-divider formula achieves the very same thing by calculating the ratio of the resistance value to the total resistance. This percentage is then multiplied by the source voltage in order to find the desired resistor's voltage drop:

$$V_{R1} = \frac{R_1}{R_T} \times V_s$$
$$= \frac{20\ \Omega}{100\ \Omega} \times V_S$$
$$= (0.2) \times 100\ V$$
$$= 20\ V$$

$$V_{R2} = \frac{R_2}{R_T} \times V_s$$
$$= \frac{30\ \Omega}{100\ \Omega} \times V_S$$
$$= (0.3) \times 100\ V$$
$$= 30\ V$$

$$V_{R3} = \frac{R_3}{R_T} \times V_s$$
$$= \frac{50\ \Omega}{100\ \Omega} \times V_S$$
$$= (0.5) \times 100\ V$$
$$= 50\ V$$

b. Voltage dropped across R_2 and R_3 = 30 + 50 = 80 V.
c. Voltage dropped across R_1, R_2, and R_3 = 20 + 30 + 50 = 100 V.

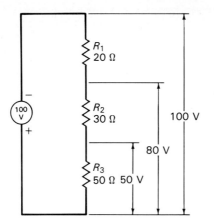

FIGURE 7-16

The voltage-divider formula can be summarized by saying that: The voltage drop across a resistor or multiple resistors in a series circuit is equal to the ratio of that resistance (R_X) to the total resistance (R_T) multiplied by the source voltage (V_S).

If three voltages of 50, 80, and 100 volts were required by a piece of equipment, we could use three individual power sources, which would be very expensive, or use one 100-V voltage source connected across the resistors from the previous problem to divide up the 100 V and supply the three required fixed voltages, as seen in Figure 7-16.

7.3.2 VARIABLE VOLTAGE DIVIDER

When discussing variable resistors in Chapter 4, we talked about a potentiometer, or variable voltage divider, which consists of a fixed value of resistance between two terminals and a wiper that can be adjusted to vary resistance between its terminal and one of the other two. Figure 7-17(a) and (b) illustrates the potentiometer's schematic symbol and physical appearance.

If the wiper is moved down, as seen in Figure 7-18, the resistance between terminals A and B will increase, while the resistance between B and C will decrease. If the wiper is moved up, as seen in Figure 7-19, the resistance between terminals A and B will decrease, while the resistance between B and C will increase. The resistances between A and B and B and C are inversely proportional to one another in that if one were to increase

FIGURE 7-17   **Potentiometer. (a) Schematic Symbol. (b) Physical Appearance.**

Series DC Circuits

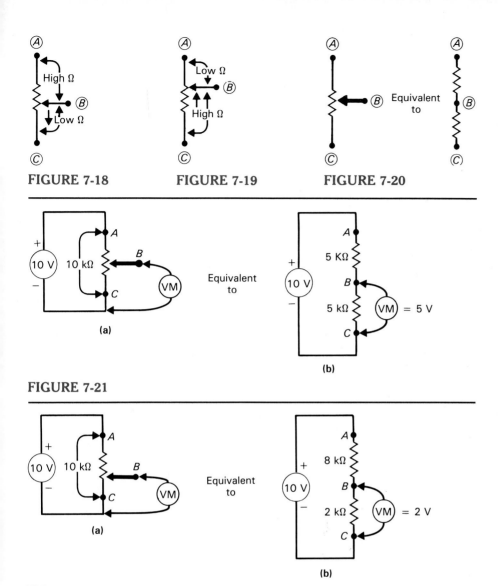

FIGURE 7-18

FIGURE 7-19

FIGURE 7-20

FIGURE 7-21

FIGURE 7-22

the other would decrease, and vice versa, and can be thought of as two separate resistors, as seen in Figure 7-20.

Figure 7-21(a) illustrates a 10-kΩ potentiometer that has been hooked up across a 10-V dc source with a voltmeter between terminals *B* and *C*. If the wiper terminals are positioned in the mid position, halfway between *A* and *C*, the potentiometer will be equivalent to two 5-kΩ resistors in series, as shown in Figure 7-21(b). Kirchhoff's voltage law states that the entire source voltage will be dropped across the resistances in the circuit, and since the resistance values are equal, each will drop half of the source voltage, that is, 5 V, so the voltmeter will indicate 5 V.

In Figure 7-22(a), the wiper has been moved down so the resistance between *A* and *B* is equal to 8 kΩ, and the resistance between *B* and *C* equals 2 kΩ. The amount of voltage drop is proportional to the resistance, and so a larger voltage will be dropped across the larger resistance. Using

Voltage in a Series Circuit **225**

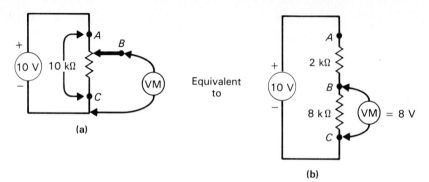

Equivalent to = 8 V

FIGURE 7-23

the voltage-divider formula, you can calculate that the 8 kΩ is 80% of the total resistance and, therefore, will drop 80% of the voltage:

$$V_{A-B} = \frac{R_{A-B}}{R_{\text{total}}} \times V_S = \frac{8\ \text{k}\Omega}{10\ \text{k}\Omega} \times 10\ \text{V} = 8\ \text{V}$$

The 2-kΩ resistance between B and C is 20% of the total resistance and consequently will drop 20% of the total voltage:

$$V_{B-C} = \frac{R_{B-C}}{R_{\text{total}}} \times V_S = \frac{2\ \text{k}\Omega}{10\ \text{k}\Omega} \times 10\ \text{V} = 2\ \text{V}$$

which will be indicated on the voltmeter.

In Figure 7-23(a), the wiper has been moved up and now 2 kΩ exists between A and B, and 8 kΩ is present between B and C. In this situation, 2 V will be dropped across the 2 kΩ between A and B, and 8 V will be dropped across the 8 kΩ between B and C.

From this discussion, you can see that the potentiometer can be adjusted to supply different voltages on the wiper. This voltage can be decreased by moving the wiper down to supply a minimum of 0 V [Figure 7-24(a)], or the wiper can be moved up to supply a maximum of 10 V [Figure 7-24(b)], or it can be placed anywhere in between, which is why the potentiometer is known as a variable voltage divider.

FIGURE 7-24

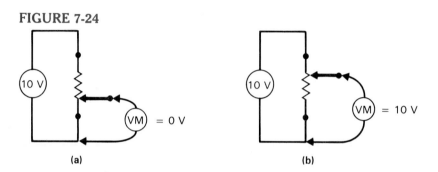

1. A series circuit is also known as a voltage-divider circuit (true or false).
2. The voltage drop across a series resistor is proportional to the value of the resistor (true or false).
3. If 6-Ω and 12-Ω resistors are connected across a dc 18-V supply, calculate I_T and the voltage drop across each.
4. State the voltage-divider formula.
5. Which component can be used as a variable voltage divider?
6. Could a rheostat be used as a variable voltage divider?

7.4
POWER IN A SERIES CIRCUIT

Power is the rate at which work is done. Work is said to have been done when energy is converted, in this case from electrical energy to heat energy. Resistors dissipate heat energy, and the rate at which they dissipate energy is called *power* and is measured in *watts* (joules per second). Resistors all have a resistive value, a tolerance, and a wattage rating. The wattage of a resistor is the amount of heat energy a resistor will dissipate per second, and this wattage is directly proportional to the resistor's size; a larger-sized resistor will be able to dissipate more heat than a smaller-sized resistor. Resistors are manufactured in several different physical sizes, and if, for example, it is calculated that for a certain value of current and voltage a 5-watt resistor is needed, and a $\frac{1}{2}$-watt resistor is put in its place, the $\frac{1}{2}$-watt resistor will burn out, because it is generating heat (5 watts) faster than it can dissipate heat ($\frac{1}{2}$ watt). A 10- or 25-watt resistor or greater could be used to replace a 5 watt; however, anything less than a 5-watt resistor will burn out.

Figure 7-25 illustrates the typical manufactured wattage resistors drawn to scale. You may remember that the formula for calculating the wattage rating of a resistor is

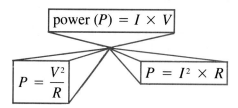

$$\text{power } (P) = I \times V$$

$$P = \frac{V^2}{R}$$

$$P = I^2 \times R$$

It is, therefore, necessary that we have some way of calculating which wattage rating is necessary for each specific application.

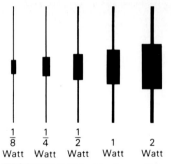

$$\frac{1}{8}$$ $$\frac{1}{4}$$ $$\frac{1}{2}$$ 1   2
Watt Watt Watt Watt Watt

FIGURE 7-25 Resistor Wattage Ratings.

A question you may be asking is why not just use large-wattage resistors everywhere. The two disadvantages with this are:

1. The larger the wattage, the greater the cost.
2. The larger the wattage, the greater the size and area the resistor occupies within the equipment.

EXAMPLE 7.8

Figure 7-26 illustrates a 20-V battery driving a 12-V/1-A television. R_1 is in series with the television and is being used to drop 8 V of the 20-V supply, so 12 V will be applied across the television.

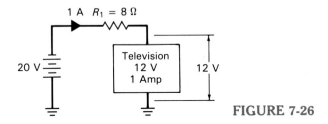

FIGURE 7-26

a. What is the wattage rating for R_1?
b. What is the series load resistance of the television?
c. What is the amount of power being consumed by the television?

SOLUTION

a. Everything is known about R_1. Its resistance is 8 Ω, it has 1 A of current flowing through it, and 8 V is being dropped across it. Consequently, any one of the three power formulas can be used to calculate the wattage rating of R_1.

$$\text{power } (P) = I \times V = 1 \text{ A} \times 8 \text{ V} = 8 \text{ W}$$

or

$$P = I^2 \times R = 1^2 \times 8 = 8 \text{ W}$$

or

$$P = \frac{V^2}{R} = \frac{8^2}{8} = 8 \text{ W}$$

The nearest commercially available wattage resistor would be a 10 watt, which would be used and could safely handle the heat.

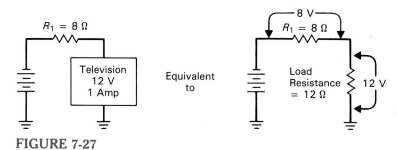

FIGURE 7-27

b. You may recall that any piece of equipment is equivalent to a load resistance. The television has 12 V across it and is pulling 1 A of current; its load resistance can be calculated by simply using Ohm's law and deriving an equivalent circuit, as seen in Figure 7-27.

$$R_L \text{ (load resistance)} = \frac{V}{I}$$
$$= \frac{12 \text{ V}}{1 \text{ A}}$$
$$= 12 \text{ } \Omega$$

c. The amount of power being consumed by the television is:
$$P = V \times I = 12 \text{ V} \times 1\text{A} = 12 \text{ W}$$

SELF-TEST REVIEW QUESTIONS (§ 7.4)

1. State the power formula.
2. Calculate the power dissipated by a 12-Ω resistor connected across a 12-V supply.
3. What fixed resistor type should probably be used for question 2, and what would be a safe wattage rating?
4. What would be the total power dissipated if R_1 dissipates 25 W and R_2 dissipates 3800 mW?

EXAMPLE 7.9

Calculate the total amount of power dissipated in the series circuit in Figure 7-28.

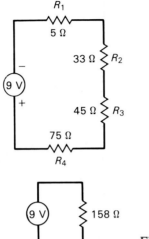

FIGURE 7-28

SOLUTION

The total power dissipated in a series circuit is equal to the sum of all the power dissipated by all the resistors. The easiest way to calculate the total power is to simplify the circuit to one resistance.

$$R_t = R_1 + R_2 + R_3 + R_4$$
$$= 5\ \Omega + 33\ \Omega + 45\ \Omega + 75\ \Omega$$
$$= 158\ \Omega$$

We now have total resistance and total voltage, so we can calculate the total power:

$$P_t = \frac{V_s^2}{R_t} = \frac{9\ V^2}{158\ \Omega} = \frac{81}{158} = 512.7 \text{ milliwatts (mW)}$$

The longer method would have been to calculate the current through the series current:

$$I = \frac{V_s}{R_t} = \frac{9\ V}{158\ \Omega} = 56.96 \text{ mA} \quad \text{or} \quad 57 \text{ mA}$$

We could then calculate the power dissipated by each separate resistor and add up all the individual values to gain a total power figure. This is illustrated in Figure 7-29.

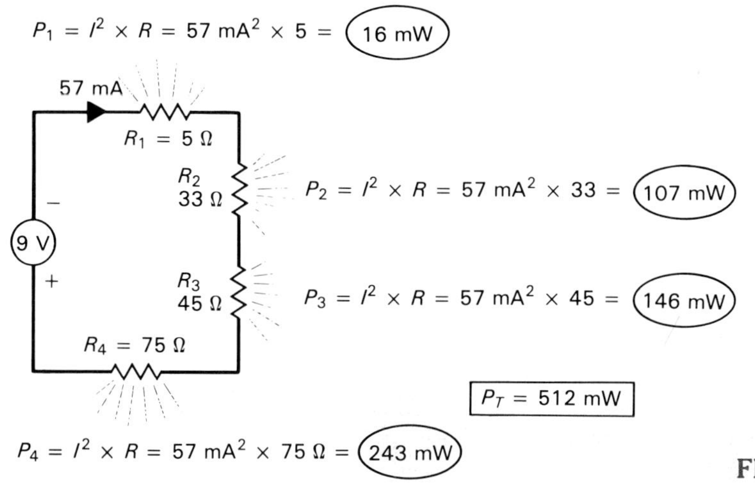

$P_1 = I^2 \times R = 57 \text{ mA}^2 \times 5 = \boxed{16 \text{ mW}}$

$P_2 = I^2 \times R = 57 \text{ mA}^2 \times 33 = \boxed{107 \text{ mW}}$

$P_3 = I^2 \times R = 57 \text{ mA}^2 \times 45 = \boxed{146 \text{ mW}}$

$P_T = 512 \text{ mW}$

$P_4 = I^2 \times R = 57 \text{ mA}^2 \times 75 \, \Omega = \boxed{243 \text{ mW}}$

FIGURE 7-29

total power = addition of all the individual power losses

$$P_t = P_1 + P_2 + P_3 + P_4$$
$$= 16 \text{ mW} + 107 \text{ mW} + 146 \text{ mW} + 243 \text{ mW}$$
$$= 512 \text{ mW}$$

7.5
TROUBLESHOOTING A SERIES CIRCUIT

A resistor will usually burn out and result in an open between its two leads when an excessive current flow occurs. This can normally, but not always, be noticed by a visual check of the resistor, which will appear charred due to the excessive heat. In some cases, you will need to use your multimeter (combined ammeter, voltmeter, and ohmeter) to check the components in the circuit to determine where a problem exists.

The voltmeter is the most useful tool when checking series circuits as it can be used to measure voltage drops by connecting the meter leads across the component or resistors. Let's now analyze a problem and see if we can solve it by logically troubleshooting the circuit and isolating the faulty component.

The two basic problems that normally exist in a series circuit are opens and shorts; however, a problem is not always as drastic as a short or an open, but may be a variation in a component's value over a long period of time, which will eventually cause a problem.

To summarize, then, we can say that one of three problems can occur to components in a series circuit:

a. A component will open (infinite resistance).

b. A component's value will change over a period of time.

c. A component will short (zero resistance).

To begin with, let us take a look at the first of the three to see the effects of an open component.

7.5.1 OPEN COMPONENT IN A SERIES CIRCUIT

A component is open when its resistance is the maximum possible (infinity).

EXAMPLE 7.10

Figure 7-30(a) illustrates a television with a load resistance of 3 Ω. The television is off because R_2 has burnt out and become an open circuit. How would you determine that the problem is R_2?

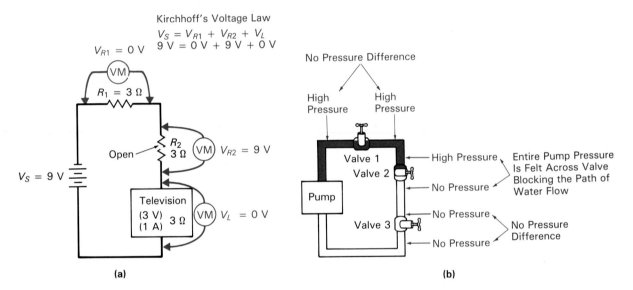

FIGURE 7-30 **Troubleshooting an Open in a Series Circuit.**

SOLUTION

If an open circuit ever occurs in a series circuit, due in this case to R_2 having burnt out, there can be no current flow, because series circuits have only one path for current to flow and that path has been broken ($I = 0$ amps). Using the voltmeter to check the amount of voltage drop across each resistor, two results will be obtained:

1. The voltage drop across a good resistor will be zero volts.
2. The voltage drop across an open resistor will be equal to the source voltage V_S.

No voltage will be dropped across a good resistor because current is zero, and if $I = 0$, then the voltage drop, which is the product of I and R, must be zero ($V = I \times R = 0 \times R = 0$ V). If no voltage is being dropped across the good resistor R_1 and the television resistance of 3 Ω, then the entire source voltage will appear across the open resistor, R_2, in order that this series circuit comply with Kirchhoff's voltage law: $V_S(9 \text{ V}) = V_{R1}(0 \text{ V}) + V_{R2}(9 \text{ V}) + V_L(0 \text{ V})$.

To further emphasize this point, refer to Figure 7-30(b), which illustrates the fluid analogy. Valve 2, like R_2, has completely blocked any form of flow (water flow = 0), and looking at the pressure differences across all three valves, you can see that no pressure difference occurs across valves 1 and 3, but the entire pump pressure is appearing across valve 2, the component that has opened the circuit.

EXAMPLE 7.11

Figure 7-31 illustrates a set of three lights connected across a 9-V battery. Bulb 3 is open and therefore there is no current flow; with no current flow, all three bulbs are off, and we need to isolate which of the three is faulty.

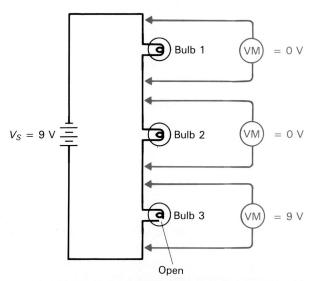

FIGURE 7-31   Troubleshooting an Open Bulb in a Series Circuit.

SOLUTION

Using the voltmeter, there will be:

1. Zero volts dropped across bulb 1; 9 V appears on both sides of bulb 1, so the potential difference or voltage drop across bulb 1 is zero. Bulb 1 is ok.
2. Zero volts dropped across bulb 2; bulb 2 is ok.
3. Nine volts dropped across bulb 3; the entire source voltage is being dropped across bulb 3. Bulb 3 is open and needs to be replaced.

7.5.2 COMPONENT VARIATION IN A SERIES CIRCUIT

Resistors will rarely go completely open, unless severely stressed because of excessive current flow. With age, resistors will normally change their resistance value. This occurs slowly and will generally always cause a decrease in the resistor's resistance and eventually cause a circuit problem. This lowering of resistance will cause an increase of current, which will cause an increase in the power dissipated; if the wattage of the resistor is exceeded, it can burn out. If they do not burn out but merely blow the circuit fuse due to this increase in current, the problem can be found by measuring the resistance values of each resistor or by measuring how much voltage is dropped across each resistor and comparing these to the calculated voltage, based on the parts list supplied by the manufacturer.

7.5.3 SHORTED COMPONENT IN A SERIES CIRCUIT

A component has gone short when its resistance is $0 \, \Omega$. Let's work out a problem using the same example of the three bulbs across the 9-V battery, as seen in Figure 7-32.

To summarize opens and shorts in series circuits, we can say that:

1. The supply voltage appears across an open component.
2. Zero volts appears across a shorted component.

EXAMPLE 7.12

Bulbs 1 and 3 are on and bulb 2 is off. A piece of wire exists between the two terminals of bulb 2, and the current is taking the lowest resistance path through the wire rather than through the filament resistor of the bulb. Bulb 2, therefore, has no current flow through it and light cannot be generated without current. How would you determine what the problem is?

SOLUTION

If bulb 2 is open (burnt-out), there cannot be any current flow in the circuit, and so bulbs 1 and 2 would be off. Bulb 2 must, therefore, be shorted. Using the voltmeter to further investigate this, you find that bulb 2 drops 0 V across it because it has no resistance except the very small wire resistance of the bypass, which means that 4.5 V must be dropped across each working bulb (1 and 3) in order to comply with Kirchhoff's voltage law. The loss of bulb 2's resistance causes the overall circuit resistance offered by bulbs 1 and 3 to decrease, which will cause an increase in current; so bulbs 1 and 3 should glow more brightly.

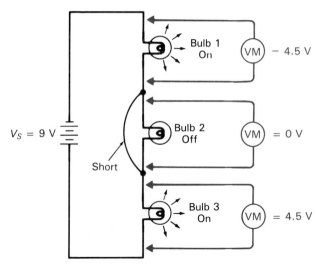

FIGURE 7-32 Troubleshooting a Short in a Series Circuit.

1. List the three basic problems that can occur with components in a series circuit.
2. How can an open component be detected in a series circuit?
3. If a series-connected resistor's value were to decrease, the voltage drop across that same resistor would increase (true or false).
4. How could a shorted component be detected in a series circuit?

SUMMARY

1. A series circuit is the connecting of components end to end in a circuit to provide only one path for current flow.
2. The current throughout a series circuit is always the same value.
3. Resistance is the opposition to current flow, and in a series resistive circuit each resistor offers opposition.
4. The total resistance in a series circuit is equal to the sum of all the individual resistances.
5. Total resistance is the only opposition the battery can sense, and based on the voltage applied and the total circuit resistance, a value of current will be produced to flow throughout the circuit.
6. A potential difference or voltage drop will occur across every resistor in a series circuit when current is flowing.
7. The amount of voltage dropped across a resistor can be calculated by Ohm's law; and since current remains the same throughout a series

circuit, the amount of voltage dropped across a resistor is proportional to its resistance ($V\updownarrow = I \times R\updownarrow$).

8. Kirchhoff's voltage law states that: The sum of the voltage drops around a series circuit is equal to the source voltage applied. A series circuit consequently is known as a voltage divider as it divides up the total source voltage among the series-connected resistors.

9. Fixed-value resistors provide a fixed voltage division, whereas the potentiometer can be used as a variable voltage divider.

10. The total power dissipated by a series circuit is equal to the sum of all the calculated individual resistor power figures.

11. A resistor will normally burn out and open the circuit in a problem situation.

12. Resistors can also, with age, change value or be accidentally shorted or bypassed.

13. The supply voltage always appears across an open component.

14. Zero volts always occurs across a shorted component.

NEW TERMS

Fixed voltage dividers

Kirchhoff's voltage law

Series circuit

Variable voltage divider

Voltage divider

NEW FORMULAS

$$\boxed{I = I_1 = I_2 = I_3 = \cdots}$$

$$\boxed{R_T = R_1 + R_2 + R_3 + \cdots}$$

$$I = \frac{V_S}{R_T}$$

$$V_{R1} = I_1 \times R_1,$$
$$V_{R2} = I_2 \times R_2,$$
$$V_{R3} = I_3 \times R_3, \text{ and so on.}$$

Voltage divider formula

$$\boxed{V_x = \left(\frac{R_x}{R_T}\right) \times V_S}$$

$$\boxed{P_T = P_1 + P_2 + P_3 + \cdots}$$

$$P_T = \frac{V_S^2}{R_T}$$

$$P_T = I^2 \times R_T$$
$$P_T = V_S \times I$$

I = series current in amperes
I_1 = current through R_1 in amperes
I_2 = current through R_2 in amperes
I_3 = current through R_3 in amperes
R_T = total circuit resistance in ohms
R_1 = resistance of R_1 in ohms
R_2 = resistance of R_2 in ohms
R_3 = resistance of R_3 in ohms
V_S = source voltage in volts
V_{R1} = voltage drop across R_1 in volts
V_{R2} = voltage drop across R_2 in volts
V_{R3} = voltage drop across R_3 in volts
V_x = voltage drop across selected resistor(s)
R_x = selected resistor(s) ohmic value
P_T = total power dissipated in watts
P_1 = power dissipated by R_1 in watts
P_2 = power dissipated by R_2 in watts
P_3 = power dissipated by R_3 in watts

Multiple Choice Questions _____

1. A series circuit:
   a. Is the connecting of components end to end
   b. Provides a single path for current
   c. Functions as a voltage divider
   d. All of the above are true

2. The total current in a series circuit is equal to:
   a. $I_1 + I_2 + I_3 + \cdots$
   b. $I_1 - I_2$
   c. $I_1 = I_2 = I_3 = \cdots$
   d. All of the above are true

3. If R_1 and R_2 are connected in series with a total current of 2 A, what will be the current flowing through R_1 and R_2, respectively?
   a. 1 A, 1 A
   b. 2 A, 1 A
   c. 2 A, 2 A
   d. All of the above could be true on some occasions.

4. The total resistance in a series circuit is equal to:
   a. The total voltage divided by the total current
   b. The sum of all the individual resistor values
   c. $R_1 + R_2 + R_3 + \cdots$
   d. All of the above are true
   e. None of the above are even remotely true

5. Which of Kirchhoff's laws applies to series circuits:
   a. His voltage law
   b. His current law
   c. His power law
   d. None of them apply to series circuits, only parallel

6. The amount of voltage dropped across a resistor is proportional to:
   a. The value of the resistor
   b. The current flow in the circuit
   c. Both (a) and (b) are true
   d. None of the above are true

7. If three resistors of 6 kΩ, 4.7 kΩ, and 330 Ω are connected in series with one another, what total resistance will the battery sense?
   a. 11.03 MΩ
   b. 11.03 Ω
   c. 6 kΩ
   d. 11.03 kΩ

8. The voltage-divider formula states that the voltage drop across a resistor or multiple resistors in a series circuit is equal to the ratio of that __ to the _____ multiplied by the _____.
   a. Resistance, source voltage, total resistance
   b. Resistance, total resistance, source voltage
   c. Total current, resistance, total voltage
   d. Total voltage, total current, resistance

9. The _____ can be used as a variable voltage divider.
   a. Potentiometer
   b. Fixed resistor
   c. SPDT switch
   d. None of the above

10. A larger physical size resistor will be able to dissipate _____ heat than a small resistor.
    a. More
    b. Less
    c. About the same
    d. None of the above

11. The _____ is the most useful tool when checking series circuits.
    a. Ammeter
    b. Wattmeter
    c. Voltmeter
    d. Both (a) and (b) are true

12. When an open component occurs in a series circuit, it can be noticed because:
    a. Zero volts appears across it
    b. The supply voltage appears across it
    c. 1.3 volts appears across it
    d. None of the above are true

13. Power can be calculated by:
    a. The addition of all the individual power figures
    b. The product of the total current and the total voltage
    c. The square of the total voltage divided by the total resistance
    d. All of the above are true

14. A series circuit is known as a:
    a. Current divider
    b. Voltage divider
    c. Current subtractor
    d. All the above

15. In a series circuit only _____ path(s) exists for current flow, while the voltage applied is distributed across all the individual resistors.
    a. Three
    b. Several
    c. Four
    d. One

Essay Questions

16. Define what is meant by a series-connected circuit.   (Introduction)
17. Describe and state mathematically what happens to current flow in a series circuit.   (7.1)
18. Describe how total resistance can be calculated in a series circuit.   (7.2)
19. Describe why voltage is dropped around a series circuit and how each voltage drop can be calculated.   (7.3)

20. Briefly describe why resistance and voltage drops are proportional to one another. (7.3)
21. Describe what is meant by a fixed and a variable voltage divider. (7.3.1) and (7.3.2)
22. How can individual and total power be calculated in a series circuit? (7.4)
23. How can you recognize shorts and opens in a series circuit when troubleshooting with a voltmeter? (7.5)
24. List the three problems that can occur in a series circuit. (7.5)
25. State Kirchhoff's voltage law. (7.3)

Practice Problems _____

26. If three resistors of 1.7 kΩ, 3.3 kΩ, and 14.4 kΩ are connected in series with one another across a 24-V source, calculate:
    a. Total resistance (R_T)
    b. Circuit current
    c. Individual voltage drops
    d. Individual and total power dissipated
27. If 40- and 35-Ω resistors are connected across a 24-V source, what would be the current flow through the resistors, and what resistance would cause half the current to flow?
28. Calculate the total resistance (R_T) of the following series-connected resistors: 2.7 kΩ, 3.4 MΩ, 370 Ω, and 4.6 MΩ.
29. Calculate the value of resistors needed to divide up a 90-V source to produce 45- and 60-V outputs, and produce a divider circuit current of 1 amp.
30. How much voltage will be dropped across R_2 if R_1 = 4.7 kΩ and R_2 = 6.4 kΩ and both are connected across a 9-V source?
31. What current would flow through R_1 if it were one-third the ohmic value of R_2 and R_3, and all were connected in series with a total current of 6.5 mA flowing out of V_S?
32. Draw a circuit showing R_1 = 2.7 kΩ, R_2 = 3.3 kΩ, and R_3 = 0.027 MΩ in series with one another across a 20-V source. Calculate:
a. I	d. P_2	g. V_{R2}
b. P_T	e. P_3	h. V_{R3}
c. P_1	f. V_{R1}	i. I_1
33. Calculate the current flowing through a 120-W, 60-W, and 200-W bulb connected in series across a 120-V source. How is the voltage divided around the series circuit?
34. If three equal-value resistors are connected across a dc power supply adjusted to supply 10 V, what percentage of the source voltage will appear across R_1?
35. Refer to the following figures and calculate:
    a. I (Figure 7-33)
    b. R_T and P_T (Figure 7-34)
    c. V_S, V_{R1}, V_{R2}, V_{R3}, V_{R4}, P_1, P_2, P_3, and P_4 (Figure 7-35)
    d. P_T, I, R_1, R_2, R_3 and R_4. (Figure 7-36)

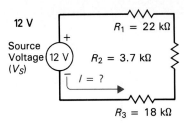

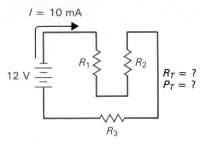

FIGURE 7-33

FIGURE 7-34

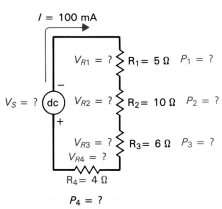

FIGURE 7-35

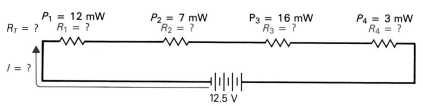

FIGURE 7-36

Troubleshooting

36. If three bulbs are connected across a 9-V battery in series, and the filament in one of the bulbs burned out causing an open in the bulb, would the other lamps be on? Explain why.

37. Using a voltmeter, how would a short be recognized in a series circuit?

38. If one of three series-connected bulbs is shorted, will the other two bulbs be on? Explain why.

39. When one resistor in a series string is open, explain what would happen to the circuit's:
    a. Current
    b. Resistance
    c. Voltage across the open component
    d. Voltage across the other components

40. When one resistor in a series string is shorted, explain what would happen to the circuit's:

    a. Current

    b. Resistance

    c. Voltage across the shorted component

    d. Voltage across the other components

8

PARALLEL DC CIRCUITS

AFTER COMPLETING THIS CHAPTER, YOU WILL BE ABLE TO:

1. Describe the difference between a series and parallel circuit.
2. Be able to recognize and determine whether circuit components are connected in series or parallel.
3. Explain why voltage measures the same across parallel-connected components.
4. State Kirchhoff's current law.
5. Describe why branch current and resistance are inversely proportional to one another.
6. Determine the total resistance of any parallel-connected resistive circuit.
7. Describe and be able to apply all formulas associated with the calculation of voltage, current, resistance, and power in a parallel circuit.
8. Describe how a short, open, or component variation will affect a parallel circuit's operation and how it can be recognized.

INTRODUCTION

Series or parallel can be determined by current. In a series circuit, there is only one path for current, whereas in parallel circuits the current has two or more paths. These paths are known as *branches*. A parallel circuit, by definition, is when two or more components are connected to the same pair of terminals from the voltage source so that the current can branch out over two or more paths. In all cases, two or more resistors are connected to the same pair of terminals from the battery source, so current splits to travel through each separate branch.

Many components, other than resistors, can be connected in parallel, and a parallel circuit can be easily identified because current is split into two or more paths. Being able to identify a parallel connection requires some practice, because they can come in many different shapes and sizes. The means for recognizing series circuits is that, if you can place your pencil at the negative terminal of the voltage source (battery) and follow the wire connections through components to the positive side of the battery and only have one path to follow, then the circuit is connected in series. If, however, you can place your pencil at the negative terminal of the voltage source and follow the wire and at some point have a choice of two or more routes, then the circuit is connected with two or more parallel branches; the number of routes determines the number of parallel branches.

EXAMPLE 8.1

Figure 8-1 illustrates five different examples of parallel resistive circuits;

FIGURE 8-1 **Parallel Circuits.**

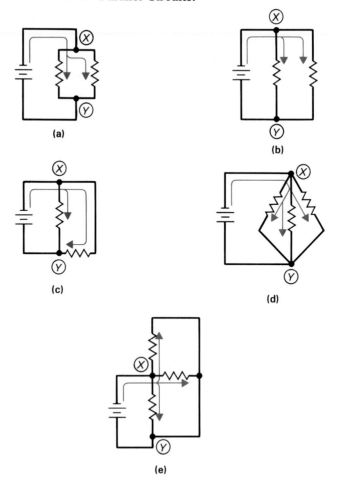

(a)

(b)

(c)

(d)

(e)

Figure 8-2 illustrates four resistors laid out on a table top:

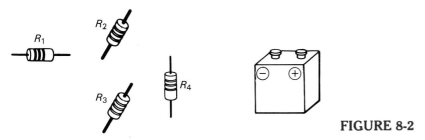

FIGURE 8-2

a. With wire leads, connect all four resistors in parallel between the negative and positive terminals of the battery.
b. Draw the schematic diagram of the parallel-connected circuit.

SOLUTION

See Figure 8-3.

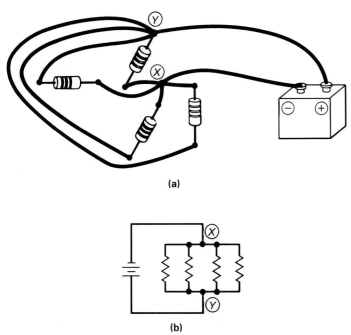

(a)

(b)

FIGURE 8-3 (a) Pictorial. (b) Schematic.

8.1
VOLTAGE IN A PARALLEL CIRCUIT

Figure 8-4(a) illustrates a simple circuit with four resistors connected in parallel across the voltage source of a 9-V battery. The current from the negative side of the battery will split between the four different paths or branches, yet the voltage drop across each branch of a parallel circuit is equal to the voltage drop across all the other branches in parallel. This means

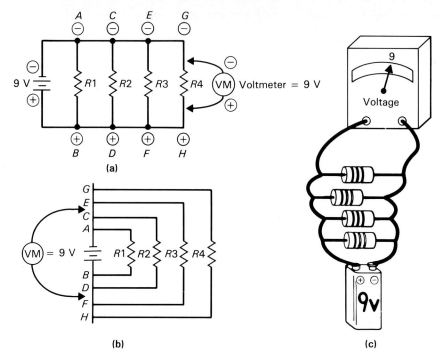

FIGURE 8-4

that, if the voltmeter were to measure the voltage across A and B or C and D or E and F or G and H, they would all be the same or, in this example, would all drop 9 V.

It is quite easy to imagine why there will be the same voltage drop across all the resistors, seeing that points A, C, E, and G are all one connection and points B, D, F, and H are all one connection. Measuring the voltage drop with the voltmeter across any of the resistors is the same as measuring the voltage across the battery, as seen in Figure 8-4(b). As long as the voltage source remains constant, the voltage drop will always be common (9 V) across the parallel resistors, no matter what value or how many resistors are connected in parallel, because the voltmeter is measuring the voltage between two common points that are directly connected to the battery, and the voltage dropped across all these parallel resistors will be equal to the source voltage.

In Figure 8-5(a) and (b), the same circuit is shown in two different

FIGURE 8-5

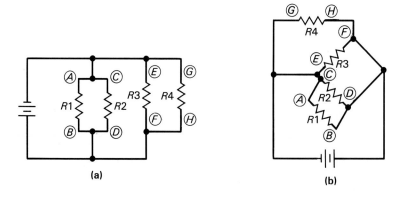

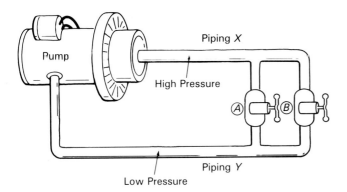

Pump

Piping X

High Pressure

Ⓐ Ⓑ

Piping Y

Low Pressure

FIGURE 8-6

ways, so you can see how the same circuit can look completely different. In both examples, the voltage drop across any of the resistors will always be the same and, as long as the voltage source is not heavily loaded, equal to the source voltage. Just as you can trace the positive side of the battery to all four resistors, you can also trace the negative side to all four resistors.

Mathematically stated, we can say that in a parallel circuit

$$V_{R1} = V_{R2} = V_{R3} = V_{R4} = V_S$$

Voltage drop across $R1$ = voltage drop across $R2$ = voltage drop across $R3$ (etc.) = source voltage

For the sake of discussion, we will relate this to a water analogy, as seen in Figure 8-6. The pressure across valves A and B will always be the

---

EXAMPLE 8.2

Refer to Figure 8-7 and calculate:
a. Voltage drop across $R1$.
b. Voltage drop across $R2$.
c. Voltage drop across $R3$.

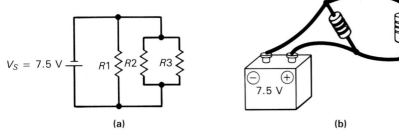

$V_S = 7.5$ V

$R1$ $R2$ $R3$

7.5 V

(a)

(b)

FIGURE 8-7 **(a) Schematic. (b) Pictorial.**

SOLUTION

Since all these resistors are connected in parallel, the voltage across every branch will be the same and equal to the source voltage applied. Therefore,

$$V_{R1} = V_{R2} = V_{R3} = V_S$$
$$7.5 \text{ V} = 7.5 \text{ V} = 7.5 \text{ V} = 7.5 \text{ V}$$

Voltage in a Parallel Circuit **247**

same, even if one offers more opposition than the other. This is because the pressure measured across either valve will be the same as checking the pressure difference between piping X and Y, which run directly back to the pump, so the pressure across A and B is the same as the pressure difference across the pump.

8.2
CURRENT IN A PARALLEL CIRCUIT

Gustav Kirchhoff, in 1847, was the first to observe and prove that, in fact, the sum of all the branch currents ($I_1 + I_2 + I_3$, etc.) was equal to the total current (I_T); in honor of his discovery, this phenomenon is known as Kirchhoff's current law. It states that the sum of all the current entering a junction is equal to the sum of all the currents leaving that same junction.

Figure 8-8(a) and (b) illustrates two examples of how this law applies. In both examples, the sum of the currents entering a junction is equal to the sum of the currents leaving that same junction. In Figure 8-8(a), the total current arrives at a junction (X) and splits to produce three branch currents, I_1, I_2, and I_3, which cumulatively equal the total current (I_T) that arrived at the junction X. The same three branch currents combine at junction Y, and

FIGURE 8-8

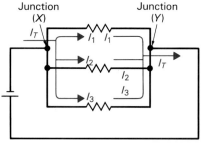

$I_T = I_1 + I_2 + I_3$ or $I_1 + I_2 + I_3 = I_T$

(a)

$I_{in} (a) + I_{in} (b) = I_{out} (a) + I_{out} (b) + I_{out} (c)$

(b)

EXAMPLE 8.3

Refer to Figure 8-9 and calculate the value of I_1.

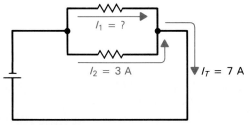

FIGURE 8-9

SOLUTION

By Kirchhoff's current law,

$$I_T = I_1 + I_2$$
$$7\,\text{A} = ? + 3\,\text{A}$$
$$I_1 = 4\,\text{A}$$

the total current (I_T) leaving that junction is equal to the sum of the three branch currents arriving at junction Y.

In Figure 8-8(b), you can see that there are two branch currents entering a junction $[I_{in}(a)$ and $I_{in}(b)]$ and three branch currents leaving that same junction $[I_{out}(a), I_{out}(b)$ and $I_{out}(c)]$, and the sum of the input currents will

EXAMPLE 8.4

Refer to Figure 8-10 and calculate the value of I_T.

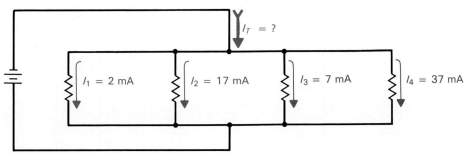

FIGURE 8-10

SOLUTION

By Kirchhoff's current law,

$$I_T = I_1 + I_2 + I_3 + I_4$$
$$? = 2\,\text{mA} + 17\,\text{mA} + 7\,\text{mA} + 37\,\text{mA}$$
$$I_T = 63\,\text{mA}$$

equal the sum of the output currents: $I_{in}(a) + I_{in}(b) = I_{out}(a) + I_{out}(b) + I_{out}(c)$.

As with series circuits, to find out how much current will flow through a parallel circuit, we need to find out how much opposition or resistance is being connected across the voltage source.

$$I_T = \frac{V_T}{R_T}$$

$$\text{total current} = \frac{\text{total or source voltage}}{\text{total resistance}}$$

When we connect resistors in parallel, the total resistance in the circuit will actually decrease; in fact, the total resistance in a parallel circuit will always be less than the value of the smallest resistor in the circuit.

Let us now further examine this point. If two sets of identical resistors (R_1, R_2, and R_3) were used to build both a series and parallel circuit, as seen in Figure 8-11, the total current flow in the parallel circuit would be larger than the total current in the series circuit, because the parallel circuit has two or more paths for current to flow, while the series circuit only has one. To explain why the total current will be larger in a parallel circuit, let us take the analogy of a freeway with only one path for traffic to flow. The freeway is equivalent to a series circuit, and a certain amount of traffic is allowed to flow along this freeway. If the freeway is expanded to accommodate two lanes, then a greater amount of traffic can flow along the freeway in the same amount of time; with more lanes, you have a greater total amount of traffic flow and, in parallel circuits, more branches will allow a greater total amount of current flow because there is less resistance in more paths than there is with only one path. This concept is summarized in Figure 8-11.

Just as a series circuit is often referred to as a voltage-divider circuit, a parallel circuit is often referred to as a current-divider circuit, because the

FIGURE 8-11

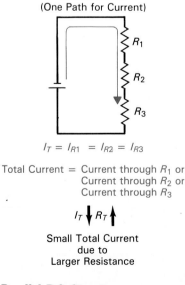

Series Circuit
(One Path for Current)

$I_T = I_{R1} = I_{R2} = I_{R3}$

Total Current = Current through R_1 or
Current through R_2 or
Current through R_3

$I_T \downarrow R_T \uparrow$

Small Total Current
due to
Larger Resistance

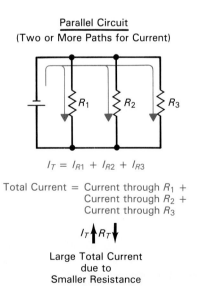

Parallel Circuit
(Two or More Paths for Current)

$I_T = I_{R1} + I_{R2} + I_{R3}$

Total Current = Current through R_1 +
Current through R_2 +
Current through R_3

$I_T \uparrow R_T \downarrow$

Large Total Current
due to
Smaller Resistance

250   **Parallel DC Circuits**

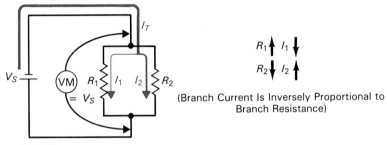

FIGURE 8-12

total current arriving at a junction will divide or split into branch currents (Kirchhoff's law), as seen in Figure 8-12.

The current division is inversely proportional to the resistance in the branch seeing that the voltage across both resistors is constant and equal to the source voltage (V_S). This means that a large branch resistance will cause

---

EXAMPLE 8.5

Calculate the following for Figure 8-13:
a. I_1
b. I_2
c. I_T

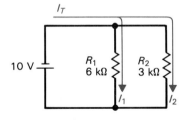

FIGURE 8-13

SOLUTION

Since R_1 and R_2 are connected in parallel across the 10-V source, the voltage across both resistors will be 10 V.

a. $I_1 = \dfrac{V_{R1}}{R_1} = \dfrac{10 \text{ V}}{6 \text{ k}\Omega} = 1.6$ mA (smaller branch current through larger branch resistance).

b. $I_2 = \dfrac{V_{R2}}{R_2} = \dfrac{10 \text{ V}}{3 \text{ k}\Omega} = 3.3$ mA (larger branch current through smaller branch resistance).

c. By Kirchoff's current law,

$$I_T = I_1 + I_2$$
$$= 1.6 \text{ mA} + 3.3 \text{ mA}$$
$$= 4.9 \text{ mA}$$

---

a small branch current ($I\downarrow = V/R\uparrow$), and a small branch resistance will cause a large branch current ($I\uparrow = V/R\downarrow$).

By rearranging Ohm's law, we can arrive at another formula, which is called the *current-divider formula* and can be used to calculate the current through any branch of a multiple-branch parallel circuit.

$$I_x = \frac{R_T}{R_x} \times I_T$$

where I_x = branch current desired
R_x = resistance in branch
R_T = total resistance
I_T = total current

EXAMPLE 8.6

Refer to Figure 8-14 and calculate the following if the total circuit resistance (R_T) is equal to 1 kΩ.
a. I_1
b. I_2
c. I_3

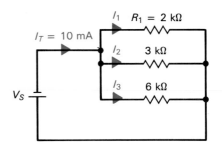

FIGURE 8-14

SOLUTION

Since the source and therefore the voltage across each branch resistor are not known, we will use the current-divider formula to calculate I_1, I_2, and I_3.

a. $I_1 = \dfrac{R_t}{R_1} \times I_T = \dfrac{1\text{ k}\Omega}{2\text{ k}\Omega} \times 10\text{ mA} = 5\text{ mA}$

(smallest branch resistance has largest branch current).

b. $I_2 = \dfrac{R_t}{R_2} \times I_T = \dfrac{1\text{ k}\Omega}{3\text{ k}\Omega} \times 10\text{ mA} = 3.33\text{ mA}$

c. $I_3 = \dfrac{R_t}{R_3} \times I_T = \dfrac{1\text{ k}\Omega}{6\text{ k}\Omega} \times 10\text{ mA} = 1.67\text{ mA}$

To double-check that the values for I_1, I_2, and I_3 are correct, you can apply Kirchhoff's current law, which is

$$I_T = I_1 + I_2 + I_3$$
$$10\text{ mA} = 5\text{ mA} + 3.33\text{ mA} + 1.67\text{ mA}$$
$$10\text{ mA} = 10\text{ mA}$$

Calculator Sequence

Step	Keypad Entry	Display Response
1.	[1] [E] [3]	1.E3
2.	[÷]	
3.	[2] [E] [3]	2.E3
4.	[×]	0.5
5.	[1] [0] [E] [3] [+/−]	10E−3
6.	[=]	5.−03

1. State Kirchhoff's current law.
2. If $I_T = 4$ A and $I_1 = 2.7$ A in a two-resistor parallel circuit, what would be the value of I_2?
3. State the current-divider formula.
4. Calculate I_1 if $R_t = 1$ kΩ, $R_1 = 2$ kΩ, and $V_t = 12$ V.

8.3
RESISTANCE IN A PARALLEL CIRCUIT

We now know that parallel circuits will have a larger current flow than an equivalent series circuit due to the smaller total resistance, as already illustrated in Figure 8-11. To calculate exactly how much total current will flow, we need to calculate the total resistance that the parallel circuit develops or contains.

The ability of a circuit to conduct current is a measure of that circuit's conductance, and you will remember from Chapter 2 that conductance (G) is equal to the reciprocal of resistance and is measured is siemens

$$G = \frac{1}{R} \quad \text{(siemens)}$$

Every resistor in a parallel circuit will have a conductance figure that is equal to the reciprocal of its resistance, and the total conductance (G_T) of the circuit will be equal to the sum of all the individual resistor conductances; therefore,

$$G_T = G_{R1} + G_{R2} + G_{R3} + \cdots$$

(Total conductance is equal to the conductance of R_1 + the conductance of R_2 + the conductance of R_3 + $\cdots$.) Once you have calculated total conductance, then the reciprocal of this figure will give you total resistance. If, for example, we have two resistors in parallel, as seen in Figure 8-15, the conductance for R_1 will equal

$$G_{R1} = \frac{1}{R_1} = \frac{1}{20\ \Omega} = 0.05\ \text{S}$$

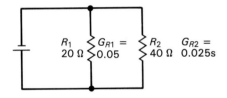

R_1 $G_{R1} =$ R_2 $G_{R2} =$
20 Ω   0.05   40 Ω   0.025s

FIGURE 8-15

The conductance for R_2 will equal

$$G_{R2} = \frac{1}{R_2} = \frac{1}{40\ \Omega} = 0.025\ \text{S}$$

The total conductance will therefore equal

$$G_{\text{total}} = G_{R1} + G_{R2}$$
$$= 0.05 + 0.025$$
$$= 0.075\ \text{S}$$

Since total resistance is equal to the reciprocal of total conductance, total resistance for the parallel circuit in Figure 8-15 will be

$$R_{\text{Total}} = \frac{1}{G_{\text{Total}}} = \frac{1}{0.075\ \text{S}} = 13.3\ \Omega$$

Combining these three steps (first calculate individual conductances, then total conductance and then total resistance) we can arrive at the following formula:

$$R_{\text{Total}} = \frac{1}{(1/R_1) + (1/R_2)}$$

Conductance of R_1 (G_{R1})
+ Conductance of R_2 (G_{R2})
= Total conductance (G_T),
and the reciprocal of total conductance is equal to total resistance

In the previous example for Figure 8-15, this combined general formula for total resistance can be verified by plugging in the example values.

$$R_T = \frac{1}{(1/R_1) + (1/R_2)}$$
$$= \frac{1}{(1/20) + (1/40)}$$
$$= \frac{1}{0.05 + 0.025}$$
$$= \frac{1}{0.075}$$
$$= 13.3\ \Omega$$

Calculator Sequence

Step	Keypad Entry	Display Response
1.	(Clear Memory)	
2.	$\boxed{2}$ $\boxed{0}$	20.
3.	$\boxed{1/x}$	5.E-2
4.	$\boxed{M+}$ (add result to memory)	5.E-2
5.	$\boxed{\text{C/CE}}$ (cancel	0.
6.	$\boxed{4}$ $\boxed{0}$	40.
7.	$\boxed{1/x}$	2.5E-2
8.	$\boxed{M+}$	2.5E-2
9.	$\boxed{\text{C/CE}}$	0.
10.	$\boxed{RM}$ (Recall memory)	7.5E-2
11.	$\boxed{1/x}$	1.3333

The formula for calculating total parallel circuit resistance for any number of resistors is

$$R_T = \cfrac{1}{(1/R_1) + (1/R_2) + (1/R_3) + (1/R_4) + \cdots}$$

EXAMPLE 8.7

Referring to Figure 8-16, calculate:
a. Total resistance
b. Voltage across R_2
c. Voltage drop across R_3

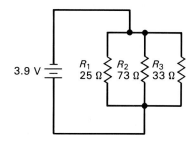

FIGURE 8-16

SOLUTION

a. $R_T = \cfrac{1}{(1/R_1) + (1/R_2) + (1/R_3)}$

$= \cfrac{1}{(1/25 \, \Omega) + (1/73 \, \Omega) + (1/33 \, \Omega)}$

$= \cfrac{1}{0.04 + 0.014 + 0.03}$

$= 11.9 \, \Omega$

With parallel resistance circuits, the total resistance is always smaller than the smallest branch resistance. In this example, the total opposition of this circuit is equivalent to 11.9 Ω, as seen in Figure 8-17.

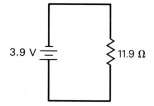

FIGURE 8-17   **Equivalent Circuit.**

b and c. With parallel resistive circuits, the voltage drop across any branch is equal to the voltage drop across each of the other branches and is equal to the source voltage, in this example 3.9 V.

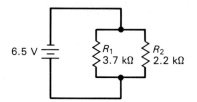

FIGURE 8-18

8.3.1 TWO RESISTORS IN PARALLEL

If only two resistors are connected in parallel, a quick and easy formula can be used to calculate total resistance.

$$R_t = \frac{R_1 \times R_2}{R_1 + R_2}$$

Total resistance is equal to product over sum.

Using this formula with the example in Figure 8-18, the total resistance can be calculated by the use of either parallel resistance formulas, (a) or (b), to make a comparison between the two;

$$\text{(a) } R_t = \frac{R_1 \times R_2}{R_1 + R_2} \qquad \text{(b) } R_t = \frac{1}{(1/R_1) + (1/R_2)}$$

$$= \frac{3.7 \text{ k}\Omega \times 2.2 \text{ k}\Omega}{3.7 \text{ k}\Omega + 2.2 \text{ k}\Omega} \qquad = \frac{1}{(1/3.7 \text{ k}\Omega) + (1/2.2 \text{ k}\Omega)}$$

$$= \frac{8.14 \text{ k}\Omega}{5.9 \text{ k}\Omega} \qquad = \frac{1}{(270.2 \times 10^{-6}) + (454.5 \times 10^{-6})}$$

$$= 1.3 \text{ k}\Omega \qquad = \frac{1}{724.7 \times 10^{-6}}$$

$$= 1.3 \text{ k}\Omega$$

8.3.2 MORE THAN TWO RESISTORS

You can see that the advantage of the product over sum parallel resistance formula (a) is its ease of use. Its disadvantage is that it can only be used for two resistors in parallel.

8.3.3 EQUAL VALUE RESISTORS IN PARALLEL

If resistors of equal value are connected in parallel, then a special case formula can be used to calculate the total resistance.

$$R_T = \frac{\text{Value of one resistor } (R)}{\text{Number of parallel resistors } (n)}$$

EXAMPLE 8.8

Refer to Figure 8-19 and calculate:
a. Total resistance in part (a).
b. Total resistance in part (b).

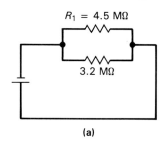

$R_1 = 4.5\ \mathrm{M\Omega}$

3.2 MΩ

(a)

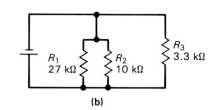

R_1 27 kΩ
R_2 10 kΩ
R_3 3.3 kΩ

(b)

FIGURE 8-19

SOLUTION

a. Figure 8-19(a) has only two resistors in parallel and therefore the two parallel resistor formula can be used; the equivalent circuit is seen in Figure 8-20.

$$R_T = \frac{R_1 \times R_2}{R_1 + R_2}$$

$$= \frac{4.5\ \mathrm{M\Omega} \times 3.2\ \mathrm{M\Omega}}{4.5\ \mathrm{M\Omega} + 3.2\ \mathrm{M\Omega}}$$

$$= \frac{14.4\ \mathrm{M\Omega}}{7.7\ \mathrm{M\Omega}}$$

$$= 1.9\ \mathrm{M\Omega}$$

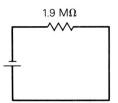

1.9 MΩ

FIGURE 8-20   Equivalent Circuit.

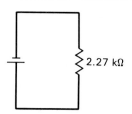

2.27 kΩ

FIGURE 8-21   Equivalent Circuit.

b. Figure 8-19(b) has more than two resistors in parallel and therefore the sum of conductances reciprocated formula must be used. The equivalent circuit can be seen in Figure 8-21.

$$R_t = \frac{1}{(1/R_1) + (1/R_2) + (1/R_3)}$$

$$= \frac{1}{(1/27\ \mathrm{k\Omega}) + (1/10\ \mathrm{k\Omega}) + (1/3.3\ \mathrm{k\Omega})}$$

$$= \frac{1}{440.0 \times 10^{-6}}$$

$$= 2.27\ \mathrm{k\Omega}$$

Calculator Sequence

Step	Keypad Entry	Display Response
1.	[4] [.] [5] [E] [6]	4.5E6
2.	[×]	
3.	[3] [.] [2] [E] [6]	3.2E6
4.	[=]	1.44E13
5.	[STO] (store in memory)	
6.	[4] [.] [5] [E] [6]	4.5E6
7.	[+]	
8.	[3] [.] [2] [E] [6]	3.2E6
9.	[=]	7.7E6
10.	[C/CE]	0.
11.	[RM] (Recall memory)	1.44E13
12.	[÷]	
13.	[7] [.] [7] [E] [6]	7.7E6
14.	[=]	1.87E6

EXAMPLE 8.9

Find the total resistance of the parallel circuit in Figure 8-22.

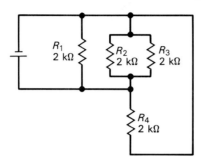

FIGURE 8-22

SOLUTION

Since all four resistors are connected in parallel and are all of the same value, the equal value resistors in parallel formula can be used.

$$R_T = \frac{R}{n} = \frac{2\text{ k}\Omega}{4} = 500\ \Omega$$

To summarize the effects of current, voltage, and resistance in parallel circuits, so far we could say:

1. Components are said to be connected in parallel when the current has to travel two or more paths between the negative and positive side of the voltage source.
2. The voltage across all the components in parallel is always the same.
3. The total current from the source is equal to the sum of all the branch currents (Kirchhoff's current law).
4. The amount of current flowing through each branch is inversely proportional to the resistance value in that branch.
5. The total resistance of a parallel circuit is always less than the value of the smallest branch resistor.

SELF-TEST REVIEW QUESTIONS (§ 8.3)

State the parallel resistance formulas for calculating R_T:

1. Two resistors
2. More than two
3. Equal value resistors

4. Calculate total resistance: $R_1 = 2.7$ kΩ, $R_2 = 24$ kΩ, and $R_3 = 1$ MΩ.

EXAMPLE 8.10

Calculate the total amount of power dissipated in Figure 8-23.

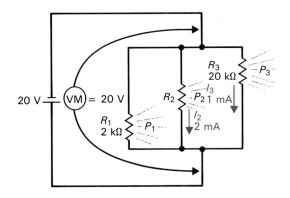

FIGURE 8-23

SOLUTION

The total power dissipated in a parallel circuit is equal to the sum of all the power dissipated by all the resistors. With P_{R1}, we only know voltage and resistance and therefore we can use the formula

$$P_{R1} = \frac{V^2}{R} = \frac{20^2}{2 \text{ k}\Omega} = 0.2 \text{ W} \quad \text{or} \quad 200 \text{ mW}$$

Calculator Sequence

Step	Keypad Entry	Display Response
1.	[2] [0]	20.
2.	[x²]	400.
3.	[÷]	
4.	[2] [E] [3]	2.E3
5.	[=]	0.2

With P_{R2}, we only know current and voltage, and therefore we can use the formula
$$P_{R2} = I \times V = 2 \text{ mA} \times 20 \text{ V} = 40 \text{ mW}$$

Calculator Sequence

Step	Keypad Entry	Display Response
1.	[2] [E] [3] [+/−]	2.E−3
2.	[x]	
3.	[2] [0]	20.
4.	[=]	40E−3

With P_{R3}, we know V, I, and R; however, we will use the third power formula:
$$P_{R3} = I^2 \times R = 1 \text{ mA}^2 \times 20 \text{ k}\Omega = 20 \text{ mW}$$

Total power (P_T) equals the sum of all the power or wattage losses for each resistor:
$$P_T = P_{R1} + P_{R2} + P_{R3}$$
$$= 200 \text{ mW} + 40 \text{ mW} + 20 \text{ mW}$$
$$= 260 \text{ mW}$$

Calculator Sequence

Step	Keypad Entry	Display Response
1.	[1] [E] [3] [+/−] [x²]	1.E−6
2.	[x]	1.E−6
3.	[2] [0] [E] [3]	20E3
4.	[=]	20.E−3

8.4

POWER IN A PARALLEL CIRCUIT

As with series circuits, the total power in a parallel resistive circuit is equal to the sum of all the power losses for each of the resistors in parallel.

$$P_T = P1 + P2 + P3 + P4 + \cdots$$
total power = addition of all the power losses

The formulas for calculating the amount of power dissipated by each resistor are

$$P = I \times V$$
$$P = \frac{V^2}{R}$$
$$P = I^2 \times R$$

EXAMPLE 8.11

In Figure 8-24, there are two ½-W (0.5-W) resistors connected in parallel. Should the wattage rating for each of these resistors be increased or decreased, or can they remain the same?

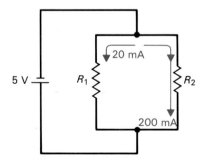

FIGURE 8-24

SOLUTION

Since current and voltage are known for both branch, the power formula used in both cases will be $P = I \times V$.

$$
\begin{aligned}
P_1 &= I \times V \\
&= 20 \text{ mA} \times 5 \text{ V} \\
&= 0.1 \text{ W}
\end{aligned}
\qquad
\begin{aligned}
P_2 &= I \times V \\
&= 200 \text{ mA} \times 5 \text{ V} \\
&= 1 \text{ W}
\end{aligned}
$$

R_1 is dissipating 0.1 W and has been designed to dissipate up to 0.5 W; it is therefore safe in this application. However, R_2 is dissipating 1 W and is only designed to dissipate 0.5 W. R_2 will consequently overheat unless it is replaced with a resistor of the same ohmic value with a 1-W or greater rating.

1. Total power in a parallel circuit can be obtained by using the same total power formula as for series circuits (true or false).
2. If $I_1 = 2$ mA and $V = 24$ V, calculate P_1.
3. If $P_1 = 22$ mW and $P_2 = 6400$ μW, $P_T = ?$
4. Is it important to observe the correct wattage ratings of resistors when they are connected in parallel?

8.5
TROUBLESHOOTING A PARALLEL CIRCUIT

In the troubleshooting discussion on series circuits, we mentioned that one of three problems can occur:

1. A component will open.
2. A component will short.
3. A component's value will change over a period of time.

Let's now discuss each of these separately with respect to parallel circuits. The example circuit in Figure 8-25 illustrates three resistors connected in parallel. Since the voltage measured across each resistor will always be the same, the voltmeter will not be very useful for isolating a faulty component.

FIGURE 8-25

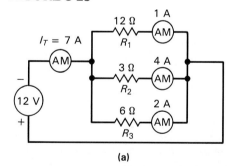

(a)

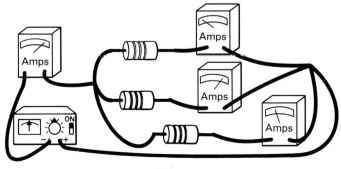

(b)

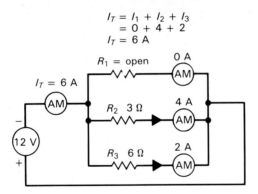

$$I_T = I_1 + I_2 + I_3$$
$$= 0 + 4 + 2$$
$$I_T = 6 \text{ A}$$

FIGURE 8-26

If you cannot isolate the faulty component by visual inspection, then the ammeter will have to be the best tool to localize the problem. Three important points should be remembered when using the ammeter:

1. Set the selector to the highest scale first and then come down in scale as necessary.
2. Ensure that the positive lead of the ammeter connects back to the positive side of the voltage source and the negative side of the ammeter connects back to the negative side of the battery (observe polarity).
3. The ammeter must be connected in series with the current path to be measured. This is the disadvantage of the ammeter, because the circuit path has to be opened and the ammeter inserted, whereas a voltmeter is merely connected across a component.

8.5.1 OPEN CIRCUIT

In Figure 8-26, R_1 has opened and so there can be no current flow through the R_1 branch. An open is equivalent to the maximum possible resistance (infinite ohms), and therefore the total resistance of this parallel circuit will increase, causing the total current flow to decrease. The total current flow, which was 7 A, will actually decrease by the amount that was flowing through the now open branch, 1 A, to a total current flow of 6 A. This problem can be noticed by one of two methods:

1. If you measure each branch current, you will isolate the problem of R_1, because there will be no current flow through R_1. However, this could take three checks.
2. If you just measure total current (one check), you will notice that total current has decreased by 1 A, and after making a few calculations you realize that the only path that had a branch current of 1 A was through R_1 and so R_1 must have gone open.

If R_1 opens, total current decreases by 1 A to 6 A ($I_T = I_1 + I_2 + I_3 = 0 + 4 + 2 = 6$ A).

If R_2 opens, total current decreases by 4 A to 3 A ($I_T = I_1 + I_2 + I_3 = 1 + 0 + 2 = 3$ A).

Parallel DC Circuits

If R_3 opens, total current decreases by 2 A to 5 A ($I_T = I_1 + I_2 + I_3 = 1 + 4 + 0 = 5$ A).

Method 2 can be used as long as the resistors are unequal. If they are equal, then method 1 may be used to locate the 0 current branch.

EXAMPLE 8.12

Figure 8-27 illustrates a set of three light bulbs connected in parallel across a 9-V battery. The filament in bulb 2 has burned out, causing an open, and so there is no current flow through that path. However, there will be current flow through the other two and so bulbs 1 and 3 will be on. How would the faulty bulb be located?

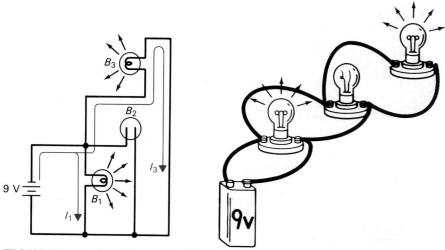

FIGURE 8-27   (a) Schematic. (b) Pictorial.

SOLUTION

As always, a visual inspection would be your first approach. Since current is flowing through both B_1 and B_3, these two bulbs will be on; and since B_2 is off, you would analyze B_2 and probably notice that the filament of the bulb has blown.

Alternatively, by using the ammeter and placing it in series with all three current branches, you would read I_1 and I_3 but not I_2, leading you to B_2 as the faulty component.

8.5.2 SHORT CIRCUIT

In Figure 8-28, R_2, which has a resistance of 3 Ω, has shorted (decreased to 1 Ω) and therefore the current through the second branch will be

$$I_2 = \frac{12 \text{ V}}{1 \text{ Ω}} = 12 \text{ A}$$

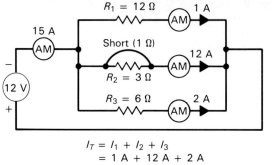

$$I_T = I_1 + I_2 + I_3$$
$$= 1\ A + 12\ A + 2\ A$$
$$I_T = 15\ A$$

FIGURE 8-28

The total resistance in the parallel circuit will decrease, causing the total flow, which was 7 A ($I_1 = 1$ A, $I_2 = 4$ A, $I_3 = 2$ A), to increase to 15 A ($I_1 = 1$ A, $I_2 = 12$ A, $I_3 = 2$ A).

This problem could be isolated by one of two methods:

1. If you check each branch current, you will notice that 12 A is flowing through R_2 and, therefore, there is not the normal resistance in that branch, and so the resistor must be shorted.

2. By just making one check of the total current, you will find that the amount of current increase will help you to isolate the faulty component.

EXAMPLE 8.13

In Figure 8-29, three resistors are connected in parallel across a 30-V source. The total current being drawn by this circuit is 35 A. Is there a fault within this circuit and, if so, what exactly is the problem?

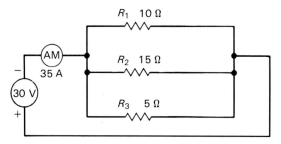

FIGURE 8-29

SOLUTION

First, let's calculate exactly what should normally occur within the circuit as shown in Figure 8-30. The 30 V will be across all three resistors, so

$$I_1 = \frac{V}{R_1} = \frac{30\ V}{10\ \Omega} = 3\ A$$

$$I_2 = \frac{V}{R_2} = \frac{30\ V}{15\ \Omega} = 2\ A$$

$$I_3 = \frac{V}{R_3} = \frac{30\ V}{5\ \Omega} = 6\ A$$

$$I_T = I_1 + I_2 + I_3 = 3 + 2 + 6 = 11\ A$$

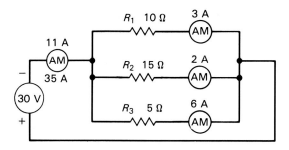

FIGURE 8-30

The circuit should normally pull 11 A; in actuality, a total current of 35 A is flowing through the defective circuit. The increased current indicates that there must be a shorted component. If R_1, R_2, or R_3 went short, a total current of 30 A would flow through the shorted branch:

$$I = \frac{V}{R} = \frac{30 \text{ V}}{1 \text{ }\Omega \text{ (short)}} = 30 \text{ A}$$

If R_1 is shorted, the total current through the R_1 branch will be

$$I_T = I_1 + I_2 + I_3 = 30 \text{ A} + 2 \text{ A} + 6 \text{ A} = 38 \text{ A}$$

If R_2 is shorted, the total current through the R_2 branch will be

$$I_T = I_1 + I_2 + I_3 = 3 \text{ A} + 30 \text{ A} + 6 \text{ A} = 39 \text{ A}$$

If R_3 is shorted, the total current through the R_2 branch will be

$$I_T = I_1 + I_2 + I_3 = 3 \text{ A} + 2 \text{ A} + 30 \text{ A} = 35 \text{ A}$$

In conclusion, therefore, we can say that there is indeed a fault in the circuit because a total current of 35 A exists instead of the required 11 A. By calculation, you can isolate the problem to be R_3 shorted, because in this situation the circuit would draw 35 A.

If R_1 is shorted, the total current will increase to 18 A ($I_T = I_1 + I_2 + I_3 = 12 + 4 + 2 = 18$ A).

If R_2 is shorted, the total current will increase to 15 A ($I_T = I_1 + I_2 + I_3 = 1 + 12 + 2 = 15$ A).

If R_3 is shorted, the total current will increase to 17 A ($I_T = I_1 + I_2 + I_3 = 1 + 4 + 12 = 17$ A).

Once again, method 2 can only be used if the resistors in the parallel circuit are all unequal. If they are equal, then method 1 may be used to locate the maximum current branch.

8.5.3 COMPONENT VARIATION

The resistive values of resistors will change with age. This increase or decrease in resistance will cause a corresponding decrease or increase in branch current and therefore total current. This can be detected by calculating the expected current flow through the branch, using Ohm's law. As long as the voltage is correct, a changed current will isolate the problem to the resistor whose value has changed from the rated value.

8.5.4 SUMMARY OF PARALLEL CIRCUIT TROUBLESHOOTING

1. An open component will cause no current flow within that branch. The total current will decrease, and the voltage across the component will be the same as the source voltage.

2. A shorted component will cause maximum current through that branch. The total current will increase and the voltage across the component will always be equal to the source voltage.

3. A change in resistance value will cause a corresponding opposite change in branch current and total current.

SELF-TEST REVIEW QUESTIONS (§ 8.5)

How would problems 1 through 3 be recognized in a parallel circuit:

1. An open component?
2. A shorted component?
3. A component's value variation?
4. An ammeter is used to troubleshoot series circuits, whereas a voltmeter is used to troubleshoot parallel circuits (true or false).

SUMMARY

1. A parallel circuit has two or more paths or branches for current to flow from the negative to the positive side of the source.

2. The voltage across all components in parallel is always the same.

3. Kirchhoff's current law states that: The sum of all the currents entering a junction is equal to the sum of all the currents leaving the same junction.

4. The amount of current flowing through each branch of a parallel circuit is inversely proportional to the resistance value in that branch, because the voltage across both resistors is constant.

5. Parallel circuits have a larger current flow than a series circuit containing the same number and value of resistors due to the smaller total resistance.

6. The total resistance in a parallel circuit is equal to the reciprocal of the circuit's total conductance and is always less than the value of the smallest branch resistor.

7. As with series circuits, the total power in a parallel resistive circuit is equal to the sum of all the power losses from each of the resistors in parallel.

8. When an open (infinite ohms) occurs in a parallel circuit, the total resistance increases, causing the total current to decrease.

9. When a short $(1\ \Omega)$ occurs in a parallel circuit, the total resistance decreases, causing the total current flow to increase.

10. Resistors will change their resistive value with age, causing a change in branch current and consequently total current.

11. An open causes no current flow through the open branch, whereas a short causes maximum current flow through that branch.

NEW TERMS

Branch current
Current divider
Fixed current divider

Kirchhoff's current law
Parallel circuit

Variable current divider

NEW FORMULAS

$$V_S = V_{R1} = V_{R2} = V_{R3} = \cdots$$

$$I_T = \frac{V_S}{R_T}$$

$$I_T = I_1 + I_2 + I_3 + \cdots$$

Current divider formula
$$I_x = \left(\frac{R_T}{R_x}\right) \times I_T$$

$$R_T = \cfrac{1}{\left(\dfrac{1}{R_1}\right) + \left(\dfrac{1}{R_2}\right) + \left(\dfrac{1}{R_3}\right) + \cdots}$$

$$R_T = \frac{R_1 \times R_2}{R_1 + R_2}$$

$$R_T = \frac{R_{EV}}{n}$$

$$P_T = P_1 + P_2 + P_3 + \cdots$$

$$P_T = \frac{V_S^2}{R_T} = I_T^2 \times R_T = V_S \times I_T$$

V_S = source voltage in volts
V_{R1} = voltage across R_1 in volts
V_{R2} = voltage across R_2 in volts
V_{R3} = voltage across R_3 in volts
I_T = total current in amps
R_T = total resistance in ohms
I_1 = current through R_1 in amps
I_2 = current through R_2 in amps
I_3 = current through R_3 in amps
I_x = branch current desired in amps
R_x = resistance in branch
R_1 = resistance of R_1 in ohms
R_2 = resistance of R_2 in ohms
R_3 = resistance of R_3 in ohms
R_{EV} = resistance of one of the equal-value resistors
n = number of parallel resistors
P_T = total power dissipated in watts
P_1 = power dissipated by R_1 in watts
P_2 = power dissipated by R_2 in watts
P_3 = power dissipated by R_3 in watts

Multiple Choice Questions

1. A parallel circuit has _____path(s) for current to flow.
   a. One
   b. Two or more
   c. Only three
   d. None of the above

2. If a source voltage of 12 V is applied across four resistors of equal value in parallel, the voltage drops across each resistor would be equal to:
   a. 12 V
   b. 3 V
   c. 4 V
   d. 48 V

3. What would be the voltage drop across two 25-Ω resistors in parallel if the source voltage were equal to 9 V?
   a. 50 V
   b. 25 V
   c. 12 V
   d. None of the above

4. If a four-branch parallel circuit has 15 mA flowing through each branch, the total current into the parallel circuit will be equal to:
   a. 15 mA
   b. 60 mA
   c. 30 mA
   d. 45 mA

5. If the total three-branch parallel circuit current is equal to 500 mA, and 207 mA is flowing through one branch and 153 mA through another, what would be the current flow through the third branch?
   a. 707 mA
   b. 653 mA
   c. 140 mA
   d. None of the above

6. A large branch resistance will cause a _____ branch current.
   a. Large
   b. Small
   c. Medium
   d. None of the above are true

7. What would be the conductance of a 1-kΩ resistor?
   a. 10 mS
   b. 1 mS
   c. 2 kΩ
   d. All of the above

8. If only two resistors are connected in parallel, the total resistance is equal to:
   a. The sum of the resistance values
   b. Three times the value of one resistor
   c. The product over the sum
   d. All of the above

9. If resistors of equal value are connected in parallel, the total resistance can be calculated by:
   a. One resistor value divided by the number of parallel resistors
   b. The sum of the resistor values
   c. The number of parallel resistors divided by one resistor value
   d. All of the above could be true

10. The total power in a parallel circuit is equal to the:
    a. The product of total current and total voltage
    b. Reciprocal of the individual power losses
    c. Sum of the individual power losses
    d. Both (a) and (b) are true
    e. Both (a) and (c) are true

Essay Questions

11. Describe the difference between a series and parallel circuit. (Introduction)

12. Explain and state mathematically the situation regarding voltage in a parallel circuit. (8.1)

13. State Kirchhoff's current law for parallel circuits. (8.2)

14. What is the current-divider formula? (8.2)

15. List the formulas for calculating the following total resistances. (8.3)
    a. Two resistors of different values
    b. More than two resistors of different values
    c. Equal value resistors

16. Describe the relationship between branch current and branch resistance. (8.2)

17. Briefly describe total and individual power in a parallel resistive circuit. (8.4)

18. Discuss troubleshooting parallel circuits as applied to: (8.5)
    a. A shorted component
    b. An open component

19. Explain why parallel circuits have a smaller total resistance and larger total current than series circuits. (8.2)

20. Briefly describe why Kirchhoff's voltage law applies to series circuits, and why Kirchhoff's current law relates to parallel circuits. (8.2)

Practice Problems

21. Calculate the total resistance of four 30kΩ resistors in parallel.

22. Find the total resistance for each of the following parallel circuits:
    a. 330 Ω and 560 Ω
    b. 47 kΩ, 33 kΩ, and 22 kΩ
    c. 2.2 MΩ, 3 kΩ, and 220 Ω

23. If 10 V is connected across three 25-Ω resistors in parallel, what will be the total and individual branch currents?

24. If a four-branch parallel circuit has branch currents equal to 25 mA, 37 mA, 220 mA, and 0.2 A, what is the total circuit current?

25. If three resistors of equal value are connected across a 14-V supply and the total resistance is equal to 700 Ω, what is the value of the branch currents?

26. If three 75-W light bulbs are connected in parallel across a 110-V supply, what is the value of each branch current? What is the branch current through the other two light bulbs if one burns out?

27. If 33-kΩ and 22-kΩ resistors are connected across a 20-V source, calculate:
    a. Total resistance
    b. Total current
    c. Branch currents
    d. Total power dissipated
    e. Individual power dissipated

28. If four parallel-connected resistors are each dissipating 75 mW, what is the total power being dissipated?

29. Calculate the branch currents through the following parallel resistor circuits when they are connected across a 10-V supply.
   a. 22 kΩ and 33 kΩ
   b. 220 Ω, 330 Ω, and 470 Ω

30. If 30-Ω and 40-Ω resistors are connected in parallel, which resistor will generate the greatest amount of heat?

31. Calculate the total conductance and resistance of the following parallel circuits:
   a. Three 5 Ω resistors
   b. Two 200-Ω resistors
   c. 1 MΩ, 500 MΩ, 3.3 MΩ
   d. 5 Ω, 3 Ω, 2 Ω

32. Connect the three resistors in Figure 8-31 in parallel across a 12-V battery and then calculate the following:
   a. V_{R1}, V_{R2}, V_{R3} **d.** P_T
   b. I_1, I_2, I_3 **e.** P_1, P_2, P_3
   c. I_T **f.** G_{R1}, G_{R2}, G_{R3}

33. Calculate R_T in Figure 8-32 (a), (b), (c), and (d).

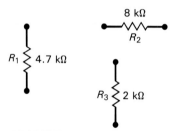

FIGURE 8-31

FIGURE 8-32

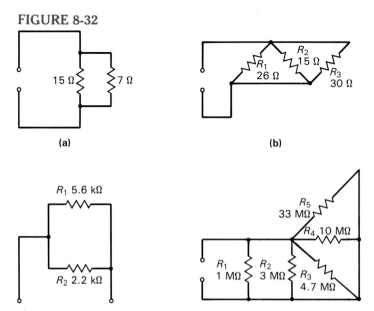

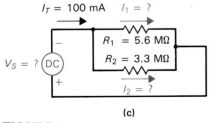

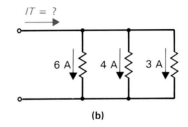

(a)

(b)

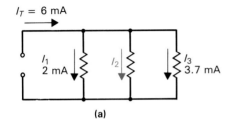

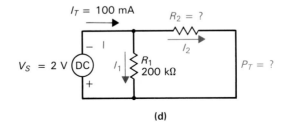

(c)

(d)

FIGURE 8-33

34. Calculate the branch currents through four 60-W bulbs connected in parallel across 110 V. How much is the total current and what would happen to the total current if one of the bulbs were to burn out? What change would occur in the remaining branch currents?

35. Calculate the following in Figure 8-33.
   a. I_2 **c.** V_S, I_1, I_2
   b. I_T **d.** R_2, I_1, I_2, P_T

Troubleshooting ―――――――――――――――――――――――――――

36. An open component in a parallel circuit will cause _____ current flow within that branch, which will cause the total current to _____.
   a. Maximum, increase
   b. Zero, decrease
   c. Maximum, decrease
   d. Zero, increase

37. A shorted component in a parallel circuit will cause _____ current through a branch, and consequently the total current will _____.
   a. Maximum, increase
   b. Zero, decrease
   c. Maximum, decrease
   d. Zero, increase

38. If a 10-kΩ and two 20-kΩ resistors are connected in parallel across a 20-V supply, and the total current measured is 2 mA, determine whether a problem exists in the circuit and, if it does, isolate the problem.

39. What situation would occur and how would we recognize the problem if one of the 20-kΩ resistors in question 38 were to short?

40. With age, the resistance of a resistor will _____, resulting in a corresponding but opposite change in _____.
   a. Increase, branch current
   b. Change, source voltage
   c. Decrease, source resistance
   d. Change, branch current

9

SERIES–PARALLEL DC CIRCUITS

AFTER COMPLETING THIS CHAPTER, YOU WILL BE ABLE TO:

1. Identify the difference between a series, parallel, and series–parallel circuit.
2. Describe how to use a three-step procedure to determine total resistance.
3. Describe for the series–parallel circuit how to use a five-step procedure to calculate:
   a. Total resistance
   b. Total current
   c. Voltage division
   d. Branch current
   e. Total power dissipated
4. Explain what loading effect a piece of equipment will have when connected to a voltage divider.
5. Identify and describe the Wheatstone bridge circuit in both the balanced and unbalanced condition.
6. Describe the R–$2R$ ladder circuit used for digital-to-analog conversion.
7. Describe the differences between a voltage and current source.
8. Analyze series–parallel networks using:
   a. The superposition theorem
   b. Thevenin's theorem
   c. Norton's theorem
9. Explain how to identify the following problems in a series–parallel circuit:
   a. Open series resistor
   b. Open parallel resistor
   c. Shorted series resistor
   d. Shorted parallel resistor
   e. Resistor value variation

INTRODUCTION

Very rarely are we lucky enough to run across straightforward series or parallel circuits. In general, all electronic equipment is composed of many components that are interconnected to form a combination of series and parallel circuits.

Figure 9-1(a) through (f) illustrates six different examples of series–parallel resistive circuits. The most important point to learn is how to distinguish between the resistors that are connected in series and the resistors that are connected in parallel, which will take a little practice.

One thing that you may not have noticed when examining Figure 9-1 is that:

Circuit 9-1(a) is equivalent to 9-1(b)

Circuit 9-1(c) is equivalent to 9-1(d)

Circuit 9-1(e) is equivalent to 9-1(f)

When analyzing these series–parallel circuits, always remember that current flow determines whether the resistor is connected in series or parallel. Begin at the negative side of the battery and apply these two rules:

1. If the total current has only one path to follow through a component, then that component is connected in series.

2. If the total current has two or more paths to follow through two or more components, then those components are connected in parallel.

PEACEFUL PROGRAMMING

Top-selling software programs for computers can originate from a diverse spectrum of sources. Some are developed by large research corporations in the center of major cities, while others are created by people in mountaintop cabins in the middle of nowhere.

Paul Lutus, who had created many best-selling software programs for the Apple computer, was fiercely independent. For a long period of time, Lutus lived and labored in his isolated, handbuilt 12-foot by 16-foot cabin in the Oregon mountains, developing a word-processsing program. Published in 1979, the highly acclaimed Apple Writer almost instantly became a best-seller, and the peace, tranquillity, and hard work finally paid off in the form of royalty payments of more than $7500 per day.

Referring to Figure 9-1, you can see that series or parallel resistor networks are easier to identify in parts (a), (c), and (e) than in parts (b), (d), and (f). Redrawing the circuit so that the components are arranged from left to right or from top to bottom is your first line of attack in your quest to identify series- and parallel-connected components.

FIGURE 9-1

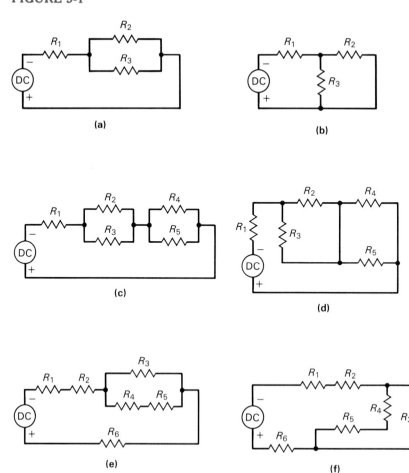

Series–Parallel DC Circuits

EXAMPLE 9.1

Refer to Figure 9-2 and identify which resistors are connected in series and which are in parallel.

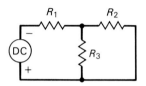

FIGURE 9-2

SOLUTION

First, let's redraw the circuit so the components are aligned either from left to right [Figure 9-3(a)] or from top to bottom [Figure 9-3(b)]. Placing your pencil at the negative terminal of the battery on whichever figure you prefer, either Figure 9-3(a) or (b), trace the current path through the circuit toward the positive side of the battery, as illustrated in Figure 9-4.

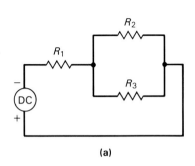

(a)

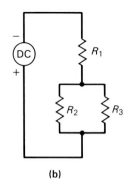

(b)

FIGURE 9-3

The total current arrives first at R_1. There is only one path for current to flow, which is through R_1 and, therefore, R_1 is connected in series. The total current proceeds on past R_1 and arrives at a junction where current divides and travels through two branches, R_2 and R_3. Since current had to split into two paths, R_2 and R_3 are therefore connected in parallel. After the parallel connection of R_2 and R_3, total current combines and travels to the positive side of the battery.

To summarize, R_1 is in series with the parallel combination of R_2 and R_3.

FIGURE 9-4

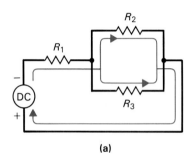

(a)

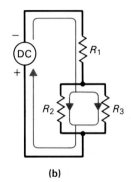

(b)

EXAMPLE 9.2

Refer to Figure 9-5 and identify which resistors are connected in series and which are connected in parallel.

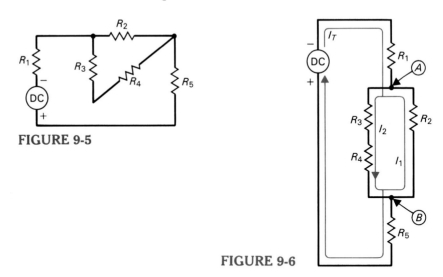

FIGURE 9-5

FIGURE 9-6

SOLUTION

Figure 9-6 illustrates the simplified, redrawn schematic. Total current leaves the negative terminal of the battery, and all of this current has to travel through R_1, which is therefore a series-connected resistor. Total current will split at junction A, and consequently R_3 and R_4 with R_2 make up a parallel combination. The current that flows through R_3 (I_2) will also flow through R_4 and have only one path to follow; therefore, R_3 is in series with R_4. I_1 and I_2 branch currents combine at junction B to produce total current, which has only one path to follow through the series resistor R_5, and finally to the positive side of the battery.

To summarize, R_3 and R_4 are in series with one another and both are in parallel with R_2, and this parallel combination is in series with R_1 and R_5.

9.1
TOTAL RESISTANCE IN A SERIES– PARALLEL CIRCUIT

No matter how complex or involved the series–parallel circuit, there is a simple three-step method to simplify the circuit to a single equivalent total resistance. Figure 9-7 illustrates an example of a series–parallel circuit. Once

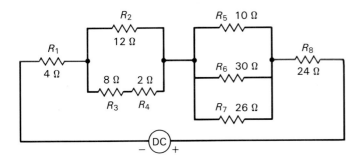

FIGURE 9-7

you have analyzed and determined the series–parallel relationship, we can proceed to solve for total resistance.

The three-step method is:

Step A: Determine the equivalent resistances of all series-connected resistors.

Step B: Determine the equivalent resistances of all parallel-connected combinations.

Step C: Determine the equivalent resistances of the remaining series-connected resistances.

Let's now put our theory to work with the example circuit in Figure 9-7.

STEP A   Solve for all series-connected resistors. In our examples, this only applies to R_3 and R_4, and since this is a series connection, we have to use the series resistance formula.

$$R_{3,4} = R_3 + R_4 = 8 + 2 = 10\ \Omega \quad \text{(series resistance formula)}$$

With R_3 and R_4 solved, the circuit appears as indicated in Figure 9-8.

STEP B   Solve for all parallel combinations. In this example, they are the two parallel combinations of (a) R_2 and $R_{3,4}$ and (b) R_5 and R_6 and R_7. Since these are parallel connections, use the parallel resistance formulas.

$$R_{2,3,4} = \frac{R_2 \times R_{3,4}}{R_2 + R_{3,4}} = \frac{12 \times 10}{12 + 10} = 5.5\ \Omega \quad \text{(two parallel resistances formula)}$$

$$R_{5,6,7} = \frac{1}{(1/R_5) + (1/R_6) + (1/R_7)} = 5.8\ \Omega$$

(more than two parallel resistances formula)

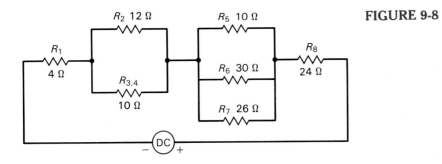

FIGURE 9-8

Total Resistance in a Series–Parallel Circuit

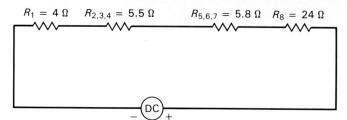

$R_1 = 4\ \Omega \quad R_{2,3,4} = 5.5\ \Omega \qquad R_{5,6,7} = 5.8\ \Omega \quad R_8 = 24\ \Omega$

FIGURE 9-9

With $R_{2,3,4}$ and $R_{5,6,7}$ solved, the circuit now appears as illustrated in Figure 9-9.

STEP C Solve for the remaining series resistances. There are now four remaining series resistances, which can be reduced to one equivalent resistance (R_{eq}), as seen in Figure 9-10, by using the series resistance formula:

$$R_{eq} = R_1 + R_{2,3,4} + R_{5,6,7} + R_8$$
$$= 4\ \Omega + 5.5\ \Omega + 5.8\ \Omega + 24\ \Omega$$
$$= 39.3\ \Omega$$

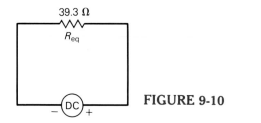

39.3 Ω

R_{eq}

FIGURE 9-10

EXAMPLE 9.3

Find the total resistance of the circuit in Figure 9-11.

FIGURE 9-11

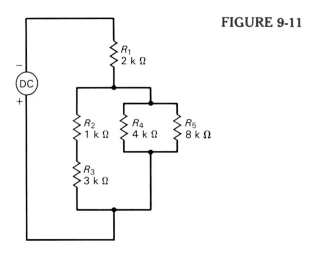

R_1
2 k Ω

R_2
1 k Ω

R_4
4 k Ω

R_5
8 k Ω

R_3
3 k Ω

Series–Parallel DC Circuits

SOLUTION

STEP A   Solve for all series resistors in parallel combinations. This applies to R_2 and R_3 (series connection):

$$R_{2,3} = R_2 + R_3 = 1\text{ k}\Omega + 3\text{ k}\Omega = 4\text{ k}\Omega$$

The resulting circuit, after completing step A, is illustrated in Figure 9-12(a).

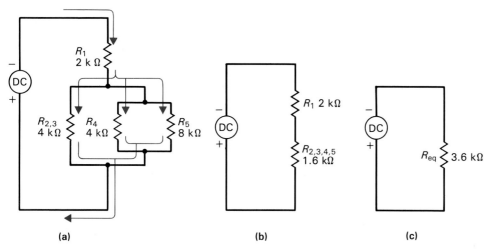

(a)   (b)   (c)

FIGURE 9-12

STEP B   Solve for all parallel combinations. Observing the circuit resulting from step A [Figure 9-12(a)], you can see that current branches into three paths, and so the following formula can be used:

$$R_{2,3,4,5} = \cfrac{1}{\left(\dfrac{1}{R_{2,3}}\right) + \left(\dfrac{1}{R_4}\right) + \left(\dfrac{1}{R_5}\right)}$$

$$= \cfrac{1}{\left(\dfrac{1}{4\text{ k}\Omega}\right) + \left(\dfrac{1}{4\text{ k}\Omega}\right) + \left(\dfrac{1}{8\text{ k}\Omega}\right)} = 1.6\text{ k}\Omega$$

The resulting circuit, after completing step B, is illustrated in Figure 9-12(b).

STEP C   Solve for the remaining series resistances. Observing the circuit resulting from step B [Figure 9-12(b)], you can see that there are two remaining series resistances. The equivalent resistance (R_{eq}) is equal to

$$R_{eq} = R_1 + R_{2,3,4,5} = 2\text{ k}\Omega + 1.6\text{ k}\Omega = 3.6\text{ k}\Omega$$

The total equivalent resistance, after completing all three steps, is illustrated in Figure 9-12(c).

1. Given a series–parallel circuit, how can we determine which resistors are connected in series and which are connected in parallel?
2. Calculate the total resistance if two series-connected 12-kΩ resistors are connected in parallel with a 6-kΩ resistor.
3. State the three-step procedure used to determine total resistance in a circuit made up of both series and parallel resistors.
4. Sketch a series–parallel resistor network made up of three resistors. R_1 and R_2 are in series with each other and are connected in parallel with R_3. If R_1 = 470 Ω, R_2 = 330 Ω, and R_3 = 270 Ω, what is R_t?

9.2
VOLTAGE DIVISION IN A SERIES–PARALLEL CIRCUIT

There is a simple three-step procedure for finding the voltage drop across each part of the series–parallel circuit. Figure 9-13 illustrates an example of a series–parallel circuit to which we will apply the three-step method for determining voltage drop.

STEP 1   Determine the circuit's total resistance. This is achieved by following the three-step method used previously for calculating total resistance.

$$\text{Step A} \quad R_{3,4} = 4 + 8 = 12 \ \Omega$$

$$\text{Step B} \quad R_{2,3,4} = \frac{1}{(1/R_2) + (1/R_{3,4})} = 6 \ \Omega$$

$$R_{5,6,7} = \frac{1}{(1/R_5) + (1/R_6) + (1/R_7)} = 12 \ \Omega$$

Figure 9-14 illustrates the equivalent circuit up to this point. We end up with one series resistor (R_1) and two series equivalent resistors ($R_{2,3,4}$ and $R_{5,6,7}$).

FIGURE 9-13

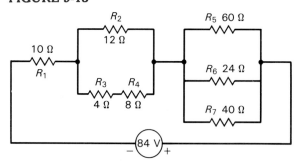

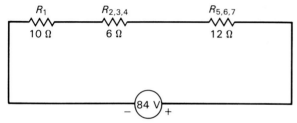

FIGURE 9-14

STEP 2   Determine the circuit's total current. This step is achieved by simply utilizing Ohm's law.

$$I_t = \frac{V_t}{R_t} = \frac{84 \text{ V}}{28 \text{ }\Omega} = 3 \text{ A}$$

STEP 3   Determine the voltage across each series resistor and each parallel combination (series equivalent resistor) in Figure 9-14. Since these are all in series, the same current (I_t) will flow through all three.

$$V_{R1} = I_t \times R_1 = 3 \text{ A} \times 10 = 30 \text{ V}$$

$$V_{R2,3,4} = I_t \times R_{2,3,4} = 3 \text{ A} \times 6 = 18 \text{ V}$$

$$V_{R5,6,7} = I_t \times R_{5,6,7} = 3 \text{ A} \times 12 = 36 \text{ V}$$

The voltage drops across the series (R_1) and series equivalent resistors ($R_{2,3,4}$ and $R_{5,6,7}$) are illustrated in Figure 9-15.

Kirchhoff's voltage law states that the sum of all the voltage drops is equal to the voltage source applied. This law can be used to confirm that our calculations are all correct:

$$
\begin{aligned}
V_t &= V_{R1} + V_{R2,3,4} + V_{R5,6,7} \\
&= 30 \text{ V} + 18 \text{ V} + 36 \text{ V} \\
&= 84 \text{ V}
\end{aligned}
$$

To summarize, refer to Figure 9-16:

30 V is dropped across R_1.

18 V is dropped across R_2.

18 V is dropped across both R_3 and R_4.

FIGURE 9-15

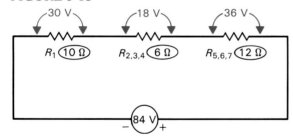

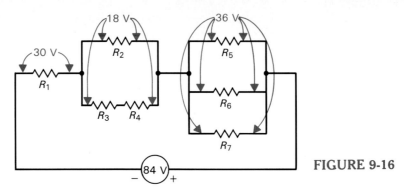

FIGURE 9-16

36 V is dropped across R_5.

36 V is dropped across R_6.

36 V is dropped across R_7.

SELF-TEST REVIEW (§ 9.2)

1. State the three-step procedure used to calculate the voltage drop across each part of a series-parallel circuit.
2. Referring to Figure 9-13, double the values of all the resistors. Would the voltage drops calculated previously change, and if so what would they be?

9.3
BRANCH CURRENTS IN A SERIES–PARALLEL CIRCUIT

In the previous example, step 2 calculated the total current flowing in a series–parallel circuit. The next step is to find out exactly how much current is flowing through each parallel branch. This is Step 4. Figure 9-17 shows the previously calculated data inserted in the appropriate places.

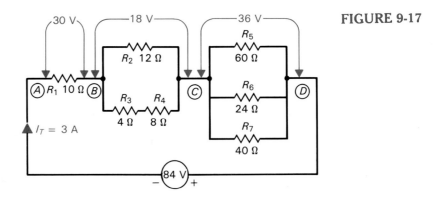

FIGURE 9-17

Total current (I_T) will exist at points A, B, C, and D. Between A and B, current has only one path to flow, which is through R_1. R_1 is therefore a series resistor and so $I_{R1} = I_t = 3$ A. Between points B and C, current has two paths, either through R_2 (12 Ω) or through R_3 and R_4 (12 Ω).

$$I_{R2} = \frac{V_{R2}}{R_2} = \frac{18 \text{ V}}{12 \text{ Ω}} = 1.5 \text{ A}$$

$$I_{R3,4} = \frac{V_{R3,4}}{R_{3,4}} = \frac{18 \text{ V}}{12 \text{ Ω}} = 1.5 \text{ A}$$

Not surprisingly, the total current of 3 A is split equally due to both branches having equal resistance.

The two 1.5-A branch currents will combine at point C to produce once again the total current of 3 A. Between points C and D, current has three paths to flow through, R_5, R_6, and R_7.

$$I_{R5} = \frac{V_{R5}}{R_5} = \frac{36 \text{ V}}{60 \text{ Ω}} = 0.6 \text{ A}$$

$$I_{R6} = \frac{V_{R6}}{R_6} = \frac{36 \text{ V}}{24 \text{ Ω}} = 1.5 \text{ A}$$

$$I_{R7} = \frac{V_{R7}}{R_7} = \frac{36 \text{ V}}{40 \text{ Ω}} = 0.9 \text{ A}$$

All three branch currents will combine at point D to produce the total current of 3 A ($I_T = I_{R5} + I_{R6} + I_{R7} = 0.6 + 1.5 + 0.9 = 3$ A).

9.4
POWER IN A SERIES–PARALLEL CIRCUIT

If resistors are in series or in parallel, the total power in a series–parallel circuit is

$$P_t = P_1 + P_2 + P_3 + P_4 + \cdots$$

total power = addition of all power losses

The formulas for calculating the amount of power lost by each resistor are

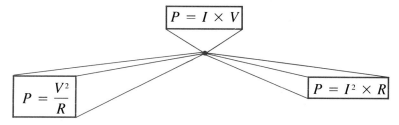

Let us calculate the power dissipated by each resistor; this final calculation will be called Step 5. Since resistance, voltage, and current are

known, either of the three formulas for power can be used to determine power.

$$P_{R1} = \frac{V_{R1}^2}{R_1} = \frac{30 \text{ V}^2}{10 \text{ }\Omega} = 90 \text{ W},$$

$$P_{R2} = \frac{V_{R2}^2}{R_2} = \frac{18 \text{ V}^2}{12 \text{ }\Omega} = 27 \text{ W},$$

$$P_{R3} = I_{R3}^2 \times R_3 = 1.5 \text{ A}^2 \times 4 \text{ }\Omega = 9 \text{ W},$$

$$P_{R4} = I_{R4}^2 \times R_4 = 1.5 \text{ A}^2 \times 8 = 18 \text{ W}$$

$$P_{R5} = \frac{V_{R5}^2}{R_5} = \frac{36 \text{ V}^2}{60 \text{ }\Omega} = 21.6 \text{ W}$$

$$P_{R6} = \frac{V_{R6}^2}{R_6} = \frac{36 \text{ V}^2}{24 \text{ }\Omega} = 54 \text{ W}$$

$$P_{R7} = \frac{V_{R7}^2}{R_7} = \frac{36 \text{ V}^2}{40 \text{ }\Omega} = 32.4 \text{ W}$$

$$P_t = P_{R1} + P_{R2} + P_{R3} + P_{R4} + P_{R5} + P_{R6} + P_{R7}$$
$$= 90 + 27 + 9 + 18 + 21.6 + 54 + 32.4$$
$$= 252 \text{ W}$$

or

$$P_t = \frac{V_t^2}{R_t} = \frac{84 \text{ V}^2}{28 \text{ }\Omega}$$
$$= 252 \text{ W}$$

The total power dissipated in this example circuit is 252 W. All the information can now be inserted in a final diagram for the example, as seen in Figure 9-18.

FIGURE 9-18

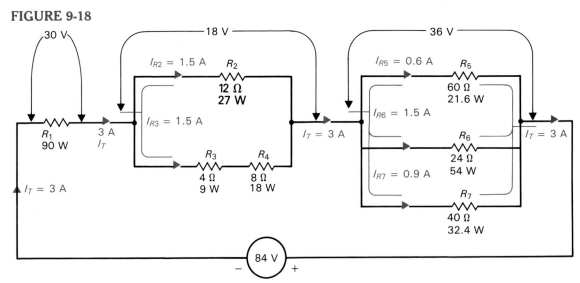

9.5

FIVE-STEP METHOD FOR SERIES–PARALLEL CIRCUIT ANALYSIS

Let's now combine and summarize all the steps for calculating resistance, voltage, current, and power in a series–parallel circuit by solving another problem.

EXAMPLE 9.4

Referring to Figure 9-19, calculate:
a. Total resistance
b. Voltage drop across all resistors
c. Current through each resistor
d. Total power dissipated by the circuit

SOLUTION

The problem has basically asked us to calculate everything about this series–parallel circuit seen in Figure 9-19. Let's first go through the steps we will follow to solve this problem.

Solving for Resistance, Voltage, Current, and Power in a Series–Parallel Circuit

STEP 1   Determine the circuit's total resistance.

   Step A Solve for series resistors in all parallel combinations.
   Step B Solve for all parallel combinations.
   Step C Solve for remaining series resistances.

STEP 2   Determine the circuit's total current.

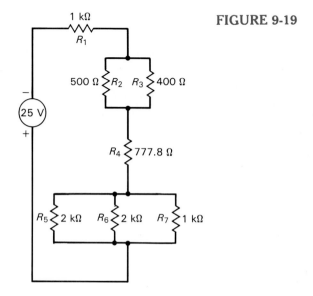

FIGURE 9-19

STEP 3 Determine the voltage across each series resistor and each parallel combination (series equivalent resistor).

STEP 4 Determine the value of current through each parallel resistor in every parallel combination.

STEP 5 Determine the total and individual power dissipated by the circuit.

Applying this procedure to our problem in Figure 9-19, we will first proceed with Step 1, which is calculating the circuit's total resistance.

STEP 1 Determine the circuit's total resistance.
  Step A There are no series resistors within parallel combinations.
  Step B There are two-resistor (R_2, R_3) and three-resistor (R_5, R_6, R_7) parallel combinations in this circuit.

$$R_{2,3} = \frac{1}{(1/R_2) + (1/R_3)} = 222.2\ \Omega$$

$$R_{5,6,7} = \frac{1}{(1/R_5) + (1/R_6) + (1/R_7)} = 500\ \Omega$$

Figure 9-20 illustrates the circuit resulting after Step B.
  Step C Solve for the remaining four resistances to gain the circuit's total resistance (R_t) or equivalent resistance (R_{eq}).

$$\begin{aligned}
R_{eq} &= R_1 + R_{2,3} + R_4 + R_{5,6,7} \\
&= 1000 + 222.2 + 777.8 + 500 \\
&= 2500\ \Omega \quad \text{or} \quad 2.5\ \text{k}\Omega
\end{aligned}$$

Figure 9-21 illustrates the circuit resulting after Step C.

STEP 2 Determine the circuit's total current.

$$I_t = \frac{V_S}{R_t} = \frac{25\ \text{V}}{2.5\ \text{k}\Omega} = 10\ \text{mA}$$

FIGURE 9-20

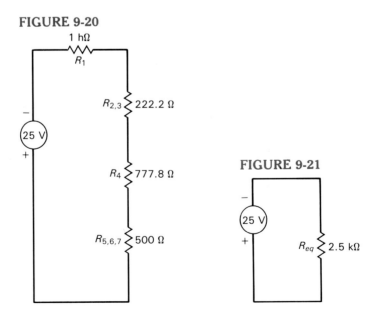

FIGURE 9-21

Series–Parallel DC Circuits

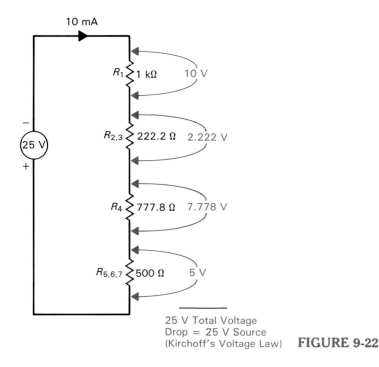

10 mA

R_1 1 kΩ   10 V

$R_{2,3}$ 222.2 Ω   2.222 V

R_4 777.8 Ω   7.778 V

$R_{5,6,7}$ 500 Ω   5 V

25 V

25 V Total Voltage
Drop = 25 V Source
(Kirchoff's Voltage Law)   **FIGURE 9-22**

STEP 3   Determine the voltage across each series resistor and each series equivalent resistor. To achieve this, we utilize the diagram obtained after completing Step B (Figure 9-20):

$$V_{R1} = I_t \times R_1 = 10\text{ mA} \times 1\text{ k}\Omega = 10\text{ V}$$
$$V_{R2,3} = I_t \times R_{2,3} = 10\text{ mA} \times 222.2 = 2.222\text{ V}$$
$$V_{R4} = I_t \times R_4 = 10\text{ mA} \times 777.8 = 7.778\text{ V}$$
$$V_{R5,6,7} = I_t \times R_{5,6,7} = 10\text{ mA} \times 500 = 5\text{ V}$$

Figure 9-22 illustrates the results after Step 3.

STEP 4   Determine the value of current through each parallel resistor (Figure 9-23). R_1 and R_4 are series-connected resistors and, therefore, their current will equal 10 mA.

$$I_{R1} = 10\text{ mA}$$
$$I_{R4} = 10\text{ mA}$$

The current through the parallel resistors is calculated by Ohm's law.

$$I_{R2} = \frac{V_{R2}}{R_2} = \frac{2.222\text{ V}}{500\text{ }\Omega} = 4.4\text{ mA}$$

$$I_{R3} = \frac{V_{R3}}{R_3} = \frac{2.222\text{ V}}{400\text{ }\Omega} = 5.6\text{ mA}$$

$$\left.\begin{array}{c} I_t = I_{R2} + I_{R3} \\ 10\text{ mA} = 4.4\text{ mA} + 5.6\text{ mA} \end{array}\right\} \text{Kirchhoff's current law}$$

$$I_{R5} = \frac{V_{R5}}{R_5} = \frac{5\text{ V}}{2\text{ k}\Omega} = 2.5\text{ mA}$$

Five Step Method for Series–Parallel Circuit Analysis   **287**

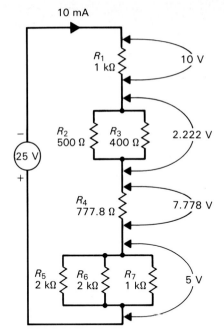

10 mA

R_1
1 kΩ

10 V

R_2
500 Ω

R_3
400 Ω

2.222 V

25 V

R_4
777.8 Ω

7.778 V

R_5
2 kΩ

R_6
2 kΩ

R_7
1 kΩ

5 V

FIGURE 9-23

$$I_{R6} = \frac{V_{R6}}{R_6} = \frac{5\text{ V}}{2\text{ k}\Omega} = 2.5\text{ mA}$$

$$I_{R7} = \frac{V_{R7}}{R_7} = \frac{5\text{ V}}{1\text{ k}\Omega} = 5\text{ mA}$$

$$\left.\begin{array}{l} I_t = I_{R5} + I_{R6} + I_{R7} \\ 10\text{ mA} = 2.5\text{ mA} + 2.5\text{ mA} + 5\text{ mA} \end{array}\right\} \text{Kirchhoff's current law}$$

STEP 5   Determine the total power dissipated by the circuit.

$$P_t = P_{R1} + P_{R2} + P_{R3} + P_{R4} + P_{R5} + P_{R6} + P_{R7}$$

or

$$P_t = \frac{V_t^2}{R_t}$$

Each resistor's power figure can be calculated and the sum would be the total power dissipated by the circuit. Since the problem does not ask for the power dissipated by each individual resistor, but for the total power dissipated, it will be easier to use the formula:

$$P_t = \frac{V_t^2}{R_t}$$

$$= \frac{25\text{ V}^2}{2.5\text{ k}\Omega}$$

$$= 0.25\text{ W}$$

9.6
SERIES–PARALLEL CIRCUITS

9.6.1 LOADING OF VOLTAGE-DIVIDER CIRCUITS

The straightforward voltage divider was discussed in Chapter 7, but at that point we did not explore some changes that will occur if a load resistance is connected to the voltage divider's output.

Figure 9-24 illustrates a voltage divider and, as you can see, the advantage of a voltage-divider circuit is that it can be used to produce several different voltages from one main voltage source by the use of a few chosen resistor values.

In our discussion on load resistance, we mentioned that every circuit or piece of equipment offers a certain amount of resistance, and this resistance represents how much a circuit or piece of equipment will load down the source supply.

Figure 9-25 illustrates an example voltage-divider circuit that is being used to develop a 10-V source from a 20-V battery supply. Figure 9-25(a) illustrates this circuit in the unloaded condition, and by making a few calculations you can understand everything about the circuit.

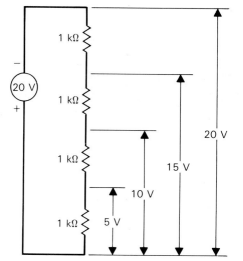

FIGURE 9-24

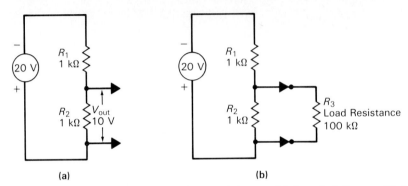

(a)   (b)

FIGURE 9-25 (a) Unloaded Output Voltage. (b) Loaded Output Voltage.

STEP 1 $R_t = R_1 + R_2 = 1 \text{ k}\Omega + 1 \text{ k}\Omega = 2 \text{ k}\Omega$

STEP 2 $I_t = \dfrac{V_t}{R_t} = \dfrac{20 \text{ V}}{2 \text{ k}\Omega} = 10 \text{ mA}$

The current that flows through a voltage divider, without a load connected, is called the *bleeder current*. In this example the current is equal to 10 mA. It is called the bleeder current because it is continually drawing or bleeding this current from the voltage source.

STEP 3 $V_{R1} = V_{R2}$ (as resistors are the same value)

         $V_{R1} = 10 \text{ V}$

         $V_{R2} = 10 \text{ V}$

In Figure 9-25(b), we have connected a piece of equipment represented as a resistance (R_3) across the 10-V supply. This automatically turns the previous series circuit of R_1 and R_2 into a series–parallel circuit made up of R_1, R_2, and the 100-kΩ load resistance. By making a few more calculations, we can discover the changes that have occurred by connecting this load resistance.

STEP 1   Total resistance (R_t)

$$\text{Step B} \quad R_{2,3} = \frac{R_2 \times R_3}{R_2 + R_3} = 990.1 \ \Omega$$

$$\text{Step C} \quad R_{1,2,3} = R_1 + R_{2,3}$$
$$= 1 \text{ k}\Omega + 990.1$$
$$= 1.99 \text{ k}\Omega$$

STEP 2   Total current (I_t)

$$V_t = \frac{20 \text{ V}}{1.99 \text{ k}\Omega} = 10.05 \text{ mA}$$

STEP 3 $V_{R1} = I_t \times R_1 = 10.05 \text{ mA} \times 1 \text{ k}\Omega = 10.05 \text{ V}$
            $V_{R2,3} = I_t \times R_{2,3} = 10.05 \text{ mA} \times 990.1 = 9.95 \text{ V}$

STEP 4 $I_{R1} = I_t = 10.05 \text{ mA}$

$$I_{R2} = \frac{V_{R2}}{R_2} = \frac{9.95 \text{ V}}{1 \text{ k}\Omega} = 9.95 \text{ mA}$$

$$I_{R3} = \frac{V_{R3}}{R_3} = \frac{9.95 \text{ V}}{100 \text{ k}\Omega} = 99.5 \text{ } \mu\text{A}$$

$$\left.\begin{array}{c} I_{R2} + I_{R3} = I_t \\ 9.95 \text{ mA} + 99.5 \text{ } \mu\text{A} = 10.05 \text{ mA} \end{array}\right\} \text{ Kirchhoff's current law}$$

As you can see, the load resistance is pulling 99.5 μA, and this pulls the voltage down to 9.95 V from the required 10 V that was desired and is normally present in the unloaded condition.

When designing a voltage divider, design engineers need to calculate how much current a particular load will pull and then alter the voltage-divider resistor values to offset the loading effect when the load is connected.

9.6.2 WHEATSTONE BRIDGE

In 1850, Charles Wheatstone developed a circuit to measure resistance. This circuit, which is still used widely today in a few applications, is called the Wheatstone bridge in his honor; it is illustrated in Figure 9-26(a). In Figure 9-26(b), the same circuit has been drawn differently, yet the circuit functions exactly the same.

(1) BALANCED BRIDGE

Figure 9-27 illustrates an example circuit in which four resistors are connected together to form a series–parallel arrangement. Let us now use the five-step procedure to find out exactly what resistance, current, voltage, and power exist throughout the circuit.

STEP 1   Total resistance (R_t)

Step A $R_{1,3} = R_1 + R_3 = 10 + 20 = 30$
$R_{2,4} = R_2 + R_4 = 10 + 20 = 30$

FIGURE 9-26

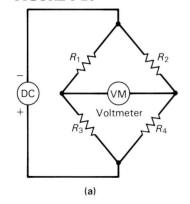

(a)

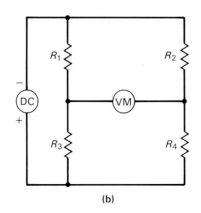

(b)

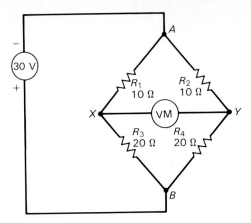

FIGURE 9-27

$$\text{Step B} \quad R_{\text{total}}: (R_{1,2,3,4}) = \frac{R_{1,3} \times R_{2,4}}{R_{1,3} + R_{2,4}}$$

$$= \frac{30 \times 30}{30 + 30} = 15\ \Omega$$

Total resistance $= 15\ \Omega$

STEP 2   Total current (I_t)

$$\frac{V_t}{R_t} = \frac{30\ \text{V}}{15\ \Omega} = 2\ \text{A}$$

STEP 3   Since $R_{1,3}$ is in parallel with $R_{2,4}$, then 30 V will appear across both $R_{1,3}$ and $R_{2,4}$.

$$V_{R1,3} = V_{R2,4} = 30\ \text{V}$$

$$V_{R1} = \frac{R_1}{R_t} \times V_t$$

$$= \frac{10}{30} \times 30 = 10\ \text{V}$$

$$V_{R2} = \frac{R_2}{R_t} \times V_t = 10\ \text{V}$$

$$V_{R3} = \frac{R_3}{R_t} \times V_t = 20\ \text{V}$$

$$V_{R4} = \frac{R_4}{R_t} \times V_t = 20\ \text{V}$$

STEP 4 $$I_{R1} = \frac{V_{R1,3}}{R_{1,3}} = \frac{30\ \text{V}}{30\ \Omega} = 1\ \text{A}$$

$$I_{R2,4} = \frac{V_{R2,4}}{R_{2,4}} = \frac{30\ \text{V}}{30\ \Omega} = 1\ \text{A}$$

$$\left.\begin{array}{l} I_{R1,3} + I_{R2,4} = I_t \\ 1\ \text{A} + 1\ \text{A} = 2\ \text{A} \end{array}\right\} \text{Kirchhoff's current law}$$

Series–Parallel DC Circuits

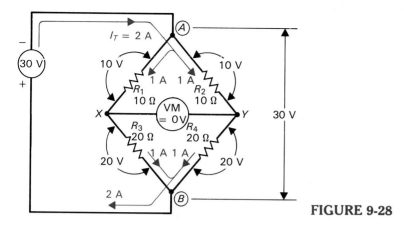

FIGURE 9-28

STEP 5   Total power dissipated $(P_t) = I_t^2 \times R_t$
$$= 2\,\text{A}^2 \times 15\,\Omega$$
$$= 4\,\text{A} \times 15\,\Omega$$
$$= 60\,\text{W}$$

Figure 9-28 illustrates these results inserted in the Wheatstone bridge example schematic. The Wheatstone bridge is said to be in the balanced condition as points X and Y have the same voltage of 20 V. This same voltage exists across R_3 and R_4 and so the voltmeter, which is measuring the voltage difference between X and Y, will indicate 0-V potential difference.

(2) UNBALANCED BRIDGE

In Figure 9-29, we have replaced R_3 with a variable resistor and set it to 10 Ω. The R_2 and R_4 resistor combination will not change its voltage drop; however, R_1 and R_3, which are now equal, will each split the 30-V supply, producing 15 V across R_3. The voltmeter will indicate the difference in potential (5 V) from the voltage across R_3 at point X (15 V) and across R_4 at point Y (20 V). The voltmeter is actually measuring the imbalance in the circuit, which is why this circuit in this condition is known as an unbalanced bridge.

FIGURE 9-29

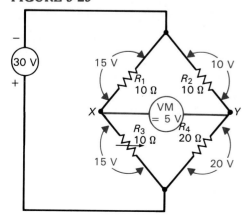

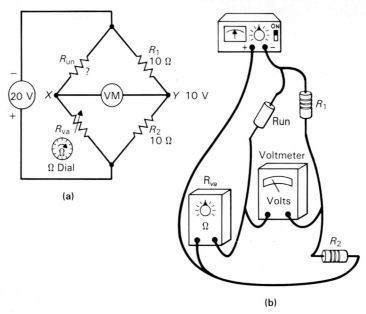

FIGURE 9-30 (a) Schematic. (b) Pictorial.

(3) DETERMINING UNKNOWN RESISTANCE

Figure 9-30 illustrates a Wheatstone bridge into which an unknown resistor (R_{un}) and variable resistor (R_{va}) have been inserted. The variable resistor is a calibrated resistor (resistance has been checked against a known, accurate resistance) in which the resistance can be adjusted and read from a calibrated dial.

The procedure to follow to find the value of the unknown resistor is as follows:

1. Adjust the variable resistor until the voltmeter indicates that the Wheatstone bridge is balanced (voltmeter indicates 0 V).
2. Read the value of the variable resistor. As long as $R_1 = R_2$, the variable resistance value will be the same as the unknown resistance value.

$$R_{va} = R_{un}$$

Since R_1 and R_2 are equal to one another, the voltage will be split across the two resistors, producing 10 V at point Y. The variable resistor must therefore be adjusted so that it equals the unknown resistance, and therefore the same situation will occur, in that the 20-V source will be split, producing 10 V at point X, indicating a balanced condition on the VOM. For example, if the unknown resistance is equal to 5 Ω, then only when the variable resistor is adjusted and equal to 5 Ω would 10 V appear at point X and the circuit be balanced. The variable resistor resistance could be read (5 Ω) and the unknown resistor resistance would be known (5 Ω).

EXAMPLE 9.5

What is the unknown resistance in Figure 9-31?

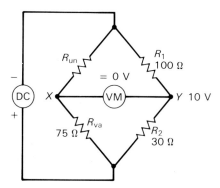

FIGURE 9-31

SOLUTION

The bridge is in a balanced condition as the voltmeter is reading a 0-V difference between points X and Y. The next step is to read off the value of the variable resistor, 75 Ω, and as long as $R_1 = R_2$, then the variable resistance value will equal the unknown resistance value.

$$R_{va} = R_{un}$$

In this case, R_1 does not equal R_2, so a variation in the formula must be applied to take into account the ratio of R_1 and R_2.

$$R_{un} = R_{va} \times \frac{R_1}{R_2}$$
$$= 75\ \Omega \times \frac{100}{30}$$
$$= 75\ \Omega \times 3.33 = 250\ \Omega$$

Calculator Sequence

Step	Keypad Entry	Display Response
1.	$\boxed{1}\ \boxed{0}\ \boxed{0}$	100
2.	$\boxed{\div}$	
3.	$\boxed{3}\ \boxed{0}$	30
4.	$\boxed{=}$	3.333
5.	$\boxed{\times}$	
6.	$\boxed{7}\ \boxed{5}$	75
7.	$\boxed{=}$	250

9.6.3 LADDER CIRCUIT

Figure 9-32 illustrates a R–$2R$ ladder circuit, which is a series–parallel circuit used for digital-to-analog conversion. To fully understand this circuit, our first step should be to find out exactly which branches will have which values of current. This can be obtained by finding out what value of resistance the current sees when it arrives at the three junctions A, B, and C. Let us begin with point C first and simplify the circuit. This is illustrated in Figure 9-33(a). No specific resistance value has been chosen, but in all cases $2R$ resistors ($2 \times R$) are twice the resistance of an R resistor.

FIGURE 9-32

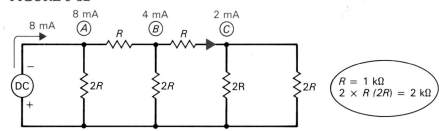

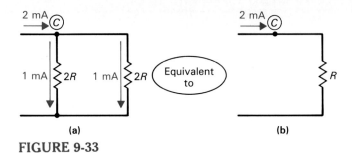

(a)   **(b)**

FIGURE 9-33

In Figure 9-33(a), if 2 mA of current arrives at point C, it sees $2R$ of resistance in parallel with a $2R$ resistance, and consequently the 2 mA of current splits and 1 mA flows through each branch. Two $2R$ resistors in parallel with one another would consequently be equivalent to one R, as seen in Figure 9-33(b).

In Figure 9-34(a), if 4 mA of current arrives at point B, it sees two series resistances to the right, which is equivalent to $2R$ [Figure 9-35(b)] and one resistance down of $2R$. The 4 mA therefore splits, causing 2 mA down one path and 2 mA down the other. The two $2R$ resistors in parallel as seen in Figure 9-34(b) are equivalent to one R, as shown in Figure 9-34(c).

In Figure 9-35(a), if 8 mA of current arrives at point A, it sees two series resistors to the right, which is equivalent to $2R$ [Figure 9-35(b)], and one resistance down of $2R$. The 8 mA of current therefore splits equally, causing 4 mA down one path and 4 mA down the other. The two $2R$ resistors in parallel [Figure 9-35(b)] are equivalent to one resistance R [Figure 9-35(c)].

Figure 9-36(a) through (f) illustrates the step by step simplification of this circuit. The question you may have at this point is: What is the main application of this circuit? The answer is as a current divider, as seen in Figure 9-37(a). The 8 mA of reference current is repeatedly divided by 2 as it moves from left to right, producing currents of 4 mA, 2 mA, and 1

FIGURE 9-34

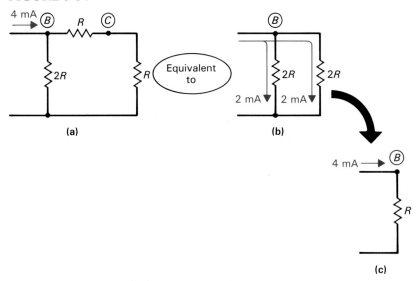

(a)   **(b)**

(c)

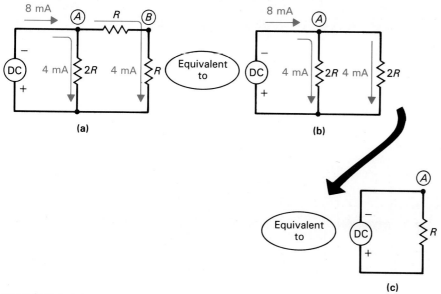

FIGURE 9-35

FIGURE 9-36

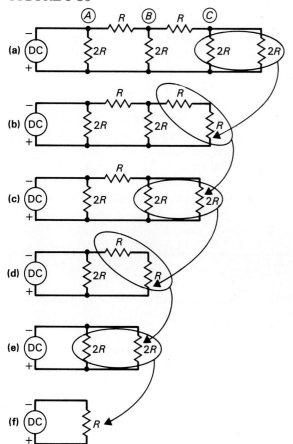

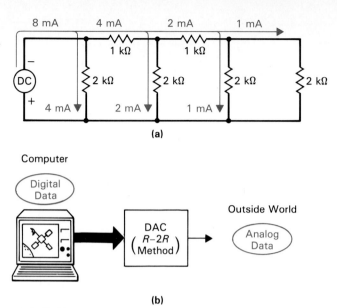

(a)

(b)

FIGURE 9-37

mA. The result of this *R–2R* current division can be used in a circuit known as a digital-to-analog converter (DAC), which is illustrated in Figure 9-37(b).

Digital data or information exists within a computer, and this system expresses numbers and letters in two discrete steps, for example, on–off, high–low, open–closed, or 0–1. Only two conditions exist within the computer, and all information is represented by this two-state system.

Analog data or information exists outside a computer in our environment, and in this system, data are expressed in many different levels, as opposed to just two with digital information. Numbers are expressed in one of ten different levels (0–9), as opposed to only two (0–1) in digital.

Due to these differences, a device is needed that will interface (convert or link two different elements) the digital information within a computer to the analog information that you and I understand. This device is called a digital-to-analog converter and uses the *R–2R* ladder circuit that we just discussed.

EXAMPLE 9.6

Determine the reference current and branch currents for the circuit in Figure 9-38.

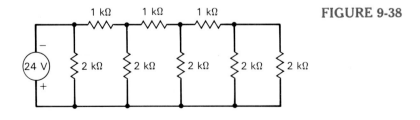

FIGURE 9-38

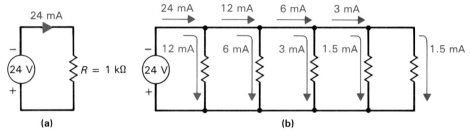

FIGURE 9-39

SOLUTION

In our simplification of the ladder circuit, we discussed previously that any R–$2R$ ladder circuit can be simplified to one resistor equal to R, as seen in Figure 9-39(a). The reference or total current supplied will therefore equal

$$I_t = \frac{V_t}{R_t} = \frac{24 \text{ V}}{1 \text{ k}\Omega} = 24 \text{ mA}$$

and the current will split through each branch, as shown in Figure 9-39(b).

SELF-TEST REVIEW (§ 9.6)

1. What is meant by loading of a voltage-divider circuit?
2. Sketch a Wheatstone bridge circuit and list an application of this circuit.
3. An R–$2R$ ladder circuit will always have a total resistance equal to?
4. For what application could the R–$2R$ ladder be used?

9.7

THEOREMS FOR DC CIRCUITS

Series–parallel circuits can become very complex in some applications, and the more help you have in simplifying and analyzing these networks, the better. The following theorems can be used as powerful analytical tools for evaluating circuits. To begin, let's discuss the differences between voltage and current sources.

9.7.1 VOLTAGE AND CURRENT SOURCES

The easiest way to understand a current source is to compare its feature to a voltage source, and so let's begin by discussing voltage sources.

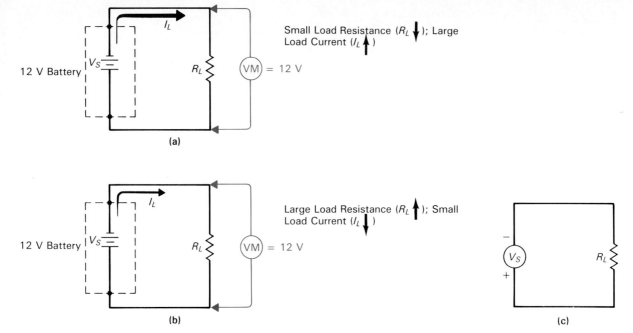

FIGURE 9-40 **Ideal Voltage Source. (a) Heavy Load. (b) Light Load. (c) Symbol.**

(1) VOLTAGE SOURCE

The battery is an example of a voltage source that in the ideal condition will produce a fixed output voltage regardless of what load resistance is connected across its terminals. This means that even if a large load current is drawn from the battery (heavy load, due to a small load resistance) or if a small load current results (light load, due to a large load resistance), the battery will always produce a constant output voltage, as seen in Figure 9-40.

In reality, every voltage source, whether a battery, power supply, or generator, will have some level of inefficiency and not only generate an output electrical dc voltage but also generate heat. This inefficiency is represented as an internal resistance, as seen in Figure 9-41(a), and in most cases this internal source resistance (R_{int}) is very low (several ohms) compared to the load resistance (R_L). In Figure 9-41(a), no load has been connected, so the

FIGURE 9-41 **Realistic Voltage Source. (a) Unloaded. (b) Loaded.**

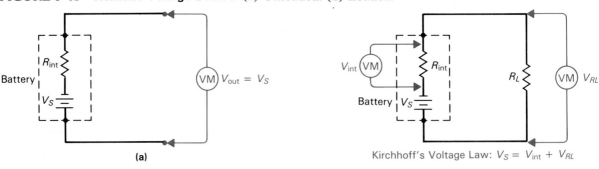

Kirchhoff's Voltage Law: $V_S = V_{int} + V_{RL}$

(b)

output or open circuit voltage will be equal to the source voltage, V_S. When a load is connected across the battery, as seen in Figure 9-41(b), R_{int} and R_L form a series circuit, and some of the source voltage appears across R_{int}; so the output or load voltage is always less than V_S. Since R_{int} is normally quite

---

EXAMPLE 9.7

Calculate the output voltage in Figure 9-42 if R_L is equal to: (a) 100 Ω; (b) 1 kΩ; (c) 100 kΩ.

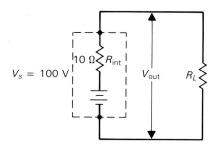

FIGURE 9-42

SOLUTION

(voltage-divider formula)
a. $R_L = 100$ Ω:

$$V_{out} = \frac{R_L}{R_T} \times V_S$$

$$= \frac{100\ \Omega}{110\ \Omega} \times 100\ V = 90.9\ V$$

b. $R_L = 1$ kΩ (1000 Ω)

$$V_{out} = \left(\frac{1000\ \Omega}{1010\ \Omega}\right) \times 100\ V = 99.0\ V$$

c. $R_L = 100$ kΩ (100,000 Ω)

$$V_{out} = \left(\frac{100,000\ \Omega}{100,010\ \Omega}\right) \times 100\ V = 99.99\ V$$

From this example, you can see that the larger the load resistance, the greater the output voltage (V_{out} or V_{RL}). To explain this in a little more detail, we can say that a large R_L is considered a light load for the voltage source as it only has to produce a small load current ($R_L\uparrow$, $I_L\downarrow$), and consequently the heat generated by the source is small ($P_{R_{int}}\downarrow = I^2\downarrow \times R$) and the voltage source is more efficient (V_{out} almost equals V_S), approaching ideal ($V_{out} = V_S$). The voltage source in this example, however, produced an almost constant output voltage (within 10% of V_S) despite the very large changes in R_L.

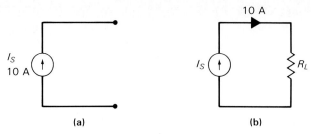

FIGURE 9-43   Ideal Current Source. (a) Unloaded. (b) Loaded.

small compared to R_L, the voltage source approaches ideal, as almost all the source voltage (V_S) appears across R_L.

In conclusion, a voltage source should have the smallest possible internal resistance so that the ouptut voltage (V_{out}) will remain constant and approximately equal to V_S independent of whether a light load (large R_L, small I_L) or heavy load (small R_L, large I_L) is connected across its output terminals.

(2) CURRENT SOURCE

Just as a voltage source has a certain voltage rating, the current source has a certain current rating; and just as a voltage source should deliver a constant output voltage, an ideal current source should deliver its constant rated current, regardless of what value of load resistance is connected across its output terminals, as seen in Figure 9-43.

A current source can be thought of as a voltage source with an extremely large internal resistance, as seen in Figure 9-44(a) and symbolized in Figure 9-44(b).

In conclusion, a current source should have a large internal resistance so that, whatever the load resistance connected across the output, it will have very little effect on the total resistance and the load current will remain constant.

The symbol for a constant current source has an arrow within a circle and this arrow points in the direction of current flow.

FIGURE 9-44   Realistic Current Source.

$$I = \frac{V}{R} = \frac{10 \text{ V}}{1 \text{ M}\Omega} = 10 \text{ }\mu\text{A}$$

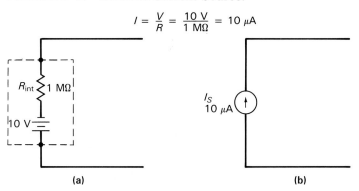

EXAMPLE 9.8

Calculate the load current supplied by the current source in Figure 9-45 if the following values of R_L are connected across the output terminals: (a) 100 Ω; (b) 1 kΩ; (c) 100 kΩ.

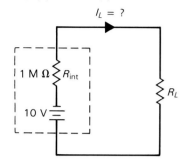

FIGURE 9-45

SOLUTION

a. $R_L = 100\ \Omega\ (R_T = R_{int} + R_L = 1{,}000{,}000 + 100\ \Omega = 1{,}000{,}100\ \Omega)$

$$I_L = \frac{V_S}{R_T}$$

$$= \frac{10\ \text{V}}{1{,}000{,}100\ \Omega} = 9.999\ \mu\text{A}$$

b. $R_L = 1\ \text{k}\Omega\ (R_T = R_{int} + R_L = 1{,}001{,}000\ \Omega)$

$$I_L = \frac{10\ \text{V}}{1{,}001{,}000\ \Omega} = 9.99\ \mu\text{A}$$

c. $R_L = 100\ \text{k}\Omega\ (R_T = R_{int} + R_L = 1{,}100{,}000\ \Omega)$

$$I_L = \frac{10\ \text{V}}{1{,}100{,}000\ \Omega} = 9.09\ \mu\text{A}$$

From this example, you can see that the current source delivered an almost constant output voltage regardless of the large load resistance change.

9.7.2 SUPERPOSITION THEOREM

This logical theorem is used not only in electronics, but also in physics and even economics. It is used to determine the net effect in a circuit that has two or more sources connected. The basic idea behind this theorem is that, if two voltage sources are both producing a current through the same circuit, the net current can be determined by first finding the individual currents and then adding them together. Stated formally: *In a network containing two or more voltage sources, the current at any point is equal to the algebraic sum of the individual source currents produced by each source acting separately.*

   The best way to fully understand the theorem is to apply it to a few examples. Figure 9-46(a) illustrates a simple series circuit with two resistors and two voltage sources. The 12-V source (V_1) is trying to produce a current in a clockwise direction, while the 24-V source (V_2) is trying to force current

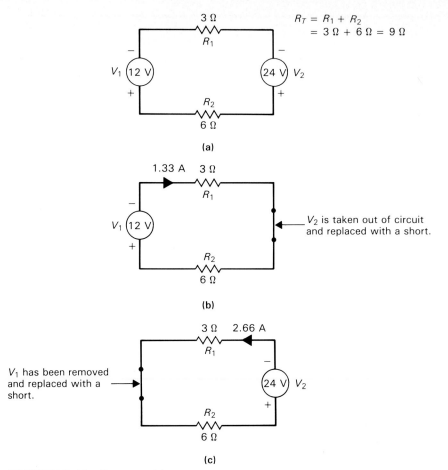

FIGURE 9-46 **Superposition.**

in a counterclockwise direction. What will be the resulting net current in this circuit?

STEP 1 To begin, let's consider what current would be produced in this circuit if only V_1 were connected, as seen in Figure 9-46(b).

$$I_1 = \frac{V_1}{R_T} = \frac{12\text{ V}}{9\text{ }\Omega} = 1.33\text{ A}$$

STEP 2 The next step is to determine how much current V_2 would produce if V_1 were not connected in the circuit, as seen in Figure 9-46(c).

$$I_2 = \frac{V_2}{R_T} = \frac{24\text{ V}}{9\text{ }\Omega} = 2.66\text{ A}$$

V_1 is attempting to produce 1.33 A in the clockwise direction, while V_2 is trying to produce 2.66 A in the counterclockwise direction. The net current will consequently be 1.33 A in the counterclockwise direction.

EXAMPLE 9.9

Calculate the current through R_2 in Figure 9-47 using the superposition theorem.

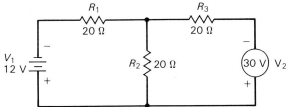

FIGURE 9-47

SOLUTION

The first step is to calculate the current through R_2 due only to the voltage source V_1. This is shown in Figure 9-48(a). R_1 is a series-connected resistor, while R_2 and R_3 are connected in parallel with one another. So:

$$R_{2,3} = \frac{R_{val}}{n} = \frac{20\ \Omega}{2} = 10\ \Omega \text{ (same value parallel resistor formula)}$$

$$R_T = R_1 + R_{2,3} = 20\ \Omega + 10\ \Omega = 30\ \Omega \text{ (total resistance)}$$

$$I_T = \frac{V_1}{R_T} = \frac{12\ V}{30\ \Omega} = 400\ mA \text{ (total current)}$$

$$I_{R2} = \frac{R_{2,3}}{R_2} \times I_T = \frac{10\ \Omega}{20\ \Omega} \times 400\ mA = 200\ mA \text{ (current divider formula)}$$

This 200 mA of current is flowing down through R_2.

The next step is to find the current flow through R_2 due only to the voltage source V_2. This is shown in Figure 9-48(b). In this instance, R_3 is a series-connected resistor, and R_1 and R_2 make up a parallel circuit. So:

$$R_{1,2} = 10\ \Omega$$

$$R_T = R_{1,2} + R_3 = 10\ \Omega + 20\ \Omega = 30\ \Omega$$

$$I_T = \frac{V_2}{R_T} = \frac{30\ V}{30\ \Omega} = 1\ A \quad (1000\ mA)$$

$$I_{R2} = \frac{R_{1,2}}{R_2} \times I_T = \frac{10\ \Omega}{20\ \Omega} \times 1000\ mA = 500\ mA$$

This 500 mA of current will flow down through R_2.

Since both V_1 and V_2 produce a current flow down through R_2, and so I_{R2} and I_{R3} have the same algebraic sign, the total current through R_2 is equal to the sum of the two currents produced by V_1 and V_2.

$$I_{R2} \text{ (total)} = I_{R2} \text{ due to } V_1 + I_{R2} \text{ due to } V_2$$
$$= 200\ mA + 500\ mA$$
$$= 700\ mA$$

FIGURE 9-48

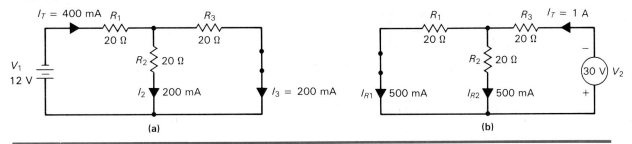

(a)

(b)

EXAMPLE 9.10

Calculate the current flow through R_1 in Figure 9-49.

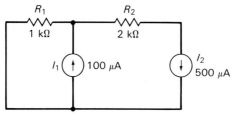

FIGURE 9-49

SOLUTION

Current sources are treated differently from voltage sources in that each current source is removed from the circuit and replaced with an open, as illustrated in the first step of the solution shown in Figure 9-50(a). In this instance, you can see that current has only one path, and so the current through R_1 is counterclockwise and equal to the I_1 source current, 100 μA.

In Figure 9-50, the current source I_1 has been removed and replaced with an open. R_1 and R_2 form a series circuit, and so the total source current from I_2 flows through R_1 (I_{R1} = 500 μA) in a clockwise direction.

Since I_1 is producing 100 μA of current through R_1 in a counterclockwise direction, and I_2 is producing 500 μA of current in a clockwise direction, the resulting current through R_1 will equal:

$$I_{R1} = I_2 \, (\text{cw}) - I_1 \, (\text{ccw})$$
$$= 500 \, \mu\text{A} - 100 \, \mu\text{A} = 400 \, \mu\text{A clockwise}$$

FIGURE 9-50

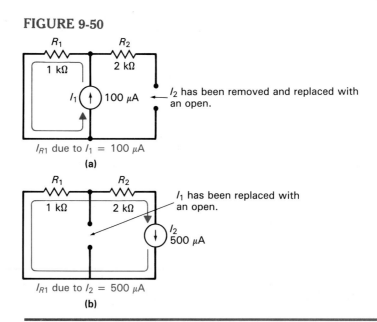

9.7.3 THEVENIN'S THEOREM

Thevenin's theorem allows us to replace the complex networks in Figure 9-51(a) with an equivalent circuit containing just one source voltage (V_{TH}) and one series-connected resistance (R_{TH}), as seen in Figure 9-51(b). Stated formally: *Any network of voltage sources and resistors can be replaced by a single equivalent voltage source (V_{TH}) in series with a single equivalent resistance (R_{TH}).*

Figure 9-52(a) illustrates an example circuit. As with any theorem, a few rules must be followed to obtain an equivalent V_{TH} and R_{TH}.

STEP 1 The first step is to disconnect the load (R_L) and calculate the voltage that will appear across points A and B, as in Figure 9-52(b). This open-circuit voltage will be the same as the voltage drop across R_2 (V_{R2}) and is called the Thevenin equivalent voltage (V_{TH}). First, let's calculate current:

$$I_T = \frac{V_S}{R_T} = \frac{12\text{ V}}{9\ \Omega} = 1.333\text{ A}$$

Therefore, V_{R2} or V_{TH} will equal:

$$V_{R2} = I_T \times R_2 = 1.333 \times 6\ \Omega = 8\text{ V}$$

and so $V_{TH} = 8$ V.

STEP 2 Now that the Thevenin equivalent voltage has been calculated, the next step is to calculate the Thevenin equivalent resistance. In this step, the source voltage is removed and replaced with a short, as seen in Figure 19-52(c), and the Thevenin equivalent resistance is equal to whatever resis-

FIGURE 9-51 Thevenin's Theorem. (a) Complex Multiple Resistors and Source Networks Are Replaced by (b) One Source Voltage (V_{TH}) and One Series-Connected Resistance (R_{TH}).

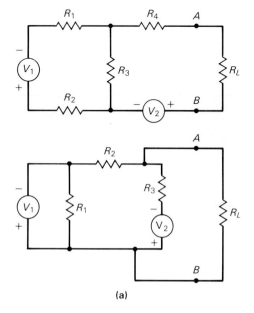

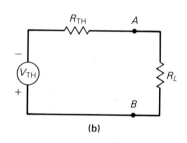

(a)

(b)

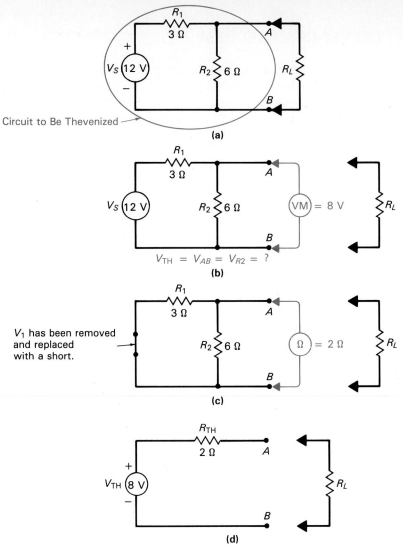

FIGURE 9-52 **Thevinin's Theorem. (a) Example Circuit. (b) Obtaining Thevenin Voltage (V_{TH}). (c) Obtaining Thevenin Resistance. (d) Thevenin Equivalent Circuit.**

EXAMPLE 9.11

Determine V_{TH} and R_{TH} for the circuit in Figure 9-53.

FIGURE 9-53

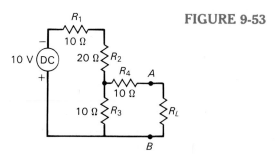

SOLUTION

The first step is to remove the load resistor R_L and calculate what voltage will appear between points A and B, as seen in Figure 9-54(a). Removing R_L will open the path for current to flow through R_4, which will consequently have no voltage drop across it. The voltage between points A and B therefore will be equal to the voltage dropped across R_3, and since R_1, R_2, and R_3 form a series circuit, the voltage-divider formula can be used:

$$V_{R3} = \left(\frac{R_3}{R_T}\right) \times V_S$$

$$= \left(\frac{10\ \Omega}{40\ \Omega}\right) \times 10\ V = 2.5\ V$$

Therefore,

$$V_{AB} = V_{R3} = V_{TH} = 2.5\ V$$

The next step is to calculate the Thevenin resistance, which will equal whatever resistance appears across the terminals A and B with the voltage source having been removed and replaced with a short, as seen in Figure

FIGURE 9-54

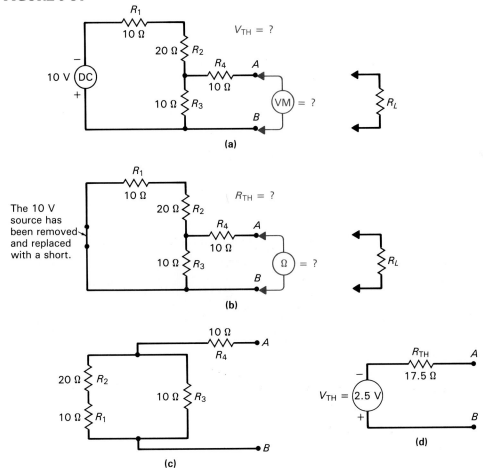

(a)

(b)

(c)

(d)

9-54(b). In Figure 9-54(c), the circuit has been redrawn so that the relationship between the resistors can be seen in more detail. As you can see, R_1 and R_2 are in series with one another and both are in parallel with R_3, and this combination is in series with R_4. Using our three-step procedure for calculating total resistance in a series–parallel circuit, the following results are obtained:

1. $R_{1,2} = R_1 + R_2 = 10\ \Omega + 20\ \Omega = 30\ \Omega$

2. $R_{1,2,3} = \dfrac{R_{1,2} \times R_3}{R_{1,2} + R_3} = \dfrac{30 \times 10}{30 + 10} = \dfrac{300}{40} = 7.5\ \Omega$

3. $R_T = R_{1,2,3} + R_4 = 7.5\ \Omega + 10\ \Omega = 17.5\ \Omega$

Figure 9-54(d) illustrates the Thevenin equivalent circuit.

---

tance exists between points A and B. In this example, R_1 and R_2 form a parallel circuit, the total resistance of which is equal to

$$R_T = \frac{R_1 \times R_2}{R_1 + R_2} = \frac{3 \times 6}{3 + 6} = \frac{18}{9} = 2\ \Omega$$

and so $R_{TH} = 2\ \Omega$.

The circuit to be Thevenized in Figure 19-52(a) can be represented by the Thevenin equivalent circuit seen in Figure 19-52(d).

The question you may have at this time is why we would need to simplify such a basic circuit, when Ohm's law could have been used just as easily to analyze the network. Thevenin's theorem has the following advantages:

1. If you had to calculate load current and load voltage (I_{R_L} and V_{R_L}) for 20 different values of R_L, it would be far easier to use the Thevenin equivalent circuit with the series-connected resistors R_{TH} and R_L, rather than applying Ohm's law to the series–parallel circuit made up of R_1, R_2, and R_L.

2. Thevenin's theorem permits you to solve complex circuits that could not easily be analyzed using Ohm's law.

9.7.4 NORTON'S THEOREM

Norton's theorem, like Thevenin's theorem, is a tool for simplifying a complex circuit into a more manageable one. Figure 9-55 illustrates the difference between a Thevenin equivalent and Norton equivalent circuit. Thevenin's theorem simplifies a complex network and uses an equivalent voltage source (V_{TH}) and an equivalent series resistance (R_{TH}). Norton's theorem, on the other hand, simplifies a complex circuit and represents it with an equivalent current source (I_N) in parallel with an equivalent Norton resistance (R_N), as seen in Figure 9-55.

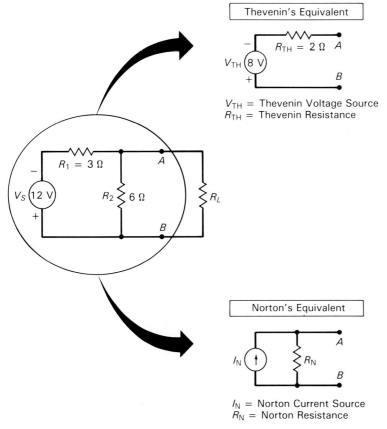

FIGURE 9-55 **Comparison of Thevenin's and Norton's Circuits.**

As with any theorem, a set of steps has to be carried out to arrive at an equivalent circuit. The example we will use is shown in Figure 9-56(a) and is the same example used in Figure 9-55.

STEP 1   Calculate the Norton equivalent current source, which will be equal to the current that would flow between terminals A and B if the load resistor was removed and replaced with a short, as seen in Figure 9-56(b). Placing a short between terminal A and B will short out the resistors R_2, and so the only resistance in the circuit will be R_1. The Norton equivalent current source in this example will therefore be equal to

$$ I_N = \frac{V_s}{R_T} = \frac{12 \text{ V}}{3 \text{ }\Omega} = 4 \text{ A} $$

STEP 2   The next step is to determine the value of the Norton equivalent resistance that will be placed in parallel with the current source, unlike Thevenin's equivalent resistance, which was placed in series. Like Thevenin's theorem though, Norton's equivalent resistance (R_N) is equal to the resistance between terminals A and B when the voltage source is removed and replaced

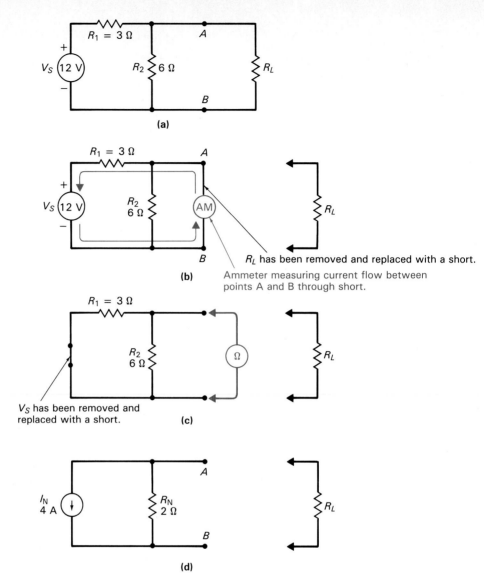

FIGURE 9-56 Norton's Theorem. (a) Example Circuit. (b) Obtaining Norton Current. (c) Obtaining Norton Resistance (R_N). (d) Norton Equivalent Circuit.

EXAMPLE 9.12

Determine I_N and R_N for the circuit in Figure 9-57.

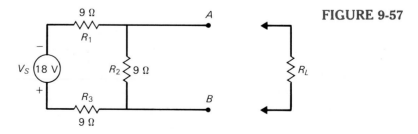

FIGURE 9-57

Series–Parallel DC Circuits

SOLUTION

If a short is placed between terminals A and B, R_2 will be shorted out, and so the current between points A and B, and therefore the Norton equivalent current, will be limited by only R_1 and R_3 and will equal [Figure 9-58(a)]

$$I_N = \frac{V_S}{R_T} = \frac{V_S}{R_1 + R_3} = \frac{18 \text{ V}}{9 + 9 \text{ } \Omega} = 1 \text{ A}$$

Replacing V_S with a short, you can see that our Norton equivalent resistance between terminals A and B is made up of R_1 and R_3 in series with one another, and both are in parallel with R_2. R_N will therefore be equal to [Figure 9-58(b)]

$$R_{1,3} = R_1 + R_3 = 9 \text{ } \Omega + 9 \text{ } \Omega = 18 \text{ } \Omega$$

$$R_T = \frac{R_{1,3} \times R_2}{R_{1,3} + R_2} = \frac{18 \times 9}{18 + 9} = 6 \text{ } \Omega$$

The Norton equivalent circuit is shown in Figure 9-58(c).

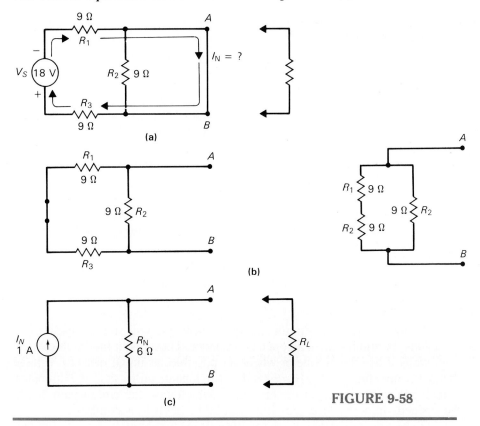

FIGURE 9-58

with a short, as seen in Figure 9-56(c). Since R_1 and R_2 form a parallel circuit, the Norton equivalent resistance will be equal to

$$R_N = \frac{R_1 \times R_2}{R_1 + R_2} = \frac{3 \text{ } \Omega \times 6 \text{ } \Omega}{3 \text{ } \Omega + 6 \text{ } \Omega} = \frac{18 \text{ } \Omega}{9 \text{ } \Omega} = 2 \text{ } \Omega$$

The Norton equivalent circuit has been determined by simply carrying out these two steps and is illustrated in Figure 9-56(d).

1. A constant voltage source will have a _____ internal resistance, while a constant current source will have a _____ internal resistance. (large or small)
2. The superposition theorem is a logical way of analyzing networks with more than one _____.
3. Thevenin's theorem represents a complex two-terminal network as a single _____ source with a series-connected single _____.
4. Norton's theorem also allows you to analyze complex two-terminal networks as a single _____ source in parallel with a single resistor.

9.8
TROUBLESHOOTING SERIES– PARALLEL CIRCUITS

Troubleshooting is defined as the process of locating and diagnosing malfunctions or breakdowns in equipment by means of systematic checking or analysis. Basically, only three problems can occur:

1. A component will open. This usually occurs if a resistor burns out or a wire or switch contact breaks.

2. A component will short. This occurs usually if a conductor, such as solder, wire, or some other conducting material, is dropped or left in the equipment, making or connecting two points that should not be connected.

3. A variation in a component's value. This occurs with age in resistors over a long period of time and can eventually cause a malfunction of the equipment.

Using the example circuit in Figure 9-59, we will step through a few problems, beginning with an open component. Throughout the troubleshooting, we will use the voltmeter whenever possible, as it can measure voltage by just connecting the leads across the component, rather than the ammeter, which has to be placed in the circuit, in which case the circuit path has to be opened. In some instances, this can be difficult to accomplish.

To begin, let's calculate the voltage drops and branch current obtained when the circuit is operating normally.

STEP 1 (A) $R_{3,4} = R_3 + R_4 = 3 + 9 = 12 \, \Omega$

(B) $R_{2,3,4} = \dfrac{R_2 \times R_{3,4}}{R_2 + R_{3,4}} = \dfrac{6 \times 12}{6 + 12} = 4 \, \Omega$

(C) $R_{1,2,3,4} = R_t = R_1 + R_{2,3,4}$
$= 2 \, \Omega + 4 \, \Omega = 6 \, \Omega$

314   Series–Parallel DC Circuits

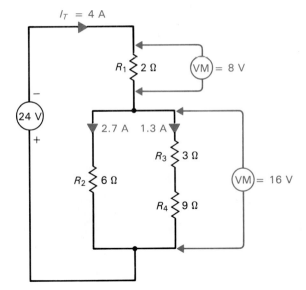

FIGURE 9-59

STEP 2 $I_t = \dfrac{V_t}{R_t} = \dfrac{24\text{ V}}{6\ \Omega} = 4\text{ A}$

STEP 3
$$V_{R1} = I_{R1} \times R_1 = 4\text{ A} \times 2 = \quad 8\text{ V}$$
$$V_{R2,3,4} = I_{R2,3,4} \times R_{2,3,4} = 4\text{ A} \times 4\ \Omega = 16\text{ V}$$
(Kirchhoff's voltage law)

STEP 4 $I_{R1} = 4\text{ A}$ (series resistor)

$$I_{R2} = \dfrac{V_{R2}}{R_2} = \dfrac{16\text{ V}}{6\ \Omega} = 2.7\text{ A}$$

$$I_{R3,4} = \dfrac{V_{R3,4}}{R_{3,4}} = \dfrac{16\text{ V}}{12\ \Omega} = 1.3\text{ A}$$

All these results have been inserted in Figure 9-59.

9.8.1 OPEN COMPONENT

(1) R_1 OPEN (FIGURE 9-60)

With R_1 open, there cannot be any current flow through the circuit as there is not a path from one side of the battery to the other. This fault can be recognized because all the 24 V will be measured across the open resistor (R_1), and 0 V will appear across all the other resistors.

(2) R_3 OPEN (FIGURE 9-61)

With R_3 open, there will be no current through the branch made up of R_3 and R_4. The current path will be through R_1 and R_2 and, therefore, the total resistance will now increase (R_t↑) from 6 Ω to

$$R_t = R_1 + R_2$$
$$= 2\ \Omega + 6\ \Omega = 8\ \Omega$$

Troubleshooting Series–Parallel Circuits   **315**

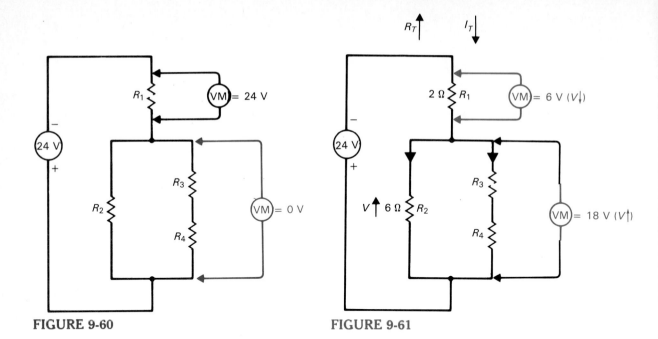

FIGURE 9-60   **FIGURE 9-61**

This 8 Ω is an increase in circuit resistance from the typical resistance which was 6 Ω, which implies an open has occurred to increase resistance. The total current will decrease ($I_t\downarrow$) from 4 A to

$$I_t = \frac{V_t}{R_t} = \frac{24\ V}{8\ \Omega} = 3\ A$$

The voltage drop across the resistors will be

$$V_{R1} = I_t \times R_1 = 3 \times 2 = 6\ V$$

$$V_{R2} = I_t \times R_2 = 3 \times 6 = 18\ V$$

If one of the parallel branches is opened, the overall circuit resistance will always increase. This increase in the total resistance will cause an increase in the voltage dropped across the parallel branch (the greater the resistance, the greater the voltage drop), which enables the technician to localize the fault area and also to determine that the fault is an open.

The voltage measured with a voltmeter will be

$$R_1 = 6\ V$$

$$R_2 = 18\ V$$

$$R_3 = 18\ V$$

$$R_4 = 0\ V$$

This identifies the problem as R_3 being open, as it drops the entire parallel circuit voltage (18 V) across itself, whereas normally the voltage would be dropped proportionally across R_3 and R_4, which are in series with one another.

9.8.2 SHORTED COMPONENT

(1) R_1 SHORTED (FIGURE 9-62)

With R_1 shorted, the total circuit resistance will decrease ($R_t \downarrow$), causing an increase in circuit current ($I_t \uparrow$). This increase in current will cause an increase in the voltage dropped across the parallel branch; however, the fault can be located once you measure the voltage across R_1, which will read 0 V, indicating that this resistor has almost no resistance (for example, 1 Ω) as it has no voltage drop across it.

(2) R_3 SHORTED (FIGURE 9-63)

With R_3 shorted, there will be a decrease in the circuit's total resistance from 6 Ω to

$$R_{2,3,4} = \frac{R_2 \times R_{3,4}}{R_2 + R_{3,4}} = \frac{6 \times 9}{6 + 9} = 3.6$$

$$R_{1,2,3,4} = R_1 + R_{2,3,4}$$
$$= 2\,\Omega + 3.6\,\Omega$$
$$= 5.6\,\Omega$$

This decrease in total resistance ($R_t \downarrow$) will cause an increase in total current ($I_t \uparrow$), which implies a short has occurred to decrease resistance. The total current will now increase from 4 A to

$$I_t = \frac{V_t}{R_t} = \frac{24\text{ V}}{5.6\,\Omega} = 4.3\text{ A}$$

FIGURE 9-62   **FIGURE 9-63**

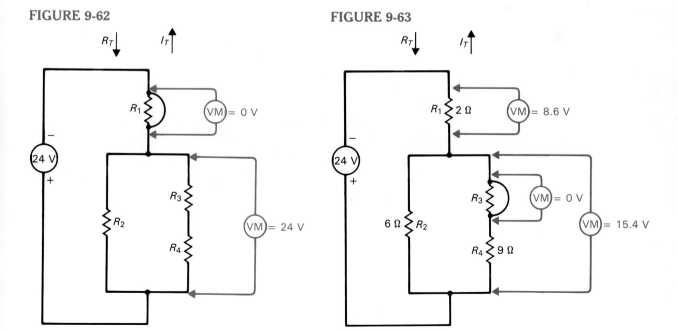

The voltage drops across the resistors will be

$$V_{R1} = I_t \times R_1 = 4.3 \, A \times 2 \, \Omega = 8.6 \, V$$

$$V_{R2,3,4} = I_t \times R_{2,3,4} = 4.3 \, A \times 3.6 \, \Omega = 15.4 \, V$$

If one of the parallel branch resistors is shorted, the overall circuit resistance will always decrease. This decrease in total resistance will cause a decrease in the voltage dropped across the parallel branch (the smaller the resistance, the smaller the voltage drop), and this enables the technician to localize the faulty area and also to determine that the fault is a short.

The voltage measured with the voltmeter (VM) will be

$$R_1 = 8.6 \, V$$

$$R_2 = 15.4 \, V$$

$$R_3 = 0 \, V$$

$$R_4 = 15.4 \, V$$

This identifies the problem as R_3 being a short, as 0 V is being dropped across it. The 15.4 V that we measure across the parallel branch should be proportionally divided across R_3 and R_4, which are in series with one another.

In summary, you may have noticed that (Figure 9-64):

1. An open component causes total resistance to increase ($R_t\uparrow$) and, therefore, total current to decrease ($I_t\downarrow$), and the open component, if in series, has the supply voltage across it, and if in a parallel branch, has the parallel branch voltage across it.

2. A shorted component causes total resistance to decrease ($R_t\downarrow$) and, therefore, total current to increase ($I_t\uparrow$), and the shorted component, if in series or parallel branches, will have 0 V across it.

FIGURE 9-64

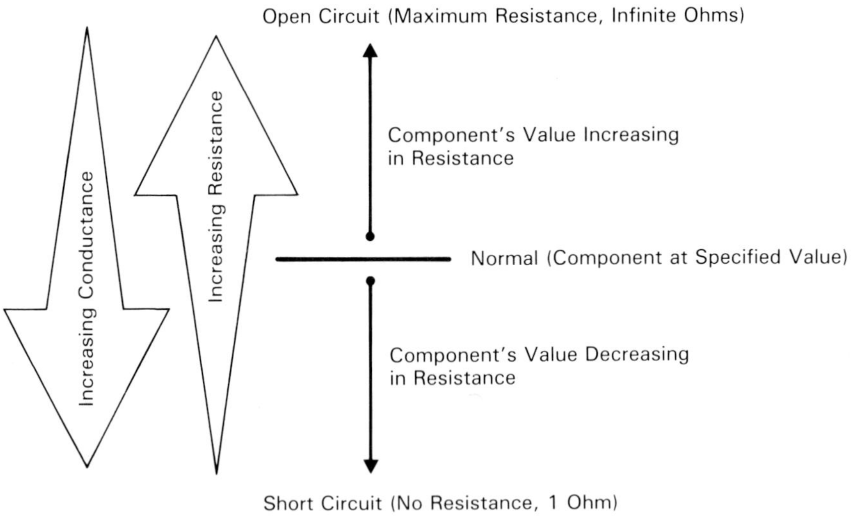

Open Circuit (Maximum Resistance, Infinite Ohms)

Increasing Conductance

Increasing Resistance

Component's Value Increasing in Resistance

Normal (Component at Specified Value)

Component's Value Decreasing in Resistance

Short Circuit (No Resistance, 1 Ohm)

Series–Parallel DC Circuits

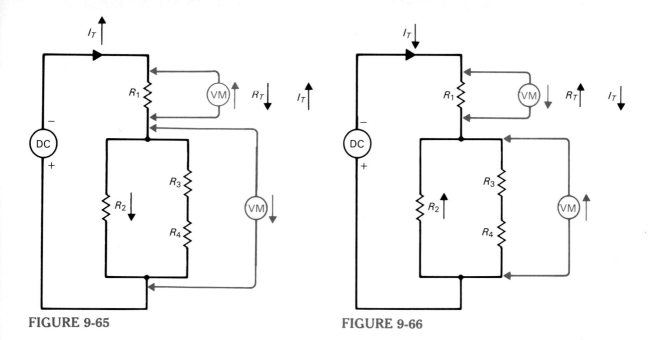

FIGURE 9-65   FIGURE 9-66

9.8.3 RESISTOR VALUE VARIATION

(1) R_2 RESISTANCE DECREASES (FIGURE 9-65)

If the resistance of R_2 decreases, the total circuit resistance will decrease and the total circuit current will increase. The result of this problem and the way in which the fault can be located is that when the voltage drop across R_1 and the parallel branch is tested, there will be an increased voltage drop across R_1 due to the increased current flow, and a decrease in the voltage drop across the parallel branch due to a decrease in the parallel branch resistance.

   With open and shorted components, the large voltage (open) or small voltage (short) drop across a component enables the technician to identify the faulty component. A variation in a component's value, however, will vary the circuit's behavior; but with this example, the symptoms could have been caused by a combination of variations. So once the area of the problem has been localized, the next troubleshooting step is to remove each of the resistors in the suspected faulty area and verify that their resistance values are correct by measuring their resistances with an ohmmeter. In this problem, we had an increase in voltage across R_1 and a decrease in voltage across the parallel branch when measuring with a voltmeter, which was caused by R_2 decreasing. The same swing in voltage readings could also have been obtained by an increase in the resistance of R_1.

(2) R_2 RESISTANCE INCREASES (FIGURE 9-66)

If the resistance of R_2 increases, the total circuit resistance will increase and the total circuit current will decrease. The voltage drop across R_1 will decrease and the voltage across the branch will increase in value. Once again, these measured voltage changes could be caused by the resistance of R_2 increasing or the resistance of R_1 decreasing.

   To reinforce your understanding let's work out a few examples of troubleshooting series–parallel circuits.

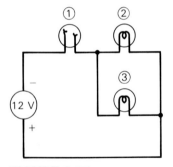

FIGURE 9-67

EXAMPLE 9.13

If bulb 1 in Figure 9-67 goes open, what effect will it have and how will the fault be recognized?

SOLUTION

A visual inspection of the circuit shows all bulbs off as there is no current path from one side of the battery to the other, since bulb 1 is opening the only path because it is connected in series. Since all bulbs are off, the faulty bulb cannot be visually isolated; however, you can easily localize the faulty bulb by using one of two methods:

1. Use the voltmeter and check the voltage across each bulb. Bulb 1 would have 12 V across it, while 2 and 3 would have 0 V across them, isolating the faulty component to bulb 1, since all the supply voltage is being measured across it.

2. By analyzing the circuit diagram, you can see that only one bulb can open and cause all the bulbs to go out, and that is bulb 1. If bulb 2 opens, 1 and 3 would still be on, and if bulb 3 opens, 1 and 2 would still remain on.

EXAMPLE 9.14

One resistor in Figure 9-68 has shorted. From the voltmeter reading shown, determine which one.

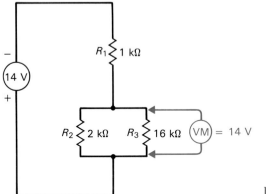

FIGURE 9-68

SOLUTION

If the supply voltage is being measured across the parallel branch of R_2 and R_3, then there cannot be any other resistance in circuit, and so R_1 must have shorted. The next step would be to locate the component, R_1, and determine what has caused it to short. If we were not told that a resistor had shorted, the same symptom could have been caused if R_2 and R_3 were both open and, therefore, the open parallel branch would allow no current to flow and maximum supply voltage would appear across it. The individual component resistance when checked will isolate the problem.

EXAMPLE 9.15

Determine if there is an open or short in Figure 9-69. If so, isolate it by the two voltage readings that are shown in the circuit diagram.

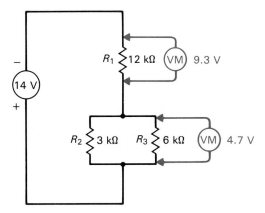

FIGURE 9-69

SOLUTION

Performing a few calculations, you should come up with a normal total circuit resistance of 14 kΩ and a total circuit current of 1 mA. This should cause 12 V across R_1 and 2 V across the parallel branch under no-fault conditions. The decrease in the voltage drop across the series resistor R_1 leads you to believe that there has been a decrease in total circuit current, which must have been caused by a total resistance increase, which points to an open component (assuming only an open or short can occur and not a component value variation).

If R_1 was open, then all the 14 V would have been measured across R_1, which did not occur.

If R_3 opened,

$$\text{total resistance} = 15 \text{ k}\Omega$$

$$\text{total current} = 0.93 \text{ mA} \left(\frac{V}{R} = \frac{14 \text{ V}}{15 \text{ k}\Omega} \right)$$

$$V_{R1} = I_t \times R_1 = 11.16 \text{ V}$$

Since the voltage dropped across R_1 was 9.3 V, R_3 is not our open.

If R_2 opened,

$$\text{total resistance} = 18 \text{ k}\Omega$$

$$\text{total current} = 0.78 \text{ mA} \left(\frac{V}{R} = \frac{14 \text{ V}}{18 \text{ k}\Omega} \right)$$

$$V_{R1} = I_t \times R_1 = 9.36 \text{ V}$$

This circuit's problem is resistor R_2, which has burned out, causing an open circuit.

Describe how to isolate the following problems in a series–parallel circuit.

1. An open component
2. A shorted component
3. A resistor value variation

SUMMARY

1. If current has only one path to follow through a component, then that component is connected in series.
2. If the total current has two or more paths to follow, then these components are connected in parallel.
3. All electronic equipment is composed of many components interconnected to form a combination of series and parallel (series–parallel) circuits.
4. The total resistance in a series–parallel circuit can be calculated by:
   Step A: Determining an equivalent resistance of all series and parallel-connected resistors in parallel combinations.
   Step B: Determining the equivalent resistance of all parallel-connected combinations.
   Step C: Determining the equivalent resistance of the remaining series-connected resistances.
5. The voltage division in a series–parallel circuit can be calculated by:
   Step 1: Determining the circuit's total resistance by applying Steps A, B, and C.
   Step 2: Determining the circuit's total current.
   Step 3: Determining the voltage drop across each series resistor and each parallel combination (series equivalent resistor).
6. The fourth step is to:
   Step 4: Determine the amount of current flowing through each branch of the parallel combinations.
7. The fifth and final step is to:
   Step 5: Determine the total power and individual power figures.
8. The Wheatstone bridge is a series–parallel circuit that was first developed in 1850 by Charles Wheatstone to measure resistance.
9. The $R-2R$ ladder circuit is a series–parallel circuit used for digital-to-analog conversion.
10. An ideal voltage source will have an internal resistance of $0 \, \Omega$ and provide a constant output voltage regardless of what load resistance is connected across its output.
11. In reality, every practical voltage source will have some level of inefficiency, and this is represented as an internal resistance, and the output voltage will remain relatively constant despite variations in load resistance.

12. An ideal current source will have an infinite internal resistance and provide a constant output current, regardless of what load resistance is connected across its output.

13. In reality, every practical current source has a relatively high internal resistance, and the output current will remain relatively constant despite variations in load resistance.

14. The superposition theorem is a useful tool when analyzing circuits with more than one voltage source.

15. Thevenin's theorem is another handy tool that can be used to represent complex series–parallel networks as a single voltage source (V_{TH}) in series with a single resistor (R_{TH}).

16. Norton's theorem can also simplify a complex series–parallel network to an equivalent form consisting of a single current source (I_N) in parallel with a single resistor (R_N).

17. Troubleshooting is the process of locating and diagnosing malfunctions or breakdowns in equipment by means of systematic checking or analysis.

18. **(a)** If a series-connected resistor in a series–parallel circuit opens, there cannot be any current flow, and the source voltage will appear across the open and 0-V will appear across all the others.
(b) If a parallel branch in a series–parallel circuit opens, the overall circuit resistance will increase and the total current will decrease. A greater parallel resistance will cause a greater voltage drop across that parallel circuit.

19. **(a)** If a series-connected resistor in a series–parallel circuit shorts, the total circuit resistance will decrease, resulting in an increase in total current. The fault can be located by measuring the voltage across the shorted resistor, which will be 0 V.
(b) If a parallel branch in a series–parallel circuit shorts, the total resistance will decrease, resulting in a total current increase. This smaller parallel circuit resistance will result in a smaller voltage drop across the parallel circuit.

20. A resistor's resistance will typically change with age, resulting in a total resistance change and therefore a total current change. The faulty component can be located due to its abnormal voltage drop.

NEW TERMS

Analog data	Interface	Superposition
Balanced bridge	Ladder circuit	theorem
Bleeder current	Network	Thevenin resistance
Calibrated	Norton resistance	Thevenin's theorem
Computer	Norton's theorem	Thevenin voltage
Current source	Norton voltage	Unbalanced bridge
Digital data	Series–parallel	Voltage source
Digital-to-analog	circuit	Wheatstone bridge
converter		

NEW FORMULAS

Wheatstone bridge

$$\boxed{R_{variable} = R_{unknown}}$$

Branch resistances are equal.

$$\boxed{R_{unknown} = R_{variable} \times \left(\frac{R_1}{R_2}\right)}$$

Branch resistors, R_1 and R_2, are not equal

REVIEW QUESTIONS

Multiple Choice Questions

1. A series–parallel circuit is a combination of:
   a. Components connected end to end
   b. Series (one current path) circuits
   c. Both series and parallel circuits
   d. Parallel (two or more current path) circuits

2. Total resistance in a series–parallel circuit is calculated by applying the _____ resistance formula to series-connected resistors and the _____ resistance formula to resistors connected in parallel.
   a. Series, parallel
   b. Parallel, series
   c. Series, series
   d. Parallel, parallel

3. Total current in a series–parallel circuit is determined by dividing the total _____ by the total _____.
   a. Power, current
   b. Voltage, resistance
   c. Current, resistance
   d. Voltage, power

4. Branch current within series–parallel circuits can be calculated by:
   a. Ohm's law
   b. The current-divider formula
   c. Kirchhoff's current law
   d. All the above

5. All voltages on a circuit diagram are with respect to _____ unless otherwise stated.
   a. The other side of the component
   b. The high-voltage source
   c. Ground
   d. All the above

6. A _____ ground has the negative side of the source voltage connected to ground, while a _____ ground has the positive side of the source voltage connected to ground.
   a. Positive, negative
   b. Chassis, earth

c. Positive, earth

d. Negative, positive

7. The output voltage will always _____ when a load or volt-meter is connected across a voltage divider.
   a. Decrease
   b. Remain the same
   c. Increase
   d. All the above could be considered true

8. A Wheatstone bridge was originally designed to measure:
   a. An unknown voltage
   b. An unknown current
   c. An unknown power
   d. An unknown resistance

9. A balanced bridge has an output voltage:
   a. Equal to the supply voltage
   b. Equal to half the supply voltage
   c. Of 0 volts
   d. Of 5 volts

10. The total resistance of a ladder circuit is best found by starting at the point _____ the source.
   a. Nearest to
   b. Farthest from
   c. Midway between
   d. All the above

11. The R–$2R$ ladder circuit finds its main application as a(an):
   a. Analog-to-digital converter
   b. Digital-to-analog converter
   c. Device to determine unknown resistance values
   d. All the above

12. The Norton equivalent resistance (R_N) is always in _____ with the Norton equivalent current source (I_N).
   a. Proportion
   b. Series
   c. Parallel
   d. Either series or parallel

13. An ideal current source has _____ internal resistance, while an ideal voltage source has _____ internal resistance.
   a. An infinite, 0 Ω of
   b. No, a large amount
   c. 0 Ω of, an infinite
   d. No, an infinite

14. The Thevenin equivalent resistance (R_{TH}) is always in _____ with the Thevenin equivalent voltage (V_{TH}).
   a. Proportion
   b. Series
   c. Parallel
   d. Either series or parallel

15. The superposition theorem is useful for analyzing circuits with:
   a. Two or more voltage sources

b. A single voltage source

c. Only two voltage sources

d. A single current source

16. A resistor, when it burns out:
    a. Decreases slightly in value
    b. Increases slightly in value
    c. Shorts
    d. Opens

17. In a series–parallel resistive circuit, an open series-connected resistor will cause _____ current, whereas an open parallel-connected resistor will result in a total current _____.
    a. An increase in, decrease
    b. A decrease in, increase
    c. Zero, decrease
    d. None of the above

18. In a series–parallel resistive circuit, a shorted series-connected resistor will cause _____ current, whereas a shorted parallel-connected resistor will result in a total current _____.
    a. An increase in, increase
    b. A decrease in, decrease
    c. An increase in, decrease
    d. A decrease in, increase

19. A resistor's resistance will typically _____ with age, resulting in a total circuit current _____.
    a. Decrease, decrease
    b. Increase, decrease
    c. Decrease, increase
    d. All of the above are true

20. The maximum power transfer theorem states that maximum power is transferred from source to a load when load resistance is equal to source resistance.
    a. True
    b. False

Essay Questions _____

21. State the five-step method for determining a series–parallel circuit's resistance, voltage, current, and power values.   (9.5)

22. Illustrate the following series–parallel circuits:
    a. R_1 in series with a parallel combination R_2, R_3, and R_4.
    b. R_1 in series with a two-branch parallel combination consisting of R_2 and R_3 in series and R_4 in parallel.
    c. R_1 in parallel with R_2, which is in series with a three-resistor parallel combination, R_3, R_4, and R_5.

23. Using the example in question 22(c), apply values of your choice and apply the five-step procedure.   (9.5)

24. Describe what is meant by ''loading of a voltage-divider circuit.''   (9.6.1)

25. Illustrate and describe the Wheatstone bridge in the: (9.6.2)
    a. Balanced condition

    b. Unbalanced condition

    c. Application of measuring unknown resistances

26. Describe how the ladder circuit acts as a current divider. (9.6.3)

27. Briefly describe the difference between a voltage source and a current source. (9.7.1)

28. Briefly describe the following theorems: (9.7)

    a. Superposition

    b. Thevenin's

    c. Norton's

    d. Maximum power transfer

29. What would be the advantages of Thevenin's and Norton's theorems to obtain an equivalent circuit? (9.7)

30. Draw the components that would exist in a: (9.7)

    a. Thevenin equivalent circuit

    b. Norton equivalent circuit

31. Describe the steps involved in obtaining: (9.7)

    a. A Thevenin equivalent circuit

    b. A Norton equivalent circuit

32. When troubleshooting series–parallel circuits, describe what effect (9.8.1)

    a. An open series

    b. An open parallel-connected resistor

    would have on total current and resistance, and how the opened resistor could be isolated.

33. When troubleshooting series–parallel circuits, describe what effect (9.8.2)

    a. A shorted series

    b. A shorted parallel-connected resistor

    would have on total current and resistance, and how the shorted resistors could be isolated.

34. Describe what effect a resistor's value variation would have and how it could be recognized. (9.8.3)

35. Give the divider formula, Ohm's law and Kirchhoff's laws used to determine: (9.5)

    a. Branch currents

    b. Voltage drops in a series–parallel circuit

    (List all six)

Practice Problems _____

36. R_3 and R_4 are in series with one another and are both in parallel with R_5. This parallel combination is in series with two series-connected resistors, R_1 and R_2. $R_1 = 2.5 \text{ k}\Omega$, $R_2 = 10 \text{ k}\Omega$, $R_3 = 7.5 \text{ k}\Omega$, $R_4 = 2.5 \text{ k}\Omega$, $R_5 = 2.5 \text{ M}\Omega$, and $V_S = 100 \text{ V}$. For these values, calculate:

    a. Total resistance

    b. Total current

    c. Voltage across series resistors and parallel combinations

    d. Current through each resistor

    e. Total and individual power figures

37. Referring to the example in question 36, calculate the voltage at every point of the circuit with respect to ground.

38. A 10-V source is connected across a series–parallel circuit made up of R_1 in parallel with a branch made up of R_2 in series with a parallel combination of R_3 and R_4. $R_1 = 100\ \Omega$, $R_2 = 100\ \Omega$, $R_3 = 200\ \Omega$, and $R_4 = 300\ \Omega$. For these values, apply the five-step procedure, and also determine the voltage at every point of the circuit with respect to ground.

39. Calculate the output voltage (V_{RL}) in Figure 9-70 if R_L is equal to (a) 25 Ω; (b) 2.5 kΩ; (c) 2.5 MΩ.

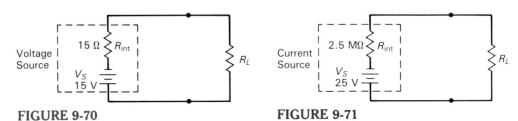

FIGURE 9-70   FIGURE 9-71

40. What load current will be supplied by the current source in Figure 9-71 if R_L is equal to (a) 25 Ω; (b) 2.5 kΩ; (c) 2.5 MΩ.

41. Use the superposition theorem to calculate total current:
   a. Through R_2 in Figure 9-72(a)
   b. Through R_3 in Figure 9-72(b)

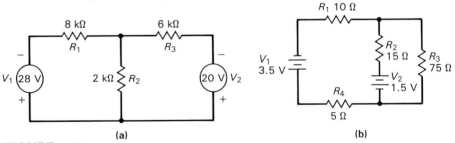

(a)   (b)

FIGURE 9-72

42. Convert the following voltage sources to equivalent current sources:
   a. $V_S = 10$ V, $R_{int} = 15\ \Omega$
   b. $V_S = 36$ V, $R_{int} = 18\ \Omega$
   c. $V_S = 110$ V, $R_{int} = 7\ \Omega$

43. Use Thevenin's theorem to calculate the current through R_L in Figure 9-73. Sketch the Thevenin and Norton equivalent circuits for Figure 9-73.

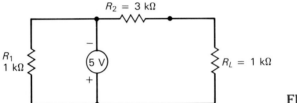

FIGURE 9-73

44. Sketch the Thevenin and Norton equivalent circuits for the networks in Figure 9-74.

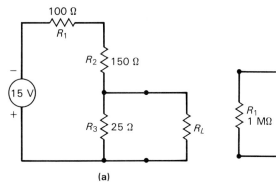

(a)

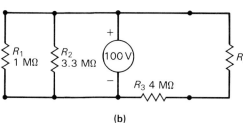

(b)

FIGURE 9-74

45. Convert the following current sources to equivalent voltage sources:
a. $I_S = 5$ mA, $R_{int} = 5$ MΩ
b. $I_S = 10$ A, $R_{int} = 10$ kΩ
c. $I_S = 0.0001$ A, $R_{int} = 2.5$ kΩ

Troubleshooting Questions

46. Referring to the example circuit in Figure 9–68, short and then open a resistor and describe the effects of both.

47. Design a simple five-resistor series–parallel circuit and insert a source voltage and resistance values. Apply the five-step series–parallel circuit procedure, and then theoretically open and short all the resistors and calculate what effect would occur and how you would recognize the problem.

48. Carbon composition resistors tend to increase in resistance with age, while most other types generally decrease in resistance. What effects would resistance changes have on their respective voltage drops?

49. Use Thevenin's theorem to simplify the circuit in Figure 9-75. What effect would the following faults have on the Thevenin equivalent circuit?
a. R_2 is shorted
b. R_2 is open

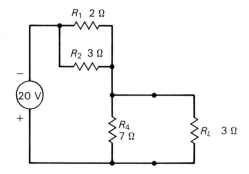

FIGURE 9-75

50. Calculate the Norton equivalent for Figure 9-75 and describe what circuit differences will occur for the same faults listed in question 49.

10

DC TROUBLESHOOTING AND MEASUREMENT EQUIPMENT

AFTER COMPLETING THIS CHAPTER, YOU WILL BE ABLE TO:

1. State the difference between an analog and digital readout multimeter.
2. Explain the construction and operation of a D'Arsonval or moving coil meter movement.
3. Describe what is meant by meter resistance and sensitivity.
4. When using the ammeter to measure current, describe why:
   a. Shunt resistors are used to achieve different range scales.
   b. Three rules must be applied.
   c. It affects the circuit in which it is connected.
5. Describe why, when using the voltmeter to measure voltage:
   a. Multiplier resistors are used to achieve different range scales.
   b. Voltmeter sensitivity determines meter accuracy.
   c. Three rules must be applied.
   d. It affects the circuit in which it is connected.
6. State the difference between an earth and chassis ground.
7. Explain why voltage measurements are usually measured with respect to ground.
8. Describe why, when using the ohmmeter to measure resistance:
   a. An internal battery source is used to supply current.
   b. The ohmmeter scale is nonlinear.
   c. The range scales of an ohmmeter are interpreted differently from those of the voltmeter and ammeter.
   d. Three rules must be applied.

9. State the advantages and disadvantages of both the digital and analog readout multimeter.
10. Explain how to use and interpret the digital multimeter when measuring:
   a. Current
   b. Voltage
   c. Resistance

INTRODUCTION

Measurement, in the field of electronics, is essential to determine the magnitude of electrical quantities such as voltage, current, and resistance.

TABULATING TIME

Born in Buffalo, New York, Herman Hollerith was the son of German immigrants. After finishing his studies at Columbia University in 1879, he began work at the census office in Washington. Hundreds of clerks were needed to tabulate by hand the 1880 census of nearly 13 million people, which would take the next seven and a half years.

Believing this task could be handled by a tabulating system, Hollerith worked for the next 10 years on a solution, which was unveiled in 1890 and took a third of the time to tabulate an increased population of more than $62\frac{1}{2}$ million people. The Hollerith tabulator used punched cards the size of dollar bills to hold data such as age, sex, country of birth, marital status, number of children, and so on.

Hollerith formed a tabulating machine company to sell to government offices his tabulating inventions, one of which even found its way to Tsarist Russia where they felt it was time for a modern census.

Over the years, the company merged with others and went through a number of name changes, the last of which occurred in 1924, five years before Hollerith died, when International Business Machines Corporation was created (now more commonly known as IBM).

10.1
THE MULTIMETER

The multimeter is a device used for measurement and troubleshooting in electronic equipment. It is called a multimeter because it is like having three meters all rolled into one, as seen in Figure 10-1. If the multimeter is switched to measure:

a. Current, it is called an ammeter.

b. Voltage, it is called a voltmeter.

c. Resistance, it is called an ohmmeter.

For many years, the analog readout type of multimeter seen in Figure 10-2(a) was the only type available. At present, these analog readout multimeters are still being used, but are slowly being replaced by digital readout multimeters, similar to the type seen in Figure 10-2(b).

Let's now take a look at each of these meters separately, beginning with the analog readout multimeter.

SELF-TEST REVIEW (§ 10.1)

1. Name the three meters incorporated within a multimeter.

2. What are the two types of readouts available for multimeters?

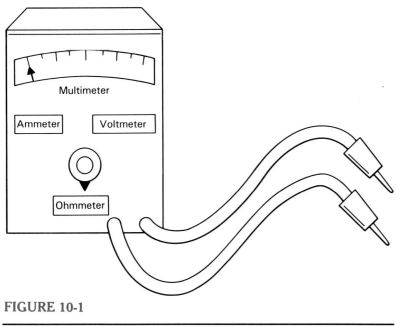

FIGURE 10-1

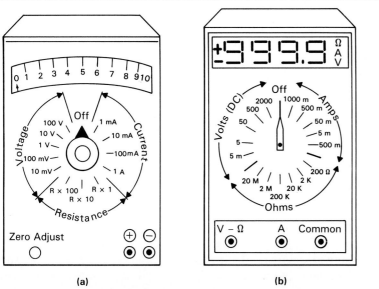

(a)   (b)

FIGURE 10-2 (a) Multimeters. (a) Analog. (b) Digital.

10.1.1 ANALOG READOUT MULTIMETER

The D'Arsonval or moving coil type of meter movement illustrated in Figure 10-3 was invented by Jacques Arsène D'Arsonval and can be used in instruments to measure current, voltage, or resistance.

(1) CONSTRUCTION

A coil of copper wire is mounted on a moving core known as an *armature*, which is free to rotate between the two poles of a stationary permanent

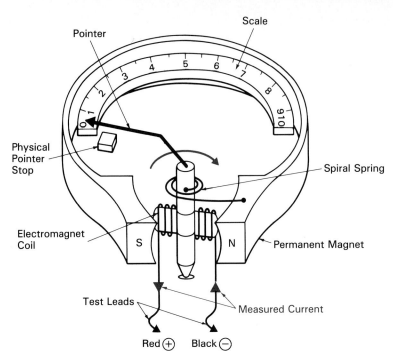

FIGURE 10-3 Moving Coil Meter.

magnet. A spiral spring is attached to the moving coil and holds the armature in a rest position so that the pointer, which is also attached to the armature, is pointing to 0 on the scale.

(2) OPERATION

This moving coil type of meter movement is used in the multimeter, and whether we are measuring current, voltage, or resistance, a current will always flow in the moving coil to represent the amount of current, voltage, or resistance being measured with our two test leads. Let's examine now what will actually happen when a current flows through the moving coil.

When a current is passed through the moving coil, a magnetic field is set up by the electromagnet, producing a north pole on the right side, and a south pole on the left side of the moving coil, as seen in Figure 10-4(a) [left-hand rule for electromagnets, Figure 10-4(b)].

The magnetic field of the electromagnet will interact with the magnetic field of the permanent magnet, forcing the armature to rotate in a clockwise direction (like poles repel), as seen in Figure 10-5.

The south pole generated in the electromagnet is repulsed by the south pole of the permanent magnet and attracted by the north pole of the permanent magnet, while the north pole generated in the electromagnet is repulsed by the north pole of the permanent magnet and attracted by the south pole of the permanent magnet. The overall effect is to overcome the tension created by the spring and move the pointer in a clockwise direction. If a greater current is passed through the electromagnet, a larger magnetic field will be generated, causing a stronger magnetic interaction between the electromagnet and permanent magnet, thus increasing the repulsion and attraction, resulting

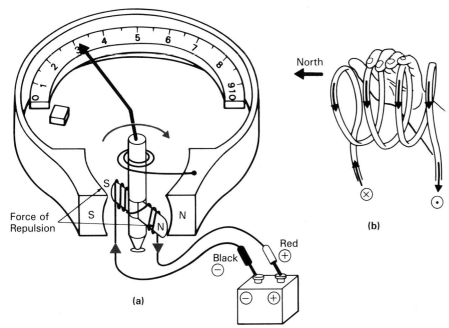

FIGURE 10-4

in an increased movement of the pointer across the scale in the clockwise direction. A greater current causes a greater meter deflection.

If current flows through the electromagnet in the opposite direction, it will generate a north pole in the left side of the electromagnet and a south pole in the right side of the electromagnet. This will result in a counter-clockwise movement of the pointer so that the pointer will strike the physical

FIGURE 10-5 Attraction and Repulsion. (a) Like Poles Repulse One Another. (b) Unlike Poles Attract One Another.

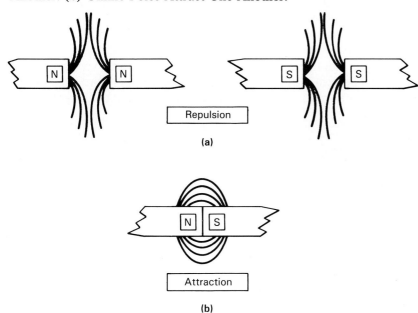

The Multimeter   **335**

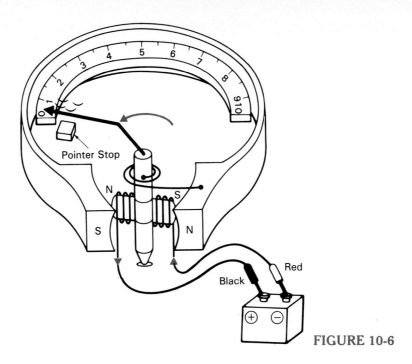

FIGURE 10-6

pointer stop, as seen in Figure 10-6. Too large a current could damage the meter, and this is why the meter leads are said to be polarity (positive and negative) sensitive. Therefore, when measuring a potential difference, the red positive test lead must always be placed on the point that is positive with respect to the other terminal, or, said another way, the negative black test lead must always probe the terminal that is negative with respect to the other point.

(3) METER RESISTANCE AND SENSITIVITY

The coil of the armature in a moving coil meter possesses some value of resistance and is called the movement's internal resistance (R_m). This internal resistance (R_m) is small and normally in the range of 1 to 500 Ω.

The wire used to make the coil is extremely thin (about the thickness of a human hair) and can therefore not, without damage, carry very much current. The typical current rating of a meter movement is therefore small, ranging from 10 μA to 10 mA. This value specifies the amount of current needed

FIGURE 10-7

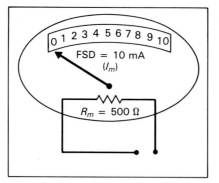

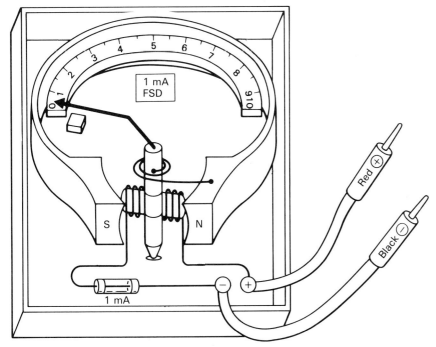

FIGURE 10-8 Physical View of 1-mA Ammeter.

to cause the meter movement to deflect the needle to its full-scale deflection (FSD) position at the far right side of the scale. The maximum current of a meter movement (I_m) will therefore tell you how sensitive a meter is; for example, if a meter movement needs only 10 μA to deflect the pointer to FSD, it is much more sensitive than a meter movement that requires 10 mA. The meter movement can be simplified to display only R_m and I_m and will then appear as in Figure 10-7.

(4) MEASURING CURRENT (DC AMMETER)

Figure 10-8 illustrates a meter with a maximum meter movement current (I_m) of 1 mA. This meter will function perfectly for any current value in the range from 0 to 1 mA. If the current being measured exceeds 1 mA, which is the current rating of the fuse, the fuse will blow and protect the meter from damage. Some modifications are necessary if we wish to measure any current greater than the FSD current (I_m). To make this modification, we must first know the meter's internal resistance (R_m) and FSD current (I_m). In this example, we will use a 1 mA, 50-Ω meter movement, as seen in Figure 10-9.

(a) Ammeter Range Scales The maximum possible current that the meter in Figure 10-9 can measure as it is presently constructed is 1 mA; however, if we wish to measure current in the range from 0 to 10 mA, we will need some other path for the 9 mA to travel through so that only 1 mA will flow through the meter movement.

A resistor is connected in parallel or shunt with the meter movement resistance (R_m) to shunt the damaging current away from the meter movement,

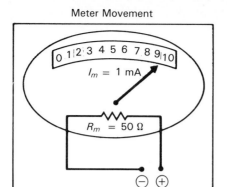

FIGURE 10-9   Symbol of 1-mA Ammeter.

as seen in Figure 10-10. The next question is—What will be the value of the shunt resistor (R_{sh}) so that it will shunt 9 mA away from the meter movement when we are measuring a current value of 10 mA? This can be calculated by first finding the voltage across R_{sh}.

$$V_{Rsh} = I_{Rm} \times R_m = 1 \text{ mA} \times 50 \, \Omega = 50 \text{ mV}$$

Since $V_{Rm} = V_{Rsm}$ (the resistors are connected in parallel), the value of R_{sh} can therefore be calculated by applying Ohm's law:

$$R_{sh} = \frac{V_{Rsh}}{I_{Rsh}} = \frac{50 \text{ mV}}{9 \text{ mA}} = 5.6 \, \Omega$$

If a maximum current of 10 mA is measured by the meter, the 10 mA will travel into the meter, where it will split through a parallel circuit made up of R_m and R_{sh}. The current split will be as follows:

FIGURE 10-10   Symbol of 10-mA Ammeter.

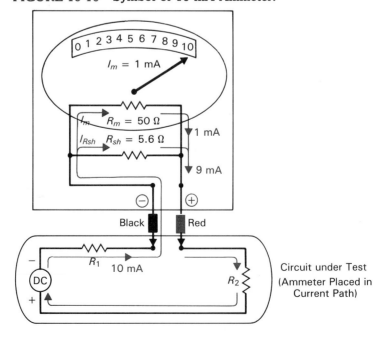

DC Troubleshooting and Measurement Equipment

$$I_{Rm} = \frac{V_{Rm}}{R_m} = \frac{50 \text{ mV}}{50 \text{ }\Omega} = 1 \text{ mA}$$

$$I_{Rsh} = \frac{V_{Rsh}}{R_{sh}} = \frac{50 \text{ mV}}{5.6 \text{ }\Omega} = 9 \text{ mA}$$

For a current of 10 mA, you will have 1 mA flowing through the moving coil and this will cause FSD.

To have a multirange ammeter, there will have to be some way of switching in different values of shunt resistors for different current ranges, as seen in Figure 10-11(a).

FIGURE 10-11   (a) Multiposition Ammeter's Internal Resistances. (b) Multiposition Ammeter's External Appearance. (c) Ammeter in circuit.

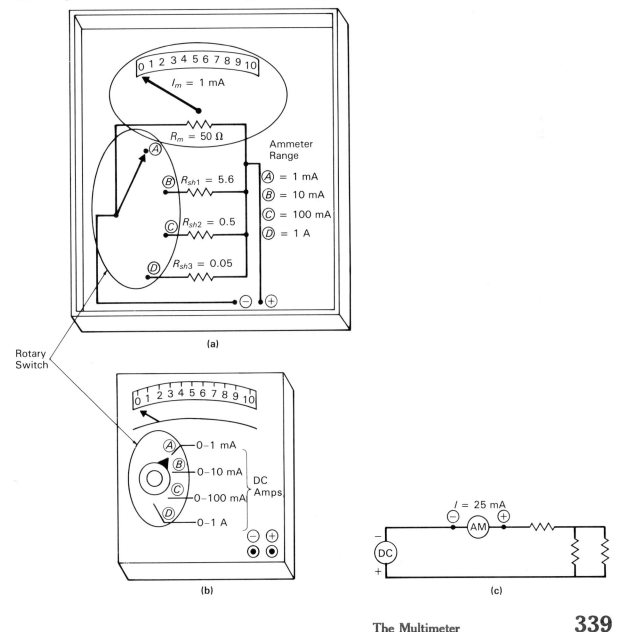

If the rotary switch is placed in position A (0 to 1 mA range), there is no need for a shunt resistor because the FSD current for the moving coil is 1 mA.

With the rotary switch in position B (0 to 10 mA range), we have switched in a shunt resistance of 5.6 Ω, and, as previously calculated, this 5.6 Ω resistor will shunt 9 mA and allow 1 mA through R_m, when a maximum of 10 mA is being measured.

In position C (0 to 100 mA), the rotary switch has switched in the shunt resistor R_{sh2}, whose job it is to shunt 99 mA away from the moving coil if 100 mA is being measured. Since we already know that 50 mV will appear across all shunt resistors, the value of R_{sh2} will equal.

$$R_{sh2} = \frac{V_{Rsh2}}{I_{Rsh2}} = \frac{50 \text{ mV}}{99 \text{ mA}} = 0.5 \text{ }\Omega$$

If the rotary switch is placed in position D (0 to 1 A or 1000 mA), R_{sh3} is switched in parallel with R_m to shunt 999 mA away from the moving coil. R_{sh3} can be calculated and is equal to

$$R_{sh3} = \frac{V_{sh3}}{I_{sh3}} = \frac{50 \text{ mV}}{999 \text{ mA}} = 0.05 \text{ }\Omega$$

Figure 10-11(b) shows the external appearance of the rotary switch that is used by the technician to select the ammeter range scale required. Figure 10-11(c) illustrates the schematic symbol of the ammeter (AM) being used to measure current in a circuit.

With multiple ranges, the scale has to be interpreted by the technician differently, depending on the range scale selected. For example, if the 1 mA range is selected, the 10 on the scale is now equivalent to 1 mA, the 9 on the scale is equivalent to 0.9 mA, the 8 to 0.8 mA, and so on. If the 10 mA range is selected, 10 on the scale is equal to 10 mA, 9 to 9 mA, 8 to 8 mA, and so on. If the 100 mA range is selected, a 2 on the scale is equivalent to 20 mA, all the way up to the 10 on the scale, which in this case is equal to 100 mA. On the 1 A (1000 mA) range scale, 6 on the scale is equivalent to 0.6 A or 600 mA, 7 on the scale is equal to 0.7 A or 700 mA, and so on.

Some range scales are not oriented only around 10, but can incorporate

FIGURE 10-12   Ammeter Scales.

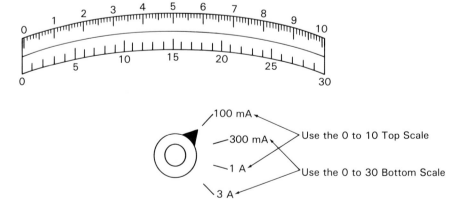

DC Troubleshooting and Measurement Equipment

a scale of 30, as well as the 10, as seen in Figure 10-12. If the 100 mA or 1 A scales are selected, the upper scale from 0 to 10 is used, with 10 on the scale representing 100 mA or 1 A. If the 300 mA or 3 A scales are selected, the lower scale from 0 to 30 is used, with 30 on the scale representing 300 mA or 3 A respectively.

(b) Using the Ammeter There are three important steps to remember when using the ammeter to measure current in a circuit, as seen in Figure 10-13.

1. Always set the selector to the higher scale first (amperes) and then reduce as needed to the milliampere or microampere ranges, just in case a larger current than anticipated is within the circuit, in which case the ammeter could be damaged.
2. The positive lead of the ammeter (red) should be connected in the circuit so that it can be traced back to the positive side of the voltage supply, and the negative lead of the meter (black) should be connected in the circuit so that it can be traced back to the negative side of the voltage supply.
3. The ammeter must always be connected in the path of the current flow and, therefore, the circuit path must be opened and the ammeter inserted into the circuit.

(c) Effect of Ammeter in a Circuit The ammeter is placed in a circuit to measure the current flowing in that circuit. The meter should ideally offer no resistance to the circuit, as any resistance in an ammeter will cause the overall circuit resistance to change, and this will change the current flow that we are trying to measure.

FIGURE 10-13   Connecting an Ammeter in Circuit to Measure Current.

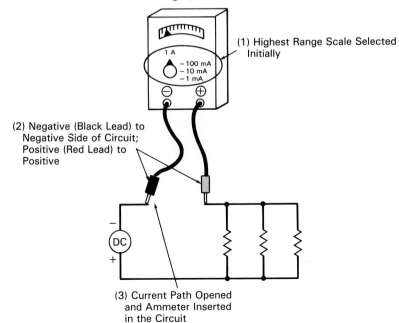

(1) Highest Range Scale Selected Initially

(2) Negative (Black Lead) to Negative Side of Circuit; Positive (Red Lead) to Positive

(3) Current Path Opened and Ammeter Inserted in the Circuit

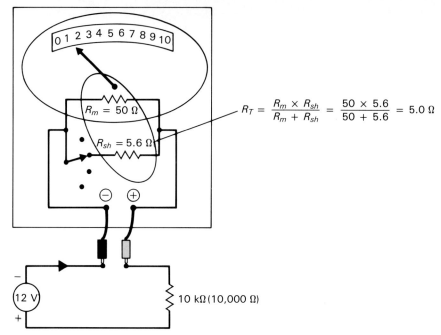

FIGURE 10-14 Effect of Ammeter in Circuit.

Figure 10-14 illustrates an example of an ammeter set up for the 10-mA range and, as you can see, the total resistance offered by the ammeter is the value of the shunt resistance in parallel with the coil resistance and is equal to 5 Ω. The resistance of the circuit (10,000 Ω) is so large compared to the internal resistance of the meter (5 Ω) that this meter resistance can be disregarded, and the measured current will be accurate.

The total circuit resistance in Figure 10-14 will equal 10,000 Ω (10 kΩ) and, therefore, the circuit current will equal 1.2 mA (12 V/10 kΩ). With the ammeter in the circuit, the total resistance will now equal 10,005 Ω and, therefore, the circuit current will equal 1.199 mA (12 V/10,005 Ω). It can be seen therefore that the ammeter has no significant effect on the circuit current it measures.

(5) MEASURING VOLTAGE (DC) VOLTMETER

A voltage increase causes a current increase and this relationship between voltage and current means that the moving coil type of meter movement discussed previously can also be used to measure voltage, as seen in Figure 10-15. In this explanation we will, for variety, use a different movement, where $I_m = 1$ mA and $R_m = 10$ Ω. If 10 mV is connected across 10 Ω of moving coil resistance, a current of 1 mA will flow:

$$I = \frac{V}{R} = \frac{10 \text{ mV}}{10 \text{ Ω}} = 1 \text{ mA}$$

This will cause full-scale deflection of the meter. With this setup, the meter can be used to measure any voltage from 0 to 10 mV.

(a) Voltmeter Range Scales To measure a voltage greater than 10 mV, we will utilize a resistor, referred to as a multiplier, to drop the extra voltage.

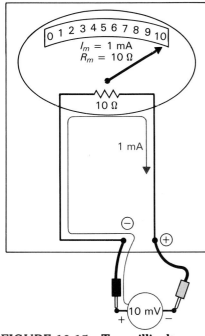

FIGURE 10-15   Ten-millivolt Voltmeter.

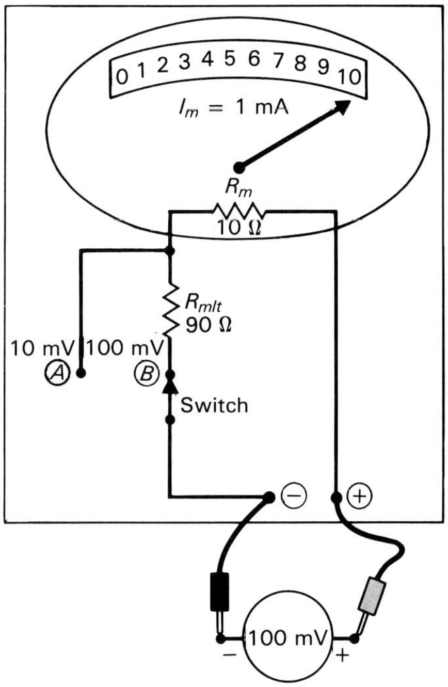

FIGURE 10-16   One Hundred-Millivolt Voltmeter.

For example, if we wished to measure a voltage in the range of 0 to 100 mV, we are already aware that a 10 mV drop across the moving coil resistance ($R_m = 10\ \Omega$) is all that is necessary to cause a full-scale deflection current of 1 mA ($I_m = 10\ \text{mV}/10 = 1\ \text{mA}$). To measure 100 mV, a multiplier resistor is connected in series with R_m to drop the extra 90 mV. The value of the multiplier resistor can be calculated by simply applying Ohm's law.

$$R_{mlt} = \frac{V_{mlt}}{I_{mlt}} = \frac{90\ \text{mV}}{1\ \text{mA}} = 90\ \Omega$$

This modification with the 90 Ω multiplier resistor connected in series with the moving coil resistance will allow us to measure a voltage range from 0 to 100 mV. When 100 mV is being measured across the positive ($+$) and negative ($-$) terminals of the meter, 90 mV will be dropped across the multiplier and 10 mV across the moving coil resistor (R_m), causing full-scale deflection.

If a switch is included, as seen in Figure 10-16, we can either (1) switch out the multiplier (position A) to convert the meter to a 10 mV range, or (2) if position B is selected, the multiplier resistor can be brought in series with R_m to obtain a 100 mV range.

Figure 10-17(a) illustrates a multirange voltmeter in which four multiplier resistors can be switched into circuit. The voltage drop across the moving coil resistance must always be 10 mV to cause FSD. The remaining voltage on whichever range has been selected will be dropped across the series-connected multiplier resistor.

POSITION A If the 10 mV range is selected, there is no need for a multiplier as 5 mV across the ($+$) and ($-$) terminals will cause one-half FSD and 10 mV across the terminals will cause FSD.

The Multimeter   **343**

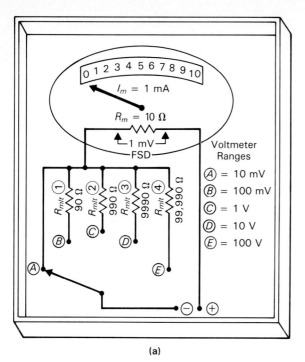

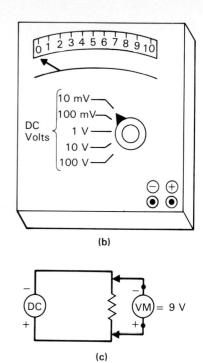

FIGURE 10-17 (a) Multiposition Voltmeter's Internal Resistances. (b) Multiposition Voltmeter's External Appearance. (c) Voltmeter Schematic Symbol.

POSITION B If the 100 mV range is selected, the difference in voltage, which in this case is 90 mV (100 mV − 10 mV) must be dropped across a multiplier resistor (R_{mlt1}) whose resistance will equal

$$R_{mlt1} = \frac{V_{mlt}}{I_{mlt}} = \frac{90 \text{ mV}}{1 \text{ mA}} = 90 \text{ }\Omega$$

POSITION C When the 1 V (1000-mV) range is selected, the voltage difference that will have to be dropped across the multiplier resistor (R_{mlt2}) will be 990 mV (1000 mV − 10 mV). The value of R_{mlt2} will therefore be

$$R_{mlt2} = \frac{V_{mlt2}}{I_{mlt2}} = \frac{990 \text{ mV}}{1 \text{ mA}} = 990 \text{ }\Omega$$

POSITION D 10 V range (10,000 mV) The voltage difference is equal to 10,000 mV − 10 mV = 9990 mV.

$$R_{mlt3} = \frac{V_{mlt3}}{I_{mlt3}} = \frac{9990 \text{ mV}}{1 \text{ mA}} = 9990 \text{ }\Omega$$

POSITION E 100 V range (100,000 mV) The voltage difference is equal to 100,000 mV − 10 mV = 99,990 mV.

$$R_{mlt4} = \frac{V_{mlt4}}{I_{mlt4}} = \frac{99,990 \text{ mV}}{1 \text{ mA}} = 99,990 \text{ }\Omega$$

Figure 10-17(b) shows the external appearance of the rotary switch used by the technician to select the voltage range scale required. Figure 10-17(c)

DC Troubleshooting and Measurement Equipment

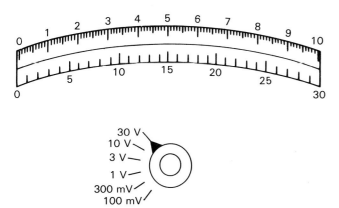

FIGURE 10-18   Voltmeter Scales.

illustrates the schematic symbol of the voltmeter being used to measure the voltage drop across a resistor.

The range scales of the voltmeter are like the ammeter in that they may not only be oriented around 10, such as 10 mV, 100 mV, 1 V, 10 V, and 100 V, but also around, for example, 3, 30, or 300 V.

Figure 10-18 illustrates a multimeter with five range scales of 100 mV, 300 mV, 1 V, 10 V, and 30 V. The top scale of 0 through 10 is used for the 100 mV, 1 V, and 10 V range scales, and the 0 through 30 scale is used for the 300 mV, 3 V, and 30 V range scales. The range scale selected refers to the FSD voltage, or maximum voltage that can be measured.

(b) Voltmeter Sensitivity (Ω/V)   The sensitivity of a voltmeter is measured in a quantity known as the ohms-per-volt rating (Ω/V). This Ω/V rating can be calculated by taking the reciprocal of the full-scale deflection current of the moving coil (I_m). For example, a 1 mA meter will have an Ω/V rating of

$$\text{voltmeter sensitivity} = \frac{1\text{ V}}{\text{meter coil current } (I_m)} = \Omega/\text{V}$$

$$= \frac{1\text{ V}}{1\text{ mA}} = 1000\ \Omega/\text{V}$$

A meter that only requires 100 μA (I_m) for FSD for the moving coil will have a sensitivity of

$$V_m \text{ sensitivity} = \frac{1\text{ V}}{I_m}$$

$$= \frac{1\text{ V}}{100\ \mu\text{A}}$$

$$= 10\ \text{k}\Omega/\text{V} \quad \text{or} \quad 10{,}000\ \Omega/\text{V}$$

It can be said, therefore, that if a meter only needs a small current to cause FSD it will have a large Ω/V rating and will be classified as a more sensitive meter.

(c) Connecting a Voltmeter in a Circuit   There are three important points to remember and apply when measuring voltage with a voltmeter, as seen in Figure 10-19.

The Multimeter   **345**

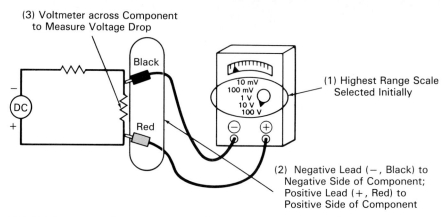

(3) Voltmeter across Component to Measure Voltage Drop

Black

10 mV
100 mV
1 V
10 V
100 V

(1) Highest Range Scale Selected Initially

DC

Red

(2) Negative Lead (−, Black) to Negative Side of Component; Positive Lead (+, Red) to Positive Side of Component

FIGURE 10-19   Connecting a Voltmeter to Measure Voltage.

1. Always set the range selector to the highest range scale first (100 V) and then reduce as needed to the lower ranges (volts or millivolts).

2. Always connect the positive lead (+, red) of the voltmeter to the component end that is connected back to the positive side of the battery or source, and the negative lead (−, black) of the meter to the component end that is connected back to the negative side of the battery or source.

3. The voltmeter must be connected across or in parallel with the component voltage to be measured.

Meter Resistance = 100 kΩ

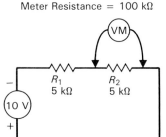

R_1
5 kΩ

R_2
5 kΩ

10 V

FIGURE 10-20

(d) Effect of Voltmeter in a Circuit A voltmeter is connected in parallel with the component to measure the voltage developed across that component. The voltmeter is, therefore, much easier to use for troubleshooting or testing purposes, as the circuit path does not have to be opened and the meter inserted in the current path, as with the ammeter.

When a voltmeter is connected across a component for voltage measurement, some of the circuit current flows through the voltmeter to deflect the meter movement. Figure 10-20 illustrates a voltmeter that has an internal resistance of 100 kΩ being used to measure the voltage drop across R_2. Without the voltmeter connected, the voltage will be dropped equally across R_1 and R_2, causing 5 V across R_2. If the voltmeter is now connected across R_2, the voltmeter resistance of 100 kΩ is now in parallel with the resistance of R_2, causing an equivalent resistance of 4.76 kΩ. The meter has loaded down the circuit and caused this change in resistance, which means that R_2 will not drop 5 V as normal, but will in fact drop 4.88 V. This error of 0.12 V caused by the loading effect of the meter is very small. As long as the internal voltmeter resistance is large with respect to the resistance of the component across which the voltage is being measured, it will be reasonably accurate. In the first example, Figure 10-20, the meter's resistance is 100 kΩ, which is very large with respect to the component's resistance of 5 kΩ, so the loading effect of the meter is negligible, causing a small error of 0.12 V.

In Figure 10-21, the same voltmeter, with a resistance of 100 kΩ, is being used to measure the voltage drop across R_2, whose resistance is also equal to 100 kΩ. Once again, the voltage of 10 V will be divided equally across R_1 and R_2 when the meter is not connected, causing a 5 V drop across R_2.

If the voltmeter is now connected across R_2 so as to measure the voltage drop across the resistor, the voltmeter's resistance of 100 kΩ is now in parallel with the resistance of R_2, causing an equivalent resistance of 50 kΩ. The voltage drop across R_2, which is normally 5 V, is now pulled down to 3.33 V due to the loading effect of the meter, which is a 1.67 V error.

These two examples illustrate the loading effect of a voltmeter on a circuit under test. However, if a voltmeter has a large internal resistance with respect to the component's resistance (10 or more times greater), the loading effect of the meter can be disregarded and the voltmeter will be more accurate.

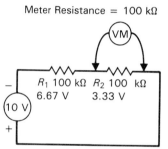

FIGURE 10-21

(e) Voltage Reference A voltmeter will always measure voltage with respect to some other point. If a resistor has a voltage drop of 30 V, for example, this means that one side of the resistor is 30 V more positive or negative than the other. This could occur if one side of the resistor was at +30 V and the other at +60 V (30 V difference in potential) or if one side was −100 V and the other was −70 V (30 V difference). Generally, when troubleshooting, a voltage measurement is taken with respect to 0 V or ground, of which there are two basic types:

1. *Earth ground*: The power lines from your local electric company have a reference point referred to as earth ground; this is because one side of the voltage source is attached to a metal rod and inserted in the ground (Figure 10-22).

2. *Circuit or chassis ground*: In electronic equipment, the metal chassis or housing of the equipment is used as a common reference point. One side of the voltage supply is connected to the metal frame or chassis of the equipment and voltage is measured with respect to this common ground reference point, referred to as a circuit, chassis, or frame ground (Figure 10-23).

Voltages are measured by placing the negative lead on the resistor and the positive lead anywhere on the chassis. This will be the same as measuring directly across the resistor, as the chassis is a common positive reference point. There is no need to run cable from the bottom of the resistor back to the positive side of battery, as the metal chassis of the equipment makes the connection.

If the voltage from the electric company is used, then the earth ground and the chassis ground are one and the same and are connected together. This ground, whether chassis or earth, is a reference point.

FIGURE 10-22 Earth Ground.

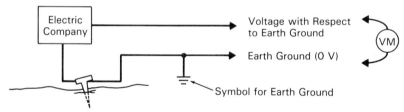

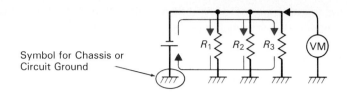

Symbol for Chassis or Circuit Ground

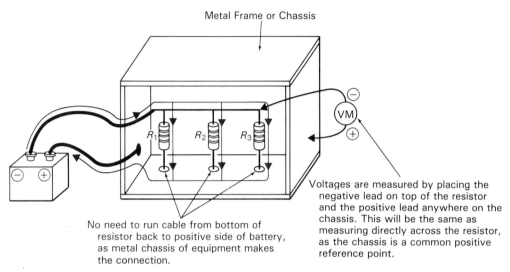

Metal Frame or Chassis

No need to run cable from bottom of resistor back to positive side of battery, as metal chassis of equipment makes the connection.

Voltages are measured by placing the negative lead on top of the resistor and the positive lead anywhere on the chassis. This will be the same as measuring directly across the resistor, as the chassis is a common positive reference point.

FIGURE 10-23 Chassis Ground.

Another term that is used with all dc voltage sources is a positive and negative ground. This refers to which polarity of the battery or source is connected to the chassis; for example, if the negative side of the battery is connected to the chassis, then positive voltages will be measured throughout the circuit with respect to the *negative ground* [Figure 10-24(a)]. On the other hand, if the positive side of the battery is connected to the chassis, then negative voltages will be measured throughout the circuit with respect to the *positive ground* [Figure 10-24(b)].

If a voltmeter is normally always used to measure voltage with respect to chassis ground (0 V), then how could you find the voltage dropped across R_1 in Figure 10-25? The voltmeter will measure 50 V at point A, but this is the voltage dropped across both R_1 and R_2. The voltage dropped across

FIGURE 10-24   (a) Negative Ground, Positive Voltage Measurements. (b) Positive Ground, Negative Voltage Measurements.

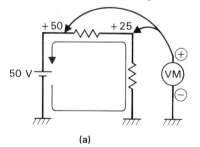

(a)

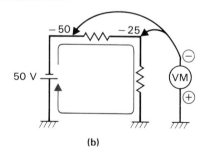

(b)

DC Troubleshooting and Measurement Equipment

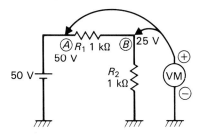

FIGURE 10-25   **Voltmeter Probing.**

R_1 can be obtained by measuring the voltage at point A with respect to chassis ground ($+50$ V) and then at point B with respect to chassis ground ($+25$ V). The difference between these two readings on your voltmeter will be the potential difference between A and B; in this case, 25 V is dropped across R_1.

The next question you are probably asking is why would we make two voltmeter checks to find out we have dropped 25 V across R_1 when we could have just hooked up the voltmeter across R_1 and measured 25 V [Figure 10-26(a) and (b)]. The reason why voltage measurements are almost always measured with respect to chassis ground rather than the other side of the resistor or component is convenience and fault prevention. If voltages are measured with respect to chassis ground, then the negative lead of the voltmeter can be permanently attached to the chassis, and you only have to worry about moving the positive meter lead to probe the desired testing points [Figure 10-26(a)].

If you measure the voltage dropped across a resistor, as seen in Figure 10-26(b), then both the positive and negative leads must be held by the technician to probe in two places, which means:

1. The technician does not have a free hand because he or she is probing with both voltmeter leads.
2. If the technician is trying to position two probes at once in two different points, it becomes twice as difficult, and the risk of slipping and causing a short and then possibly a further problem in the circuit under test is present.

For safety reasons, it is often recommended that the test leads be attached for voltage measurements with the equipment off; then turn it on to observe

FIGURE 10-26   **(a) Two Checks to Find the Voltage Drop across R_1.**
(b) One Check to Find the Voltage Drop across R_1.

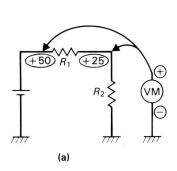

(a)

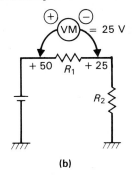

(b)

The Multimeter   **349**

EXAMPLE 10.1

Using the voltmeter with the negative lead tied to the chassis, the voltages at points A, B, and C are as shown in Figure 10-27. From these readings, calculate the voltage drops across R_1, R_2, R_3, and R_4.

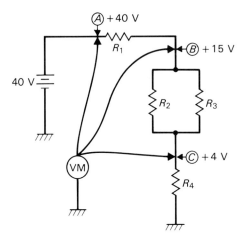

FIGURE 10-27

SOLUTION

The difference between A and B = $40 - 15 = 25$ V; therefore, 25 V is dropped across R_1.

The difference between B and C = $15 - 4 = 11$ V; therefore, 11 V is dropped across R_2 and R_3 (resistors are in parallel).

The difference between C and chassis ground = $4 - 0 = 4$ V; therefore, 4 V is dropped across R_4.

---

the reading. This is especially important in high-voltage systems of 300 V or more.

(6) MEASURING RESISTANCE (OHMMETER)

Resistance can also be measured using the same moving coil meter movement. Figure 10-28 illustrates the meter configuration for measuring resistance. The only difference is that a series 1.5 V battery source is used to supply current. When the meter leads are connected together (shorted), the zero ohms adjust resistor is adjusted to cause full-scale deflection (1 mA); this will occur when the meters total resistance is equal to

$$R = \frac{V}{I} = \frac{1.5 \text{ V}}{1 \text{ mA}} = 1500 \ \Omega$$

Since the moving coil resistance is equal to 10 Ω (R_m), then the zero ohms adjust resistance will need to be adjusted to equal 1490 Ω (1500 Ω − 10 Ω = 1490 Ω) in order to cause a total circuit resistance of 1500 Ω and, therefore, achieve full scale deflection current (I_m) of 1 mA. Adjustment

DC Troubleshooting and Measurement Equipment

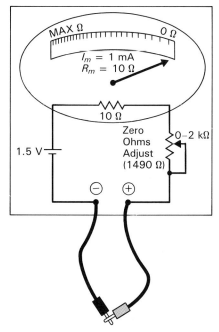

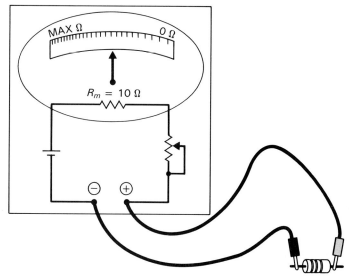

FIGURE 10-28 **Measuring Resistance.**   FIGURE 10-29

of the zero ohms adjust resistor to obtain FSD is called zeroing the meter. When the meter leads are shorted together, the meter will deflect to the full scale position (right side), indicating that there is no resistance (0 Ω) between the two meter leads. As different values of resistance are inserted between the two meter probes, different values of current will result, causing different levels of pointer deflection. A higher value of resistance will cause a smaller value of current to flow through the meter circuit, resulting in a smaller deflection of the pointer, as seen in Figure 10-29.

The reason for the ohms adjust control is due to the discharge of the internal 1.5 V ohmmeter battery. As its voltage falls after months of use, the resistance of the zero ohms adjust resistor is decreased to still cause an FSD current of 1 mA. If, for example, the battery in the circuit of Figure 10-28 has discharged to 1 V, the zero ohms adjust resistor needs to be varied to a value of 990 Ω and, therefore, the total meter circuit resistance will equal

$$\text{zero ohms resistance} + \text{moving coil resistance}$$
$$990 \ \Omega + 10 \ \Omega = 1000 \ \Omega$$

This will cause a meter circuit current of

$$I_m = \frac{V}{R} = \frac{1 \ V}{1000 \ \Omega} = 1 \ mA$$

The meter is now said to be zeroed, because 1 mA of meter circuit current is flowing through the moving coil, causing FSD.

The main function of the ohms adjust resistor is so that zeroing of the ohmmeter can be achieved even though the battery voltage will naturally decay with age, thus allowing accurate resistance measurements.

When the meter probes are shorted together, it causes FSD, which

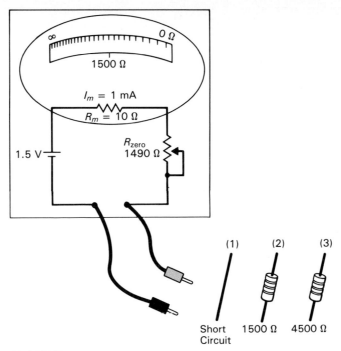

FIGURE 10-30

indicates to the technician that 0 Ω (a short circuit) exists between the two probes. On the other hand, if the two probes are open, then the meter needle remains in its resting position indicating that infinite (∞) resistance (maximum possible resistance or open circuit) exists between the two probes.

Figure 10-30 illustrates an example ohmmeter, which has a 1 mA, 10 Ω meter movement. Let's now proceed to take three examples of measured resistance and see how the ohmmeter responds to indicate these resistances.

(1) If the two leads are shorted together, the total meter circuit resistance will equal

$$R_m + R_{zero} + \text{measured resistance} = 10 + 1490 + 0$$
$$= 1500 \, \Omega$$

The moving coil current will consequently equal

$$I_m = \frac{V}{R} = \frac{1.5 \, V}{1500 \, \Omega} = 1 \, \text{mA}$$

and this will cause full-scale deflection [Figure 10-31(a)].

(2) If the two meter leads are connected across a 1500 Ω resistor, the total circuit resistance will equal

$$R_m + R_{zero} + \text{measured resistance} = 10 + 1490 + 1500$$
$$= 3000 \, \Omega$$

The moving coil current will consequently equal

$$I_m = \frac{V}{R} = \frac{1.5 \, V}{1500 \, \Omega} = 0.5 \, \text{mA}$$

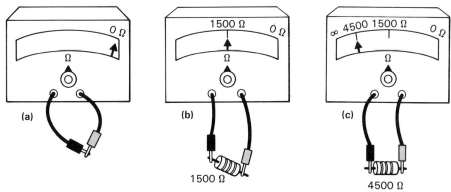

FIGURE 10-31

Since 1 mA causes full-scale deflection, then 0.5 mA will cause one half full scale deflection [Figure 10-31(b)].

(3) If we connect our ohmmeter leads across a resistor of 4500 Ω, the total meter circuit resistance will equal

$$R_m + R_{zero} + \text{measured resistance} = 10 + 1490 + 4500$$
$$= 6000 \ \Omega$$

The moving coil current consequently equals

$$I_m = \frac{V}{R} = \frac{1.5 \text{ V}}{6000 \ \Omega} = 0.25 \text{ mA}$$

Since 1 mA causes full scale deflection, 0.25 mA will cause one quarter full scale deflection [Figure 10-31(c)].

The ohmmeter scale is referred to as a nonlinear scale, as opposed to the voltmeter and ammeter scales, which are linear. Figure 10-32(a) illustrates

FIGURE 10-32 (a) Nonlinear Ohmmeter Scale. (b) Linear Voltmeter and Ammeter Scale.

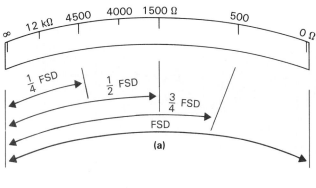

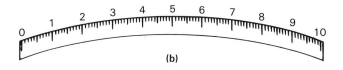

The Multimeter **353**

the nonlinearity of the ohmmeter, while Figure 10-32(b) shows how the voltmeter and ammeter scales follow a linear variation.

Maybe the best way to clearly see the difference between the two types of scales is to first analyze the ohmmeter's nonlinear scale. The difference with the ohmmeter, first, is that 0 begins on the right side. Between three quarters FSD and one-half FSD on the output scale is a change of 500 to 1500 on the input resistance being measured, which is a 1000 Ω change. Between one quarter FSD to one half FSD is an input resistance change of 3000 Ω. Between one half FSD and three quarters FSD is the same amount of distance on the output scale, yet the input change is three times greater between one quarter and one half FSD. The output scale, therefore, does not vary in direct proportion or at the same rate as the input resistance being measured and is consequently referred to as a nonlinear scale (the divisions are not uniformly spaced).

If the output scale varies in direct proportion to the input being measured, as with the voltmeter and ammeter scales, the scale is said to be linear. This can be further explained by using the same simple example.

Between one quarter and one half FSD is a change of 2.5 to 5 on the output scale and is in direct proportion to the input change of 2.5 to 5 V (or amperes). Between one half and three quarters FSD is exactly the same amount of change as one quarter to one half FSD, both on the output scale and the input being measured, which is why Figure 10-32(b) is referred to as a linear scale (all divisions are evenly spaced).

(a) Ranges of an Ohmmeter Figure 10-33(a) illustrates the external appearance of an ohmmeter with multiple ranges, and Figure 10-33(b) illustrates the internal series and shunt resistors that allow different range

FIGURE 10-33 **(a) Multiposition Ohmmeter's External Appearance. (b) Multiposition Ohmmeter's Internal Resistances.**

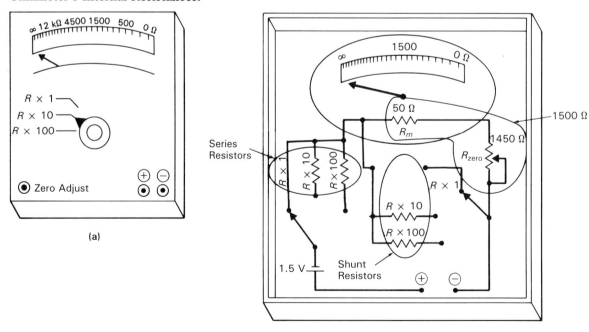

DC Troubleshooting and Measurement Equipment

EXAMPLE 10.2

If the meter leads are shorted together, a resistance of $0 \, \Omega$ exists between them, and the meter should cause full scale deflection of the pointer (1 mA of moving coil current through R_m) to the $0 \, \Omega$ position on the far right side of the scale.

R_m and R_{zero} are in series with one another, giving a combined resistance of $1500 \, \Omega$ and causing a total circuit resistance of $1500 \, \Omega$. Meter circuit current will, therefore, equal

$$I = \frac{V}{R} = \frac{1.5 \, V}{1500 \, \Omega} = 1 \, mA$$

which will flow through R_m, causing FSD of the pointer to the $0 \, \Omega$ position of the scale.

EXAMPLE 10.3

If a $1500 \, \Omega$ resistor is connected between the two meter leads, the meter should cause the pointer to deflect to the one half FSD position (0.5 mA of moving coil current through R_m). With the $1500 \, \Omega$ resistor between the two meter leads, a total circuit resistance of

$$R_t = R_m + R_{zero} + R_{measured}$$
$$= 50 + 1450 + 1500$$
$$= 3000 \, \Omega$$

This will cause a circuit current of $I = V/R = 1.5 \, V/3000 \, \Omega = 0.5 \, mA$, which will cause one half FSD of the pointer to the 1500 position on the scale.

For the $R \times 10$ and $R \times 100$ scales, an ohmmeter will switch in a different value of series and shunt resistors in order to obtain the $R \times 10$ and $R \times 100$ ohmmeter positions.

settings to be attained. The range scales of an ohmmeter are interpreted differently from those of the ammeter and voltmeter. The ohmmeter's range scales in Figure 10-33(a) of $R \times 1$, $R \times 10$, and $R \times 100$ are used as multipliers; for example, if the pointer is at 500 and the range scale selected is $R \times 10$, the resistor's resistance will be equal to

$$R \times 10 = 500 \times 10 = 5000 \quad or \quad 5 \, k\Omega$$

A typical ohmmeter will have a small battery of usually 1.5 V, which is used for the $R \times 1$, $R \times 10$, and $R \times 100$ ranges, while a larger battery of typically 9 or 15 V will be used for the $R \times 1000$ range and higher.

Figure 10-33(b) illustrates the circuit for the $R \times 1$ in which there is no shunt or series resistor connected in the Circuit.

(b) Using the Ohmmeter (Figure 10-34) There are three steps to remember when measuring resistance:

1. Connect (short) the positive and negative leads together and adjust the zero adjust control so that the pointer is at $0 \, \Omega$ on all scales.

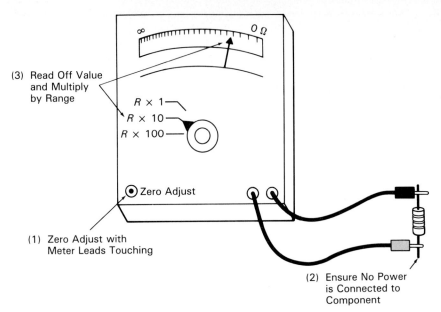

(3) Read Off Value and Multiply by Range

$R \times 1$
$R \times 10$
$R \times 100$

Zero Adjust

(1) Zero Adjust with Meter Leads Touching

(2) Ensure No Power is Connected to Component

FIGURE 10-34

This is also known as a confidence test to let you know that the meter is functioning.

2. Ensure that no power is connected to the component to be measured, as the circuit voltage and internal ohmmeter battery voltage could aid one another, causing an excessive current, which could damage the meter movement.

3. Connect the leads across the component and read off the resistance indicated by the scale, multiplying by 1, 10, 100, 1000, or more, depending on the ohms range selected.

(7) SUMMARY OF AMMETER, VOLTMETER, AND OHMMETER

After discussing the ammeter, voltmeter, and ohmmeter separately, we will now combine all three to end up with a multipurpose meter (multimeter), or, as they are sometimes called, a VOM (volt-ohm-milliammeter), as seen in Figure 10-35. Having all three meters in one package is both convenient and cheaper than having to buy three meters.

SELF-TEST REVIEW (§ 10.1.1)

1. Define meter resistance and sensitivity.
2. What is the name given to the different values of resistors used to achieve different ammeter range scales?
3. An ammeter should ideally offer the largest possible value of resistance when connected in a circuit to measure current (true or false).
4. What is the name given to the different values of resistors used to produce different voltmeter range scales?

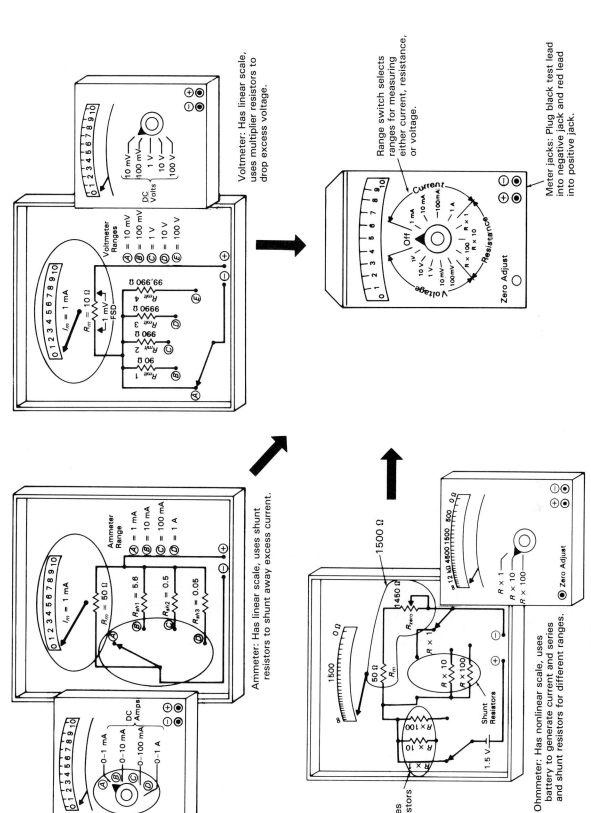

Voltmeter: Has linear scale, uses multiplier resistors to drop excess voltage.

Ammeter: Has linear scale, uses shunt resistors to shunt away excess current.

Ohmmeter: Has nonlinear scale, uses battery to generate current and series and shunt resistors for different ranges.

Range switch selects ranges for measuring either current, resistance, or voltage.

Meter jacks: Plug black test lead into negative jack and red lead into positive jack.

FIGURE 10-35   The Multimeter.

357

5. Define ohms/volt and how it relates to sensitivity.
6. What should a voltmeter internal resistance be ideally?
7. What are the major differences between an ammeter/voltmeter and an ohmmeter?
8. State the three step procedure that should be applied when measuring resistance.

10.1.2 DIGITAL READOUT MULTIMETER

The digital multimeter (DMM) illustrated in Figure 10-36 is gradually replacing the analog type multimeter previously discussed because it is easier to read and has greater accuracy.

(1) ANALOG VERSUS DIGITAL

(a) Ease of Reading One of the greatest problems with the analog type meter is the errors that occur due to the human factor when reading off

FIGURE 10-36   Digital Multimeters. (a) Handheld. (b) Benchtop.

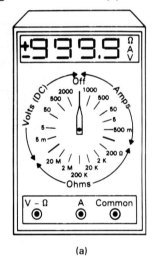

(a)

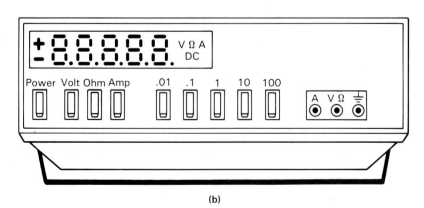

(b)

Analog

Digital

FIGURE 10-37

the value from the many different scales. The analog thin needle against a calibrated scale is similar to the hands of a clock against the number scale indicating the hours. When you look at an analog clock, you have to determine where the hands are, which number the hand is nearest to, and so on. With a digital clock, however, the time is read directly from a display, as seen in Figure 10-37.

With the analog meter, the decoding of the scale is necessary, while the digital multimeter displays the amount, polarity (+ or −), and the units (V, A, or Ω) on typically a four- or five-digit readout.

(b) Accuracy DMMs are typically accurate to ± 0.01%, have no need for a zero ohms adjust, and their loading effect when measuring voltage is negligible on all ranges, as the internal meter resistance is typically 10 to 22 MΩ. Returning to the example of the analog and digital clock, a person reading the time from the traditional analog clock would say that the time is almost 12:30. The wearer of the digital watch, however, will be unquestionably specific and give the time as 12:27. An analog reading on a meter of about 7 V becomes 7.15 V with the far more accurate digital multimeter.

Digital meters have far more complicated internal circuitry; to describe the meter properly would need a good understanding in digital circuitry, which is a subject you will more than likely encounter in your future studies of electronics. This complex circuitry causes the digital readout meters to have a more expensive price tag than the analog readout multimeters. Another disadvantage with the digital multimeter is its slow response to display the amount on the readout once it has been connected in circuit. For this reason, some technicians still prefer the analog multimeter due to its lower price and quick response, while other technicians prefer the accuracy and ease of use of the digital readout multimeter.

(2) USING THE DIGITAL MULTIMETER

When troubleshooting electronic equipment, the multimeter becomes the technician's eyes, allowing him or her to see the situation and the problem within the circuit or system. Figure 10-38 illustrates a typical digital multimeter, all of which operate in much the same way.

For the meter to be of any use, it must first be connected to the circuit or device to be tested. The two test leads for the meter, one of which is red and the other black, must be inserted into the correct meter lead jacks on

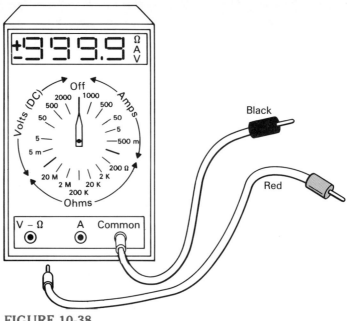

FIGURE 10-38

the meter. The black lead is connected to the meter jack marked *common*. The red lead is connected to the meter jack marked A for amperes or V–Ω for volts or ohms.

(a) Measuring Current The measurement of current is rarely performed when troubleshooting, as the circuit path has to be opened in order to insert the DMM in series with the current flow. However, if current is to be measured, the red lead is inserted into the A (ampere) jack.

(b) Measuring Voltage The measurement of voltage and resistance is where the DMM finds its greatest utilization. For voltage or resistance measurement, the red meter lead is inserted into the V–Ω (volt or ohms) meter jack.

Each setting of the meter's selector is marked with the maximum voltage that can be displayed. For example, if we set the selector for 500 V, the meter will display any measured voltage from 0 to 500 V. If the selector is set, for example, to the 5 V range and a 9 V battery is connected across it, the display will flash on and off or some meters will indicate overload by displaying OL. Both of these displays mean that the voltage being measured is greater than the range selected and that you have to switch to a higher voltage range. With the example multimeter, we can measure anywhere between the setting of 5 mV to 2000 V.

(c) Measuring Resistance The red lead is left in the V–Ω meter jack and the desired range is selected. If the lowest 200 Ω range is selected, the meter can be used to measure resistances anywhere from 0.1 to 200 Ω. Once again, if the measured resistance is greater than the range selected, the display will flash on and off or some meters will indicate overload by displaying OL.

DC Troubleshooting and Measurement Equipment

Remember that resistance measurements are carried out without power being applied to the component under test, and resistance values can vary by as much as 20% due to the tolerance of certain resistors; so do not be misled if your meter reading is slightly different to the color band value printed on the resistor. If a resistor's value is off and exceeds the tolerance, the resistor should be replaced.

A resistor will rarely short, but tyically will open. If a resistor does open, the DMM display will flash on and off or display OL even when the highest range scale is selected, because the resistors infinite resistance (open circuit) is greater than the largest scale of 20 MΩ.

SELF-TEST REVIEW (§ 10.1.2)

1. State the advantages of the DMM over the analog multimeter.
2. State the advantages of the analog multimeter over its digital readout counterpart.
3. How would a DMM indicate that the voltage being measured is greater than the range selected?
4. If the 1000 Ω range is selected on the DMM and a readout of 373 results, what is the value of the resistance being measured?

SUMMARY

1. The multimeter is a device used for the measurement of current, voltage, and resistance and is used for experimentation with and troubleshooting of electronic equipment.
2. The multipurpose meter has three meters all rolled into one package: the ammeter measures current, the voltmeter measures voltage, and the ohmmeter measures resistance.
3. The analog readout multimeter utilizes a moving coil, which is deflected by a current when measuring volts, ohms, or amperes.
4. The ammeter uses shunt resistors in parallel with the meter movement and has a very low internal resistance when placed in circuit to minimize the loading effect.
5. The voltmeter uses multiplier resistors in series with the meter movement and has a very large internal resistance when connected across a component to minimize the loading effect.
6. The ohmmeter has a combination of series and parallel resistors in and across the meter movement, and uses an internal current source in the form of a battery to deflect the pointer.
7. The digital readout type of multimeter is gradually replacing the analog type due to its accuracy and easy reading; however, the analog multimeter has a much faster response and is less costly.

NEW TERMS

Analog readout
Chassis ground
D'Arsonval meter
movement
Digital readout
Earth ground
Full-scale deflection
(FSD)

Linear scale
Meter movement
Meter resistance
Meter sensitivity
Multimeter
Multiplier resistor
Multirange meter
Negative ground

Nonlinear scale
Ohms per volt
(Ω/V)
Polarity sensitive
Positive ground
Sensitivity
Shunt resistor
Zeroing

NEW FORMULAS

$$\text{Voltmeter sensitivity } (\Omega/V) = \frac{1\ V}{\text{meter coil current } (I_m)}$$

REVIEW QUESTIONS

Multiple Choice Questions

1. The device used to measure the number of coulombs per second passing a given point is called:
   a. An ammeter
   b. A voltmeter
   c. An ohmmeter
   d. A wattmeter

2. The device used to measure the potential difference between two points is called:
   a. An ammeter
   b. A voltmeter
   c. An ohmmeter
   d. A wattmeter

3. The device used to measure the amount of opposition to current flow offered by a particular component is called:
   a. An ammeter
   b. a voltmeter
   c. An ohmmeter
   d. A wattmeter

4. The _____ or moving coil type of meter movement was named after its inventor.
   a. Joule
   b. Wattmeter
   c. D'Arsonval
   d. Guglielmo

5. A meter movement that requires only 10 μA to deflect the pointer to FSD is _____ than a meter that requires 10 mA.
   a. More sensitive
   b. Less sensitive
   c. Insensitive
   d. All of the above could be true

6. Ammeter ranges are accomplished by _____ resistors connected in _____ with the meter movement.
   a. Multiplier, series
   b. Shunt, parallel
   c. Multiplier, parallel
   d. Shunt, series

7. Voltmeter ranges are accomplished by _____ resistors connected in _____ with the meter movement.
   a. Multiplier, series
   b. Shunt, parallel
   c. Multiplier, parallel
   d. Shunt, series

8. An ammeter should ideally offer _____ resistance to the circuit in which it is connected.
   a. 100 kΩ of
   b. A very large
   c. A very low
   d. 150 Ω of

9. A voltmeter should ideally offer _____ resistance to the circuit in which it is connected.
   a. 100 kΩ of
   b. A very large
   c. A very low
   d. 150 Ω of

10. The sensitivity of a voltmeter is measured in a quantity known as the _____ rating.
   a. Ohms/cm
   b. Coulombs/second
   c. Ohms/volt
   d. Amps/volt

11. The only difference between the ohmmeter and the voltmeter or ammeter is that the ohmmeter uses a/an _____ to supply current.
   a. External voltage source
   b. Battery
   c. Internal voltage source
   d. Both (b) and (c) are true
   e. Both (a) and (b) are true

12. The two advantages of the digital multimeter are that it is:
   a. Easy to read and cheap in price
   b. Accurate and cheap in price
   c. Accurate and easy to interpret
   d. Inexpensive and has a quick response

13. The two advantages of the analog multimeter are that it is:
   a. Easy to read and cheap in price
   b. Accurate and cheap in price
   c. Accurate and easy to interpret
   d. Inexpensive and has a quick response

14. Each setting of a digital multimeter is marked with the _____ value that can be displayed.
   a. Maximum
   b. Minimum
   c. Mid-range
   d. Lowest

15. If a larger value than the range setting is measured with the DMM, the digital display will:
   a. Flash on and off
   b. Display OL for overload
   c. Both (a) and (b) above
   d. None of the above

Essay Questions

16. List the three steps that should be remembered when using the ohmmeter. (10.1.1(6))

17. List the three steps that should be remembered when using the voltmeter. (10.1.1(5))

18. List the three steps that should be remembered when using the ammeter. (10.1.1(4))

19. Briefly describe the construction and operation of the D'Arsonval or moving coil type of meter movement. (10.1.1)

20. Briefly describe how ammeters have multiple range scales by the use of shunt resistors. (10.1.1)

21. Briefly describe how voltmeters have multiple range scales by the use of multiplier resistors. (10.1.1)

22. Describe what loading effect an ammeter and voltmeter will have when measuring current and voltage in a circuit. (10.1.1)

23. How is the ohms per volt rating of a voltmeter obtained?

24. What is the difference between an earth ground and a chassis ground? (10.1.1)

25. Briefly describe the advantages and disadvantages of the digital multimeter and how it is used to measure: (10.1.2)
   a. Current
   b. Voltage
   c. Resistance

Practice Problems

26. At what position would the pointer be (1 to 10 scale) if 15 μA flows through a 60 μA meter movement?

27. If a voltmeter on the 1000 V range has a sensitivity of 60,000 Ω/V, what is its internal resistance?

28. For a 50 μA, 100 Ω ammeter meter movement, determine the value of shunt resistors for the:
    a. 100 mA range
    b. 100 μA range

29. What is the value of multiplier resistors for a 50,000-Ω/V voltmeter on the:
    a. 10 V range?
    b. 100 V range?

30. If an ohmmeter's pointer points to 60 on the $R \times 100$ range, the resistance of the resistor being measured is equal to how many ohms?

31. A voltmeter has a sensitivity of 20,000 Ω/V. Calculate the resistance of the voltmeter if the following voltage scales are being used. (a) 25 V; (b) 100 V; (c) 1.5 V.

32. Calculate the shunt resistances needed for the ammeter in Figure 10-11 if $I_m = 100$ μA and $R_m = 1$ kΩ.

33. Calculate the multiplier resistances needed for the voltmeter in Figure 10-17 if $I_m = 50$ μA and $R_m = 2000$ Ω.

34. Sketch the schematic for the following meters and indicate resistor values.
    a. Ammeter $I_m = 50$ μA, $R_m = 1500$ Ω. Four range scales: 100 μA, 100 mA, 1 A, and 10 A.
    b. Voltmeter $I_m = 0.5$ mA, $R_m = 1500$ Ω. Four range scales: 1 V, 3 V, 10 V, 30 V.
    c. Ohmmeter $I_m = 1$ mA, $R_m = 150$ Ω, Battery = 1.5 V, $R_{zero} = $?

35. To double the current and voltage ranges of the meters in question 34(a) and (b), what changes would have to occur?

11

ALTERNATING CURRENT (AC)

AFTER COMPLETING THIS CHAPTER, YOU WILL BE ABLE TO:

1. Explain the difference between alternating and direct current.
2. Describe pulsating dc.
3. State the two main applications for ac.
4. Give the three advantages that ac has over dc from a power point of view.
5. Describe basically the ac power distribution system from the electric power plant to the home or industry.
6. List the main formulas and information pertaining to the five basic wave shapes.
7. Describe the three waves used to carry information between two points.
8. Define the differences between electrical and electronic equipment.
9. Describe what is a fundamental and a harmonic frequency.
10. Explain how the three basic information carriers are used to carry many forms of information on different frequencies within the frequency spectrum.

INTRODUCTION

One of the best ways to describe anything new is to begin by redescribing something known and then discuss the unknown. The known topic is direct current. Direct current (dc) is the flow of electrons in one DIRECTion and one direction only. DC voltage is nonvarying and normally obtained from a battery or power supply unit, as seen in Figure 11-1. The only variation in voltage from a battery occurs due to the battery's discharge, but, even then, the current will still flow in only one direction, as seen in Figure 11-2.

A dc voltage of 9 or 6 V could be graphically illustrated as seen in Figure 11-3. Whether 9 or 6 V, the voltage can be seen to be constant or the same at any time.

Some power supplies supply a form of dc known as pulsating dc, which varies periodically from zero to a maximum, back to zero, and then repeats. Figure 11-4(a) illustrates the physical appearance and schematic diagram of a battery charger that is connected across two series resistors. The battery charger is generating a waveform, as shown in Figure 11-4(b), known as pulsating dc. At time 1 [Figure 11-4(c)], the power supply is generating 9 V and direct current is flowing from negative to positive. At time 2 [Figure 11-4(d)], the power supply is producing 0 V and, therefore, no current is being produced. In between time 1 and time 2, the voltage out of the power supply will decrease from 9 V to 0 V; however, no matter what the voltage, whether 8, 7, 6, 5, 4, 3, 2, or 1 V, current will only be flowing in one direction (unidirectional) and is therefore referred to as dc.

dc electrical energy into a mechanical rotation output. Whether steady or pulsating, direct current is current in only one DIRECTion.

Alternating current (ac) flows first in one direction and then in the opposite direction. This reversing current is produced by an alternating voltage source, as seen in Figure 11-5(a), which reaches a maximum in one direction (positive), decreases to zero, and then reverses itself and reaches a maximum in the opposite direction (negative). This is graphically illustrated in Figure 11-5(b).

During the time of the positive voltage alternation, the polarity of the voltage will be as seen in Figure 11-5(c), and so current will flow from negative to positive in a counterclockwise direction.

During the time of the negative voltage alternation, the polarity of the voltage will reverse, as seen in Figure 11-5(d), causing current to flow once again from negative to positive, but, in this case, in the opposite clockwise direction.

THE LASER

In 1898, H. G. Wells's famous book, *The War of the Worlds*, had Martian invaders with laserlike death rays blasting bricks, firing trees, and piercing iron as if it were paper. In 1917, Albert Einstein stated that, under certain conditions, atoms or molecules could absorb light and then be stimulated to release this borrowed energy. In 1954, Charles H. Townes, a professor at Columbia University, conceived and constructed with his students the first "maser" (acronym for microwave amplification by stimulated emission of radiation). In 1958, Townes and Arthur L. Shawlow wrote a paper showing how stimulated emission could be used to amplify light waves as well as microwaves, and so the race was on to develop the first "laser." In 1960, Theodore H. Maiman, a scientist at Hughes Aircraft Company, directed a beam of light from a flash lamp into a rod of synthetic crystal, which responded with a burst of crimson light so bright that it outshone the sun.

An avalanche of new lasers emerged, some as large as football fields, while others were no bigger than a pinhead. They can be made to produce invisible infrared or ultraviolet light or any visible color in the rainbow, and the high-power lasers can vaporize any material a million times faster and more intensely than a nuclear blast, while the low-power lasers are safe to use in children's toys.

At present, the laser is being used by the FBI to detect fingerprints that are 40 years old, in the Star Wars defense program, in compact disk players, in underground fiber-optic communication to transmit hundreds of telephone conversations, to weld car bodies, to drill holes in baby-bottle nipples, to create three-dimensional images called holograms, and as a surgeon's scalpel in the operating room. Not a bad beginning for a device that when first developed was called "a solution looking for a problem."

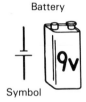

Battery

Symbol

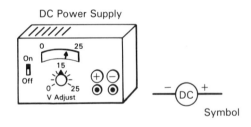

DC Power Supply

Symbol

FIGURE 11-1 DC Sources.

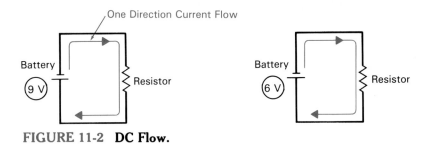

FIGURE 11-2 **DC Flow.**

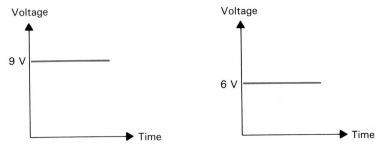

FIGURE 11-3 **Graphical Representation of DC.**

FIGURE 11-4 **Pulsating DC.**

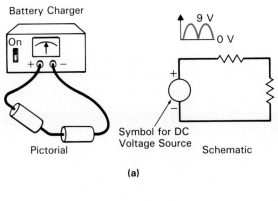

(a)

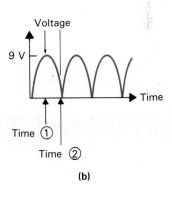

(b)

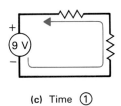

(c) Time ①

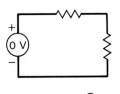

(d) Time ②

Introduction

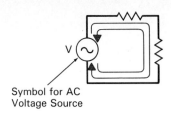

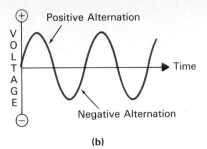

Symbol for AC
Voltage Source

(a)

(b)

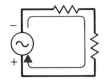

(c) Positive Voltage Alternation,
CCW Current Flow

(d) Negative Voltage Alternation,
CW Current Flow

FIGURE 11-5   Alternating Current.

11.1
WHY ALTERNATING CURRENT?

The question that may be troubling you at this point is: if we have been managing fine for the past chapters with dc, why do we need ac?

There are two main applications for ac:

1. Power transfer: To supply electrical power for lighting, heating, cooling, appliances, and machinery in both home and industry.

2. Information transfer: To communicate or carry information, such as radio music and television pictures between two points.

SELF-TEST REVIEW (INTRODUCTION AND § 11.1)

1. Give the full names of the following abbreviations: (a) ac; (b) dc.
2. Is the pulsating waveform generated by a battery charger considered to be ac or dc? Why?
3. The polarity of a/an _____ voltage source will continually reverse, and therefore so will the circuit current.
4. The polarity of a/an _____ voltage source will remain constant, and therefore current will only flow in one direction.
5. List the two main applications of ac.
6. State briefly the difference between ac and dc.

To begin with, let us discuss the first of these applications, power transfer.

11.1.1 POWER TRANSFER

There are three advantages that ac has over dc from a power point of view. Let us now discuss these in detail.

(1) Flashlights, radios, and portable televisions all use batteries (dc) as a source of power. In these applications where a small current is required, batteries will last a good length of time before there is a need to recharge or replace them. Many appliances and most industrial equipment need a large supply of current, and in this situation a generator would have to be used to generate this large amount of current. Generators operate in the opposite way to motors, in that a generator converts a mechanical rotation input into an electrical output. Generators can be used to generate either dc or ac, but ac generators can be larger, less complex internally, and cheaper to operate, which are the first reasons why we use ac instead of dc for supplying power. Figure 11-6 illustrates a typical ac generator.

(2) From a power point of view, ac is always used by electric companies when transporting power over long distances to supply both the home and industry with electrical energy. Recalling the power formula, you will remember that power is proportional to either current or voltage squared ($P \propto I^2$ or $P \propto V^2$), which means, to supply power to the home or industry, we would supply either a large current or voltage. As you can see in Figure 11-7, between the electric power plant and home or industry, are power lines carrying the power. The amount of power lost (heat) in these power lines can be calculated by using the formula $P = I^2 \times R$, where I is the current flowing through the line and R is the resistance of the power lines. This means that the larger the current, the greater the amount of power lost in the lines in the form of heat and, therefore, the less the amount of power supplied to the home or industry. For this reason, power companies transport electric energy at a very high voltage between 200,000 to 600,000 V. Since the voltage is high, the current can be low and provide the same amount of power to the consumer. Yet, by keeping the current low, the amount of heat loss generated in the power lines is minimal.

Now that we have discovered why it is more efficient to transport high voltages over long distance, than high current, what does this have to do

FIGURE 11-6 **AC Generator.**

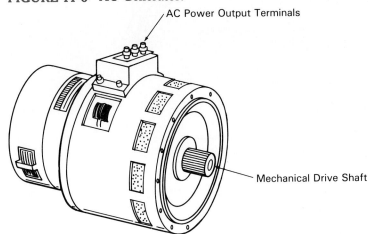

AC Power Output Terminals

Mechanical Drive Shaft

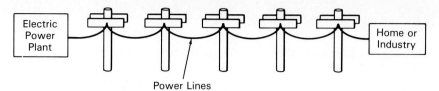

FIGURE 11-7 **Power Lines Connect Power from the Electric Company to Home and Industry.**

with ac? An ac voltage can easily and efficiently be transformed up or down to a higher or lower voltage by utilizing a device known as a *transformer*, and even though dc voltages can be stepped up and down, the method is inefficient and more complex.

(3) Nearly all electronic circuits and equipment are powered by dc voltages, which means that once the ac power arrives at the home or industry, in most cases, it will have to be converted into dc power to operate electronic

FIGURE 11-8 **AC Power Distribution.**

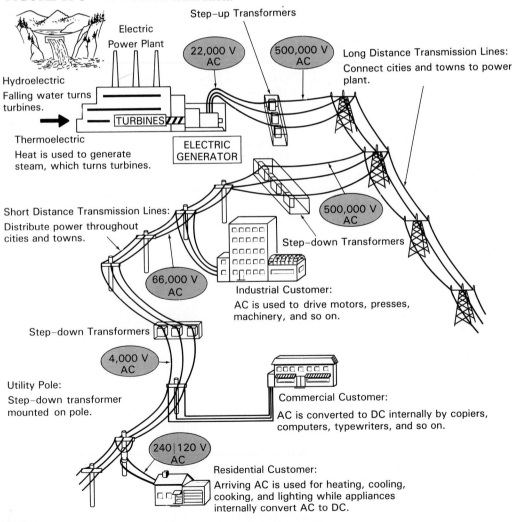

Alternating Current (AC)

FIGURE 11-9 AC Power Distribution. (a) Hydroelectric Dam. (b) Electric Power Plant.
(c) Generator. (d) Control Room. (e) Step-down Transformer. (f) Utility Pole.

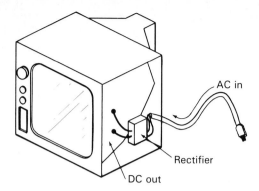

AC in

Rectifier

DC out

FIGURE 11-10 **Rectification (Converting AC to DC) by Appliances.**

equipment. It is a relatively simple process to convert ac to dc; however, conversion from dc to ac is a complex and comparatively inefficient process.

Figure 11-8 illustrates ac power distribution from the electric power plant to the home and industry. The ac power distribution system begins at the electric power plant, which has the powerful large generators driven by turbines to generate large ac voltages. The turbines can be driven by either falling water (hydroelectric), or from steam, which is produced with intense heat by burning either coal, gas, or oil or from a nuclear reactor (thermo-electric). The turbine supplies the mechanical energy to the generator, to be transformed into ac electrical energy.

The generator generates an ac voltage of approximately 22,000 V, which is stepped up by transformers to approximately 500,000 V. This voltage is applied to the long distance transmission lines, which connect the power plant to the city or town. At each city or town, the voltage is tapped off the long distance transmission lines and stepped down to approximately 66,000 V and is distributed to large scale industrial customers. The 66,000 V is stepped down again to approximately 4000 V and distributed throughout the city or town by short distance transmission lines. This 4000 V is used by small scale industrial customers and residential customers who receive the ac power via step down transformers on utility poles, which step down the 4000 V to 240 V and 120 V. The appearance of some of the devices that make up the power distribution system can be seen in Figure 11-9.

A large amount of equipment and devices within industry and the home will run directly from the ac power, such as heating, lighting, and cooling. Some equipment that runs on dc, such as televisions and computers, will accept the 120 V ac and internally convert it to the dc voltages required. Figure 11-10 illustrates a television that is operating from the 120 V ac from the wall outlet and internally converting it to dc so it can power the circuits necessary to produce both audio (sound) and video (picture) information. This unit that converts ac to dc is called a *rectifier*.

SELF-TEST REVIEW (§ 11.1.1)

1. In relation to power transfer, what three advantages does ac have over dc?

2. A generator converts an electrical input into a mechanical output. (True/False)

3. What formula is used to calculate the amount of power lost in a transmission line?
4. What is a transformer?
5. What voltage is provided to the wall outlet in the home?
6. Most appliances internally convert the _____ input voltage into a _____ voltage.

11.1.2 INFORMATION TRANSFER

Information, by definition, is the property of a signal or message that conveys something meaningful to the recipient. Communication, which is the transfer of information between two points, began with speech and progressed to handwritten words in letters and printed words in newspapers and books. To achieve greater distances of communication, face to face communications evolved into telephone and radio communications.

A simple communication system can be seen in Figure 11-11(a). The

FIGURE 11-11 Information Transfer. (a) Wire Communication System. (b) Sound Wave. (c) Electrical Wave.

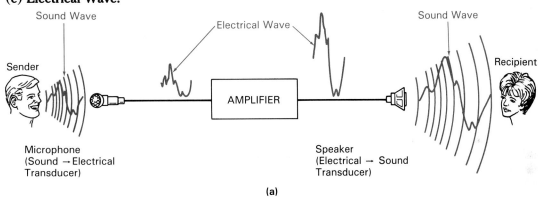

(a)

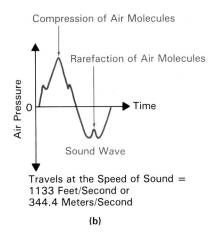

Travels at the Speed of Sound = 1133 Feet/Second or 344.4 Meters/Second

(b)

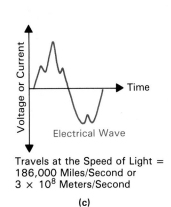

Travels at the Speed of Light = 186,000 Miles/Second or 3×10^8 Meters/Second

(c)

Why Alternating Current?   **375**

audio or sound waves generated when the person on the left speaks are converted to an electrical signal or wave by the microphone. This electrical wave is fed into an amplifier, which boosts the magnitude of the electrical signal. The enlarged electrical signal emerging out of the amplifier circuit is then converted back to the original sound waves of larger amplitude by the speaker, where they are then received by the recipient.

The voice information or sound wave produced by the sender is a variation in air pressure, as seen in Figure 11-11(b), and travels at the speed of sound. Sound waves or sounds are normally generated by a vibrating reed or plucked string in the case of musical instruments. In this example, the sender's vocal chords vibrate backward and forward, producing a rarefaction or decreased air pressure, where few air molecules exist, and a compression or increased air pressure, where many air molecules exist. Like the ripples produced by a stone falling in a pond, the sound waves produced by the sender are constantly expanding and traveling outward.

The microphone is in fact a transducer (energy converter), because it

FIGURE 11-12 Information Transfer. (a) Wireless Communication System. (b) Simplified Electromagnetic Wave. (c) Components of an Electromagnetic Wave.

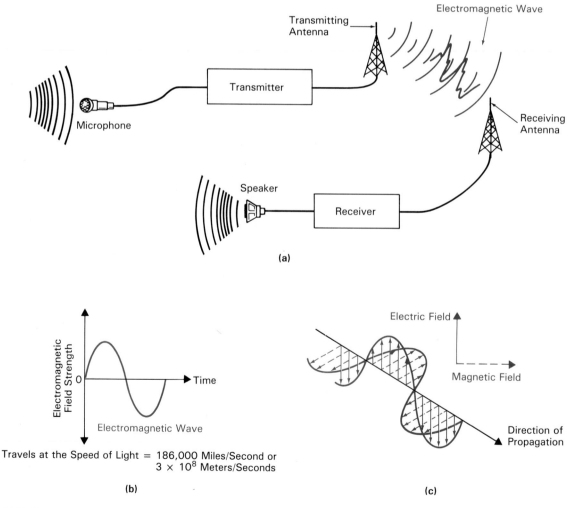

Alternating Current (AC)

converts the sound wave (which is a form of mechanical energy) into electrical energy in the form of voltage and current, which varies in the same manner as the sound wave and therefore contains the sender's message or information.

The electrical wave, shown in Figure 11-11(b), is a variation in voltage or current and can only exist in a wire conductor or circuit. This electrical signal travels at the speed of light, and after being increased in magnitude by the amplifier, it is applied to a loudspeaker.

The speaker, like the microphone, is also an electroacoustical transducer that converts the electrical energy input into a mechanical sound wave output. These sound waves strike the outer eardrum, causing the ear diaphragm to vibrate, and these mechanical vibrations actuate nerve endings in the ear, which convert the mechanical vibrations into electrochemical impulses that are sent to the brain. The brain decodes this information by comparing these impulses with a library of previous sounds and so provides the sensation of hearing.

To communicate between two distant points, a wire must be connected between the microphone and speaker. However, if an electrical wave is applied to an antenna, the electrical wave is converted into a radio or elecromagnetic wave, as seen in Figure 11-12(a), and communication is established without the need of a connecting wire; hence the term wireless communication. Antennas are designed to radiate and receive electromagnetic waves, which vary in field strength, as seen in Figure 11-12(b), and can exist in either air or space. These radio waves, as they are also known, travel at the speed of light and allow us to achieve great distances of communication.

More specifically, they are composed of two basic components, as seen in Figure 11-12(c). The electrical voltage applied to the antenna is converted into an electric field and the electrical current into a magnetic field. This electromagnetic (electric-magnetic) wave is used to carry a variety of information, such as speech, radio broadcasts, television signals, and so on.

EXAMPLE 11.1

How long will it take the sound wave produced by a rifle shot to travel 9630.5 feet?

SOLUTION

This problem makes use of the following formula: Distance = velocity × time, or

$$D = \text{distance}$$
$$v = \text{velocity}$$
$$t = \text{time}$$

If someone travels at 20 mph for 2 hours, the person will travel 40 miles ($D = v \times t = 20$ mph $\times$ 2 hours $= 40$ miles). In this problem, the distance (9630.5) and the sound wave's velocity (1133 ft/sec) are known. By rearranging the formula, we can find time:

$$\text{time} = \frac{\text{distance}}{\text{velocity}} = \frac{9630.5 \text{ ft}}{1133 \text{ ft/s}} = 8.5 \text{ s}$$

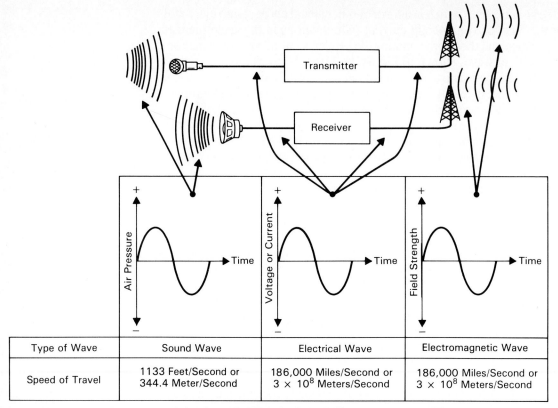

FIGURE 11-13 **Summary of the Sound, Electrical, and Electromagnetic Waves.**

In summary, the sound wave is a variation in air pressure, the electrical wave is a variation of voltage or current, and the electromagnetic wave is a variation of electric and magnetic field strength, as seen in Figure 11-13.

EXAMPLE 11.2

How long will it take an electromagnetic (radio) wave to reach a receiving antenna that is 2000 miles away from the transmitting antenna?

SOLUTION

In this problem, both distance (2000 miles) and velocity (186,000 miles/s) are known, and time has to be calculated:

$$\text{time} = \frac{\text{distance}}{\text{velocity}} = \frac{2000 \text{ miles}}{186{,}000 \text{ miles/s}}$$
$$= 1.075 \times 10^{-2} \quad \text{or} \quad 10.8 \text{ ms}$$

1. Define information and communication.
2. The _____ wave is a variation in air pressure, the _____ wave is a variation in _____ strength, and the _____ wave is a variation of voltage or current.
3. Which of the three waves described in question 2:
   a. Can only exist in the air?
   b. Can exist in either air or a vacuum?
   c. Exists in a wire conductor?
4. Sound waves travel at the speed of sound, which is _____, while electrical and electromagnetic waves travel at the speed of light, which is _____.
5. A human ear is designed to receive _____ waves, an antenna is designed to transmit or receive _____ waves, and an electronic circuit is designed to pass only _____ waves.
6. Give the names of the following energy converters or transducers:
   a. Sound wave (mechanical energy) to electrical wave
   b. Electrical wave to sound wave
   c. Electrical wave to electromagnetic wave
   d. Sound wave to electrochemical impulses

11.1.3 AC WAVE SHAPES

In all fields of electronics, whether medical, industrial, consumer, or data processing, different types of information are being conveyed between two points, and electronic equipment is managing the flow of this information.

Let's now discuss the basic types of ac wave shapes. The way in which a wave varies in magnitude with respect to time describes its wave shape. All ac electrical waves can be classified into one of six groups, and these are illustrated in Figure 11-14.

(1) SINE WAVE

The sine wave is the most common type of waveform. It is the natural output of a generator that converts a mechanical input, in the form of a rotating shaft, into an electrical output in the form of a sine wave. In fact, for one cycle of the input shaft, the generator will produce one sinusoidal ac voltage waveform, as seen in Figure 11-15. When the input shaft of the generator is at 0°, the ac output is 0 V. As the shaft is rotated through 360°, the ac output voltage will rise to a maximum positive voltage at 90°, fall back to 0 V at 180°, and then reach a maximum negative voltage at 270°, and finally return to 0 V at 360°. If this ac voltage is applied across a closed circuit, it produces a current that continually reverses or alternates in each direction.

Figure 11-16 illustrates the sine wave, with all the characteristic in-

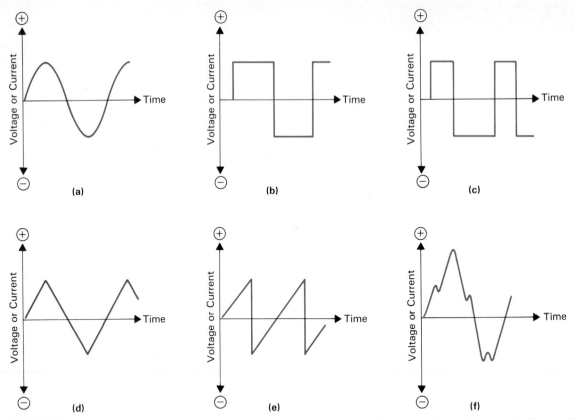

FIGURE 11-14 AC Wave Shapes. (a) Sine Wave. (b) Square Wave. (c) Pulse Wave. (d) Triangular Wave. (e) Sawtooth Wave. (f) Irregular Wave.

FIGURE 11-15 Degrees of a Sine Wave.

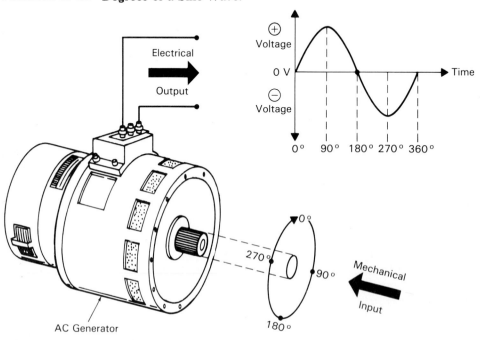

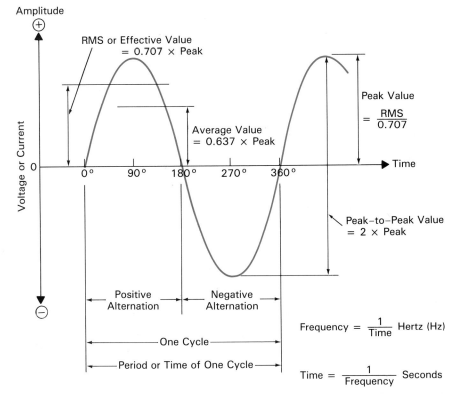

FIGURE 11-16 Sine Wave.

formation inserted, which at first glance looks a bit ominous. Let's analyze and discuss each piece of information individually, beginning with the sine wave's amplitude.

(a) Amplitude Figure 11-17 plots direction and amplitude against time. The amplitude or magnitude of a wave is often represented by a vector arrow, also illustrated in Figure 11-17. The vector's length indicates the magnitude of the current or voltage, while the arrow's point is used to show the direction, or polarity.

FIGURE 11-17 Sine Wave Amplitude.

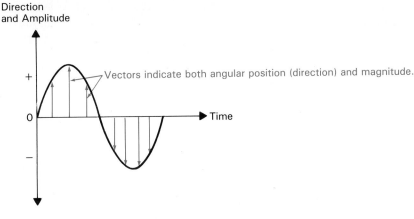

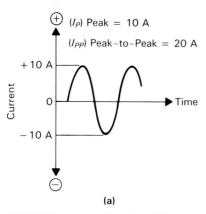

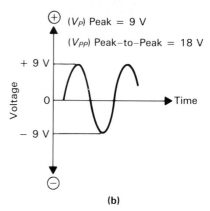

(a)   (b)

FIGURE 11-18   **Peak and Peak-to-Peak of a Sine Wave.**

(b) Peak Value The peak of an ac wave occurs on both the positive and negative alternation, but is only at the peak (maximum) for an instant. Figure 11-18(a) illustrates an ac current waveform rising to a positive peak of 10 A, falling to zero, and then reaching a negative peak of 10 A in the reverse direction; Figure 11-18(b) shows an ac voltage waveform reaching positive and negative peaks of 9 V.

(c) Peak-to-Peak Value The peak-to-peak value of a sine wave is the value of voltage or current between the positive and negative maximum values of a waveform. For example, the peak-to-peak value of the current waveform in Figure 11-18(a) is equal to $I_{pp} = 2 \times I_p = 20$ A. In Figure 11-18(b), it would be equal to $V_{pp} = 2 \times V_p = 18$ V.

(d) RMS or Effective Value Both the positive and negative alternation can accomplish the same amount of work; but the ac waveform is only at its maximum value for an instant in time, spending most of its time between peak currents and, therefore, cannot supply the same amount of power as a dc value of 10 A or 9 V.

The effective value of a sine wave is equal to 0.707 of the peak value. Let's now see how this value was obtained. Power is equal to either $P = I^2 \times R$ or $P = V^2/R$; said another way, power is proportional to the voltage or current squared. If every instantaneous value of either the positive or negative half-cycle of any voltage or current sinusoidal waveform is squared, as shown in Figure 11-19, and then averaged out to obtain the mean value, the square root of this mean value would be equal to 0.707 of the peak.

For example, if the process is carried out on the 10-A current waveform previously considered, the result would equal 7.07 A (0.707 × 10 A = 7.07 A), which is 0.707 of the peak value of 10 A.

$$\boxed{\text{rms} = 0.707 \times \text{peak}}$$

This root-mean-square (rms) result of 0.707 can always be used to tell us how effective an ac sine wave will be. For example, a 10 A dc source would be 10 A effective because it is continually present and always delivering power to the circuit to which it is connected, while a 10 A ac source would

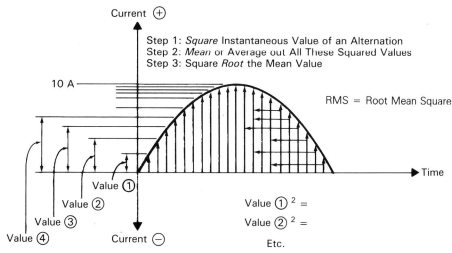

Step 1: *Square* Instantaneous Value of an Alternation
Step 2: *Mean* or Average out All These Squared Values
Step 3: Square *Root* the Mean Value

RMS = Root Mean Square

Value ① ² =
Value ② ² =
Etc.

FIGURE 11-19 **Obtaining the RMS Value of 0.707 for a Sine Wave.**

only be 7.07 A effective, as seen in Figure 11-20, because it is only at 10 A for a short period of time.

A 10 V ac sine wave alternation would be as effective or supply the same amount of power to a circuit as a 7.07 V dc source.

Unless otherwise stated, ac values of voltage or current are always given in rms, so the peak value can be calculated by transposing the original rms formula of rms = peak × 0.707, and ending up with

$$\boxed{\text{peak} = \frac{\text{rms}}{0.707}}$$

Since $1/0.707 = 1.414$, then the peak can also be calculated by

$$\boxed{\text{peak} = \text{rms} \times 1.414}$$

(e) Average The average value of the positive or negative alternation is found by taking either the positive and negative alternation, and listing the amplitude or vector length of current or voltages at 1° intervals, as shown in Figure 11-21(a). The sum of all these values is then divided by the total number of values (averaging), which for all sine waves will calculate out to

FIGURE 11-20 **Effective Equivalent.**

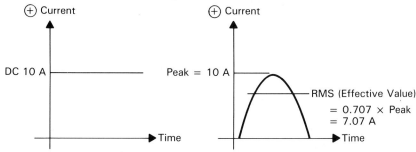

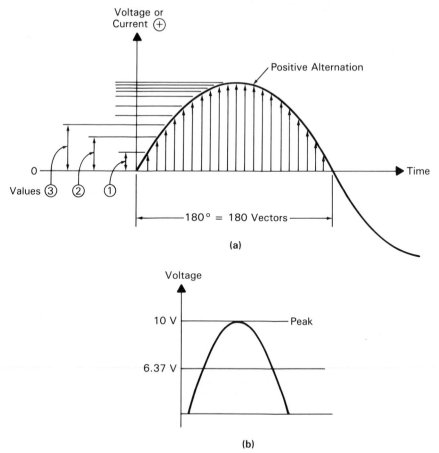

(a)

(b)

FIGURE 11-21   **Average Value of a Sine-Wave Alternation = 0.637 × Peak.**

be 0.637 of the peak voltage or current. For example, the average sine wave alternation with a peak of 10 V, as seen in Figure 11-21(b), is equal to

$$\boxed{\text{Average} = 0.637 \times \text{peak}}$$
$$= 0.637 \times 10 \text{ V}$$
$$= 6.37 \text{ V}$$

The average of a positive or negative alternation (half-cycle) is equal to 0.637 × peak; however, the average of the complete cycle, including both the positive and negative half-cycles, is mathematically zero as the amount of voltage or current above the zero line is equal but opposite to the amount of voltage or current below the zero line, as shown in Figure 11-22.

(f) Frequency and Period   The period (T) is the time required for one complete cycle (positive and negative alternation) of the sinusoidal current or voltage waveform. A *cycle* by definition, is the change of an alternating wave from zero to a positive peak, to zero, and then to a negative peak, and finally back to zero (see Figure 11-23).

Frequency is the number of repetitions of a periodic wave in a unit of time. It is symbolized by f and is given the unit hertz (cycles per second), in honor of the German physicist, Heinrich Hertz.

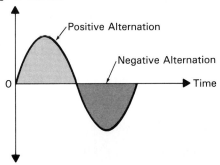

Voltage or Current

Positive Alternation

Negative Alternation

Time

0

Positive Alternation = Negative Alternation

Area = Area

and mathematically Cancel to Equal Zero

FIGURE 11-22 **Average Value of a Complete Sine-Wave Cycle = 0.**

Sinusoidal waves can take a long or a short amount of time to complete one cycle. This time is related to frequency in that period is equal to the reciprocal of frequency, and vice versa.

$$f \text{ (hertz)} = \frac{1}{T}$$

$$T \text{ (seconds)} = \frac{1}{f}$$

where T = period and f = frequency.

FIGURE 11-23 **Frequency and Period.**

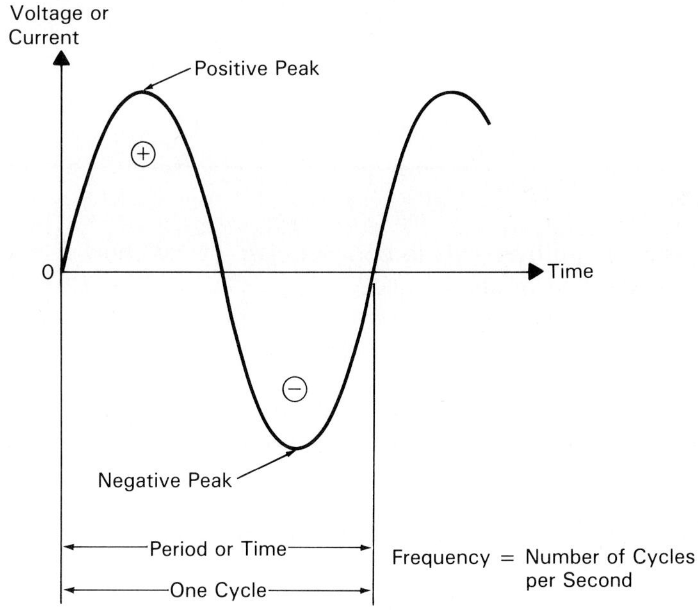

Voltage or Current

Positive Peak

$\oplus$

0 → Time

$\ominus$

Negative Peak

Period or Time

One Cycle

Frequency = Number of Cycles per Second

EXAMPLE 11.3

Calculate V_p, V_{p-p}, V_{rms} and V_{avg} of a 16 V peak sine wave.

SOLUTION

$$V_p = 16 \text{ V}$$
$$V_{p-p} = 2 \times V_p = 2 \times 16 \text{ V} = 32 \text{ V}$$
$$V_{rms} = 0.707 \times V_p = 0.707 \times 16 \text{ V} = 11.3 \text{ V}$$
$$V_{avg} = 0.637 \times V_p = 0.637 \times 16 \text{ V} = 10.2 \text{ V}$$

EXAMPLE 11.4

Calculate V_p, V_{p-p}, and V_{avg} of a 120 V (rms) ac main supply.

SOLUTION

$$V_p = \text{rms} \times 1.414 = 120 \text{ V} \times 1.414 = 169.68 \text{ V}$$
$$V_{p-p} = 2 \times V_p = 2 \times 169.68 \text{ V} = 339.36 \text{ V}$$
$$V_{avg} = 0.637 \times V_p = 0.637 \times 169.68 \text{ V} = 108.09 \text{ V}$$

EXAMPLE 11.5

If a sine wave has a period of 400 µs, what is its frequency?

SOLUTION

$$\text{frequency } (f) \frac{1}{\text{time } (T)} = \frac{1}{400 \text{ µs}} = 2.5 \text{ kHz}$$

EXAMPLE 11.6

If it takes a sine wave 25 milliseconds to complete two cycles, how many of the cycles will be received in one second?

SOLUTION

If the period of two cycles is 25 milliseconds, then one cycles period will equal 12.5 milliseconds. The number of cycles per second or frequency will equal

$$f = \frac{1}{t} = \frac{1}{12.5 \text{ ms}} = 80 \text{ Hz} \quad \text{or} \quad 80 \text{ cycles/second}$$

EXAMPLE 11.7

Calculate the period of the following:
a. 100 MHz
b. 40 cycles every 5 seconds
c. 4.2 kilocycles/second
d. 500 kHz

SOLUTION

$$f = \frac{1}{t}$$

a. $f = \dfrac{1}{100 \text{ MHz}} = 10$ nanoseconds (ns)

b. 40 cycles/5 seconds = 8 cycles/second (8 Hz).

$f = \dfrac{1}{8 \text{ Hz}} = 125$ ms

c. $f = \dfrac{1}{4.2 \text{ kHz}} = 238$ μs

d. $f = \dfrac{1}{500 \text{ kHz}} = 2$ μs

For example, the ac voltage of 120 V (rms) arrives at the household electrical outlet alternating at a frequency of 60 hertz (Hz). This means that 60 cycles arrive at the household electrical outlet in 1 second. If 60 cycles occur in 1 second, as seen in Figure 11-24(a), then it is actually taking 1/60th of a second for one of the 60 cycles to complete its cycle, which calculates out to be

$$\text{1/60th of 1 second} = \frac{1 \text{ sec}}{60 \text{ cycles}} \times 1 \text{ second} = 16.67 \text{ milliseconds (ms)}$$

FIGURE 11-24   A 120-Volt, 60-Hz AC Supply.

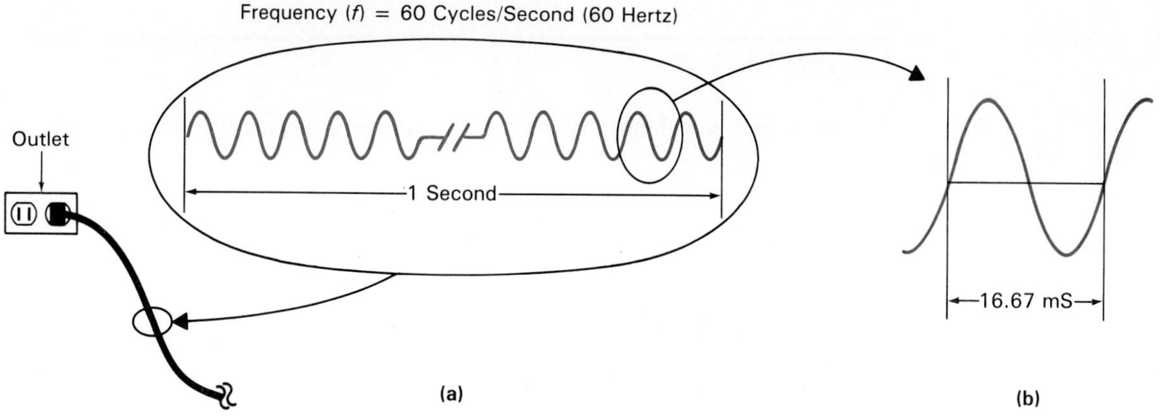

Frequency (*f*) = 60 Cycles/Second (60 Hertz)

Outlet

1 Second

16.67 mS

(a)

(b)

So the time or period of one cycle can be calculated by using the formula period $(T) = 1/f = 1/60$ Hz $= 16.67$ ms, as seen in Figure 11-24(b).

If the period or time of a cycle is known, then frequency can be calculated; for example

$$\text{frequency } (f) = \frac{1}{\text{period}} = 1/16.67 \text{ ms} = 60 \text{ Hz}$$

As previously illustrated in Figure 11-8, all homes in the United States receive at their wall outlets an ac voltage of 120 V rms at a frequency of 60 Hz. This frequency was chosen for convenience, as a lower frequency would require larger transformers, and if the frequency were too low, the slow switching (alternating) current through the light bulb would cause them to flicker. A higher frequency than 60 Hz was found to cause an increase in the amount of heat generated in the core of all power distribution transformers due to eddy currents and hysteresis losses. Consequently, 60 Hz was chosen in the United States; however, other countries, such as England and most of Europe, use an ac power line frequency of 50 Hz (240 V).

(g) Wavelength Wavelength, as its name states, is the physical length of one complete cycle and is generally measured in meters. The wavelength (λ, lambda) of a complete cycle is dependent on the frequency and velocity of the transmission

$$\boxed{\lambda = \frac{\text{velocity}}{\text{frequency}}}$$

Electromagnetic Waves Radio waves travel at the speed of light in air or a vacuum, which is 3×10^8 meters/second or 3×10^{10} cm/second.

$$\boxed{\lambda \text{ (cm)} = \frac{3 \times 10^{10} \text{ cm/s}}{\text{frequency (Hz)}}} \quad \text{or} \quad \boxed{\lambda \text{ (m)} = \frac{3 \times 10^8 \text{ m/s}}{f \text{ (Hz)}}}$$

Subsequently, the higher the frequency, the shorter the wavelength, which is why a shortwave radio receiver is designed to receive high frequencies ($\lambda\downarrow = 3 \times 10^8/f\uparrow$).

EXAMPLE 11.8

Calculate the wavelength of the electromagnetic waves illustrated in Figure 11-25.

SOLUTION

(a) $\lambda = \dfrac{3 \times 10^8}{f \text{ (Hz)}} \text{ m/s} = \dfrac{3 \times 10^8}{10 \text{ kHz}} = 30,000 \text{ m} \quad \text{or} \quad 30 \text{ km}$

(b) $\lambda = \dfrac{3 \times 10^{10}}{f \text{ (Hz)}} \text{ cm/s} = \dfrac{3 \times 10^{10}}{2182 \text{ kHz}} = 13,748.9 \text{ cm} \quad \text{or} \quad 137.489 \text{ m}$

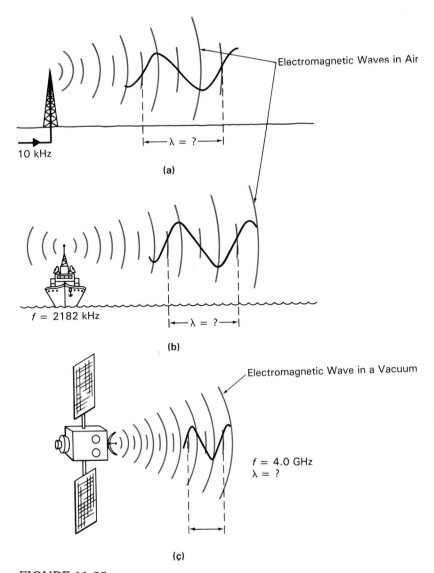

Electromagnetic Waves in Air

$\lambda = ?$

10 kHz

(a)

$f = 2182$ kHz

$\lambda = ?$

(b)

Electromagnetic Wave in a Vacuum

$f = 4.0$ GHz
$\lambda = ?$

(c)

FIGURE 11-25

(c) $\lambda = \dfrac{3 \times 10^{10}}{f\,(\text{Hz})}$ cm/s $= \dfrac{3 \times 10^{10}}{4.0\ \text{GHz}} = \dfrac{3 \times 10^{10}}{4 \times 10^{9}} = 7.5$ cm or 0.075m

As frequency increases, wavelength decreases.

Sound Waves Sound waves travel at a slower speed than electromagnetic waves, as their mechanical vibrations depend on air molecules, which offer resistance to the traveling wave. For sound waves, the wavelength formula will be equal to

$$\lambda\,(\text{ft}) = \frac{1133\ \text{ft/s}}{f\,(\text{Hz})} \quad \text{or} \quad \lambda\,(\text{m}) = \frac{344.4\ \text{m/s}}{f\,(\text{Hz})}$$

Why Alternating Current?   **389**

EXAMPLE 11.9

Calculate the wavelength of the sound waves illustrated in Figure 11-26.

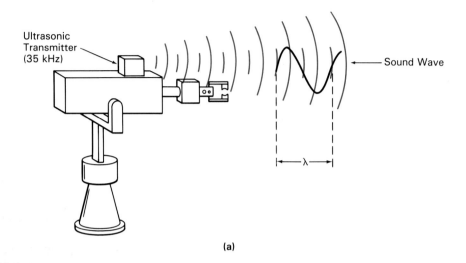

Ultrasonic Transmitter (35 kHz)

Sound Wave

(a)

Frequency Range = 300 Hz to 3 kHz
Wavelength Range = ? to ?

(b)

FIGURE 11-26

SOLUTION

(a) $\lambda \text{ (m)} = \dfrac{344.4 \text{ m/s}}{f \text{ (Hz)}} = \dfrac{344.4}{35 \text{ kHz}} = 9.8 \times 10^{-3\text{m}} = 9.8 \text{ mm}$

(b) 300 Hz: $\lambda \text{ (ft)} = \dfrac{1133 \text{ ft/s}}{f \text{ (Hz)}} = \dfrac{1133}{300 \text{ Hz}} = 3.78 \text{ ft}$

3000 Hz: $\lambda \text{ (ft)} = \dfrac{1133 \text{ ft/s}}{f \text{ (Hz)}} = \dfrac{1133}{3000 \text{ Hz}} = 0.378 \text{ ft}$

(h) Phase Relationships The phase of a sine wave is always relative to another sine wave of the same frequency. Figure 11-27(a) illustrates two sine waves that are in phase with one another, while Figure 11-27(b) shows two sine waves that are out of phase with one another. Sine wave A is our reference, since the positive going zero crossing is at 0°, its positive peak is at 90°, its negative going zero crossing is at 180°, its negative peak is at

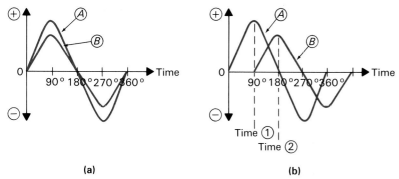

FIGURE 11-27 **Phase Relationships. (a) In phase. (b) Out of Phase.**

270°, and the cycle completes at 360°. In Figure 11-27(a), sine wave *B* is in phase with *A* since its peaks and zero crossings occur at the same time as sine wave *A*. In Figure 11-27(b), sine wave *B* has been shifted to the right by 90° with respect to the reference sine wave *A*. This *phase shift* or *phase angle* of 90° means that sine wave *A leads B* by 90°, or *B lags A* by 90°. Sine wave *A* is said to lead *B* as its positive peak, for example, occurs first at time 1, while the positive peak of *B* occurs later at time 2.

EXAMPLE 11.10

What are the phase relationships between the two waveforms illustrated in Figure 11-28(a) and (b)?

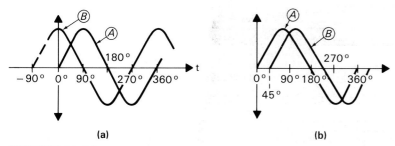

(a)   **(b)**

FIGURE 11-28

SOLUTION

a. The phase shift or angle is 90°. Sine wave *B* leads sine wave *A* by 90°, or *A* lags *B* by 90°.

b. The phase shift or angle is 45°. Sine wave *A* leads sine wave *B* by 45°, or *B* lags *A* by 45°.

(i) The Meaning of Sine The square, rectangular, triangular, and sawtooth waveform shapes are given their names because of their waveform shapes. The sine wave is the most common type of waveform shape, and the name given to this wave needs to be further explained.

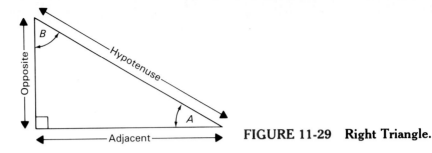

FIGURE 11-29 Right Triangle.

Figure 11-29 illustrates a right angle triangle. Sine is a trigonometric function that relates to either angle A or B in the right angle triangle. If we concentrate on angle A, we can say that

$$\text{sine of angle } A = \frac{\text{opposite side}}{\text{hypotenuse}}$$

This means that if we decrease the length of the opposite side (while keeping the hypotenuse the same), then the sine of angle A would also decrease, as seen in Figure 11-30.

On the other hand, an increase in the length of the opposite side (while keeping the hypotenuse constant) will cause an increase in angle A and its sine, and so it can be said that the sine of angle A is proportional to the length of the opposite side, which represents the instantaneous magnitude of the sinusoidal variation.

Angle A can vary anywhere from 0° to 90°, which corresponds to sine values of 0 to 1, respectively, as seen in Figure 11-31. Table 11-1 lists a few of the sine values, and Appendix 0 has a more complete standard table of trigonometric functions, which are also stored in your scientific calculator.

Sine, therefore, is simply a trigonometric function that relates to any of the two angles (excluding the right angle) in a right angle triangle. The sine wave is called the sine wave as it changes in value at the same rate as the trigonometric function known as sine, as seen in Figure 11-32.

TABLE 11-1

Angle	Sine
0°	0.000
15	0.259
30	0.500
45	0.707
60	0.866
75	0.966
90	1.000

(2) SQUARE WAVE

The square wave is a periodic (repeating) wave that alternates from a positive peak value to a negative peak value, and vice versa, for equal lengths of time.

In Figure 11-33 you can see an example of a square wave that is at a frequency of 1 kHz and has a peak of 10 V. If the frequency of a wave is known, then its period or time of one cycle can be calculated by using the formula $T = 1/f = 1/1 \text{ kHz} = 1 \text{ ms}$ or 1/1000th of a second. One complete

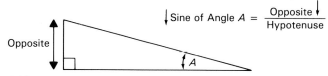

FIGURE 11-30 **Sine of Angle _A_ Is Proportional to Opposite Length.**

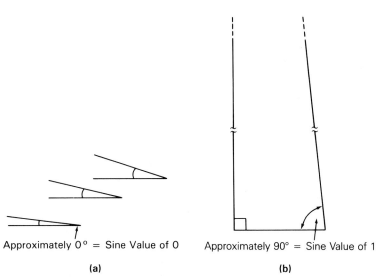

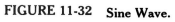

Approximately 0° = Sine Value of 0   Approximately 90° = Sine Value of 1

(a)   (b)

FIGURE 11-31 **Sine of Angle _A_ Is Inversely Proportional to the Hypotenuse Length.**

FIGURE 11-32 **Sine Wave.**

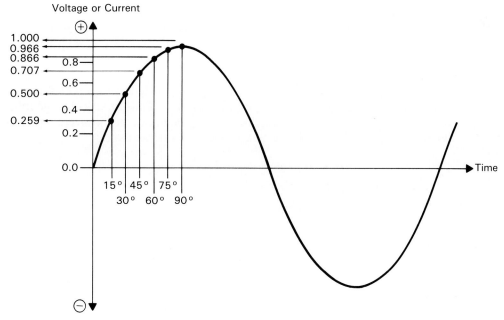

Why Alternating Current?   **393**

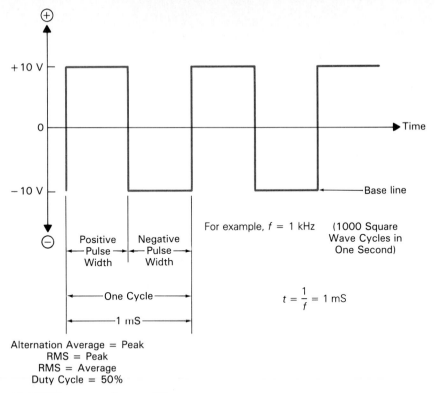

For example, $f = 1$ kHz   (1000 Square Wave Cycles in One Second)

$$t = \frac{1}{f} = 1 \text{ mS}$$

Alternation Average = Peak
RMS = Peak
RMS = Average
Duty Cycle = 50%

FIGURE 11-33 Square Wave.

cycle will take 1 ms to complete, and so the positive and negative alternations will each last for 0.5 ms.

If the peak of the square wave is equal to 10 V, then the peak-to-peak value of this square wave will equal $V_{p\text{-}p} = 2 \times V_p = 20$ V.

To summarize, the square wave alternates from a positive peak value of $+10$ V to a negative peak value of -10 V for equal time lengths (half-cycles) of 0.5 ms.

(a) Duty Cycle Duty cycle is an important relationship, which has to be considered when discussing square waveforms. The *duty cycle* is the ratio of a pulse width (positive or negative pulse or cycle) to the overall period or time of the wave and is normally given as a percentage.

$$\boxed{\text{duty cycle } (\%) = \frac{\text{pulse width } (P_w)}{\text{period } (t)} \times 100\%}$$

The duty cycle of the example square wave in Figure 11-23 will equal

$$
\begin{aligned}
\text{duty cycle } (\%) &= \frac{\text{pulse width } (P_w)}{\text{period } (p)} \times 100\% \\
&= \frac{0.5 \text{ ms}}{1 \text{ ms}} \times 100\% \\
&= 50\%
\end{aligned}
$$

Since a square wave always has a positive and negative alternation that are equal in time, the duty cycle of all square waves is equal to 50%, which actually means that the positive cycle lasts for 50% of the time of one cycle.

(b) Average The average or mean value of a square wave can be calculated by using the formula

$$\boxed{V \text{ or } I \text{ average} = \text{base line} + (\text{duty cycle} \times \text{peak to peak})}$$

The average of the complete square wave cycle in Figure 11-33 should calculate out to be zero, as the amount above the line equals the amount below. If we apply the formula to this example, you can see that

$$
\begin{aligned}
V_{avg} &= \text{base line} + (\text{duty cycle} \times \text{peak to peak}) \\
&= -10 \text{ V} + (0.5 \times 20 \text{ V}) \\
&= -10 \text{ V} + 10 \text{ V} \\
&= 0 \text{ V}
\end{aligned}
$$

However, a square wave does not always alternate about 0; for example, Figure 11-34 illustrates a 16 $V_{p\text{-}p}$ square wave that rests on a base line of 2 V. The average value of this square wave is equal to

$$
\begin{aligned}
V_{avg} &= \text{base line} + (\text{duty cycle} \times \text{peak to peak}) \\
&= 2 \text{ V} \times (0.5 \times 16 \text{ V}) \\
&= (+2) + (+8 \text{ V}) \\
&= 10 \text{ V}
\end{aligned}
$$

EXAMPLE 11.11

Calculate the duty cycle and V_{avg} of a 0 to 5 V square wave.

SOLUTION

The duty cycle of a square wave is always 0.5 or 50%.

$$
\begin{aligned}
V_{avg} &= \text{baseline} + (\text{duty cycle} \times V_{p\text{-}p}) \\
&= 0 \text{ V} + (0.5 \times 5 \text{ V}) \\
&= 0 \text{ V} + 2.5 \text{ V} = 2.5 \text{ V}
\end{aligned}
$$

FIGURE 11-34

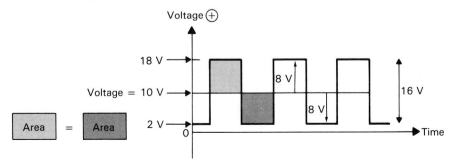

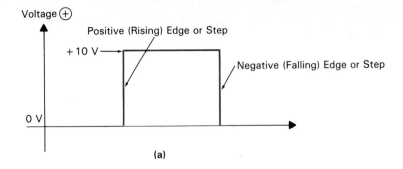

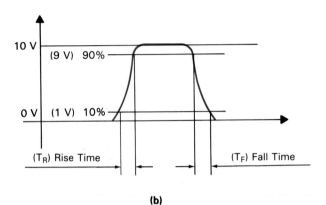

(b)

FIGURE 11-35 **(a) Ideal and (b) Actual Square Waves.**

Up to this point, we have seen the ideal square wave, which has instantaneous transition from the negative to the positive values, and vice versa, as seen in Figure 11-35(a). In fact, the transitions from negative to positive (positive or leading edge) and from positive to negative (negative or trailing edge) are not as ideal as seen here. It takes a small amount of time for the wave to increase to its positive value (the *rise time*) and consequently an equal amount of time for a wave to decrease to its negative value (the *fall time*). Rise time (t_r), by definition, is the time it takes for an edge to rise from 10% to 90% of its full amplitude, while fall time (t_f) is the time it takes for an edge to fall from 90% to 10% of its full amplitude, as seen in Figure 11-35(b).

With a waveform such as that in Figure 11-35(b), it is difficult, unless a standard is used, to know exactly what points to use when measuring the width of either the positive or negative alternation. The standard width is always measured between the two 50% amplitude points, as seen in Figure 11-36.

(c) Frequency-Domain Analysis A periodic wave is a wave that repeats the same wave shape from one cycle to the next. Figure 11-37(a) is a *time domain* representation of a periodic sine wave, which is the same way it would appear on an oscilloscope display as it plots the sine wave's amplitude against time. Figure 11-37(b) is a *frequency domain* representation of the same periodic sine wave, and this graph, which shows the wave as it would appear on a spectrum analyzer, plots the sine wave's amplitude against

396   **Alternating Current (AC)**

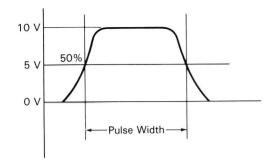

FIGURE 11-36   Pulse Width of a Square Wave.

frequency instead of time. This graph shows all the frequency components contained within a wave, and since, in this example, the sine wave has a period of 1 ms and therefore a frequency of 1 kHz, there is one bar at the 1 kHz point of the graph and its size represents the sine wave's amplitude. Pure sine waves have no other frequency components.

Other periodic wave shapes, such as square, pulse, triangular sawtooth, or irregular, are actually made up of a number of sine waves having a particular frequency, amplitude, and phase. To produce a square wave, for instance, you would start with a sine wave, as seen in Figure 11-38(a), whose frequency is equal to the resulting square-wave frequency. This sine wave is called the *fundamental* frequency, and all the other sine waves that will be added to this fundamental are called *harmonics* and will always be lower in amplitude and higher in frequency. These harmonics or multiples are harmonically related to the fundamental, in that the second harmonic is twice the fundamental frequency, the third harmonic is three times the fundamental frequency, and so on.

Square waves are composed of a fundamental frequency and an infinite number of odd harmonics (third, fifth, seventh and so on). If you look at the progression in Figure 11-38(a) through (d), you see that by continually adding these odd harmonics the waveform comes closer to a perfect square wave, as seen in Figure 11-38(e).

Figure 11-39 plots the frequency domain of a square wave, with the bars representing the odd harmonics of decreasing amplitude.

FIGURE 11-37   Analysis of a 1-Kilohertz Sine Wave. (a) Time Domain. (b) Frequency Domain.

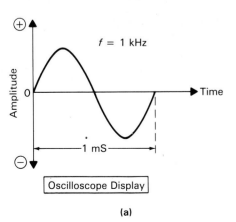

(a)

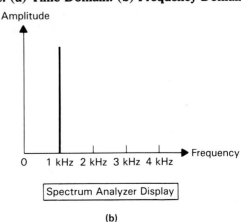

(b)

Why Alternating Current?   **397**

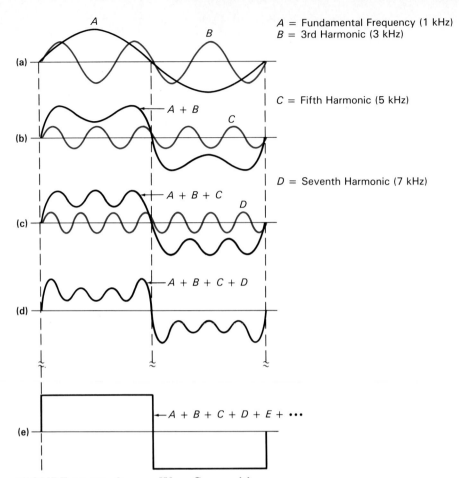

A = Fundamental Frequency (1 kHz)
B = 3rd Harmonic (3 kHz)

C = Fifth Harmonic (5 kHz)

D = Seventh Harmonic (7 kHz)

FIGURE 11-38 Square Wave Composition.

(3) RECTANGULAR OR PULSE WAVE

The rectangular wave is similar to the square wave in many respects, in that it is a periodic wave that alternately switches between one of two fixed values. The difference in the rectangular wave is that it does not remain at the two peak values for equal lengths of time, as seen in the examples in Figure 11-40(a) and (b).

FIGURE 11-39   Frequency Domain Analysis of a Square Wave.

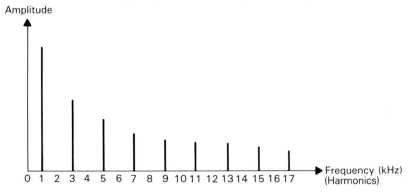

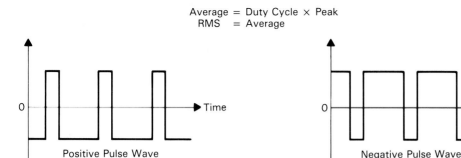

Average = Duty Cycle × Peak
RMS = Average

Positive Pulse Wave

Negative Pulse Wave

(a)

(b)

FIGURE 11-40   **Rectangular or Pulse Wave.**

In Figure 11-40(a), the rectangular wave remains at its negative level for a longer period than its positive, while the rectangular wave in Figure 11-29(b) stays at its positive value for the longer period of time and is only momentarily at its negative value.

(a) PRF, PRT, and Pulse Length (P_l) When discussing a rectangular wave, a few terms change. Instead of stating the cycles per second as frequency, it is called *pulse repetition frequency* (PRF), which is far more descriptive. The reciprocal of frequency is time, and with rectangular pulse waveforms the reciprocal of the PRF is pulse repetition time (PRT), as summarized in Table 11-2. Frequency is equivalent to PRF and time to PRT; the only difference is the name.

Let us look at the example in Figure 11-41 of a 5 V rectangular wave at a frequency of 1 kHz and a pulse width of 1 μs, and practice with these new terms. With a pulse repetition frequency of 1 kHz, the time between the leading edges of pulses (PRT) will be 1/1 kHz = 1 ms. Pulse width (P_w), pulse duration (P_d), or pulse length (P_l) are all terms that describe the length of time for which the pulse lasts, and in this example it is equal to 1 μs, which means that 999 μs exists between the end of one pulse and the beginning of the next.

TABLE 11-2

Square and Sine Wave	Rectangular Wave
Equivalent to	
Frequency $(f) = \dfrac{1}{\text{time }(t)}$	pulse repetition frequency (PRF) $= \dfrac{1}{\text{pulse repetition time (PRT)}}$
Time $(t) = \dfrac{1}{\text{frequency }(f)}$	pulse repetition time (PRT) $= \dfrac{1}{\text{pulse repetition frequency (PRF)}}$

Why Alternating Current?   **399**

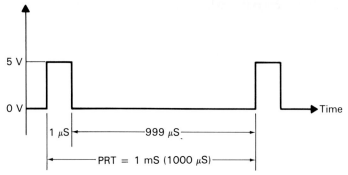

FIGURE 11-41

(b) Duty Cycle The duty cycle is calculated in exactly the same way as the square wave and is a ratio of the pulse width to the overall time (PRT). In our example, this will be equal to

$$\text{duty cycle } (\%) = \frac{\text{pulse width } (P_w)}{\text{PRT}} \times 100\%$$
$$= \frac{1 \ \mu s}{1000 \ \mu s}$$
$$= \text{duty cycle figure of } 0.001$$
$$= 0.001 \times 100\%$$
$$= 0.1\%$$

The result tells us that the positive pulse lasts for 0.1% of the total time (PRT).

(c) Average The average or mean value of this waveform is calculated by using the same formula;

$$V \text{ or } I \text{ average} = \text{base line} + (\text{duty cycle} \times \text{peak to peak})$$
$$V_{\text{avg}} = 0 \text{ V} + (0.001 \times 5 \text{ V})$$
$$= 0 \text{ V} + (5 \text{ mV})$$
$$= 5 \text{ mV}$$

FIGURE 11-42

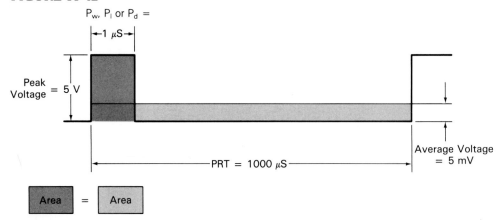

Alternating Current (AC)

EXAMPLE 11.12

Calculate the duty cycle and average voltage of the following radar pulse waveform:

$$\text{peak voltage, } V_p = 20 \text{ kV}$$
$$\text{pulse length, } P_L = 1 \text{ } \mu s$$
$$\text{baseline voltage} = 0V$$
$$\text{PRF} = 3300 \text{ pulses per second (pps)}$$

SOLUTION

$$\text{duty cycle} = \frac{\text{pulse length } (P_L)}{\text{PRT}} \times 100\%$$

$$= \frac{1 \text{ } \mu s}{303 \text{ } \mu s} \times 100\% \quad \left(\text{PRT} = \frac{1}{\text{PRF}} = \frac{1}{3300} = 303 \text{ } \mu s \right)$$
$$= (3.3 \times 10^{-3}) \times 100\%$$
$$= 0.33\%$$

$$V_{avg} = \text{baseline} + (\text{duty cycle} \times V_{p\text{-}p})$$
$$= 0 \text{ V} + [(3.3 \times 10^{-3}) \times 20 \times 10^3$$
$$= 66 \text{ V}$$

Figure 11-42 illustrates the average value of this rectangular waveform. If the voltage and width of the positive pulse are taken and spread out over the entire PRT, they will have a mean level equal, in this example, to 5 mV.

(d) Frequency-Domain Analysis The pulse or rectangular wave is closely related to the square wave, as seen in Figure 11-43; however, some changes occur in its harmonic content. One is that even-number harmonics are present and their amplitudes do not fall off as quickly as do those of the square wave. The amplitude and phase of these sine wave harmonics are determined by the pulse width and pulse repetition frequency, and the narrower the pulse, the greater the number of harmonics present.

(4) TRIANGULAR WAVE

A triangular wave consists of a positive and negative ramp of equal values, as seen in Figure 11-44. Both the positive and negative ramps have a linear increase and decrease, respectively. Linear, by definition, is the relationship between two quantities that exists when a change in a second quantity is directly proportional to a change in the first quantity. The two quantities in this case are voltage or current and time. As seen in Figure 11-44, if the increment of change of voltage ΔV (pronounced delta vee) is changing at the same rate as the time increment, Δt (delta tee), then the ramp is said to be linear.

With Figure 11-45(a), the voltage has risen 1 V in 1 second (time 1) and maintains that rise through to time 4 and, consequently, is known as a linearly rising slope. In Figure 11-45(b), the voltage is falling first from 6 to 5 V, which is a 1 V drop in 1 second, and in time 2 from 6 to 2 V, which

Why Alternating Current?

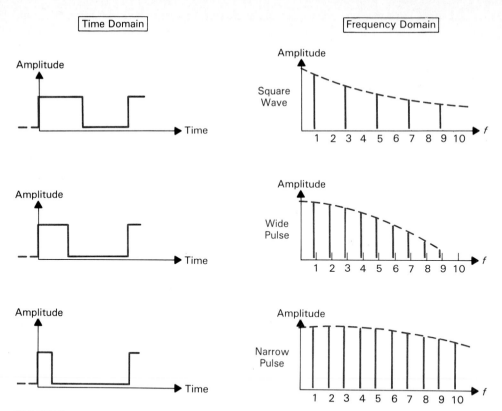

FIGURE 11-43   Time and Frequency Domain Analysis of a Pulse Waveform.

FIGURE 11-44   Triangular Wave.

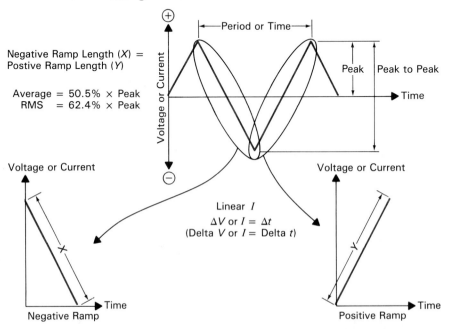

Time ①
 Slope (V/S) = $\frac{\Delta V}{\Delta t}$ = $\frac{1\,V}{1\,s}$ = 1 Volt/Second

Time ②
 Slope (V/S) = $\frac{\Delta V}{\Delta t}$ = $\frac{4\,V}{4\,s}$ = 1 Volt/Second

ΔV = Increment of Voltage Change
Δt = Increment of Time Change

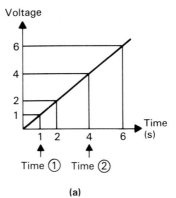

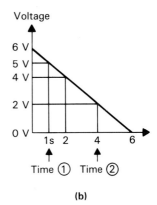

(a)   (b)

FIGURE 11-45   Linear Triangular Wave Rise and Fall.

is a 4 V drop in 4 seconds. The rate of fall still remains the same, and so the waveform is referred to as a linearly falling slope.

This formula for slope will also apply to a current waveform; the formula will now be

$$\text{slope (A/s)} = \frac{\Delta I}{\Delta t}$$

ΔI = increment of current change
Δt = increment of time change

With triangular waves, frequency and time, seen in Figure 11-46, apply as usual with

$$\text{frequency} = \frac{1}{\text{time}} \quad (\text{Hz})$$

$$\text{time} = \frac{1}{\text{frequency}} \quad (\text{s})$$

(a) Frequency-Domain Analysis The frequency domain of the triangular wave is shown in Figure 11-47. It is frequently used to test electronic

FIGURE 11-46   Triangular Wave Period and Frequency.

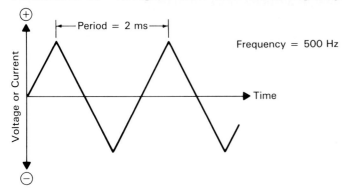

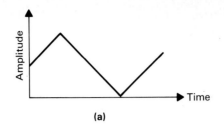

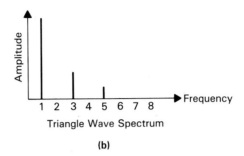

Triangle Wave Spectrum

(b)

FIGURE 11-47 **Analysis of a Triangular Wave. (a) Time Domain. (b) Frequency Domain.**

equipment as it tends to highlight circuit problems that would normally not show up when a sine-wave signal is applied.

(5) SAWTOOTH WAVE

On an oscilloscope display (time-domain presentation), the sawtooth wave is very similar to a triangular wave, in that a sawtooth wave has a linear ramp. However, unlike the triangular wave, which reverses and has an equal but opposite ramp back to its starting level, the sawtooth "flies" back to its starting point immediately and then repeats the previous ramp, as seen in Figure 11-48, which shows both a positive and negative ramp sawtooth. The sawtooth waveform *B* contains both odd and even harmonics, as illustrated in Figure 11-49, which shows both the time- and frequency-domain analysis of a negative-going ramp.

(6) OTHER WAVEFORMS

The waveforms discussed so far are some of the more common types; however, since every waveform shape (except a pure sine wave) is composed of a large number of sine waves combined in an infinite number of ways, any waveform shape is possible. Figure 11-50 illustrates a variety of waveforms that can be found in all fields of electronics.

FIGURE 11-48 **Sawtooth Wave. (a) Positive Ramp. (b) Negative Ramp.**

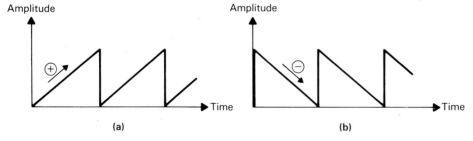

404 Alternating Current (AC)

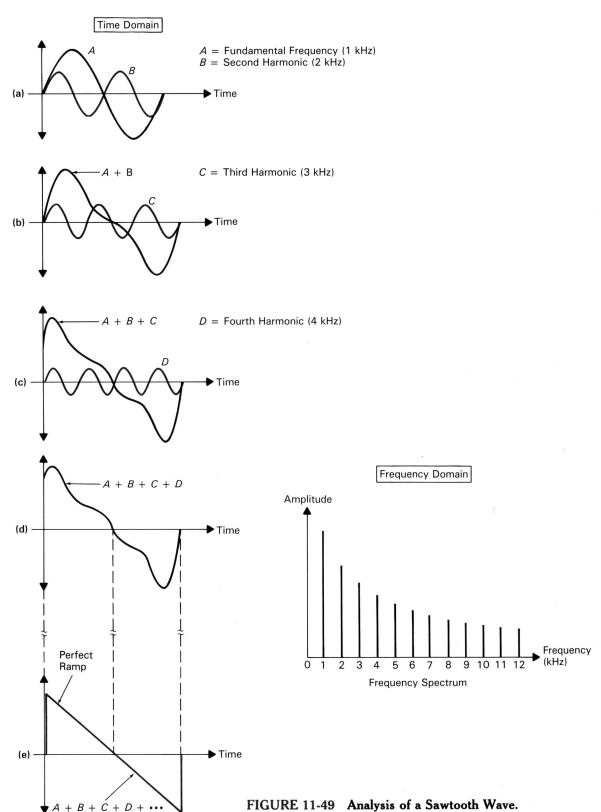

FIGURE 11-49   Analysis of a Sawtooth Wave.

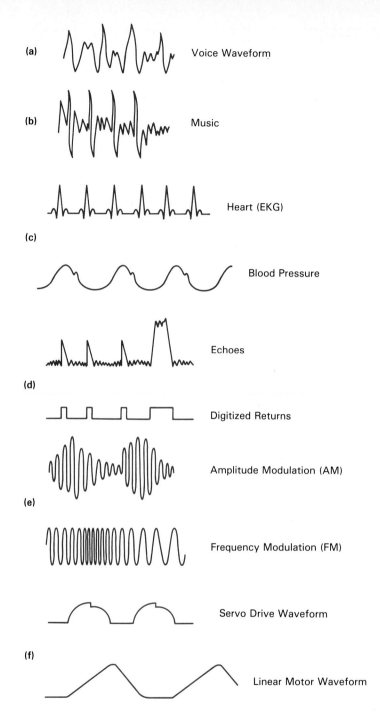

FIGURE 11-50   Waveforms. (a) Telephone Communications. (b) Radio Broadcast. (c) Medical. (d) Radar/Sonar. (e) Communications. (f) Industrial.

1. Sketch the following waveforms:
   a. Sine wave
   b. Square wave
   c. Rectangular wave
   d. Triangular wave
   e. Sawtooth wave
2. An oscilloscope gives a _____ domain representation of a periodic wave, while a spectrum analyzer gives a _____ domain representation.
3. What are the wavelength formulas for sound waves and electromagnetic waves, and why are they different?
4. A square wave is composed of an infinite number of _____ harmonics.

11.1.4 ELECTRICITY AND ELECTRONICS

In the beginning of this chapter, it was stated that ac is basically used in two applications, for (1) power transfer and (2) information transfer. These two uses for ac help define the difference between electricity and electronics. Electronic equipment manages the flow of information, while electrical equipment manages the flow of power. In summary;

Equipment	Manages
Electrical	Power (large values of V and I)
Electronic	Information (small values of V and I)

1. _____ equipment manages the flow of information, and these ac waveforms normally have small values of current and voltage.
2. _____ equipment manages the flow of power, and these ac waveforms normally have large values of current and voltage.

SUMMARY

1. Direct current (dc) is the flow of electrons in only one DIRECTion.
2. Pulsating dc rises from zero to maximum, and then from maximum to zero, and repeats; however, current flow is still only in one direction.

3. Alternating current (ac) flows first in one direction and then reverses to flow in the opposite in response to a corresponding change in voltage.

4. Alternating current is produced by an alternating voltage source that reaches a maximum in one direction (positive), decreases to zero, and then reverses itself and reaches a maximum in the opposite direction (negative).

5. Power and information are the two applications of ac.

6. Alternating current is best suited for delivering power because:
   a. AC generators, which are less complex internally and cheaper than dc generators, can supply a far larger amount of continuous power than a battery of the same size.
   b. AC can be stepped up by transformers to a large voltage, so current and consequently line power losses are minimum.
   c. In the home and industry, both ac and dc are needed to power equipment, and if we begin with ac, dc is easy to obtain; however, changing from dc to ac is much more difficult.

7. Information is the property of a signal or message that converys something meaningful to the recipient.

8. Communication is the transfer of information between two points.

9. The sound wave is a variation in air pressure, the electrical wave is a variation of voltage or current, and the electromagnetic wave is a variation of electric and magnetic field strength.

10. The sine wave is the most common type of waveform, and is called a sine wave because it changes in value at the same rate as the trigonometric function known as sine.

11. The amplitude of a waveform is its magnitude and is often represented by a vector.

12. The peak or maximum of an ac wave occurs on both the positive and negative alternation.

13. The peak-to-peak value of an ac wave is the value between the positive and negative maximum values.

14. The rms or effective value of a sine wave is equal to 0.707 of the peak value and is the dc equivalent voltage.

15. The average value of a complete sine wave cycle is zero. The average value of the positive or negative alternation is equal to 0.637 of the peak value.

16. The frequency of a periodic wave is equal to the reciprocal of its period.

17. The wavelength of either an electromagnetic or sound wave is dependent on the frequency and velocity of the transmission.

18. The phase of a periodic wave is always relative to another periodic wave of the same frequency. If both waveforms are in phase, the phase angle or phase shift is said to be 0. If out of phase, one waveform will lag or lead the other by some number of degrees.

19. The duty cycle of a waveform is the ratio of pulse width (P_w) to the overall period and is normally given as a percentage.

20. The time domain representation of a wave (as seen on an oscilloscope) plots amplitude against time, while the frequency domain representation

of a wave (seen on a spectrum analyzer) plots amplitude against frequency.

21. Waveshapes, other than a pure sine wave, are actually made up of a number of sine waves having a particular frequency, amplitude, and phase relationship to the fundamental frequency. Harmonics are multiples of the fundamental frequency.

22. Electronic equipment manages the flow of information, while electrical equipment manages the flow of power.

NEW TERMS

Adjacent
Alternating current (ac)
Amplifier
Amplitude
Antenna
Audio
Average value
Battery charger
Communication
Delta (Δ)
Duty cycle
Effective value
Electrical wave
Electric field
Electroacoustical transducer
Electromagnetic wave
Fall time
Frequency
Frequency domain
Fundamental frequency
Harmonics
Hypotenuse
Information

Lag
Lead
Leading edge
Microphone
Opposite
Oscilloscope
Peak-to-peak value (p-p)
Peak value (p)
Period (t)
Periodic
Phase
Power transfer
Pulsating dc
Pulse duration (PD)
Pulse length (PL)
Pulse repetition frequency (PRF)
Pulse repetition time (PRT)
Pulse waveform
Pulse width (PW)
Rectangular waveform
Rectifier

Rise time
Root mean square (rms)
Sawtooth waveform
Sine
Sine wave
Sinusoidal
Slope
Sound wave
Speaker
Spectrum analyzer
Speed of light
Speed of sound
Square wave
Time domain
Trailing edge
Transformer
Transmission lines
Triangular wave
Utility pole
Video
Wavelength

NEW FORMULAS

Distance = velocity × time

$$\frac{D}{V \mid t}$$

Sine of angle $A = \dfrac{\text{opposite}}{\text{hypotenuse}}$

$$f = \frac{1}{t}, \quad t = \frac{1}{f}, \quad \begin{array}{l} f = \text{frequency in hertz} \\ t = \text{period or time in seconds} \end{array}$$

Sine wave: $V_{p\text{-}p} = 2 \times V_p$ $V_{p\text{-}p}$ = peak-to-peak voltage
V_p = peak voltage

$I_{p\text{-}p} = 2 \times I_p$ $I_{p\text{-}p}$ = peak-to-peak current
I_p = peak current

$$\text{RMS} = 0.707 \times \text{peak}$$

$$\text{Peak} = \frac{\text{RMS}}{0.707} \quad \text{or} \quad 1.414 \times \text{rms}$$

$$\text{Average} = 0.637 \times \text{peak}$$

Square wave: Duty cycle (%) $= \dfrac{\text{pulse width } (P_w)}{\text{period } (t)} \times 100\%$

V or *I* average $=$ baseline $+$ (duty cycle $\times$ peak-to-peak)

Rectangular wave: PRF $= \dfrac{1}{\text{PRT}}$ PRF = pulse repetition frequency in pulses per second (pps)

PRT $= \dfrac{1}{\text{PRF}}$ PRT = pulse repetition time

Triangular wave: Slope (amps/sec) $= \dfrac{\Delta I}{\Delta t}$ delta I (ΔI) = increment of current change
delta t (Δt) = increment of time change

Wavelength (λ) $= \dfrac{\text{velocity}}{\text{frequency (Hz)}}$

Electromagnetic wave: λ (cm) $= \dfrac{3 \times 10^{10} \text{ cm/s}}{f}$ or λ (m) $= \dfrac{3 \times 10^8 \text{ m/s}}{f}$

Sound wave: λ (ft) $= \dfrac{1133 \text{ ft/s}}{f}$ or λ (m) $= \dfrac{344.4 \text{ m/s}}{f}$

REVIEW QUESTIONS

Multiple Choice Questions

1. A current that rises from zero to maximum positive, returns to zero, and then repeats is known as:
   a. Alternating current   **c.** Pulsating DC
   b. AC   **d.** Steady DC

2. A current that rises from zero to maximum positive, decreases to zero, and then reverses to reach a maximum in the opposite direction (negative) is known as:
   a. Alternating current   **c.** Steady direct current
   b. Pulsating direct current **d.** All the above

3. The advantage(s) of ac over dc from a power distribution point of view is/are:

410   Alternating Current (AC)

a. Generators can supply more power than batteries
b. AC can be transformed to a high or low voltage easily, minimizing power loss
c. AC can easily be converted into dc
d. All the above
e. Only (a) and (c)

4. The approximate voltage appearing on long distance transmission lines in the ac distribution system is:
a. 250 V
b. 2500 V
c. 500,000 V
d. 250,000 V

5. The most common type of alternating wave shape is the:
a. Square wave
b. Sine wave
c. Rectangular wave
d. Triangular wave

6. _____ equipment manages the flow of information.
a. Electronic
b. Electrical
c. Discrete
d. Intergrated

7. _____ equipment manages the flow of power.
a. Electronic
b. Electrical
c. Discrete
d. Integrated

8. The peak-to-peak value of a sine wave is equal to:
a. Twice the rms value
b. 0.707 times the rms value
c. Twice the peak value
d. 1.14 × the average value

9. The rms value of a sine wave is also known as the:
a. Effective value
b. Average value
c. Peak value
d. All the above

10. The peak value of a 115 V (rms) sine wave is:
a. 115 V
b. 230 V
c. 162.7 V
d. Two of the above could be true

11. The mathematical average value of a sine wave cycle is:
a. 0.637 × peak
b. 0.707 × peak
c. 1.414 × rms
d. Zero

12. The frequency of a sine wave is equal to the reciprocal of _____.
a. The period
b. One cycle
c. Time
d. All the above
e. None of the above

13. What is the period of a 1-MHz sine wave?
a. 1 millisecond
b. One millionth of a second
c. 10 milliseconds
d. 100 microseconds

14. The sine of 90° is:
a. 0
b. 0.5
c. 1
d. Any of the above

15. What is the frequency of a sine wave that has a cycle time of 1 ms?
a. 1 MHz
b. 1 kHz
c. 200 m
d. 10 kHz

16. The pulse width (P_w) is the time between the _____ points on the positive and negative edges of a pulse.
a. 10%
b. 90%
c. 50%
d. All the above

17. The duty cycle is the ratio of _____ to period.
    a. Peak
    b. Average power
    c. Pulse length
    d. Both (a) and (c) are true

18. With a pulse waveform, PRF can be calculated by taking the reciprocal of:
    a. The duty cycle
    b. PRT
    c. Pd
    d. Pl

19. The sound wave exists in _____ and travels at approximately _____.
    a. Space, 1130 feet/second
    b. Wires, 186,282.397 miles/second
    c. Air, 3×10^6 meters/second
    d. None of the above

20. The electrical and electromagnetic waves travel at a speed of:
    a. 186,000 miles/second
    b. 3×10^8 meters/second
    c. 162,000 nautical miles/second
    d. All of the above

Essay Questions

21. Describe the three advantages that ac has over dc from a power point of view.   (11.1.1)

22. Describe briefly the ac power distribution system.   (11.1.1)

23. What is the difference between the words electricity and electronics?   (11.1.4)

24. What are the five basic ac information wave shapes?   (11.1.3)

25. Describe briefly the following terms as they relate to the sine wave: (11.1.3)
    a. RMS
    b. Peak
    c. Peak to peak
    d. Average
    e. The name sine
    f. Frequency
    g. Period
    h. Wavelength
    i. Phase

26. Describe briefly the following terms as they relate to the square wave.   (11.1.3)
    a. Duty cycle
    b. Average

27. Describe briefly the following terms as they relate to the rectangular wave.   (11.1.3)
    a. PRT   c. Duty cycle
    b. PRF   d. Average

28. Briefly describe the meaning of the terms fundamental frequency and harmonics.   (11.1.3)
29. List and describe all the pertinent information relating to the following information carriers: (11.1.2)
    a. Sound wave
    b. Electrical wave
    c. Electromagnetic wave
30. Describe the difference between frequency- and time-domain analysis.   (11.1.3)

Practice Problems

31. Calculate the periods of the following sine-wave frequencies:
    a. 27 kHz   b. 3.4 MHz   c. 25 Hz   d. 365 Hz
    e. 60 Hz   f. 200 kHz
32. Calculate the frequency for each of the following values of time:
    a. 16 ms   b. 1 s   c. 15 μs   d. 0.05 s
    e. 200 μs   f. 350 ms
33. A 22 V peak sine wave will have the following values:
    a. RMS voltage =
    b. Average voltage =
    c. Peak to peak voltage =
34. A 40 mA rms sine wave will have the following values:
    a. Peak current =
    b. Peak-to-peak =
    c. Average =
35. How long would it take an electromagnetic wave to travel 60 miles?
36. A 10 kHz rectangular pulse, with a pulse width of 10 μs, will have a duty cycle of _____%.
37. Calculate the PRT of a 400 kHz pulse waveform.
38. Calculate the average current of the pulse waveform in question 36 if its peak current is equal to 15 A.
39. What is the duty cycle of a 10 V peak square wave at a frequency of 1 kHz?
40. Considering a fundamental frequency of 1 kHz, calculate the frequency of its:
    a. Third harmonic
    b. Second harmonic
    c. Seventh harmonic

12

AC TROUBLESHOOTING AND MEASUREMENT EQUIPMENT

AFTER COMPLETING THIS CHAPTER, YOU WILL BE ABLE TO:

1. Describe how the multimeter is able to measure ac as well as dc.
2. Explain some of the multimeter accessories, such as:
   a. Current clamps.
   b. RF probes.
   c. High voltage probes.
3. Define the function and basic controls of a frequency counter.
4. Explain, in relation to the oscilloscope:
   a. Application and function
   b. Control operation
   c. Measuring voltage, time, and frequency
   d. The single and dual trace scopes
5. Define the function and basic controls of:
   a. An audio-frequency generator.
   b. A radio frequency generator.
   c. A function generator.

INTRODUCTION

As a technician or engineer, you are going to be required to diagnose and repair a problem in the shortest amount of time possible. To aid in the efficiency of this fault finding and repair process, you can make use of certain pieces of test equipment. Humans have five kinds of sensory systems: touch, taste, sight, sound, and smell. Four can be used for electronic troubleshooting: sight, sound, touch, and smell.

Electronic test equipment can be used either to *sense* a circuit's condition or to *generate* a signal to see the response of the component or circuit to that signal. For example, the dc multimeter can be used to measure voltage, current, or resistance. When voltage or current is selected, the meter senses the power present in the circuit and gives an indication of that power as voltage or current on either an analog or digital readout display. When resistance is being measured, the multimeter utilizes an internal battery to generate a current that flows out of the meter. The amount of current flow determines the amount of meter deflection, and this current is dependent on the amount of resistance connected to the meter to be measured.

Let's begin by discussing the "sense" types of test equipment, starting with a device for sensing dc voltages and current and that, with a small modification, can be used to sense ac voltages and current.

LOGGING ON

During the seventeenth-century, European thinkers were obsessed with any device that could help in mathematical calculation. Scottish mathematician John Napier decided to meet this need, and in 1614 he published his new discovery of logarithms. In this book, consisting mostly of tediously computed tables, Napier stated that a logarithm is the exponent of a base number, for example, 100 is 10^2, 27 is $10^{1.43136}$, 10 is 10^1, 6 is $10^{0.77815}$, and any number no matter how large or small can be represented in this manner. He also outlined how the multiplication of two numbers could be achieved by simple addition. For example, if the logarithm of one number, 2, which is 0.30103, is added to the logarithm of another number, 4, which is 0.60206. The result is equal to the logarithm of 8, which is 0.90309. Therefore, the multiplication of two large numbers could now be achieved by looking up the logarithms of the two numbers in a log table, adding them together and then finding the number that corresponds to that sum in an antilog (reverse log) table. In this example, the antilog of 0.9039 is 8.

Napier's table of logarithms were used by William Oughtred, who, just ten years after Napier's death in 1617, developed a handy mechanical device that could be used for rapid calculation. This device, considered the first pocket calculator, was the slide rule.

As well as a brilliant mathematician, Napier was also interested in designing military weapons. One such unfinished project was a death ray system consisting of an arrangement of mirrors and lenses that, when aligned, would produce a concentrated lethal beam of sunlight.

12.1

SIGNAL SENSING EQUIPMENT

AC METER

Figure 12-1 illustrates the dc analog and digital readout multimeters discussed in Chapter 10. These meters are limited in their capabilities in that they can only measure direct (one direction) current because direct current is needed to deflect the needle from left to right. Most multimeters can be used to measure either ac or dc. When the technician or engineer wishes to measure ac, the ac current or voltage is converted to dc internally by a circuit known as a *rectifier*, as seen in Figure 12-2, before passing on to the meter movement.

The dc produced by the rectification process is in fact pulsating, as seen in Figure 12-3, so the current through the meter movement is a series of pulses rising from zero to maximum (peak) and from maximum back to zero. Frequencies below 10 Hz (lower limit) will cause the needle of the meter movement to deflect back and forth on the scale as the meter follows the pulsating dc. This makes it difficult to read the meter. From 10 Hz to approximately 2 kHz, the meter movement will not be able to follow the fluctuation, and the meter will remain in a position equal to the average value of the ac sine wave being measured (0.637 of peak). Most meters are normally calibrated internally to indicate rms values (0.707 of peak) rather than average values, because this effective value is most commonly used when expressing ac voltage or current. The upper limit of the ac meter is approximately 2 to 8 kHz, and beyond this limit the meter becomes pro-

AC Troubleshooting and Measurement Equipment

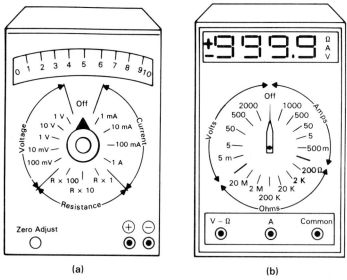

(a)   (b)

FIGURE 12-1   Multimeters. (a) Analog. (b) Digital.

gressively inaccurate due to reactance. Reactance will be discussed further in the following chapters; however, the moving coil (which displays a value of inductance) and the rectifier (which contains capacitance) will both have some value of reactance; and this reactance will result in an inaccurate indication.

(1) CURRENT CLAMPS

The voltmeter is probably the most frequently used setting on the multimeter. A meter reading can be obtained by just connecting the probes across the

FIGURE 12-2   AC Meter Uses a Rectifier to Produce DC.

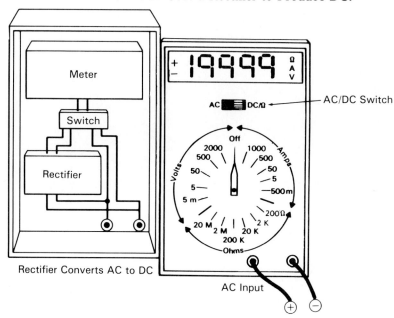

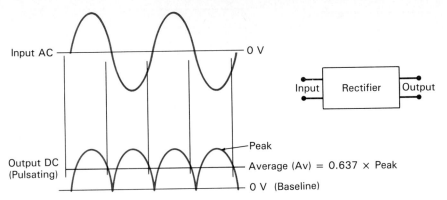

FIGURE 12-3 Rectifier.

component or source to be measured, unlike the ammeter, which requires that the circuit current path be opened, and the ammeter inserted in the path of current flow. If a current measurement is required, a clamp can be used, as seen in Figure 12-4, which allows us to sense the amount of current flow through the conductor without opening the current path.

The alternating current flowing through the conductor produces an expanding and collapsing magnetic field, which cuts across the coil of wire wound around the core of the clamp, and induces an alternating voltage in the coil (1 mA induces 1 mV). The induced alternating voltage causes an alternating current to flow, which is converted to dc by the rectifier and used to operate the meter movement. The larger the current flowing in the conductor, the larger the magnetic field surrounding the conductor, which results in a greater induced voltage, current, and consequent meter movement. These clamps are generally ineffective at measuring smaller currents (microamps), because the magnetic field produced by the current is too weak.

FIGURE 12-4 Current Clamp.

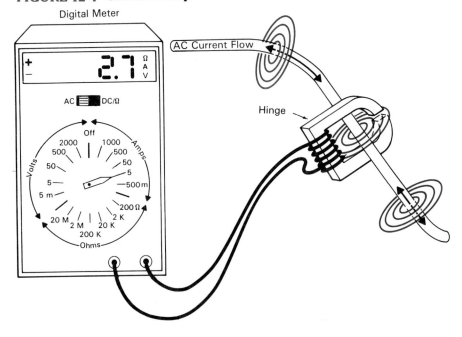

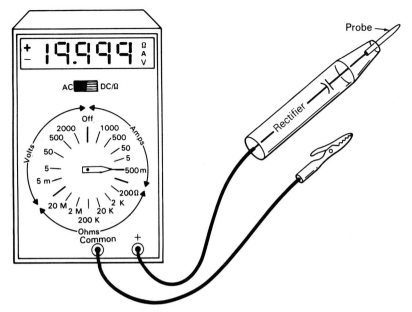

FIGURE 12-5   Radio-Frequency Probe.

(2) RADIO FREQUENCY (RF) PROBE

The meter, as previously mentioned, has a high range limit of approximately 2 kHz. If higher-frequency (radio frequencies) electrical waves are to be measured, an RF probe, as seen in Figure 12-5, can be used. The probe picks up the high-frequency ac voltage from a conducting point on the circuit and passes or couples it to a capacitor, which blocks any dc that could be

FIGURE 12-6   High-Voltage Probe.

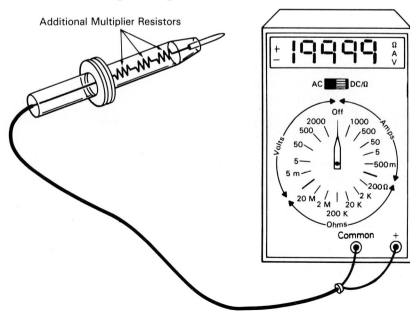

EXAMPLE 12.1

A DMM indicates 3.9 V on its display when a test point is probed by a × 1000 high-voltage probe. What is the voltage at this test point?

SOLUTION

A × 1000 probe will divide the voltage by 1000, and so the displayed voltage must be multiplied by 1000 to obtain the correct value:

$$3.9 \text{ V} \times 1000 = 3.9 \text{ kV}$$

---

present at the test point, as we only want to measure the high-frequency ac. The rectifier, within the probe will convert this ac input into a dc output, which will be displayed as an rms value on the meter display.

(3) HIGH-VOLTAGE PROBE

The typical multimeter can handle voltages up to approximately 1000 V. If you wish to measure voltages higher than this, another component, known as a high voltage (HV) probe, can be used, such as the one seen in Figure 12-6. The high-voltage probe has additional multiplier resistors to drop the extra voltage. Most high-voltage probes are designed so that 1/100th of voltage at the probe tip from the test point will appear out of the probe and be applied to the meter. For example, if 10 kV is being measured, 100 V will appear out of the probe and be applied to the meter (1/100th × 10,000 V = 100 V). This probe would be called a × 100 probe because the 100 V shown on the meter display would now have to be multiplied by 100 for the operator to determine the voltage (100 V × 100 = 10 kV).

The high-voltage probe is especially well insulated to protect its user, who should apply all safety precautions and exercise extreme caution.

(4) ANALOG AND DIGITAL

The digital meter is superior to the analog meter in basically two ways:

1. The digital multimeter (DMM) has an easy to read display, with decimal points and polarity, while with the analog multimeter the value has to be interpreted by the needle position and range selected, which may result in human errors.

2. Some of the best analog multimeters have accuracies of 1%, whereas a DMM will typically have an accuracy of 0.01%.

The analog multimeter, however, has one advantage over the DMM: When measuring low frequency ac signals of several hertz, the analog multimeter will deflect between zero and some value due to the pulsating dc. The operator will see the signal is pulsating at a low frequency due to the back and forth meter needle movement, whereas with the DMM this continual change will cause the digits on the display to continually change and therefore not allow the operator to take a reading.

When it comes to a choice between an analog or digital multimeter, it seems that most people prefer the DMM because of its easy to read display and accuracy.

1. What is the difference between sensing and generating test equipment?
2. The current clamp does not require the circuit current path to be broken in order to measure current (true or false). Can it be used to measure dc amps?
3. What is the difference between an RF and HV probe?
4. List the pros and cons of a DMM.

12.1.2 FREQUENCY COUNTER

It is very important that anyone involved in the design, manufacture, and servicing of electronic equipment be able to accurately measure the frequency

FIGURE 12-7 Frequency Counter.

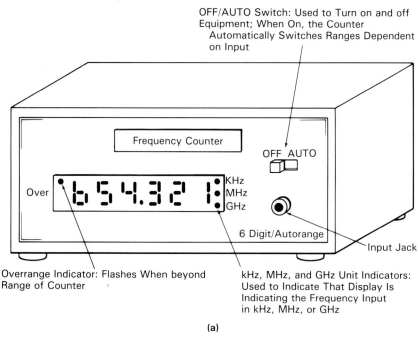

OFF/AUTO Switch: Used to Turn on and off Equipment; When On, the Counter Automatically Switches Ranges Dependent on Input

Overrange Indicator: Flashes When beyond Range of Counter

kHz, MHz, and GHz Unit Indicators: Used to Indicate That Display Is Indicating the Frequency Input in kHz, MHz, or GHz

(a)

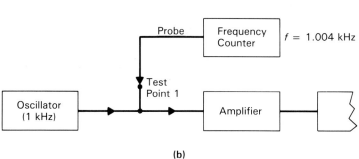

(b)

of a periodic wave. Without the ability to accurately measure frequency, there could be no communications, home entertainment, or a great number of other systems.

The frequency meter or counter, seen in Figure 12-7(a), analyzes the frequency of the periodic wave being applied to the input jack and provides a readout on the display of its frequency. In Figure 12-7(b), the frequency counter is being used to determine the accuracy of an oscillator's output of 1 kHz.

Most frequency counters on the market today can measure frequencies anywhere from hertz to gigahertz.

SELF-TEST REVIEW (§ 12.1.2)

1. What is the function of a frequency meter?
2. Is a frequency counter considered to be a sensing or generating piece of equipment?

12.1.3 OSCILLOSCOPE

Figure 12-8 illustrates a typical oscilloscope (sometimes abbreviated to scope), which is primarily used to display the shape and spacing of electrical signals. The oscilloscope displays the actual sine, square, rectangular, triangular, or sawtooth wave shape that is occurring at any point in a circuit on a cathode ray tube (CRT), which is also used in televisions for displaying video information. From the display on the CRT, we can measure or calculate

FIGURE 12-8 **Oscilloscope.**

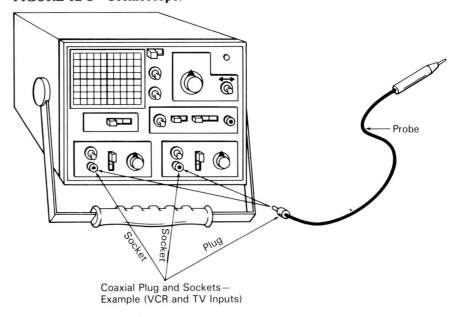

Coaxial Plug and Sockets—
Example (VCR and TV Inputs)

time, frequency, and amplitude characteristics such as rms, average, peak, and peak-to-peak.

The oscilloscope allows us to see what is happening at every point through a circuit. In Figure 12-9, you can see the different waveforms at different points on the circuit. There are also voltage test points that can be tested with a voltmeter if a scope is not available. A voltmeter, as you can well imagine, does not supply the technician or engineer with as much information as the oscilloscope.

(1) CONTROLS

Oscilloscopes come with a wide variety of features and functions; however, the basic operational features are almost identical. Figure 12-10 illustrates the front panel of a typical oscilloscope; we will now discuss the various control functions. Some of these controls are difficult to understand without practice and experience, and so practical experimentation is very necessary if you hope to gain a clear understanding of how to operate the oscilloscope.

FIGURE 12-9   Schematic Diagram with Voltage and Waveform Test Points.

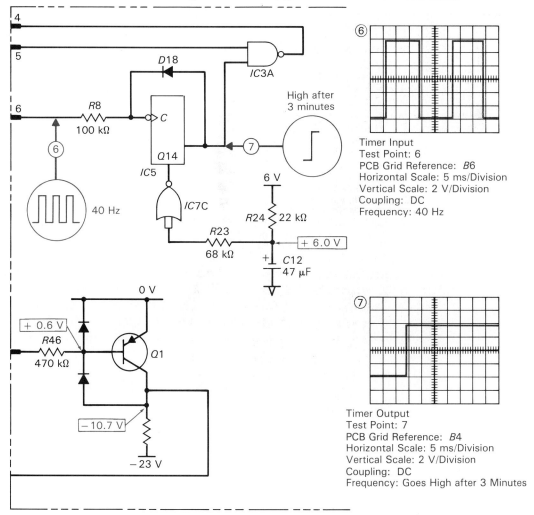

Timer Input
Test Point: 6
PCB Grid Reference: *B*6
Horizontal Scale: 5 ms/Division
Vertical Scale: 2 V/Division
Coupling: DC
Frequency: 40 Hz

Timer Output
Test Point: 7
PCB Grid Reference: *B*4
Horizontal Scale: 5 ms/Division
Vertical Scale: 2 V/Division
Coupling: DC
Frequency: Goes High after 3 Minutes

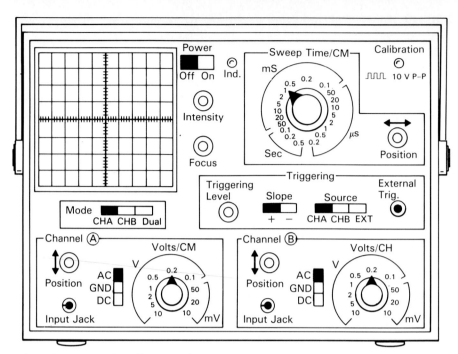

FIGURE 12-10   Oscilloscope Controls.

General Controls

Intensity control: Controls the brightness of the trace, which is the pattern produced on the screen of a CRT.

Focus control: Used to focus the trace.

Power OFF/ON: Switch will turn on oscilloscope while indicator shows when oscilloscope is turned on.

Some oscilloscopes have the ability to display more than one pattern or trace on the CRT screen, as seen by the examples in Figure 12-11. A dual-trace oscilloscope can produce two traces or patterns on the CRT screen at the same time, whereas a single trace oscilloscope can only trace out one pattern on the screen. The dual trace oscilloscope is very useful as it allows us to make comparisons between the phase, amplitude, shape, and timing of two signals from two separate test points. One signal or waveform is applied to the channel *A* input of the oscilloscope, while the other waveform is applied to the channel *B* input.

Mode Switch: This switch allows us to select which channel input should be displayed on the CRT screen.

CHA: The input arriving at channel *A*'s jack is only displayed on the screen as a single trace.

CHB: The input arriving at channel *B*'s jack is only displayed on the screen as a single trace.

Dual: Both the inputs arriving at jacks *A* and *B* are displayed on the screen as a dual trace.

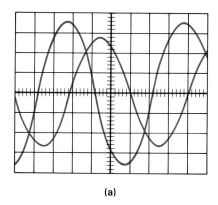

(a)

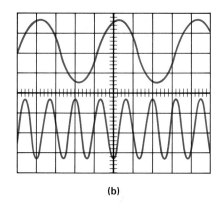

(b)

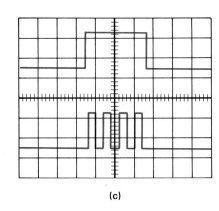

(c)

FIGURE 12-11   Dual-Trace
Oscilloscopes for Comparison.

(a) **Calibration Output**   This output connection provides a point where a fixed 1 V peak-to-peak square-wave signal can be obtained at a frequency of 1 kHz. This signal is normally fed into either channel *A* or *B*'s input to test probes and the oscilloscope operation.

(b) **Channel A and B Horizontal Controls** ↔ *Position control*: This control will move the position of the one (single trace) or two (dual trace) waveforms horizontally (left or right) on the CRT screen.

Sweep time/cm switch: The oscilloscope contains circuits that produce a beam of light that is swept continually from the left to the right of the CRT screen. When no input signal is applied, this sweep will produce a straight horizontal line in the center of the screen. When an input signal is present, this horizontal sweep is influenced by the input signal, which moves it up and down to produce a pattern on the CRT screen the same as the input pattern (sine, square, sawtooth, and so on). This sweep time/cm switch selects the speed of the sweep from left to right, and it can be either fast (0.2 microseconds per centimeter; 0.2 μs/cm) or slow (0.5 second per centimeter; 0.5 s/cm). A low-frequency input signal (long cycle time or period) will require a long time setting (0.5 s/cm) so that the sweep can capture and display one or more cycles of the input. A number of settings are available, with lower time settings displaying fewer cycles and higher time settings showing more cycles of an input.

(c) Triggering Controls These provide the internal timing control between the sweep across the screen and the input waveform.

Triggering level control: This determines the point where the sweep starts.

Slope switch (+): Sweep is triggered on positive-going slope.
(−): Sweep is triggered on negative-going slope.

Source switch, *CHA*: The input arriving into channel *A* jack triggers the sweep.
CHB: The input arriving into channel *B* jack triggers the sweep.
EXT: The signal arriving at the external trigger jack is used to trigger the sweep.

(d) Channel A and B Vertical Controls Both *A* and *B* channel controls are identical.

Volts/cm switch: This switch sets the number of volts to be displayed by each major division on the vertical scale of the screen.

↕ *Position control*: Moves the trace up or down for easy measurement or viewing.

AC-DC-GND switch: In the AC position, a capacitor on the input will pass the ac component entering the input jack, but block any dc components.
In the GND position, the input is grounded (0 V) so that the operator can establish a reference.
In the DC position, both ac and dc components are allowed to pass on to and be displayed on the screen.

(2) MEASUREMENTS

The oscilloscope is probably the most versatile of test equipment as it can be used to test:

DC voltage
AC voltage
Waveform duration
Waveform frequency
Waveform shape

(a) Voltage Measurement The screen is divided into eight vertical and ten horizontal divisions, as seen in Figure 12-12. This 8 × 10 cm grid is called the *graticule*. Every vertical division has a value depending on the setting of the volts/cm control. For example, if the volts/cm control is set to 5 V, then the waveform in Figure 12-13(a), which rises up four major divisions, will have a peak positive alternation value of 20 V (4 div × 5 V/div = 20 V).

As another example, look at the positive alternation in Figure 12-13(b).

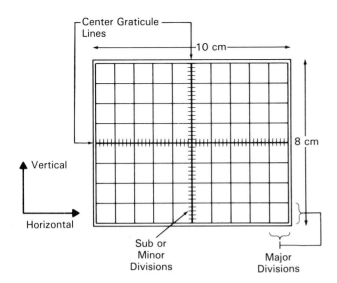

FIGURE 12-12   Oscilloscope Grid.

The positive alternation rises up three major divisions and then extends another four subdivisions, which are each equal to 1 V because five subdivisions exist within one major division, and one major division is, in this example, equal to 5 V. The positive alternation in Figure 12-13(b) has a peak therefore of three major divisions (3×5 V/cm $= 15$ V), plus three subdivisions (3×1 V $= 3$ V), which equals 18 volts peak.

In Figure 12-13(c), we have selected the 10 volt/cm position, which means that each major division is equal to 10 V and each subdivision is equal to 2 V. In this example, the waveform peak will be equal to two major divisions ($2 \times 10 = 20$ V), plus four subdivisions (4×2 V $= 8$ V), which is equal to 28 V. Once the peak value of a sine wave is known, the peak to peak, average, and rms can be calculated mathematically.

When measuring a dc voltage with the oscilloscope, the volts/cm is applied in the same way, as seen in the example in Figure 12-14. A positive dc voltage in this situation will cause deflection toward the top of the screen, whereas a negative voltage will cause deflection toward the bottom of the screen.

To determine the dc voltage, count the number of major divisions and then to this add the number of minor divisions. In the example in Figure 12-14, a major division equals 1 V/cm and, therefore, a minor division equals 0.2 V/cm, so the dc voltage being measured is interpreted as $+2.6$ V.

(b)   Time and Frequency Measurement   The frequency of an alternating wave, such as that seen in Figure 12-15(a), is inversely proportional to the amount of time it takes to complete one cycle ($f = 1/t$). Consequently, if time can be measured, frequency can be determined.

The time/cm control relates to the horizontal line on the oscilloscope graticule and is used to determine the period of a cycle so that frequency can be calculated. For example, in Figure 12-15(b), a cycle lasts five major horizontal divisions, and since the 20 μs/division setting has been selected, the period of the cycle will equal 5×20 μs/division $= 100$ μs. If the period is equal to 100 μs, the frequency of the waveform will be equal to $f = 1/t = 1/100$ μs $= 10$ kHz.

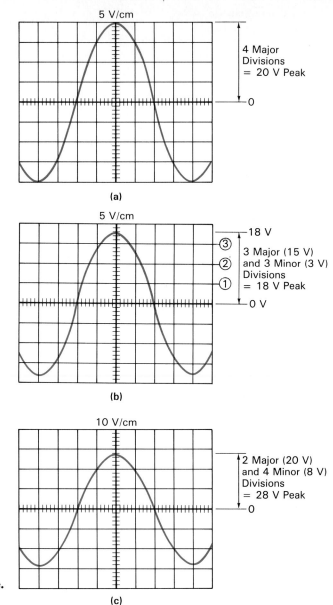

5 V/cm

4 Major
Divisions
= 20 V Peak

0

(a)

5 V/cm

18 V

③
② 3 Major (15 V)
and 3 Minor (3 V)
① Divisions
= 18 V Peak

0 V

(b)

10 V/cm

2 Major (20 V)
and 4 Minor (8 V)
Divisions
= 28 V Peak

0

(c)

FIGURE 12-13 Measuring AC Voltage.

FIGURE 12-14 Measuring DC Voltage.

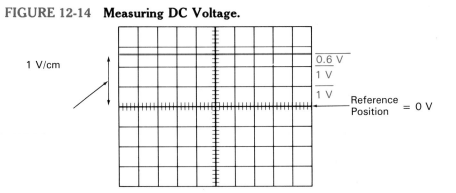

1 V/cm

0.6 V
1 V
1 V

Reference
Position = 0 V

AC Troubleshooting and Measurement Equipment

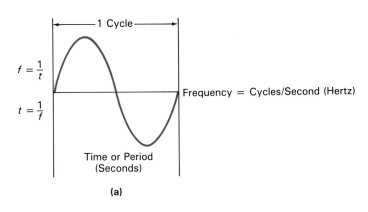

$$f = \frac{1}{t}$$

$$t = \frac{1}{f}$$

Frequency = Cycles/Second (Hertz)

Time or Period
(Seconds)

(a)

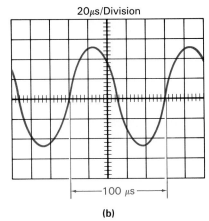

20μs/Division

—100 μs—

(b)

FIGURE 12-15 Time and Frequency Measurement.

EXAMPLE 12.2

A complete sine-wave cycle occupies four horizontal divisons and four vertical divisions from peak to peak. If the oscilloscope is set on the 20 ms/cm and 500 mV/cm, calculate:

(a) $V_{p\text{-}p}$ (e) V_{avg}
(b) t (f) V_{rms}
(c) f
(d) V_p

SOLUTION

$$4 \text{ horizontal divisions} \times 20 \text{ ms/div} = 80 \text{ ms}$$
$$4 \text{ vertical divisions} \times 500 \text{ mV/div} = 2 \text{ V}$$

(a) $V_{p\text{-}p} = 2$ V   (e) $V_{avg} = 0.637 \times V_p = 0.637$ V
(b) $t = 80$ ms   (f) $V_{rms} = 0.707 \times V_p = 0.707$ V

(c)
$$f = \frac{1}{t} = 12.5 \text{ Hz}$$

(d) $V_p = 0.5 \times V_{p\text{-}p} = 1$ V

1. Name the device within the oscilloscope on which the waveforms can be seen.
2. On the 20 μs/div time setting, a full cycle occupies four major divisions. What is the waveform's frequency and period?
3. On the 2 V/cm voltage setting, the waveform swings up and down two major divisions (total 4 cm vertical swing). Calculate the waveform's peak and peak to peak voltage.
4. What are the advantages of the dual trace oscilloscope?

12.2
SIGNAL GENERATING EQUIPMENT

Signal generators are used to produce many kinds of electrical waves for testing or controlling the operation of different circuits. As with the oscilloscope, the best way to become very familiar with the operation of these instruments is with practical experimentation involving your available test equipment.

12.2.1 AUDIO-FREQUENCY GENERATOR

Audio is a Greek word meaning "I hear," and although we can only hear sound wave frequencies from about 20 to 20,000 Hz, audio-frequency gen-

FIGURE 12-16 **Audio-Frequency Generator.**

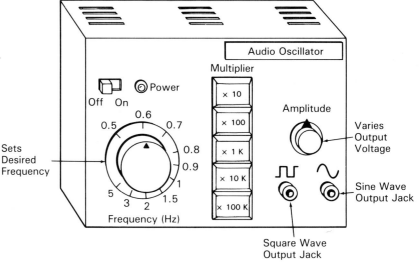

erators will generally produce electrical sine or square waves from 1 Hz to 1 MHz. An audio-frequency (AF) generator or oscillator can be seen in Figure 12-16.

12.2.2 RADIO-FREQUENCY GENERATORS

Figure 12-17(a) illustrates a typical RF generator that can be used to generate any frequency from 3 KHz to 3 GHz. The electrical wave produced can be either a constant amplitude continuous wave, as seen in Figure 12-17(b), or a varying amplitude signal, known as an amplitude modulated (AM) wave, as seen in Figure 12-17(c).

FIGURE 12-17 **Radio-Frequency Generator.**

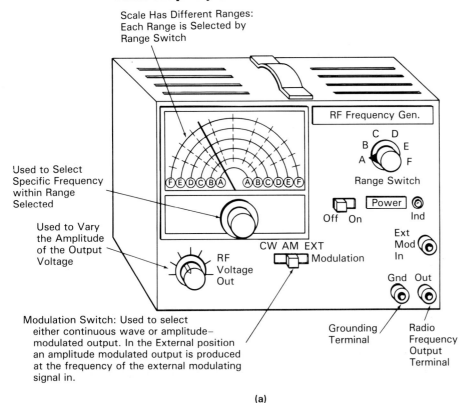

(a)

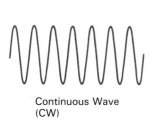

Continuous Wave
(CW)

(b)

Amplitude Modulated
(AM)

(c)

Signal Generating Equipment   431

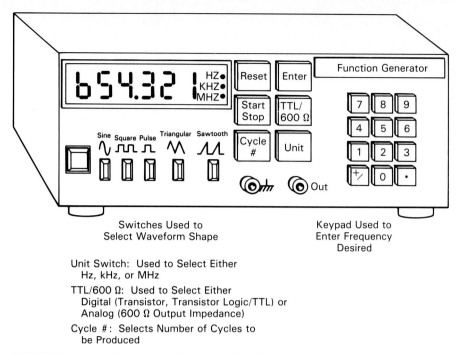

Switches Used to
Select Waveform Shape

Keypad Used to
Enter Frequency
Desired

Unit Switch: Used to Select Either
Hz, kHz, or MHz

TTL/600 Ω: Used to Select Either
Digital (Transistor, Transistor Logic/TTL) or
Analog (600 Ω Output Impedance)

Cycle #: Selects Number of Cycles to
be Produced

FIGURE 12-18 **Programmable Function Generator.**

12.2.3 FUNCTION GENERATORS

Figure 12-18 illustrates a typical function generator, which can perform or function as a sine, square, rectangular, triangular or sawtooth waveform generator.

SELF-TEST REVIEW (§ 12.2)

1. What is the difference between the audio and radio frequency generator?
2. In relation to the RF generator, what is the difference between a constant amplitude wave and amplitude modulated wave?
3. List the types of waveform shapes that a function generator can produce.
4. Do any of the generators mentioned in this section generate sound or electromagnetic waves?

SUMMARY

1. Test equipment can be used to either sense a circuit's condition or generate a signal to see the response of the component or circuit to that signal.

2. Most multimeters can be used to measure either ac or dc.

3. When multimeters are selected to measure ac, a rectifier is used to convert the ac input into a dc voltage internally.

4. Multimeters are normally calibrated to indicate rms values of the ac being measured.

5. A current clamp allows the technician or engineer to measure current without opening the current path.

6. The RF probe can be used with the multimeter to more accurately measure higher frequencies above 2 kHz.

7. The high voltage probe can be used to measure voltages in the kilovolt range by connecting additional multiplier resistors.

8. The analog multimeter unlike the DMM can be used to measure low frequency ac.

9. The digital multimeter is generally more popular because of its easy to read display and accuracy.

10. The frequency counter measures and displays the number of cycles per second (hertz) on a digital display.

11. The oscilloscope can be used to display waveform shapes, and from this presentation we can calculate the waveform's time, frequency, and amplitude characteristics.

12. A dual trace oscilloscope can produce two traces or waveforms on the screen, which allows the technician or engineer to make comparisons between the phase, amplitude, shape, and timing.

13. The audio frequency, radio frequency, and function generator are all used to produce a wide range of electrical waves for testing or controlling the operation of different circuits.

14. A function generator may function or perform as a sine, square, triangular, pulse, or sawtooth waveform generator.

NEW TERMS

AC meter

Audio frequency (AF) generator

Current clamp

Frequency counter

Function generator

High-voltage (HV) probe

Instrument

Oscilloscope

Radio frequency (RF) generator

Radio frequency (RF) probe

Signal generator

REVIEW QUESTIONS

Multiple Choice Questions

1. Which of the following would be considered a sensing instrument?
   a. Ohmmeter
   b. Voltmeter
   c. Audio oscillator
   d. Two of the above are correct

2. Which of the following would be considered a generating instrument?
   a. Voltmeter
   b. Ammeter
   c. Ohmmeter
   d. Two of the above are correct

3. When the analog multimeter is selected to measure ac:
   a. A rectifier is connected internally in circuit.
   b. A pulsating dc waveform is applied to the meter movement.
   c. The meter will indicate the rms or dc equivalent of the ac input.
   d. All the above

4. What is the frequency limit associated with accurately measuring ac with a multimeter?
   a. 10 Hz to 2 kHz
   b. 1000 Hz to 15,000 Hz
   c. 15 MHz to 10 GHz
   d. All three are true

5. The current clamp, when connected to a multimeter:
   a. Allows you to measure ac voltage with the same ease as current
   b. Indicates current flow in a conductor based on magnetic field strength
   c. Allows you to measure ac current with the same ease as voltage
   d. Both (a) and (b) are true
   e. Both (b) and (c) are true

6. Radio frequency probes are used to detect high frequency _____ waves and display a rms value on the meter.
   a. Sound
   b. Electrical
   c. Electromagnetic
   d. None of the above

7. A × 1000 high voltage probe means that _____ of the voltage at the probe tip will appear out of the probe and be applied to the meter.
   a. One tenth
   b. 1000%
   c. One thousandth
   d. Most

8. Which of the two meter types has the highest accuracy?
   a. The AMM (analog multimeter)
   b. The DMM (digital multimeter)
   c. The VOM (volt-ohm-milliammeter)
   d. Both (a) and (c)

9. What are the two advantages of the analog multimeter?
   a. Able to read low frequency ac and has low cost
   b. Accuracy and able to read low frequency ac
   c. Easy to read display and accuracy
   d. Low cost and accuracy

10. The oscilloscope can be used to measure:
    a. AC and DC voltage
    b. Frequency
    c. Duration
    d. Rise and fall times
    e. All the above

434   AC Troubleshooting and Measurement Equipment

Essay Questions

11. How can a multimeter be used to measure ac current or voltage? (12.1.1)
12. Briefly describe the following multimeter accessories: (12.1.1)
    a. The current clamp
    b. The RF probe
    c. The high-voltage probe
13. Describe the pros and cons of the analog and the digital readout multimeters. (12.1.1)
14. Explain the purpose of the frequency counter in Figure 12-7. (12.1.2)
15. If a multimeter indicates the rms value of a sine wave, what does an oscilloscope indicate? (12.1.3)
16. How can the oscilloscope be used to measure: (12.1.3)
    a. DC voltage
    b. Frequency
17. What advantage does a dual trace oscilloscope have over a single trace? (12.1.3)
18. Briefly describe the function of:
    a. The AF generator controls in Figure 12-16
    b. The RF generator controls in Figure 12-17
    c. The function generator controls in Figure 12-18
19. In relation to the oscilloscope, describe: (12.1.3)
    a. The sweep time/cm switch
    b. The volts/cm switch
20. Describe the difference between generating and sensing test equipment. (Intro.)

Practice Problems

21. If a × 100 high voltage probe's measure is 4.2 kV, what does the meter indicate?
22. If one cycle of a sine wave occupies 4 cm on the oscilloscope horizontal grid and 5 cm from peak to peak on the vertical grid, calculate frequency, period, rms, average, and peak for the following control settings:
    a. 0.5 V/cm, 20 µs/cm
    b. 10 V/cm, 10 ms/cm
    c. 50 mV/cm, 0.2 µs/cm
23. Assuming the same graticule and switch settings of the oscilloscope in Figure 12-10, what would be the lowest setting of the volts/cm and time/division switches to fully view a 6 V rms, 350 kHz sine wave?
24. If the volts/cm switch is positioned to 10 V/cm and the waveform extends 3.5 divisions from peak to peak, what is the peak-to-peak value of this wave?
25. If a square wave occupies 5.5 horizontal cm on the 1 µs/cm position, what is its frequency?

13

ELECTROMAGNETISM (AC)

AFTER COMPLETING THIS CHAPTER, YOU WILL BE ABLE TO:

1. Describe the difference between dc and ac electromagnetism.
2. Explain the relationship between flux density and magnetizing force.
3. Describe the cycle known as the hysteresis loop.
4. Define electromagnetic induction.
5. State Faraday's and Lenz's laws relating to electromagnetic induction.
6. Describe the following applications of electromagnetic induction:
   a. The generator
   b. The moving coil microphone

INTRODUCTION

In Chapter 6, we discussed electromagnetism as it relates to direct current (dc). In this chapter, we will see what differences relate to electromagnetism with alternating current (ac). The word electromagnetism can be broken into two words: electro (electrical) and magnetism. Electromagnetism is the magnetism resulting from the flow of an electrical current.

AT THE CORE OF APPLE

Stephen Wozniak and Steven Jobs met at their Los Altos, California, high school and due to their common interest, electronics, became friends.

Wozniak was a conservative youth whose obsession with technology left little room for social relationships or studying; in fact, after one year at the University of Colorado, he dropped out with his academic record littered with F's. In contrast to this serious nature, Wozniak was reknowned for his high-tech pranks for which, on one occasion, he spent a night in juvenile hall for wiring up a fake bomb in a friend's locker. In another incident, he devised a way to place a free telephone call to the Pope at the Vatican, identifying himself as then Secretary of State Henry Kissinger.

Jobs, on the other hand, had other interests outside of electronics. He searched for intellectual, emotional, and spiritual stimulation. After one semester at Reed College, he dropped out to pursue an interest in Eastern religions, which led him to temples in India, searching for the meaning of life.

The first enterprise built by Wozniak and sold by Jobs was an illegal device called a blue box that could crash the telephone system. It generated a set of tones that fooled the computerized telephone switching systems and opened up free, long-distance circuits; these systems allowed "phone phreaks" to take long and illegal joy rides throughout the worlds's telephone networks.

By selling a Volkswagen van and a programmable calculator, Wozniak and Jobs raised the initial capital of $1300 to start a business in April 1976 called Apple Computer, since Jobs had a passion for the Beatles, who recorded under the Apple record label.

In just five years, Apple Computer grew faster than any other company in history, from a two-man assetless partnership building computers in a family home to a publicly traded corporation earning its youthful entrepreneurs, Jobs (27) and Wozniak (24), fortunes of nearly $200 million each.

13.1

DC VERSUS AC ELECTROMAGNETISM

A magnetic field results when current flows through any piece of conductor or wire, as shown in Figure 13-1. If a conductor is wound to form a spiral, as illustrated in Figure 13-2(a), the conductor, which is now referred to as

FIGURE 13-1 Conductor and Coil Electromagnetism.

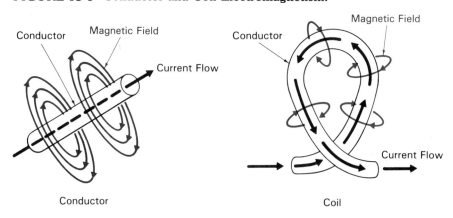

Electromagnetism (AC)

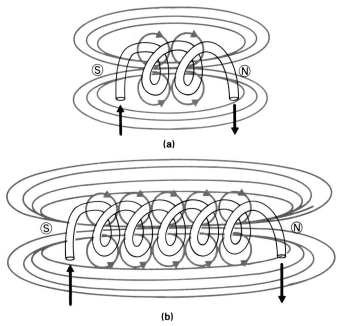

(a)

(b)

FIGURE 13-2   Electromagnet. (a) Small Number of Turns. (b) Large
Number of Turns.

a *coil*, will develop, as a result of current flow, a magnetic field, which will sum or intensify within the coil, and the net effect of all the conductor magnetic fields is a stronger resultant magnetic field, as seen in Figure 13-2(b). With electrical current, a magnetic field is produced, and this is why the component is referred to as an electromagnet.

The left-hand rule can be applied to electromagnets to determine the magnetic polarity (north and south poles), as seen in Figure 13-3. If the fingers of your left hand point in the direction of current flow, then your thumb will be pointing to the north end of the electromagnet.

Up till now, we have only been discussing current flow through a coil in one direction (dc). A dc voltage produces current in one direction and

FIGURE 13-3   **Left-Hand Rule for Coils.**

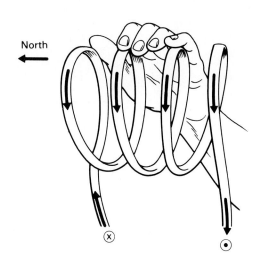

North

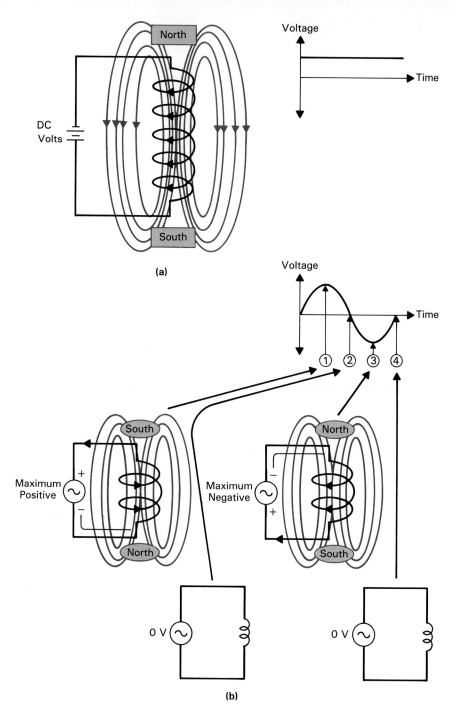

FIGURE 13-4 Electromagnets.

therefore generates a magnetic field in a coil of fixed polarity, as seen in Figure 13-4(a), and as determined by the left-hand rule.

Alternating current (ac) is continually varying, and as the polarity of the magnetic field is dependent on the direction of current flow (left-hand rule), the magnetic field will also be alternating in polarity, as seen in Figure 13-4(b).

440   Electromagnetism (AC)

Time 1. The alternating voltage has risen to a maximum positive level and causes current flow as seen in the circuit. This will cause a magnetic field with a south pole above and a north pole below.

Time 2. Between positions 1 and 2, the voltage, and therefore current, will decrease from a maximum positive value to zero. This will cause a corresponding collapse of the magnetic field from maximum (position 1) to zero (position 2).

Time 3. Voltage and consequently current increases from zero to maximum negative between positions 2 and 3. The increase in current flow causes a similar increase or buildup of magnetic flux, producing a north pole above and south pole below.

Time 4. From 3 to 4, the current within the circuit diminishes to zero, and the magnetic field once again collapses.

In summary:

1. Direct current (dc) produces a constant magnetic field of a fixed polarity, for example, north–south.
2. An alternating current (ac) produces an alternating magnetic field, which continuously switches polarity, for example, north–south, south–north, north–south, and so on.

The magnetic field strength of an electromagnet can be increased by increasing the number of electromagnetic turns, increasing the current or decreasing the length of the coil (magnetizing force, H, or field intensity = current × number of turns/length of coil). This field strength can be further increased by placing an iron core within the electromagnetic coils, as illustrated in Figure 13-5.

The iron core has less reluctance (opposition to the magnetic lines of force) than air, and so the flux density B is increased. Another way of saying this is that the permeability (conductance) of magnetic lines of force in iron is greater than that of air, and if the permeability of iron is large, then the reluctance (opposition) must be small.

FIGURE 13-5 Electromagnet. (a) Physical Appearance. (b) Symbol.

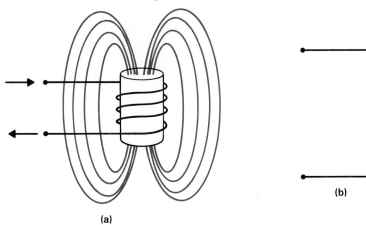

(a)

(b)

1. A magnetic field is only produced when ac is passed through a conductor (true or false).
2. Describe how the left-hand rule applies to electromagnetism.
3. What form of current flow produces a magnetic field that continuously switches in polarity?
4. What factors of an electromagnet can be changed to increase magnetic field strength?

13.2
FLUX DENSITY (*B*) VERSUS MAGNETIZING FORCE (*H*)

The *B–H* curve in Figure 13-6 illustrates the relationship between the two most important magnetic properties: flux density (*B*) and magnetizing force (*H*). Figure 13-6(a) illustrates the *B–H* curve, while Figure 13-6(b) illustrates the positive rising portion of the ac current that is being applied to the iron core electromagnetic circuit in Figure 13-6(c).

FIGURE 13-6   B–H Curve. (b) Current Applied to Circuit. (c) Circuit.

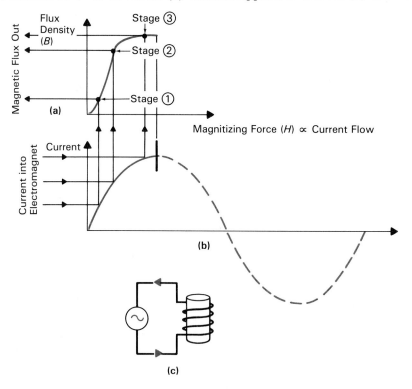

Electromagnetism (AC)

The magnetizing force is actually equal to $H = I$ (current) $\times N$ (number of turns)/l (length of coil); but since the number of turns and length of coil are fixed for the coil being used, the magnetizing force (H) is proportional to the current (I) applied, which is shown in Figure 13-6(b). The positive rise of the current from zero to maximum positive is applied through the electromagnet and will produce a corresponding bloom or buildup in magnetic flux at a rate indicated by the B–H curve shape in Figure 13-6(a).

It is important to note that, as the magnetizing force (current) is increased, there are three distinct stages in the change of flux density or magnetic flux.

Stage 1. Up to this point, the increase in flux density is slow, as a large amount of force is required to commence alignment of the molecule magnets.

Stage 2. Increase in flux density is now rapid and almost linear as the molecule magnets are aligning easily.

Stage 3. In this state, they cannot be magnetized any further because all the molecule magnets are fully aligned and no more flux density can be easily obtained. This is called the *saturation point*, and it is the state of magnetism beyond which an electromagnet is incapable of further magnetic strength, that is, the point beyond which the B–H curve is a straight, horizontal line, indicating no change.

Saturation can easily be described by a simple analogy of a sponge. A dry sponge can only soak up a certain amount of water. As it continues to absorb water, a point will be reached where it will have soaked up the maximum amount of water possible. At this point, the sponge is said to be saturated with water, and no matter how much extra water you supply, it cannot hold any more.

The electromagnet is saturated at stage 3 and cannot produce any more magnetic flux, even though more magnetizing force is supplied as the sine wave continues on to its maximum positive level.

Looking at these three stages and the B–H curve that is produced, you can see that, in fact, the magnetization (setting up of the magnetic field, B) lags the magnetizing force (H) because of molecular friction. This lag or time difference between magnetizing force (H) and flux density is known as *hysteresis*.

13.2.1 HYSTERESIS LOOP

Figure 13-7(a) illustrates what is known as a hysteresis loop, which is formed when you plot magnetizing force (H) against flux density (B) through a complete cycle of alternating current, as seen in Figure 13-7(b). Initially, when the electric circuit switch is open, the iron core is unmagnetized; therefore, both H and B are zero at point a. When the switch is closed, as seen in Figure 13-7(c), the current [Figure 13-7(b)] is increased and flux density [Figure 13-7(a)] increases until saturation point b is reached. This part of the waveform is exactly the same as the B–H curve that was previously

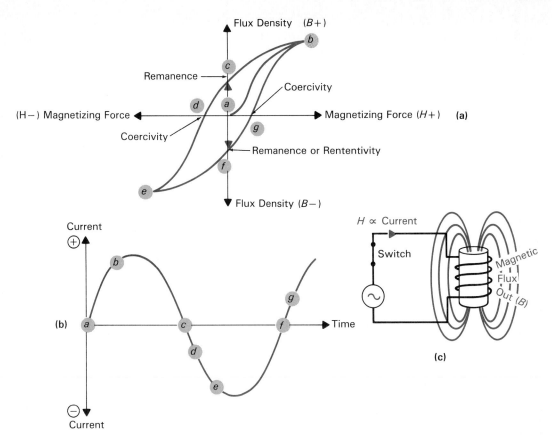

FIGURE 13-7 Hysteresis Loop.

drawn and discussed. The current continues on beyond saturation point b; however the flux density cannot increase beyond saturation.

At point c, the magnetizing force (current) is zero and B (flux density) falls to a value c that is the positive magnetic flux remaining after the removal of the magnetizing force (H). This particular value of flux density is termed remanence or retentivity.

The current or magnetizing force now reverses, and the amount of current in the reverse direction that causes flux density to be brought down from c to zero d after the core has been saturated is termed the *coercive force*.

The current and so magnetizing force continue on toward a maximum negative until saturation in the opposite magnetic polarity occurs at point e.

At point f, the magnetizing force (current) is zero and B falls to remanence f, which is the negative magnetic flux remaining after the removal of the magnetizing force H. The value of current between f and g is the coercive force needed in the reverse direction to bring flux density down to zero.

In Figure 13-8, it is easier to see how the current variation corresponds to the magnetizing force variation.

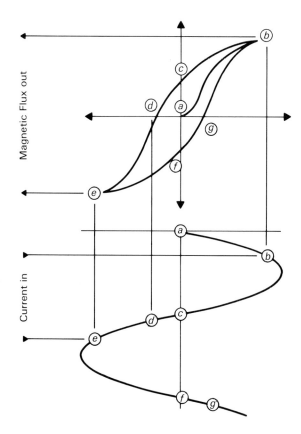

FIGURE 13-8 **Hysteresis Loop.**

1. The hysteresis loop is formed when you plot H against B through a complete cycle of ac (true or false).
2. What is magnetic saturation?

13.3
ELECTROMAGNETIC INDUCTION

Electromagnetic induction is the voltage induced or produced in a coil as the magnetic lines of force link with the turns of a coil. Flux linkage occurs when magnetic lines of force link a coil of wire. Whenever the flux linkage changes, an emf is generated in the coil.

13.3.1 FARADAY'S LAW OF ELECTROMAGNETIC INDUCTION

In 1831, Michael Faraday carried out an experiment including the use of a coil, a zero center ammeter (galvanometer), and a bar permanent magnet, as seen in Figure 13-9. Faraday discovered that [Figure 13-10(a) through (f)]:

(a) When the magnet is moved into a coil so that the magnetic lines of flux cut the turns of the coil, an emf is induced in the coil known as an induced voltage, which causes current to flow within the circuit and the meter to deflect in one direction, for example, to the right.

(b) When the magnet is stationary within the solenoid, the magnetic lines are no longer cutting the coils of the solenoid, and so there is no induced voltage and the meter returns to zero.

(c) When the magnet is pulled out of the coil, a voltage is induced that causes current to flow in the opposite direction to that of (a) and the meter deflects in the opposite direction, for example, to the left.

(d) If the magnet is moved into or out of the coil at a greater speed, the voltage induced also increases.

(e) If the size of the magnet and therefore the magnetic flux strength are increased, the induced voltage also increases.

(f) If the number of turns in the coil is increased, the induced voltage also increases.

In summary: When the magnetic flux linking a coil is changing, an emf is induced, the magnitude of which depends on the number of turns in the coil, rate of change of flux linkage, and flux density.

13.3.2 LENZ'S LAW

About the same time a German physicist, Heinrich Lenz, performed a similar experiment along the same lines as Faraday. His law states that: The current induced in a coil due to the change in the magnetic flux is such as to oppose the cause producing it.

To further explain this law, refer to Figure 13-11. When the magnet moves into the coil, a voltage is induced such that the current flows in the coil and produces a pole at the face of the coil (left hand rule), which opposes

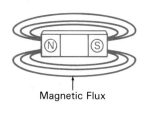

Magnetic Flux

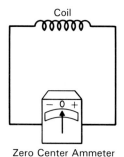

Coil

Zero Center Ammeter

FIGURE 13-9 **Electromagnetic Induction.**

Electromagnetism (AC)

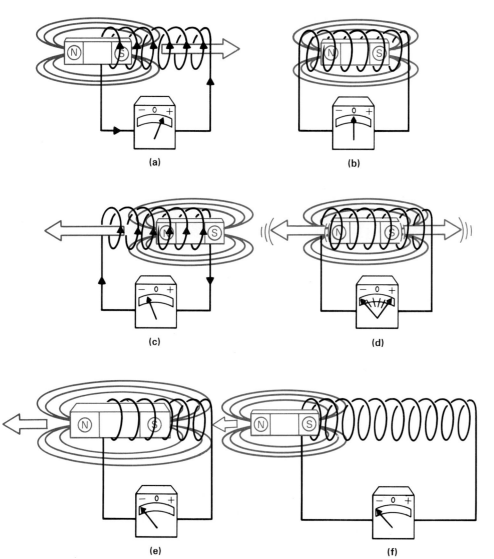

(a)

(b)

(c)

(d)

(e)

(f)

FIGURE 13-10   Farraday's Electromagnetic Induction Discoveries.

FIGURE 13-11   Lenz's Law.

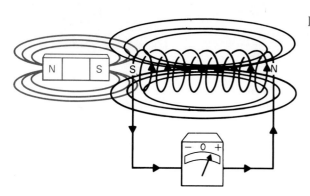

the entry of the magnet. In the example in Figure 13-11, we can see that, as the permanent magnet moves into the coil, its magnetic lines of flux cut the turns of the coil and induce a voltage (electromagnetic induction), which causes current to flow in the coil as indicated by the meter movement. If you apply the left hand rule to the coil, you can see that the current flow has produced a south pole on the left-hand side to oppose the entry or motion of the magnet that is producing the current.

13.3.3 THE WEBER AND ELECTROMAGNETIC INDUCTION

Consideration of Faraday's law enables us to take a closer look at the unit of flux (ϕ). The weber is equal to 10^8 magnetic lines of force, and from the electromagnetic induction point of view, if 1 weber of magnetic flux cuts a conductor for a period of 1 second, a voltage of 1 volt will be induced.

13.3.4 APPLICATIONS OF ELECTROMAGNETIC INDUCTION

(1) AC GENERATOR (ALTERNATOR)

The ac generator or alternator is an example of a device that uses electromagnetic induction to generate electricity. If you stroll around your city or town during the day or night and try to spot every piece of equipment, appliance, or device that is running from the ac electricity supplied by generators from the electric power plant, it begins to make you realize how inactive, dark, and difficult our modern society might become without ac power.

When discussing Faraday's discoveries of electromagnetic induction,

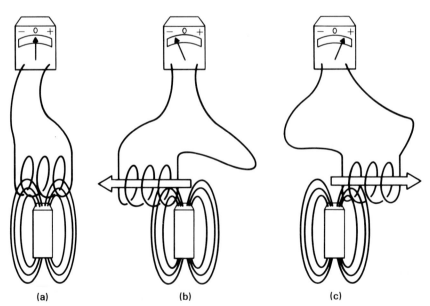

(a)   (b)   (c)

FIGURE 13-12   (a) Stationary Conductor, No Induced Voltage. (b) Moving Conductor, Induced Voltage. (c) Moving Conductor, Induced Voltage.

448   Electromagnetism (AC)

the coil or conductor remained stationary and the magnet was moved. It can be operated in the opposite manner so that the magnetic field remains stationary and the coil or conductor is moved. As long as magnetic lines of force have relative motion with respect to the conductor, an emf will be induced in the coil. Figure 13-12(a) illustrates a piece of wire wound to form a coil and attached to a galvanometer; its needle rests in the center position, indicating zero current. If the conductor remains stationary within the magnetic lines of flux being generated by the permanent magnet, there is no emf or voltage induced into the wire and so no current flow through the circuit and meter.

If the conductor is moved past the permanent magnet so that it cuts the magnetic lines of flux, as seen in Figure 13-12(b), then an emf is generated within the conductor, which is known as an induced voltage, and this will cause current to flow through the wire in one direction and be indicated on the meter by the deflection of the needle to the left.

If the conductor is moved in the opposite direction past the magnetic field, it will induce a voltage of the opposite polarity and cause the meter to deflect to the right, as seen in Figure 13-12(c).

The value or amount of induced voltage is indicated by how far the meter deflects and this voltage is dependent on three factors:

1. The speed at which the conductor passes through the magnetic field.
2. The strength or flux density of the magnetic field.
3. The number of turns in the coil.

If the speed at which the conductor passes through the magnetic field or the strength of the magnetic field or the number of turns of the coil is increased, then the induced voltage will also increase. This is merely a repetition of Faraday's law, but in this case we moved the conductor instead of the magnetic field; however, the results were the same.

(a) Basic Generator The basic generator action makes use of Faraday's discoveries of electromagnetic induction. In Chapter 10, the physical appearance of a large 700,000 kW power plant generator was shown. Figure 13-13 illustrates some smaller generators that are mobile and can therefore be used in remote locations.

Figure 13-14 illustrates the basic generator's construction. The mechanical drive energy input will produce ac electrical energy out by means of electromagnetic induction.

A loop of conductor, known as an *armature*, is rotated continually through 360° by a mechanical drive. This armature resides within a magnetic field produced by an electromagnet. Voltage will be induced into the armature and will appear on slip rings, which are also being rotated. A set of stationary brushes rides on the rotating slip rings and picks off the generated voltage and applies this voltage across the load. This voltage will cause current to flow within the circuit and be indicated by the zero center ammeter.

Let's now take a closer look at the armature as it sweeps through 360°, or one complete revolution.

Figure 13-15 illustrates four positions of the armature as it rotates through 360° in the clockwise direction.

(a)

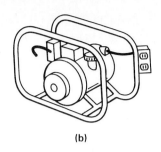

(b)

FIGURE 13-13 AC Generator. (a) 250-Kilowatt Diesel Generator. (b) Small (2 kW) Camping Generator.

Position 1: At this instant, the armature is in a position such that it does not cut any magnetic lines of force. The induced voltage in the armature conductor is equal to 0 V and there is no current flow through the circuit.

Position 2: As the conducting armature moves from position 1 to 2, you can see that more and more magnetic lines of flux will be cut, and the induced emf in the armature (being coupled off by the brushes from the slip rings) will also increase to a maximum value. The current flow throughout the circuit will rise to a maximum as the voltage increases, and this can be seen by the zero center ammeter deflection to the right. From the waveform, you can see the sinusoidal increase from zero to a maximum positive as the armature is rotated from 0° to 90°.

Position 3: The armature continues its rotation from 90° to 180°, cutting through a maximum quantity and then fewer magnetic lines of force. The induced voltage decreases from the maximum positive at 90° to

FIGURE 13-14 Basic Generator Construction.

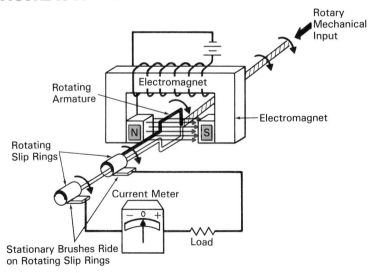

450   Electromagnetism (AC)

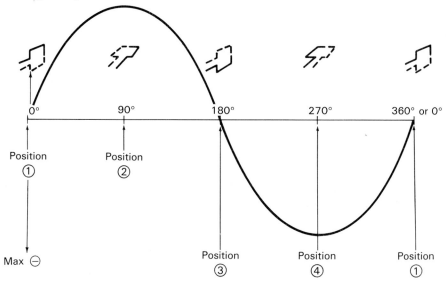

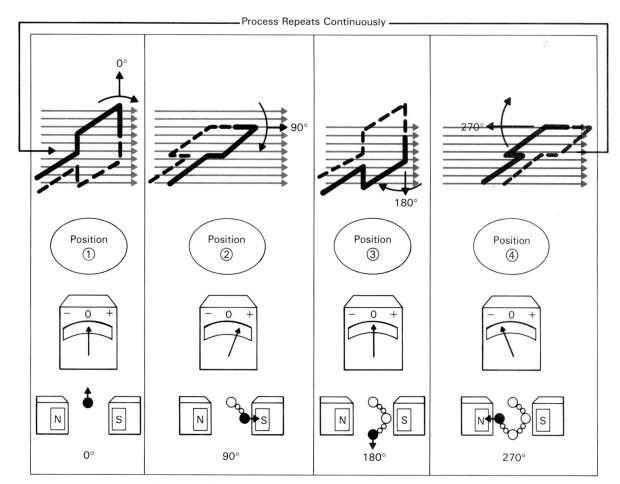

FIGURE 13-15 **360° Generator Operation.**

0 V at 180°. At the 180° position, as with the 0° position, the armature is once again perpendicular to the magnetic field, and so no lines are cut and the induced voltage is equal to 0 V.

Position 4: From 180° to 270°, the armature is still moving in a clockwise direction, and as it travels toward 270°, it cuts more and more magnetic lines of force. The direction of the cutting action between 0° and 90° causes a positive induced voltage; and since the cutting position between 180° to 270° is the reverse, between 180° to 270° a negative induced voltage will result in the armature causing current flow in the opposite direction, as indicated by the deflection of the zero center ammeter to the left. The voltage induced when the armature is at 270° will be equal to the voltage generated when the armature was at the 90° position, but of opposite polarity. The current will therefore also be equal in value but opposite in its direction of flow.

From position 4 (270°), the armature turns to the 360° or 0° position, which is equivalent to position 1, and the induced voltage decreases from maximum negative to zero.

To summarize, in Figure 13-15, one complete revolution of the mechanical energy input causes one complete cycle of the ac electrical energy output.

(2) MOVING COIL MICROPHONE

The moving coil microphone is an example of how electromagnetic induction is used to convert information carrying sound waves to information carrying electrical waves.

Sound is the movement of pressure waves in the air. To create these pressure waves, a device such as a string, reed, or stretched membrane or the human vocal cords must be vibrated to compress and expand the nearby air molecules. Figure 13-16 illustrates a taut string that is vibrating back and forth and generating maximum (A) and minimum (C) pressure regions.

The frequency or pitch of the sound wave is determined by the number of complete vibrations per second (hertz or cycles/second), while the amplitude or intensity of the sound wave is determined by the amount the string shifts from left to right from its normal position (B).

FIGURE 13-16 Sound Wave.

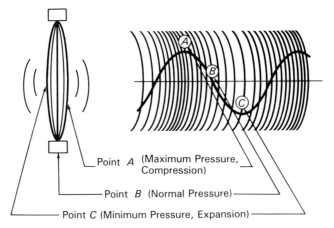

Point A (Maximum Pressure, Compression)

Point B (Normal Pressure)

Point C (Minimum Pressure, Expansion)

Electromagnetism (AC)

A moving coil microphone converts mechanical sound waves into an electrical replica by use of elecromagnetic induction. Figure 13-17 illustrates the physical appearance and construction of a moving coil type of microphone. A coil of wire is suspended in an air gap between magnetic poles and attached to a delicate diaphragm (flexible membrane). A strong magnetic field from a permanent magnet surrounds the coil, and a perforated protecting cover or shield is included to protect the delicate diaphragm.

Sound waves strike the diaphragm, causing it to vibrate back and forth. Since the coil is attached to the diaphragm, it will also be moved back and forth. This movement will cause the coil to cut the magnetic lines of force,

FIGURE 13-17 **Moving Coil Microphone. (a) Physical Appearance. (b) Construction.**

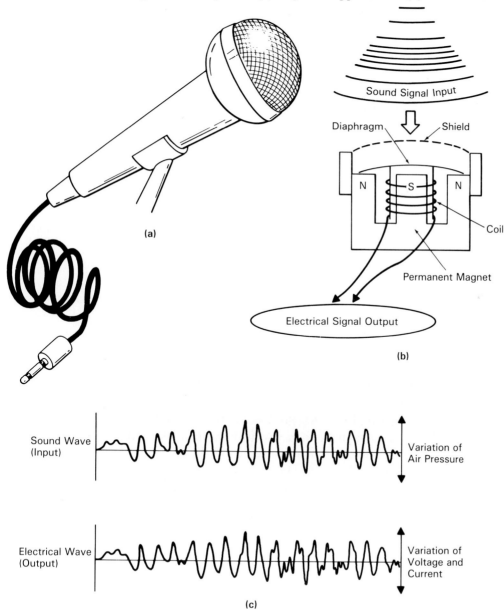

from the permanent magnet, and a resulting alternating voltage will be induced in the coil (electromagnetic induction). The electrical voltage produces an alternating current, which will have the same waveform shape (and consequently information) as the sound wave that generated it, as seen in Figure 13-17(c).

SELF-TEST REVIEW (§ 13.3)

1. Define electromagnetic induction.
2. Briefly describe Faraday's and Lenz's laws in relation to electromagnetic induction.
3. What waveform shape does the ac generator produce?
4. Why is the microphone called an electroacoustical transducer?

SUMMARY

1. The word electromagnetism can be broken into two parts: electro or electrical and magnetism, and, by definition, it is the magnetism resulting from the flow of an electrical current.
2. Direct current (dc) produces a constant magnetic field of unchanging or constant magnetic polarity.
3. Alternating current (ac) produces an alternating magnetic field that continuously switches magnetic polarity.
4. The B–H curve illustrates the relationship between flux density (B) and magnetizing force (H).
5. Magnetic saturation is the state of magnetism beyond which an electromagnet is incapable of further magnetic strength.
6. The hysteresis loop plots magnetizing force (H) against flux density (B) through a complete cycle of alternating current.
7. Remanence is the amount of flux density remaining in the core after the removal of the magnetizing force.
8. The coercive force is the amount of current needed in the reverse direction to reduce the flux density remaining in the core to zero.
9. Electromagnetic induction is the voltage induced or produced in a coil as the magnetic lines of force cut the coil.
10. Faraday's law states that, when the magnetic flux linking a conductor is changing, an emf is induced, the magnitude of which depends on the number of coil turns, rate of change of flux linkage, and the flux density.
11. Lenz's law states that, when a magnet moves into a coil, a voltage is induced such that the current flows in the coil and produces a pole at the face of the coil that opposes the entry of the magnet.
12. An ac generator or alternator is an example of a device that uses electromagnetic induction to generate ac electricity.

Development of an Electronic Product

The following 16 pages will acquaint you with the people involved in an electronics company. The flow chart on this page indicates the order, from top to bottom, in which an electronic product is developed from conception to shipping. As we proceed through this photographic tour, we will see that many people are required to keep a company productive. However let us first concentrate on the people who come in direct contact with the electronic equipment: the technician, the engineer, and the assembler.

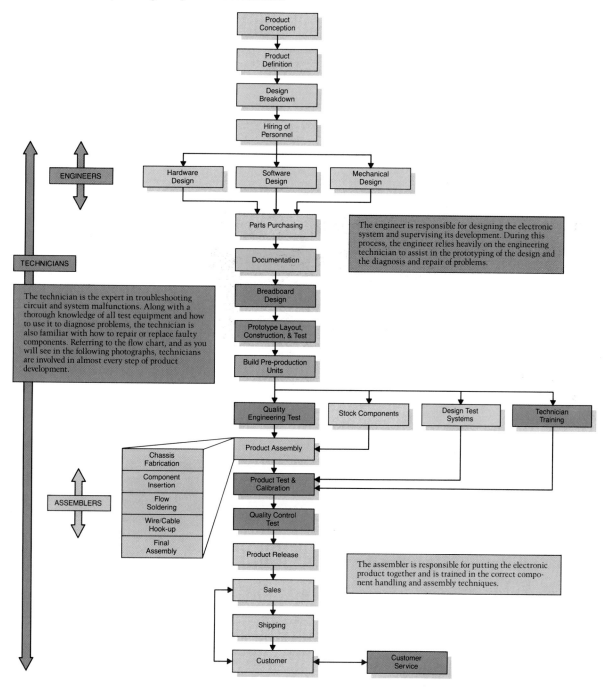

The engineer is responsible for designing the electronic system and supervising its development. During this process, the engineer relies heavily on the engineering technician to assist in the prototyping of the design and the diagnosis and repair of problems.

The technician is the expert in troubleshooting circuit and system malfunctions. Along with a thorough knowledge of all test equipment and how to use it to diagnose problems, the technician is also familiar with how to repair or replace faulty components. Referring to the flow chart, and as you will see in the following photographs, technicians are involved in almost every step of product development.

The assembler is responsible for putting the electronic product together and is trained in the correct component handling and assembly techniques.

The first step in the process can be seen in this photograph. The key company managers meet to define the new product. This new product will have to meet the needs of the ever-expanding electronics industry. During this meeting, the new product's budget, target dates, and key features will be outlined.

Once the product has been defined, it is the job of engineering to specify and design all major components of the new equipment. In this photograph, the design of the new product is being broken-up into smaller tasks by an engineering manager, and these task projects are being assigned to different engineers and engineering technicians.

```
Product
Conception
    ↓
Product
Definition
    ↓
Design
Breakdown
    ↓
Hiring of
Personnel
```

New personnel may be needed by the company to develop and produce this new product. Here, a prospective employee is receiving a technical and personnel interview for a technician position.

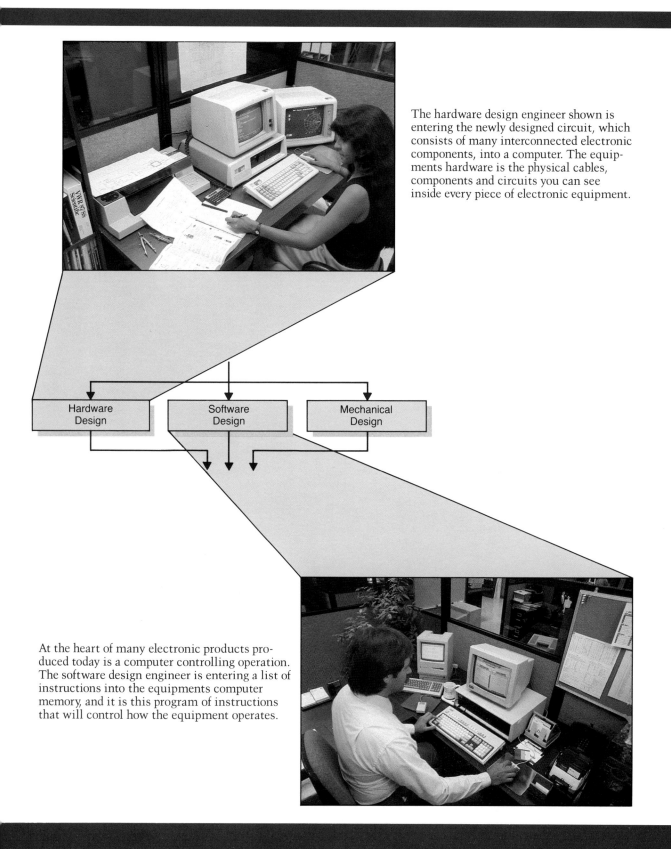

The hardware design engineer shown is entering the newly designed circuit, which consists of many interconnected electronic components, into a computer. The equipments hardware is the physical cables, components and circuits you can see inside every piece of electronic equipment.

| Hardware Design | Software Design | Mechanical Design |

At the heart of many electronic products produced today is a computer controlling operation. The software design engineer is entering a list of instructions into the equipments computer memory, and it is this program of instructions that will control how the equipment operates.

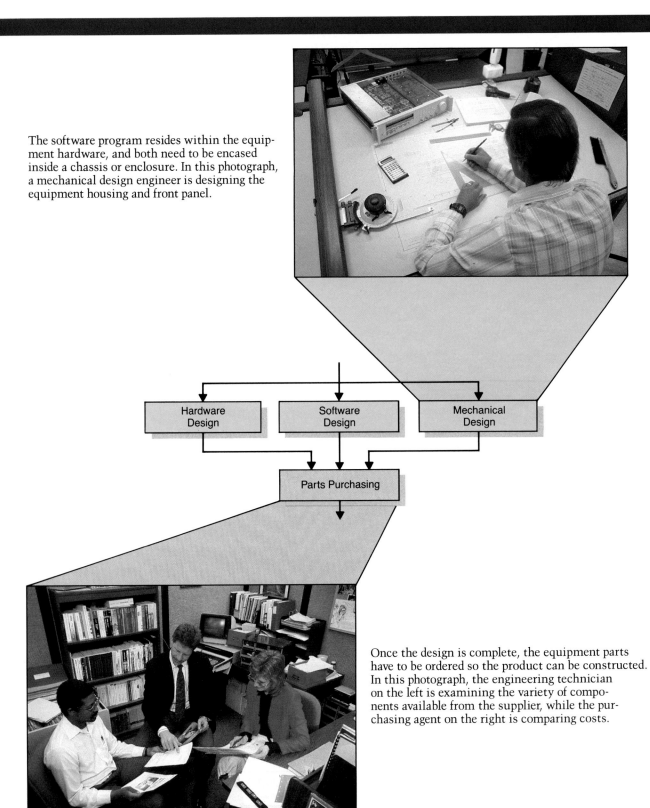

The software program resides within the equipment hardware, and both need to be encased inside a chassis or enclosure. In this photograph, a mechanical design engineer is designing the equipment housing and front panel.

Hardware Design

Software Design

Mechanical Design

Parts Purchasing

Once the design is complete, the equipment parts have to be ordered so the product can be constructed. In this photograph, the engineering technician on the left is examining the variety of components available from the supplier, while the purchasing agent on the right is comparing costs.

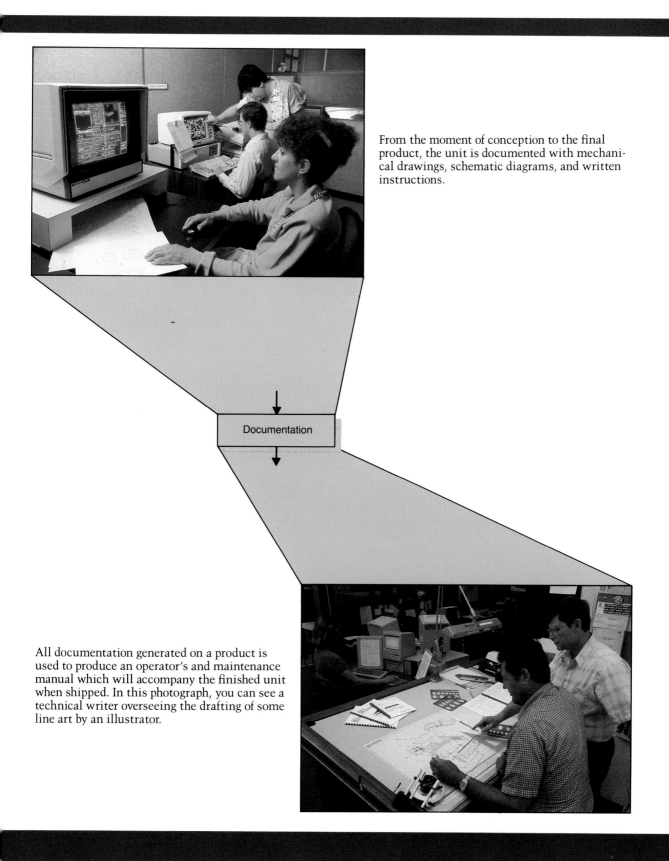

From the moment of conception to the final product, the unit is documented with mechanical drawings, schematic diagrams, and written instructions.

Documentation

All documentation generated on a product is used to produce an operator's and maintenance manual which will accompany the finished unit when shipped. In this photograph, you can see a technical writer overseeing the drafting of some line art by an illustrator.

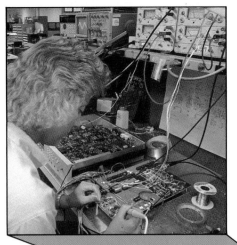

From sketches supplied by the engineers, a breadboard model of the design is constructed. The breadboard model is an experimental arrangement of a circuit in which the components are temporarily attached to a flat board. In this arrangement, the components can be tested to prove the feasibility of the circuit. A breadboard facilitates making easy changes when they are necessary. Here you can see an engineering technician breadboarding the design.

Breadboard Design

Prototype Layout, Construction, & Test

The engineer does not consider the final location of the components in constructing the breadboard model. At the next stage, however, a prototype working model completely representative of the final, mass-produced product is hand-assembled. The breadboard is replaced by a printed circuit board (PCB). In this scene, an engineering technician is producing a PCB layout from the design schematic diagrams.

The newly constructed prototype seen here is undergoing a complete evaluation of its mechanical and electrical form, design, and performance.

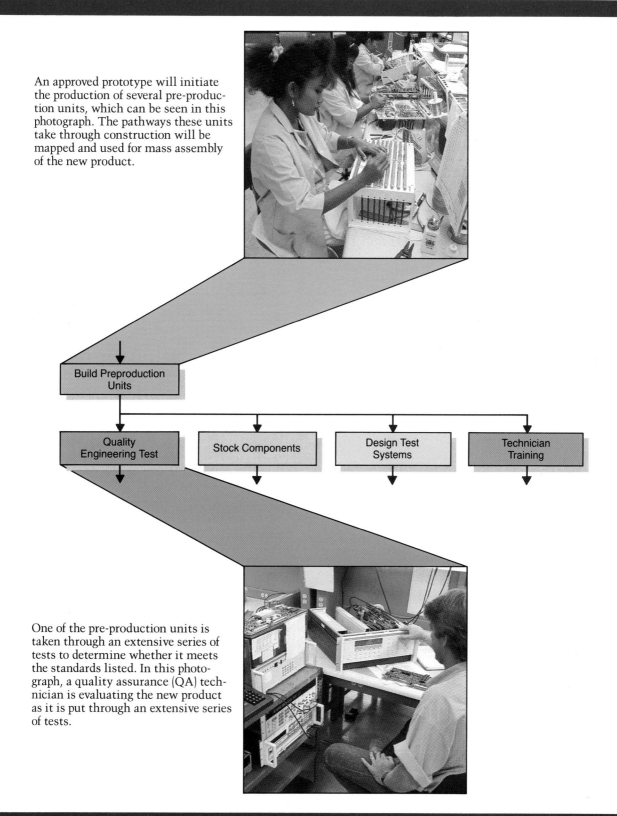

An approved prototype will initiate the production of several pre-production units, which can be seen in this photograph. The pathways these units take through construction will be mapped and used for mass assembly of the new product.

Build Preproduction Units

Quality Engineering Test

Stock Components

Design Test Systems

Technician Training

One of the pre-production units is taken through an extensive series of tests to determine whether it meets the standards listed. In this photograph, a quality assurance (QA) technician is evaluating the new product as it is put through an extensive series of tests.

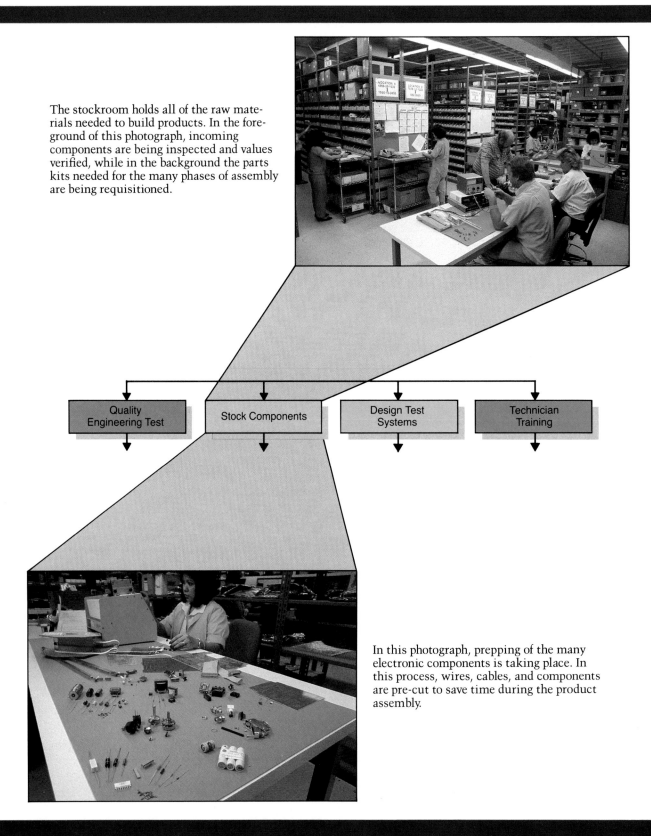

The stockroom holds all of the raw materials needed to build products. In the foreground of this photograph, incoming components are being inspected and values verified, while in the background the parts kits needed for the many phases of assembly are being requisitioned.

| Quality Engineering Test | Stock Components | Design Test Systems | Technician Training |

In this photograph, prepping of the many electronic components is taking place. In this process, wires, cables, and components are pre-cut to save time during the product assembly.

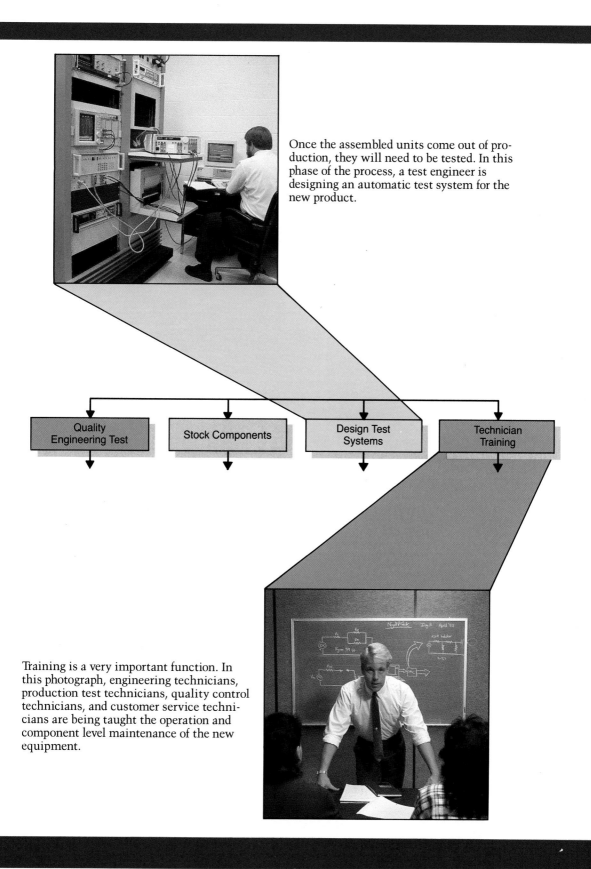

Once the assembled units come out of production, they will need to be tested. In this phase of the process, a test engineer is designing an automatic test system for the new product.

Quality Engineering Test	Stock Components	Design Test Systems	Technician Training

Training is a very important function. In this photograph, engineering technicians, production test technicians, quality control technicians, and customer service technicians are being taught the operation and component level maintenance of the new equipment.

The components which make up the unit are grouped into several areas where complex precision assembly takes place. This photograph of the machine shop shows the chassis or housing for the equipment being fabricated.

In this photograph, all of the electronic components are being inserted into their respective positions within the printed circuit boards.

Chassis
Fabrication

Component
Insertion

Flow
Soldering

Wire/Cable
Hook-up

Final
Assembly

Product Assembly

Once the boards have been filled with components, they are run through a flow- or wave-soldering machine. This machine solders the components to the board by moving the printed circuit board over a flowing wave of molten solder in a solder bath.

The wires and cables that interconnect all of the separate boards and units within the equipment are added at this stage.

Product Assembly

Chassis Fabrication

Component Insertion

Flow Soldering

Wire/Cable Hook-up

Final Assembly

Final assembly of the product is taking place in this photograph. The equipment will have the remaining units inserted, and its front and rear panels will be mounted and connected.

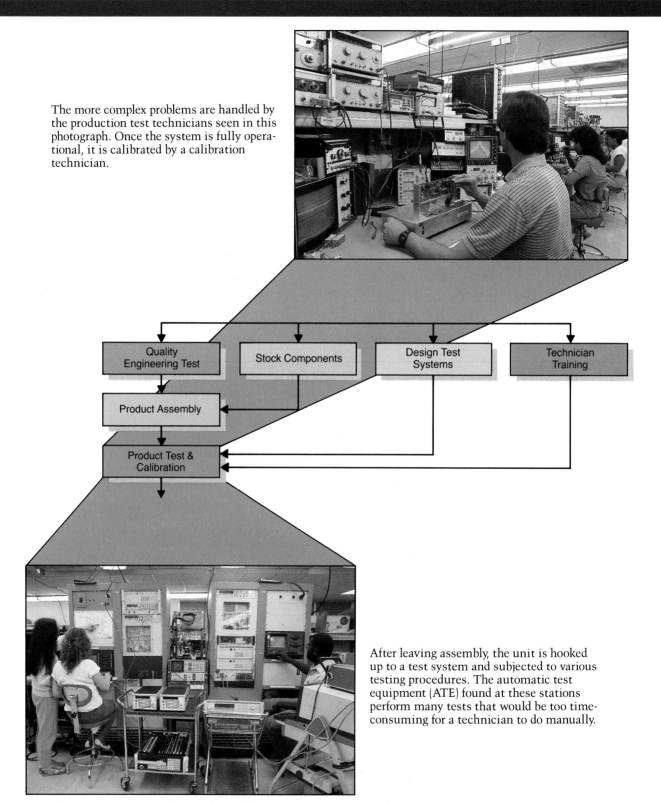

The more complex problems are handled by the production test technicians seen in this photograph. Once the system is fully operational, it is calibrated by a calibration technician.

Quality Engineering Test

Stock Components

Design Test Systems

Technician Training

Product Assembly

Product Test & Calibration

After leaving assembly, the unit is hooked up to a test system and subjected to various testing procedures. The automatic test equipment (ATE) found at these stations perform many tests that would be too time-consuming for a technician to do manually.

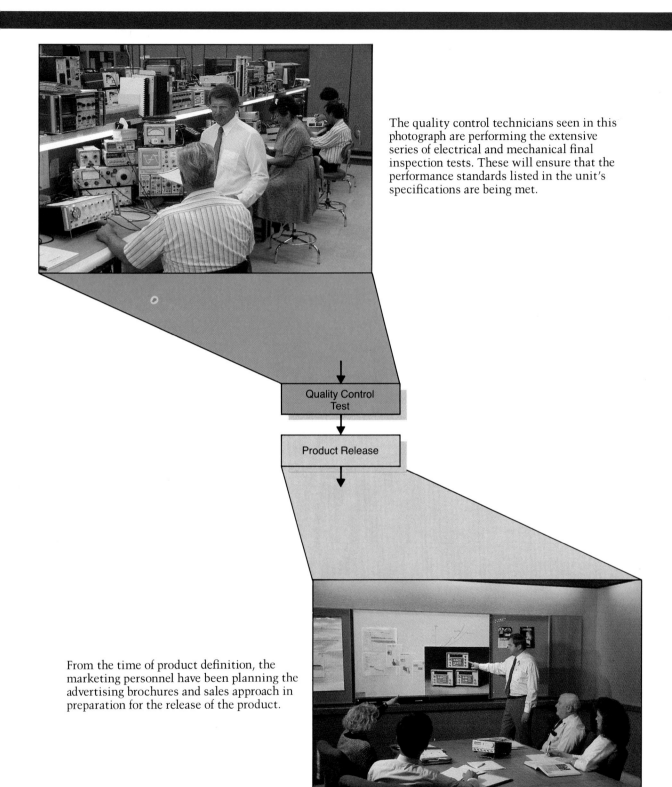

The quality control technicians seen in this photograph are performing the extensive series of electrical and mechanical final inspection tests. These will ensure that the performance standards listed in the unit's specifications are being met.

Quality Control Test

Product Release

From the time of product definition, the marketing personnel have been planning the advertising brochures and sales approach in preparation for the release of the product.

In this photograph, an applications engineer is making a sales call and will demonstrate to the customer the new product with all of its features and possible applications.

In the background of this photograph, you can see all of the equipment ready to be shipped. In the foreground, a delicate electronic unit is being carefully packaged to prevent damage during transit.

Sales

Shipping

Customer

Once the customer has received the electronic equipment, customer service provides assistance in maintenance and repair of the unit through direct in-house service or at service centers throughout the world. This photograph shows some in-house service technicians troubleshooting problems on returned units.

Sales

Shipping

Customer

Customer Service

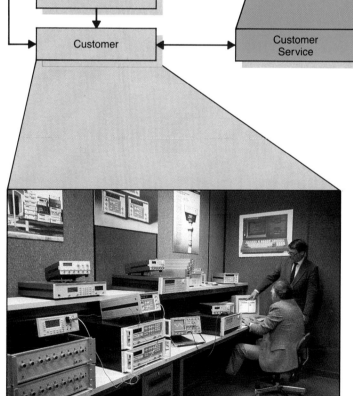

In this photograph, a customer is receiving instruction on the operation of the purchased unit.

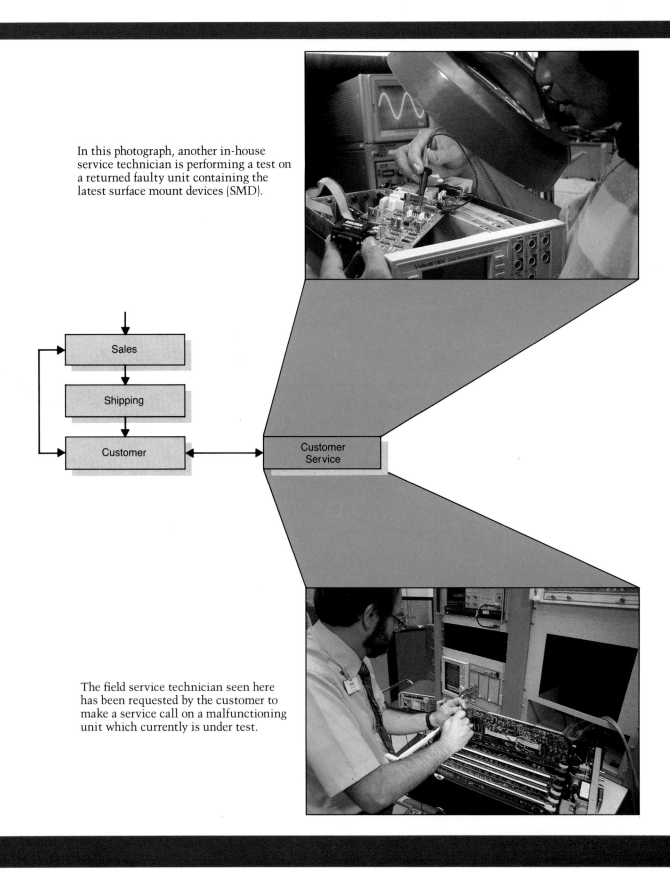

In this photograph, another in-house service technician is performing a test on a returned faulty unit containing the latest surface mount devices (SMD).

Sales

Shipping

Customer

Customer Service

The field service technician seen here has been requested by the customer to make a service call on a malfunctioning unit which currently is under test.

13. The moving coil microphone is an example of how electromagnetic induction is used to convert information-carrying sound waves into information-carrying electrical waves.

Alternating magnetic field	**Electromagnetic induction**	**Lenz's law**
Armature	**Faraday's law**	**Moving coil microphone**
Coercive force	**Flux linkage**	**Remanence**
Coercivity	**Galvanometer**	**Retentivity**
Diaphragm	**Hysteresis loop**	**Saturation**

REVIEW QUESTIONS

Multiple Choice Questions

1. Direct current produces a _____ magnetic field of _____ polarity.
   a. Alternating, unchanging
   b. Constant, alternating
   c. Constant, unchanging
   d. Both (a) and (c) are true
2. Magnetizing force (H) is equal to:
   a. $I \times N \times l$
   b. $I \times N + l$
   c. $I \times N/l$
   d. $N \times l/I$
3. Electromagnetism:
   a. Is the magnetism resulting from electrical current flow
   b. Is the electrical voltage resulting in a coil from the relative motion of a magnetic field
   c. Both (a) and (b) could be true
   d. None of the above
4. Electromagnetic induction:
   a. Is the magnetism resulting from electrical current flow
   b. Is the electrical voltage resulting in a coil from the relative motion of magnetic field
   c. Both (a) and (b) could be true
   d. None of the above
5. The _____ plots magnetizing force against flux density through a complete cycle of alternating current.
   a. B–H curve
   b. Coercive force
   c. Power curve
   d. Hysteresis loop

6. When the magnetic flux linking a conductor is changing, an emf is induced, the magnitude of which depends on the number of coil turns, rate of change of flux linkage change, and flux density. This law was discovered by:
   a. Heinrich Lenz
   b. Guglielmo Marconi
   c. Michael Faraday
   d. Joseph Henry

7. The current induced in a coil due to the change in the magnetic flux is such as to oppose the cause producing it. This law was discovered by:
   a. Heinrich Lenz
   b. Guglielmo Marconi
   c. Michael Faraday
   d. Joseph Henry

8. An ac generator uses _____ to generate ac electricity.
   a. Electromagnetism
   b. Electromagnetic induction
   c. Magnetism
   d. None of the above

9. The generator converts _____ energy into _____ energy.
   a. Electrical, electrical
   b. Mechanical, electrical
   c. Chemical, electrical
   d. None of the above

10. The loop of conductor rotated through 360° in a generator is known as a(an):
    a. Electromagnet
    b. Field coil
    c. Armature
    d. Both (a) and (b)

11. Sound waves are a form of:
    a. Electrical energy
    b. Chemical energy
    c. Magnetic energy
    d. Mechanical energy

12. The moving coil microphone converts _____ waves into _____ waves.
    a. Sound, electrical
    b. Electrical, sound
    c. Electromagnetic, sound
    d. Sound, radio

13. There is a reciprocal relationship between permeability and _____.
    a. Flux density
    b. Magnetizing force
    c. Remanence
    d. Reluctance

14. The amount of current in the reverse direction needed to reduce the flux density (B) to zero is termed the:

a. Coercive force
b. Magnetizing force
c. Electromotive force
d. All the above

Essay Questions

15. What effect does current have when it is passed through a coil of conductor? (13.1)
16. What effect does a magnet have when it is moved into and out of a coil? (13.3)
17. State Faraday's law. (13.3.1)
18. State Lenz's law. (13.3.2)
19. Describe the different effects when ac and dc are passed through a coil. (13.1)
20. Illustrate and describe all the different points on a hysteresis curve. (13.2.1)
21. Describe the meaning of the following terms: (13.2)
    a. Flux density
    b. Magnetizing
    c. Remanence
    d. Coercive force
    e. Electromagnetism
    f. Electromagnetic induction
22. Briefly describe the operation of the generator. (13.3.4)
23. Briefly describe the operation of the moving coil microphone. (13.3.4)
24. Illustrate and describe the operation of the generator through 360°. (13.3.4)
25. List some other applications of ac electromagnetism.

14

CAPACITANCE AND CAPACITORS

AFTER COMPLETING THIS CHAPTER, YOU WILL BE ABLE TO:

1. Define the term capacitance.
2. Describe the basic capacitor construction.
3. Explain the charging and discharging process and its relationship to electrostatics.
4. State the unit of capacitance and explain how it relates to charge and voltage.
5. List and explain the factors determining capacitance.
6. Define the terms:
   a. Capacitance breakdown
   b. Capacitor leakage
7. Calculate total capacitance in parallel and series capacitance circuits.
8. Describe the advantages and differences between the five basic types of fixed capacitors.
9. Describe the advantages and differences between the four basic types of variable capacitors.
10. Explain the characteristics and new techniques used to create the 1 farad capacitor.
11. Describe the coding of capacitor values on the body by use of alphanumerics or color.
12. Explain the capacitor time constant as it relates to dc charging and discharging a capacitor.
13. Explain how the capacitor charges and discharges when ac is applied.

INTRODUCTION

Up to this point, we have concentrated on circuits containing only resistance, which oppose the flow of current and then convert or dissipate power in the form of heat. Capacitance and inductance are two circuit properties that act differently from resistance in that they will charge or store the supplied energy and then return almost all the stored energy back to the circuit, rather than lose it in wasted heat. Inductance will be discussed in the following chapter.

Capacitance is the ability of a circuit or device to store electrical charge. A device or component specifically designed to have this capacity or capacitance is called a capacitor. A capacitor stores an electrical charge similar to a bucket holding water. Capacitors store electrons, and basically the amount of electrons stored is a measure of the capacitor's capacitance.

14.1
CAPACITOR CONSTRUCTION

Figure 14-1 illustrates the main parts and schematic symbol of the capacitor. Several years ago, capacitors were referred to as condensers; however, that term is very rarely used today. Two leads are connected to two parallel metal conductive plates which are separated by an insulating material known as a *dielectric*. The conductive plates are normally made of metal foil, while the dielectric can be paper, air, glass, ceramic, mica, or some other form of insulator.

FIGURE 14-1   Capacitor. (a) Symbol. (b) Basic Construction.

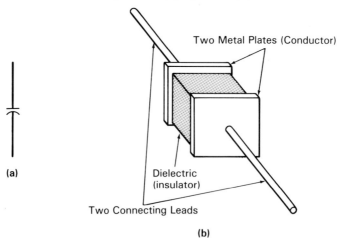

(a)

Two Metal Plates (Conductor)

Dielectric
(insulator)

Two Connecting Leads

(b)

1. What are the main parts of a capacitor?
2. What is the name given to the insulating material between the two conductive plates?

14.2
CHARGING AND DISCHARGING A CAPACITOR

Like a secondary battery, a capacitor can be made to charge or discharge, and when discharging it will return all of the energy it consumed during charge.

14.2.1 CHARGING A CAPACITOR

Capacitance is the ability of a capacitor to store an electrical charge. Figure 14-2 illustrates how the capacitor stores an electric charge. This capacitor is shown as two plates with air acting as the dielectric.

In Figure 14-2(a), the switch is open and so no circuit current results. An equal number of electrons exists on both plates, and so the voltmeter (VM) indicates zero, or no potential difference exists across the capacitor.

In Figure 14-2(b), the switch is now closed and electrons travel to the positive side of the battery, away from the right hand plate of the capacitor.

FIGURE 14-2   Charging a Capacitor.

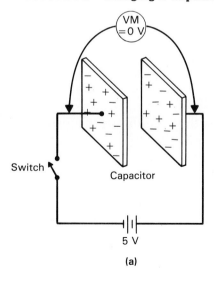
(a)

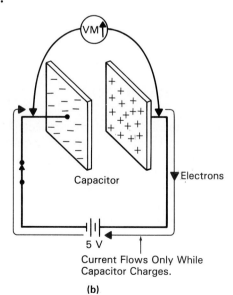
(b)

Current Flows Only While Capacitor Charges.

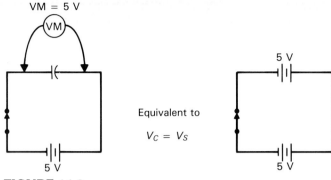

FIGURE 14-3

This creates a positive right hand capacitor plate, which results in an attraction of free electrons from the negative side of the battery to the left hand plate of the capacitor. In fact, for every electron that leaves the right hand capacitor plate and is attracted into the positive battery terminal, another electron leaves the negative side of the battery and travels to the left hand plate of the capacitor. Current appears to be flowing from the negative side of the battery, around to the positive, through the capacitor. This is not really the case, because no electrons can flow through the insulator or dielectric. There appears to be one current flowing throughout the circuit, when in fact there are actually two separate currents, each flowing in opposite directions (one from the battery to the capacitor and the other from the capacitor to the battery).

A voltmeter across the capacitor will indicate an increase in the potential difference between the plates, and the capacitor is said to be charging toward, in this example, 5 volts. This potential difference builds up across the two plates until the potential difference (voltage) across the capacitor is equal to the potential difference (voltage) of the battery. In the example, when the capacitor reaches a potential difference or charge of 5 V, then the capacitor will be equivalent to a 5 V battery, as seen in Figure 14-3. In this condition no potential difference exists between the battery and capacitor, and therefore no current flow can exist without a potential difference and the capacitor is said to be charged.

14.2.2 DISCHARGING A CAPACITOR

If the capacitor is now disconnected from the circuit by opening switch 1, as seen in Figure 14-4(a), it will remain in its charged condition. If switch 2 is now closed, a path exists across the charged capacitor, as seen in Figure 14-4(b), and the excess of electrons on the left plate will flow through the conducting wire to the positive plate on the right side. The capacitor is now said to be discharging. When equal numbers of electrons exist on both sides, the capacitor is said to be discharged; and since both plates have equal charge, the potential difference across the capacitor will be zero, as seen in Figure 14-5.

Figure 14-6(a) and (b) illustrates the charge and discharge currents, which flow in opposite directions. In both cases, the current flow is always from one plate to the other and never exists through the dielectric insulator.

Capacitance and Capacitors

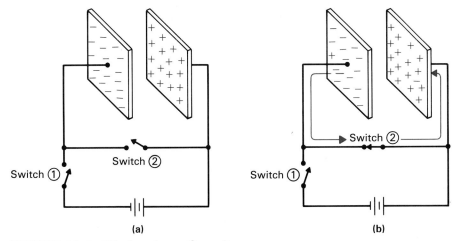

(a)   (b)

FIGURE 14-4 Discharging a Capacitor.

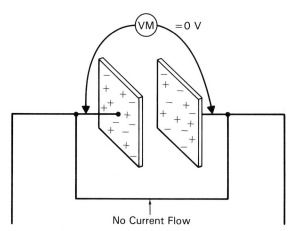

FIGURE 14-5 Discharged Capacitor.

FIGURE 14-6 (a) Charging Current Path. (b) Discharging Current Path.

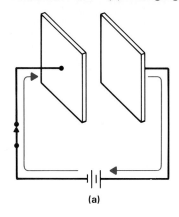

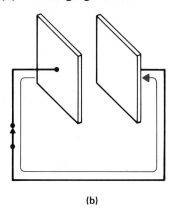

(a)   (b)

1. When both plates of a capacitor have an equal charge, the capacitor is said to be charged (true or false).
2. If 2 μA of current flows into one plate of a capacitor and 2 μA flows out of the other plate, how much current is flowing through the dielectric of this working capacitor?

14.3
ELECTROSTATICS

Just as a magnetic field is produced by the flow of current, an electric field is produced by voltage.

Current generates a magnetic field

Voltage generates an electric field

Figure 14-7 illustrates an example capacitor circuit with the capacitor charged and the switch open. In this condition, the capacitor retains its charge and an invisible electric or electrostatic (voltage) field will be produced by nonmoving or static electrical (electrostatic) charges of different polarities. You will remember that like charges repel and unlike charges attract. Invisible electrostatic lines of flux or force can be illustrated to show this electrostatic force of attraction or repulsion. These lines are polarized away from a positive electrostatic (stationary electrical) charge and toward the negative electrostatic charge, as seen in Figure 14-8(a). If two like charges are in close proximity to one another, the electrostatic lines organize themselves into a pattern, as shown in Figure 14-8(b).

The charged capacitor, seen in Figure 14-9, has an electric or electro-

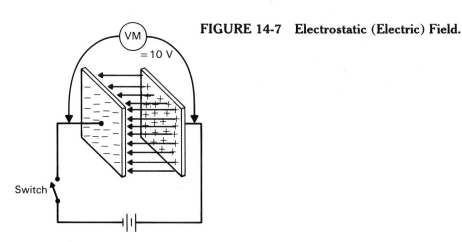

FIGURE 14-7 Electrostatic (Electric) Field.

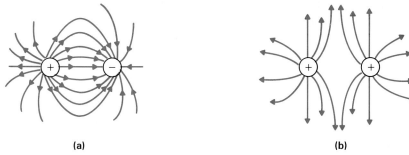

(a)　　　　　　　　　　　　　　　　(b)

FIGURE 14-8   Electrostatic Field. (a) Electrostatic Lines of Attraction. (b) Electrostatic Lines of Repulsion.

static field existing between the positively charged and negatively charged plates. The strength of the electrostatic field is proportional to the charge or potential difference on the plates and inversely proportional to the distance between the plates.

$$\text{field strength (V/m)} = \frac{\text{charge difference } (V), \text{ volts}}{\text{distance between plates } (d), \text{ meters}}$$

The dielectric or insulator between the plates, like any other material, has its own individual atoms, and although the dielectric electrons are more tightly bound to their atoms than conductor electrons, stresses are placed on the atoms within the dielectric, as seen in Figure 14-10. The electrons in orbit around the dielectric atoms are displaced or distorted by the electric field existing between the positive and negative plate. If the charge potential across the capacitor is high enough and the distance between the plates is small enough, the attraction and repulsion exerted on the dielectric atom can be large enough to free the dielectric atom's electrons. The material then becomes ionized, and a chain reaction of electrons jumping from one atom to the next in a right to left movement occurs. If this occurs, a large number of electrons will flow from the negative to the positive plate, and the dielectric is said to have broken down. This situation occurs if the capacitor is placed in a circuit where the voltages within the circuit exceed the voltage rating of the capacitor.

If the voltage rating of the capacitor is not exceeded, an electrostatic or electric field still exists between the two plates and causes this pulling of

FIGURE 14-9   Electric Field between the Plates of a Capacitor.

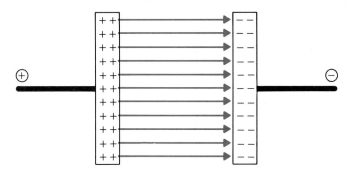

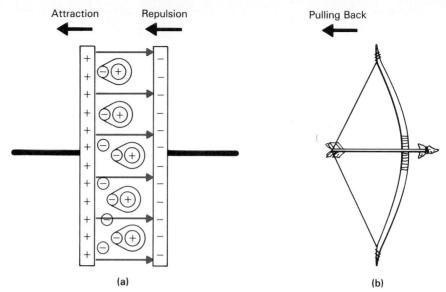

Attraction   Repulsion   Pulling Back

(a)   (b)

FIGURE 14-10   **Electric Polarization.**

the atoms's electrons within the dielectric toward the negative plate. This displacement is known as electric polarization and is similar to a pulling back effect on a bow, as seen in Figure 14-10(b). When the capacitor is given a path for discharge, as seen in Figure 14-11, the electric field in the dielectric, which is causing the distortion, is the force field that drives the electrons, like the bow drives the arrow.

To summarize electrostatics and capacitors, the charges on the plates of a capacitor produce an electric field, the electric field causes the distortion of the atoms known as electric polarization, and this pulling back or distortion, which is held there by the electric field, is the electron moving force (emf) that drives the electrons when a discharge path is provided. The energy in a capacitor is actually stored in the electric or electrostatic field within the dielectric.

Now that we understand these points, we can see where the word dielectric comes from. The dielectric is the insulating material that exists

FIGURE 14-11

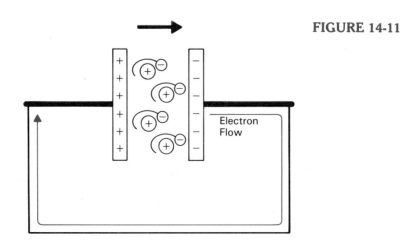

Electron Flow

between two (di) plates and undergoes electric polarization when an electric field exists within it (dielectric).

SELF-TEST REVIEW (§ 14.3)

1. An _____ field is generated by the flow of current, while an _____ field is produced by voltage.
   a. Electric, magnetic
   b. Magnetic, electric
2. State the formula for calculating field strength.
3. What is electric polarization?
4. Describe why the term dielectric is used to indicate the insulating material between the plates of a capacitor.

14.4
THE UNIT OF CAPACITANCE

Capacitance is the ability of a capacitor to store an electrical charge, and the unit of capacitance is the farad (F), in honor of Michael Faraday's work in 1831 in the field of capacitance. A capacitor with the capacity of 1 farad (1 F) can store 1 coulomb of electrical charge (6.24×10^{18} electrons) if 1 volt is applied across the capacitor's plates, as seen in Figure 14-12.

A 1 F capacitor is a very large value and not typically found in electronic equipment. Most values of capacitance found in electronic equipment are in the units between the microfarad ($\mu F = 10^{-6}$) and picofarad ($pF = 10^{-12}$). A microfarad is 1-millionth of a farad (10^{-6}). So if a 1 F capacitor can store 6.24×10^{18} electrons with 1 V applied, then a 1 μF capacitor, which has 1 millionth the capacity of a 1 F capacitor, can only store 1 millionth of a coulomb, or $(6.24 \times 10^{18}) \times (1 \times 10^{-6}) = 6.24 \times 10^{12}$ electrons when 1 V is applied, as seen in Figure 14-13.

FIGURE 14-12   One Farad of Capacitance.

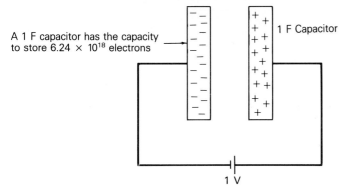

A 1 F capacitor has the capacity to store 6.24×10^{18} electrons

1 F Capacitor

1 V

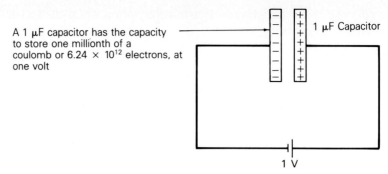

A 1 μF capacitor has the capacity to store one millionth of a coulomb or 6.24 × 10¹² electrons, at one volt

1 μF Capacitor

1 V

FIGURE 14-13   One-millionth of a Farad.

EXAMPLE 14.1

Convert the following to either microfarads or picofarads (whichever is more appropriate):

a. 0.00002 F
b. 0.00000076 F
c. 0.00047 × 10⁻⁷ F

SOLUTION

a. 20 μF
b. 0.76 μF
c. 47 pF

Since there is a direct relationship between capacitance, charge, and voltage, there must be a way of expressing this relationship in a formula.

$$\text{capacitance, } C \text{ (farads)} = \frac{\text{charge, } Q \text{ (coulombs)}}{\text{voltage, } V \text{ (volts)}}$$

C = capacitance in farads
V = voltage in volts
Q = charge in coulombs

By transposition of the formula, we arrive at the following combinations for the same formula:

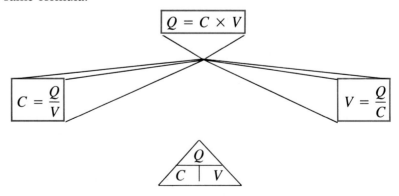

$$Q = C \times V$$

$$C = \frac{Q}{V}$$

$$V = \frac{Q}{C}$$

EXAMPLE 14.2

If a capacitor has the capacity to hold 36 C (2.25×10^{20} electrons) when 12 V is applied across its plates, what is the capacitance of the capacitor?

SOLUTION

$$
\begin{aligned}
C &= \frac{Q}{V} \\
&= \frac{36}{12} \\
&= 3 \text{ F}
\end{aligned}
$$

EXAMPLE 14.3

How many electrons could a 3 μF capacitor store when 5 V is applied across it?

SOLUTION

$$
\begin{aligned}
Q &= C \times V \\
&= 3 \ \mu F \times 5 \text{ V} \\
&= 15 \ \mu C
\end{aligned}
$$

(15 microcoulombs is 15 millionths of a coulomb). Since 1 C = 6.24×10^{18} electrons, 15 μC = $(15 \times 10^{-6}) \times 6.24 \times 10^{18}$ = 9.36×10^{13} electrons.

EXAMPLE 14.4

If a capacitor of 2 F has stored 42 C of charge (2.63×10^{20} electrons), what is the voltage across the capacitor?

SOLUTION

$$
\begin{aligned}
V &= \frac{Q}{C} \\
&= \frac{42 \ C}{2 \ F} \\
&= 21 \text{ V}
\end{aligned}
$$

1. What is the unit of capacitance?
2. State the formula for capacitance in relation to charge and voltage.
3. Convert 30,000 μF to farads.
4. If a capacitor holds 17.5 C of charge when 9 V is applied, what is the capacitance of the capacitor?

14.5

DETERMINING CAPACITANCE

The capacitance of a capacitor is determined by three factors:

1. The plate area of the capacitor
2. The distance between the plates
3. The type of dielectric used

Let's now discuss these three factors in more detail, beginning with the plate area.

(1) PLATE AREA (A)

The capacitance of a capacitor is directly proportional to the plate area. This area in square inches is the area of only one plate and is calculated by multiplying length by width. This is illustrated in Figure 14-14(a) and (b). In these two examples, the (b) capacitor plate is twice as large as the (a) capacitor plate, and since capacitance is proportional to plate area ($C \propto A$), the capacitor in example (b) will have double the capacity or capacitance of (a). Since the energy of a charged capacitor is in the electric field between the plates and the plates of capacitor (b) are double those of (a), there is twice as much area for the electric field to exist, and this doubles the capacitor's capacitance.

(2) DISTANCE BETWEEN THE PLATES (d)

The distance or separation between the plates is dependent on the thickness of the dielectric used. The capacitance of a capacitor is inversely proportional to this distance between plates, in that an increase in the distance ($d\uparrow$) causes a decrease in the capacitor's capacitance ($C\downarrow$). In Figure 14-15(a), a large distance between the capacitor plates, results in a small capacitance, whereas in Figure 14-15(b) the dielectric thickness and the plate separation is half that of capacitor (a). This illustrates also how the capacitance of a capacitor can be doubled, in this case by halving the space between the plates. The gap across which the electric lines of force exist is halved in capacitor (b), and this doubles the strength of the electric field, which consequently doubles capacitance. Simply stated, an electric line of force (Z)

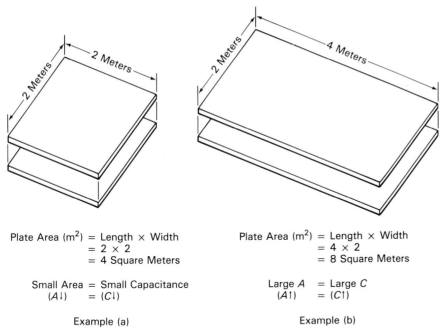

Plate Area (m²) = Length × Width
                = 2 × 2
                = 4 Square Meters

Small Area = Small Capacitance
   (A↓)   = (C↓)

Example (a)

Plate Area (m²) = Length × Width
                = 4 × 2
                = 8 Square Meters

Large A = Large C
  (A↑) = (C↑)

Example (b)

FIGURE 14-14 Capacitance Is Proportional to Plate Area.

in Figure 14-15(a) can be used to produce two electric lines of force (X and Y) in Figure 14-15(b) if the distance is half.

(3) DIELECTRIC

The insulating dielectric of a capacitor concentrates the electric lines of force between the two plates. Consequently, different dielectric materials can change the capacitance of a capacitor by being able to concentrate or establish an electric field with greater ease than other dielectric insulating materials.

FIGURE 14-15 Capacitance Is Inversely Proportional to Plate Separation of Distance (d).

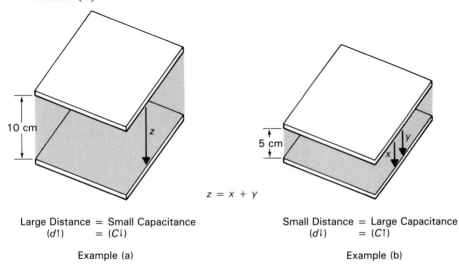

$$z = x + y$$

Large Distance = Small Capacitance
   (d↑)   = (C↓)

Example (a)

Small Distance = Large Capacitance
   (d↓)   = (C↑)

Example (b)

Determining Capacitance

471

TABLE 14-1
DIELECTRIC CONSTANTS

Material	Dielectric Constant (K)
Vacuum	1.0
Air	1.0006
Teflon	2.0
Wax	2.25
Paper	2.5
Amber	2.65
Rubber	3.0
Oil	4.0
Mica	5.0
Ceramic (low)	6.0
Bakelite	7.0
Glass	7.5
Water	78.0
Ceramic (high)	8000.

The different material compositions can cause different values of K.

The dielectric constant (K) is the ease with which an insulating material can establish an electrostatic (electric) field. A vacuum is the least effective dielectric and has a dielectric constant of 1, as seen in Table 14-1. All the other insulators listed in this table will support electrostatic lines of force more easily than a vacuum. The vacuum is used as a reference. All the other materials have dielectric constant values that are relative to the vacuum dielectric constant of 1. For example, mica has a dielectric constant of 5.0, which means that mica can cause an electric field five times the intensity of a vacuum; and, since capacitance is proportional to the dielectric constant ($C \propto K$), the mica capacitor will have five times the capacity of the same-size vacuum dielectric capacitor. In another example, we can see that the capacitance of a capacitor can be increased by a factor of almost 8000 by merely using ceramic rather than air as a dielectric between the two plates.

Thus, plate area, separation, and the dielectric used are the three factors that change the capacitance of a capacitor. The formula that combines these three factors is

$$C = \frac{(8.85 \times 10^{-12}) \times K \times A}{d}$$

where C = capacitance in farads (F)
A = plate area in square meters (m²)
K = dielectric constant
8.85×10^{-12} is a constant
d = distance between the plates in meters (m)

This formula summarizes what has been said in relation to capacitance. The capacitance of a capacitor is directly proportional to the dielectric constant (K) and the plates' area and is inversely proportional to the dielectric thickness or distance between the plates.

Capacitance and Capacitors

EXAMPLE 14.5

What is the capacitance of a capacitor of ceramic, 0.3 m² plate area whose dielectric thickness is 0.0003 m?

SOLUTION

$$C = \frac{(8.85 \times 10^{-12}) \times K \times A}{d}$$

$$= \frac{(8.85 \times 10^{-12}) \times 6 \times 0.3 \text{ m}^2}{0.0003 \text{ m}} = 5.31 \times 10^{-8}$$

$$= 0.0531 \text{ } \mu\text{F}$$

SELF-TEST REVIEW (§ 14.5)

1. List the three variable factors that determine the capacitance of a capacitor.
2. State the formula for capacitance.
3. If a capacitor's plate area is doubled, the capacitance will _____?
4. If a capacitor's dielectric thickness is halved, the capacitance will _____?

14.6
CAPACITOR BREAKDOWN

Capacitors store a charge just as a container or tank stores water. The charge is proportional to the voltage applied across the capacitor ($Q = C \times V$). The amount of charge a capacitor can hold is not only proportional to the capacitance of the capacitor, because a fixed value capacitor can be made to hold a greater charge by increasing the voltage across the plates, as seen in Figure 14-16(a). If the voltage across the capacitor is increased further, the charge held by the capacitor will increase until the dielectric between the two plates of the capacitor breaks down and a spark jumps or arcs between the plates.

Using the water analogy shown in Figure 14-16(b), if the pressure of the water being pumped in is increased, the amount of water stored in the tank will also increase until a time is reached when the tank's seams at the bottom of the tank cannot contain the large amount of pressure and break down under strain. The amount of water stored in the tank is proportional

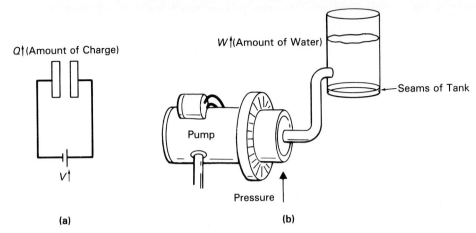

Q↑(Amount of Charge)

V↑

(a)

W↑(Amount of Water)

Seams of Tank

Pump

Pressure

(b)

FIGURE 14-16   Breakdown. (a) Voltage Increase, Charge Increases until Dielectric Breakdown. (b) Pressure Increase, Water Increase until Seams Break Down.

to the pressure applied, just as the amount of charge stored in a capacitor is proportional to the amount of voltage applied.

The breakdown voltage of a capacitor is determined by the strength of the dielectric used. Table 14-2 illustrates some of the different strengths of many of the common dielectrics. As an example, let's consider a capacitor that uses 1 mm of air as a dielectric between its two plates. This particular capacitor can withstand any voltage up to 787 V. If the voltage is increased further, the dielectric will break down and current will flow between the plates, destroying the capacitor (air capacitors, however, can recover from ionization).

SELF-TEST REVIEW (§ 14.6)

1. Which dielectric material has the best breakdown voltage figure?
2. Current flows through the dielectric at and below the breakdown voltage of the material (true or false).

TABLE 14-2
DIELECTRIC STRENGTHS

Material	Dielectric Strength (V/mm)
Air	787.
Oil	12,764.
Ceramic	39,370.
Paper	49,213.
Teflon	59,055.
Mica	59,055.
Glass	78,740.

14.7

CAPACITOR LEAKAGE

The ideal or perfect insulator should have a resistance equal to infinite ohms. Insulators or the dielectric used to isolate the two plates of a capacitor are not perfect and therefore have some very high value of resistance. This means that some value of resistance exists between the two plates, as seen in Figure 14-17; although this value of resistance is very large, it will still allow a small amount of current to flow between the two plates (in most applications a few nanoamps or picoamps). This small current is referred to as leakage current and causes any charge in a capacitor to slowly, over a long period of time, discharge between the two plates. A capacitor should have a large leakage resistance to ensure the smallest possible leakage current.

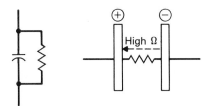

FIGURE 14-17   Capacitor Leakage.

14.8

CAPACITORS IN COMBINATION

Like resistors, capacitors can be connected in either series or parallel. The rules for determining total capacitance for parallel and series connected capacitors are opposite to series and parallel connected resistors.

14.8.1 PARALLEL

In Figure 14-18(a), you can see a 2 μF and 4 μF capacitor connected in parallel with one another. As the top plate of capacitor A is connected to the top plate of capacitor B with a wire, and a similar situation occurs with the bottom plates, you can see that this is the same as if the top and bottom plates were touching one another, as seen in Figure 14-18(b). When drawn so that the respective plates are touching, the dielectric constant and plate separation is the same as in Figure 14-18(a); however, now we can easily see that the plate area is actually increased. Consequently, if capacitors are connected in parallel, the effective plate area is increased; and since capacitance is proportional to plate area [$C = (8.85 \times 10^{-12}) \times K \times (A/d)$], the capacitance will also increase. Total capacitance is actually calculated by adding the plate areas, so total capacitance is equal to the sum of all the individual capacitances in parallel.

$$C_t = C_1 + C_2 + C_3 + C_4 + \cdots$$

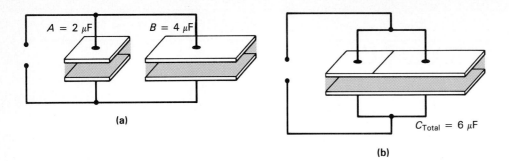

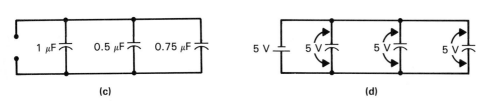

FIGURE 14-18 **Capacitors in Parallel.**

EXAMPLE 14.6

Determine the total capacitance of the circuit in Figure 14.18(c).

SOLUTION

$$C_t = C_1 + C_2 + C_3$$
$$= 1 \ \mu F + 0.5 \ \mu F + 0.75 \ \mu F$$
$$= 2.25 \ \mu F$$

As with any parallel-connected circuit, the source voltage appears across all the components. If, for example, 5 V is connected to the circuit of Figure 14-18(d), all the capacitors will charge to the same voltage of 5 V because the same voltage always exists across each section of a parallel circuit.

14.8.2 SERIES

In Figure 14-19(a), we have taken the two capacitors of 2 μF and 4 μF and connected them in series. Since the bottom plate of the A capacitor is connected to the top plate of the B capacitor, they can be redrawn so that they are touching, as seen in Figure 14-19(b).

The top plate of the A capacitor is connected to a wire into the circuit, and the bottom plate of B is connected to a wire into the circuit. This connection creates two center plates that are isolated from the circuit and can therefore be disregarded, as seen in Figure 14-19(c). The first thing you will notice is that the dielectric thickness ($d\uparrow$) has increased, causing a greater separation between the plates. The effective plate area of this capacitor has decreased, as it is just the area of the top plate only. Even though the bottom plate extends out further, the electric field can only exist between the two plates, and so the surplus metal of the bottom plate has no metal plate opposite for the electric field to exist.

Consequently, when capacitors are connected in series the effective plate area is decreased ($A\downarrow$) and the dielectric thickness increased ($d\uparrow$),

FIGURE 14-19   Capacitors in Series.

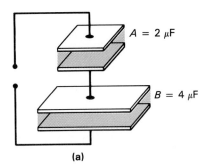

(a)

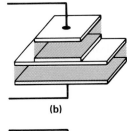

(b)

(c)

(d)

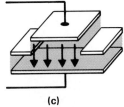

$$V_{C1} = \frac{C_T}{C_1} \times V_T$$

$$= \frac{1\ \mu F}{2\ \mu F} \times 6\ V = 3\ V$$

$$V_{C2} = \frac{C_T}{C_2} \times V_T$$

$$= \frac{1\ \mu F}{2\ \mu F} \times 6\ V = 3\ V$$

(e)

$$V_{C1} = \frac{C_T}{C_1} \times V_T = \frac{1.33\ \mu F}{2\ \mu F} \times 6\ V = 4\ V$$

$$V_{C2} = \frac{C_T}{C_2} \times V_T = \frac{1.33\ \mu F}{4\ \mu F} \times 6\ V = 2\ V$$

(f)

and both of these effects result in an overall capacitance decrease ($C\downarrow\downarrow$ = $(8.85 \times 10^{-12}) \times K \times A\downarrow/d\uparrow$).

The plate area is actually decreased to the smallest individual capacitance connected in series, which in this example is the plate area of A. If the plate area were the only factor, then capacitance would always equal the smallest capacitor value, but the dielectric thickness is always equal to the sum of all the capacitor dielectrics, and this factor always causes the total capacitance (C_t) to be less than the smallest individual capacitance when connected in series.

The total capacitance of two or more capacitors in series is calculated by using the following formulas: For two capacitors in series,

$$\boxed{C_t = \frac{C_1 \times C_2}{C_1 + C_2}} \quad \text{(product over sum)}$$

For more than two capacitors in series,

$$\boxed{C_t = \frac{1}{(1/C_1) + (1/C_2) + (1/C_3) + \cdots}}$$

EXAMPLE 14.7

Determine the total capacitance of the circuit in Figure 14-19(d).

SOLUTION

$$
\begin{aligned}
C_t &= \frac{1}{(1/C_1) + (1/C_2) + (1/C_3)} \\
&= \frac{1}{(1/4\ \mu\text{F}) + (1/2\ \mu\text{F}) + (1/1\ \mu\text{F})} \\
&= \frac{1}{1.75 \times 10^6} = 5.7143 \times 10^{-7} \\
&= 0.6\ \mu\text{F}
\end{aligned}
$$

The total capacitance for capacitors in series is calculated in the same way as total resistance when resistors are in parallel.

As with series-connected resistors, all the series-connected capacitor voltages will equal the voltage applied (Kirchhoff's voltage law). With capacitors connected in series, the charged capacitors act as a voltage divider, and a voltage-divider formula can be applied to capacitors in series.

$$\boxed{V_{cx} = \frac{C_t}{C_x} \times V_t}$$

where V_{cx} = voltage across desired capacitor

C_x = desired capacitor's value

C_t = total capacitance

V_t = total supplied voltage

If the capacitor values are the same, as seen in Figure 14-19(e), then the voltage is divided equally across each capacitor, as each capacitor has an equal amount of charge and therefore has half of the applied voltage or 3 V across it.

When the capacitor values are different, the smaller value of capacitor will actually charge to a higher voltage than the larger capacitor. In the example in Figure 14-19(f), the smaller capacitor is actually half the size of the other capacitor, and it has charged to twice the voltage. Since Kirchhoff's voltage law has to apply to this and every series circuit, you can easily calculate that the voltage across C_1 will equal 4 V and is twice that of C_2, which is 2 V. To fully understand this, we must first understand that, although the capacitance is different, both capacitors have an equal value of coulomb charge held within them, which in this example is 8 μC.

$$Q_1 = C_1 \times V_1$$
$$= 2 \ \mu F \times 4 \ V = 8 \ \mu C$$
$$Q_2 = C_2 \times V_2$$
$$= 4 \ \mu F \times 2 \ V = 8 \ \mu C$$

This equal charge occurs because the same amount of current flow exists throughout a series circuit, and so both capacitors are being supplied with the same number or quantity of electrons. The charge held by C_1 is large with respect to its small capacitance, whereas the same charge held by C_2 is small with respect to its larger capacitance.

If the charge remains the same and the capacitance is small, then the voltage drop across the capacitor will be large, because the charge is large with respect to the capacitances.

$$V\uparrow = \frac{Q}{C\downarrow}$$

On the other hand, for a constant charge, a large capacitance will have a small charge voltage because the charge is small with respect to the capacitance:

$$V\downarrow = \frac{Q}{C\uparrow}$$

We can apply the water analogy once more and imagine two series-connected buckets, one of which is twice the size of the other. Both are being supplied by the same series pipe, which has an equal flow of water throughout, and are consequently each holding an equal amount of water, for example, 1 gallon. The one gallon of water in the small bucket is large with respect to the size of the bucket, and a large amount of pressure exists within that bucket. The one gallon of water in the large bucket is small with respect to the size of the bucket, so a small amount of pressure exists within this bucket. The pressure within a bucket is similar to the voltage across a

EXAMPLE 14.8

Using the voltage-divider formula, calculate the voltage dropped across each of the capacitors in Figure 14-19(d) if $V_s = 24$ V.

SOLUTION

$$V_{C1} = \frac{C_t}{C_1} \times V_s = \frac{0.5714\ \mu F}{4\ \mu F} \times 24\ V \qquad = 3.4\ V$$

$$V_{C2} = \frac{C_t}{C_2} \times V_s = \frac{0.5714\ \mu F}{2\ \mu F} \times 24\ V \qquad = 6.9\ V$$

$$V_{C3} = \frac{C_t}{C_3} \times V_s = \frac{0.5714\ \mu F}{1\ \mu F} \times 24\ V \qquad = 13.7\ V$$

$$V_s = V_{C1} + V_{C2} + V_{C3} = 3.4 + 6.9 + 13.7 = 24\ V$$
(Kirchhoff voltage law)

capacitor, and a smaller bucket or capacitor will have a greater pressure or voltage charge associated with it, as compared to a large capacitance with an equal amount of charge.

To summarize capacitors in series, all the series-connected components will have the same charging current throughout the circuit, and, because of this, two or more capacitors in series will always have equal amounts of coulomb charge. If the charge (Q) is equal, then the voltage across the capacitor is determined by the value of the capacitor. A small capacitance will charge to a larger voltage ($V\uparrow = Q/C\downarrow$), whereas a large value of capacitance will charge to a smaller voltage ($V\downarrow = Q/C\uparrow$).

SELF-TEST REVIEW (§ 14.7 and § 14.8)

1. A capacitor should have a _____ leakage resistance to ensure a _____ leakage current (large or small)
2. If 2 μF, 3 μF, and 5 μF capacitors are connected in series, what will be the total circuit capacitance?
3. If 7 pF, 2 pF, and 14 pF capacitors are connected in parallel, what will be the total circuit capacitance?
4. State the voltage divider formula as it applies to capacitance.
5. With resistors, the large value of resistor will drop a larger voltage, whereas with capacitors the smaller value of capacitor will actually charge to a higher voltage (true or false).

14.9

TYPES OF CAPACITORS

Capacitors come in a variety of shapes and sizes and can be either fixed or variable in their values of capacitance. Within these groups, capacitors are generally classified by the dielectric used between the plates.

A *fixed* value capacitor is a capacitor whose capacitance value remains constant and cannot be altered. Fixed capacitors normally come in a disc or tubular package, as seen in Figure 14-20 and consist of metal foil plates separated by one of the following types of insulators (dielectric), which is the means by which we classify them.

1. Mica
2. Ceramic
3. Paper
4. Plastic
5. Electrolytic

A *variable* value capacitor is a capacitor whose capacitance value can be changed by rotating a shaft. The variable capacitor normally consists of one electrically connected movable plate and one electrically connected stationary plate. There are basically four types of variable capacitors, which are also classified by the dielectric used.

1. Air
2. Mica
3. Ceramic
4. Plastic

First, let's take a closer look at the different types of fixed capacitors.

14.9.1 FIXED CAPACITORS

(1) MICA

Figure 12-21 illustrates the physical appearance and construction of the mica capacitor. In Figure 14-21(a), which illustrates the construction of the mica

FIGURE 14-20 **Fixed Capacitors.**

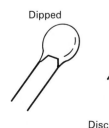
Dipped

Molded

Disc

Tubular

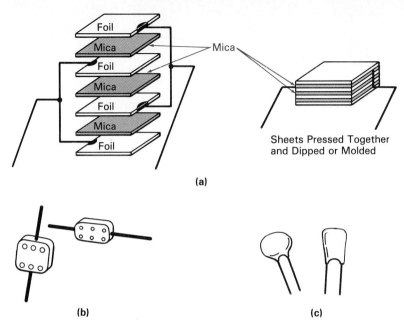

FIGURE 14-21 Mica Capacitors. (a) Construction. (b) Molded. (c) Dipped.

capacitor, you can see that thin foil plates (normally aluminum) are alternatively stacked to form the plates of the capacitor, all of which are isolated from one another by a thin layer of mica dielectric. Every other plate is connected to one lead on the right, while the other metal foil plates are attached to the other connecting lead on the left. This arrangement of stacked plates provides an increase in the capacitance due to the larger overall plate area.

The complete assembly of plates and mica is then sealed inside a protective casing. Figure 14-21(b) and (c) illustrate the two disc type casings. If molding equipment is used to protect and seal the assembly seen in Figure 14-21(b), the capacitor is referred to as a molded mica capacitor; however, if the plate and mica assembly is sealed by a dipping process, as seen in Figure 14-21(c), the capacitor is referred to as a dipped mica capacitor.

(2) CERAMIC

Figure 14-22 illustrates the physical appearance and construction of the molded and dipped types of ceramic capacitor. In their construction, you will notice the stacking of plates and isolation by ceramic, similar to the mica capacitor. Generally, the low dielectric constant ceramic ($K = 6.0$) is placed within the molded package, while the high dielectric constant ceramic (different composition) ($K = 8000$) is placed within the dipped package to obtain higher voltage ratings and high values of capacitance in small packages.

Figure 14-22(c) shows the ceramic chip surface mount capacitors that are now available on the market for both discrete and integrated types of circuits. Values typically range from 16 to 1600 pF. Different values of capacitance can be obtained by using different types of ceramic, thereby varying the dielectric constant.

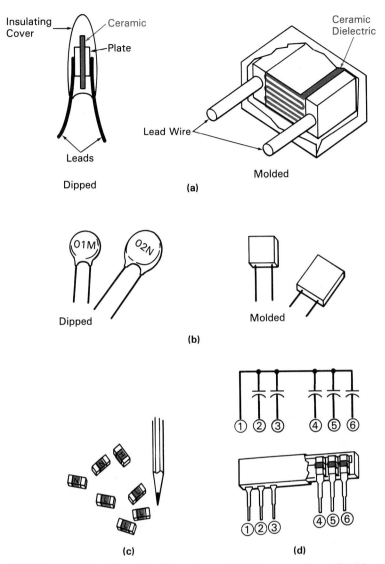

FIGURE 14-22 **Ceramic Capacitors. (a) Construction. (b) Physical Appearance of Dipped and Molded Type. (c) Chip Ceramic Casing. (d) SIP Casing.**

Figure 14-22(d) shows how the chip ceramic capacitors can be mounted in a SIP (single in line package) casing.

(3) PAPER

Figure 14-23 illustrates the construction and physical appearance of tubular paper capacitors. Two long strips of foil are separated by a paper dielectric that has been saturated with paraffin to ensure that the paper dielectric is a good insulator. The two foil plates separated by the paper dielectric are rolled into a tubular shape and placed in either a molded or dipped case

The leads of the capacitor can be either axial lead, as in Figure 14-23(b) (leads coming out of either end), or radial lead, as in Figure 14-23(c) (leads coming out of a single end).

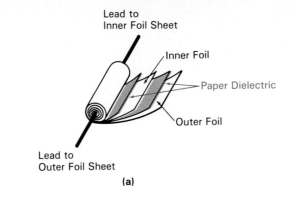

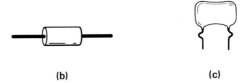

FIGURE 14-23 Paper Capacitors.
(a) Construction. (b) Axial Lead Wrapped. (c) Radial Lead Dipped.

(4) PLASTIC

Plastic film capacitors have almost completely replaced the older types of paper capacitors, and their method of construction is illustrated in Figure 14-24(a). The construction is identical to the paper type; however, in this instance a plastic film dielectric strip is used as the dielectric and then rolled into a cylinder. Mylar, polycarbonate, Teflon, and polypropylene are all examples of plastic that are used as a dielectric.

The final cylindrical package can be either encased in an outer plastic layer, as seen in Figure 14-24(b), or dipped, as seen in Figure 14-24(c).

(5) ELECTROLYTIC

Figure 14-25 shows the construction and physical appearance of typical electrolytic capacitors. These types of capacitors are constructed in a similar

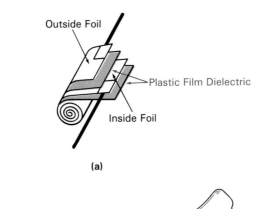

FIGURE 14-24 Plastic Capacitors.
(a) Construction. (b) Axial Lead. (c) Radial Lead.

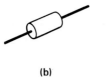

Capacitance and Capacitors

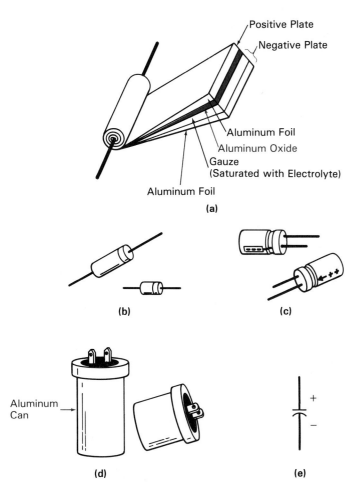

FIGURE 14-25 **Electrolytic Capacitors. (a) Construction. (b) Axial Lead.
(c) Single Ended. (d) Can Electrolytic. (e) Schematic Symbol for Electrolytic.**

manner as both paper and plastic types of capacitors, the difference being
that the foil plates are separated by a strip of gauze that has been saturated
with a conductive fluid known as an electrolyte. In the manufacturing process,
a dc voltage causes a current flow in one direction, which causes the electrolyte to chemically interact with the aluminum foil and create a coating of
aluminum oxide on the surface of the positive aluminum foil plate, as seen
in Figure 14-25(a), which causes it to change chemically. This oxide, which
is formed by an electrochemical reaction, becomes the dielectric for the
electrolytic capacitor and, because it is extremely thin, the electrolytic capacitor can have a large capacitance for a small size. The chemical change
in the positive aluminum plate during this electrochemical process makes
the electrolytic capacitor polarized. Thus the electrolytic capacitor must
always have a positive charge applied to its positive plate and a negative
charge to its negative plate. If this rule is not followed, the electrolytic
capacitor becomes a safety hazard, because it can in the worse case condition
explode violently.

Figure 14-25(e) shows the schematic symbols for electrolytic capacitors,
which are always marked with a ($+$) or ($-$) sign to indicate polarity. In

Name	Construction	Approximate Range of Values and Tolerances	Characteristics	
Mica	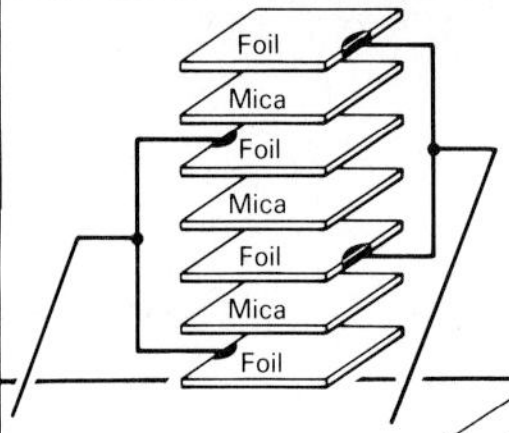 Foil Mica Foil Mica Foil Mica Foil	1 pF–0.1 μF ±1% to ±5%	Lower voltage rating than other capacitors of the same size	Small Capacitor Values
Ceramic	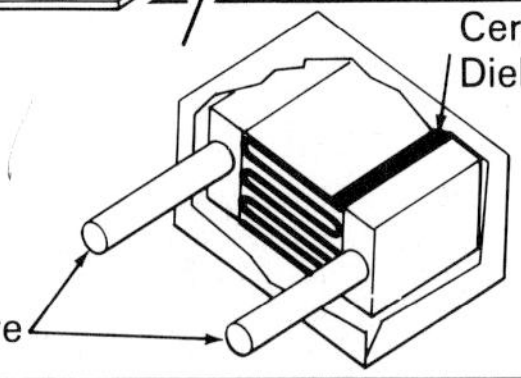 Ceramic Dielectric Lead Wire	*Low Dielectric K*: 1 pF–0.01 μF ±0.5% to ±10% *High Dielectric K*: 1 pF–0.1 μF ±10% to ±80%	Most popular small value capacitor due to lower cost than mica, and its ruggedness	Small Capacitor Values
Paper	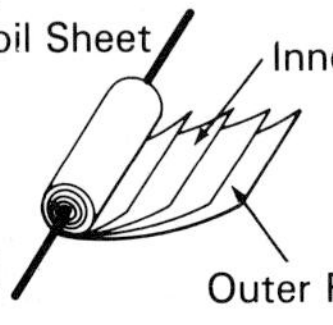 Lead to Inner Foil Sheet Inner Foil Lead to Outer Foil Sheet Outer Foil	1 pF–1μF ±10%	Has a large plate area and therefore large capacitance for a small size	Large Capacitor Values
Plastic	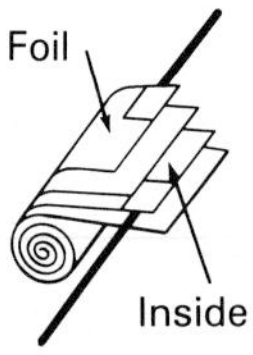 Outside Foil Inside Foil	1 pF–10 μF ±5% to ±10%	Has almost completely replaced paper capacitors; has large capacitance values for small size and high voltage ratings	Large Capacitor Values
Electrolytic (Aluminum and Tantalum)	(Tantalum has tantalum rather than aluminum foil plates) 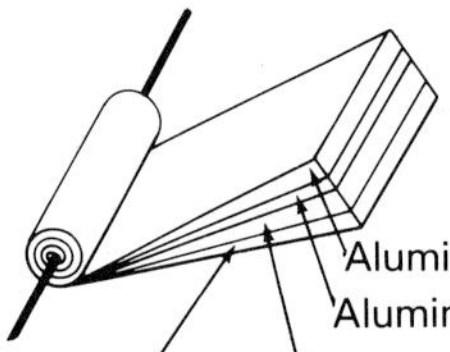 Aluminum Foil Aluminum Oxide Aluminum Foil Gauze (Saturated With Electrolyte)	1 μF–1F ±10% to ±50%	Most popular large value capacitor: large capacitance into small area, wide range of values. Disadvantages are: cannot be used in AC-circuits as they are polarized; poor tolerances; low leakage resistance and so high leakage current. *Tantalum* advantages over aluminum include smaller size, longer life than aluminum, which has an approximate lifespan of 12 years. Disadvantages: 4 to 5 times the price.	Large Capacitor Values

FIGURE 14-26 Fixed Capacitors.

Figure 14-25(c), you can see the manufacturer's method of polarity indication.

Figure 14-25(d) shows a can electrolytic capacitor. With this type of electrolytic capacitor, the negative outside foil plate is connected to the metal can casing, which acts as the negative capacitor lead.

Tantalum instead of aluminum is used in some electrolytic capacitors for the plates and has advantages over the aluminum type, which are:

1. Higher capacitance per volt for a given unit volume.
2. Longer life and excellent shelf life (storage).

3. Able to operate in a wider temperature range.
4. Temperature stability is better.
5. Construction features make them more rugged.

The cost of tantulum electrolytics, however, is almost four to five times that of aluminum electrolytics, and their operating voltages are much lower.

Figure 14-26 illustrates, reviews, and includes added information on the five basic fixed capacitors previously discussed.

14.9.2 VARIABLE CAPACITORS

Variable capacitors are the second basic type, and within this group, four types are commercially available. Like the fixed type of capacitors, they are classified by dielectric.

Variable

1. Air
2. Mica

The variable type uses a hand-rotated shaft to vary the effective plate area, and so capacitance, as seen in Figure 14-27(a).

Adjustable

3. Ceramic
4. Plastic

The adjustable type uses screw-in or -out mechanical adjustment to vary the distance between the plates, and so capacitance, as seen in Figure 14-27(b).

Capacitance is dependent on effective plate area and can therefore be varied by either changing the effective plate area or distance between the plates. The dielectric constant is fixed and depends on the particular type being used. The schematic symbol for a variable capacitor is shown in Figure 14-27(c).

FIGURE 14-27   (a) Variable Capacitor. (b) Adjustable Capacitor. (c) Schematic Symbol for Variable or Adjustable Capacitor.

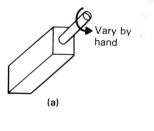

(a)

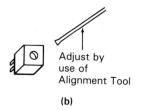

(b)

(c)

(1) AIR (VARIABLE)

Figure 14-28(a) illustrates the construction of a typical air dielectric variable capacitor. With this type of capacitor, the effective plate area is adjusted to vary capacitance by causing a set of rotating plates (rotor) to mesh with a set of stationary plates (stator). When the rotor plates are fully out, the capacitance is minimum, and when the plates are fully in, the capacitance is maximum, because the maximum amount of rotor plate area is now opposite the stator plate, creating the maximum value of capacitance.

The plates are usually made of aluminum to prevent corrosion and the dielectric between the plates is air. The capacitor has to be carefully manufactured to ensure that the rotor plates do not touch the stator plates when the shaft is rotated and the plates interweave. Figure 14-28(b) illustrates two typical variable air type capacitors.

In radio equipment, it is sometimes necessary to have two or more variable capacitors that have been constructed in such a way that a common shaft (rotor) runs through all the capacitors and varies their capacitance simultaneously, as seen in Figure 14-28(c). If you mechanically couple (gang) two or more variable capacitors so that they can all be operated from a single control, the component is known as a *ganged capacitor*, and the symbol for this arrangement is also shown in Figure 14-28(c).

FIGURE 14-28 Variable Air-type Capacitors. (a) Construction. (b) Physical Appearance. (c) Ganged.

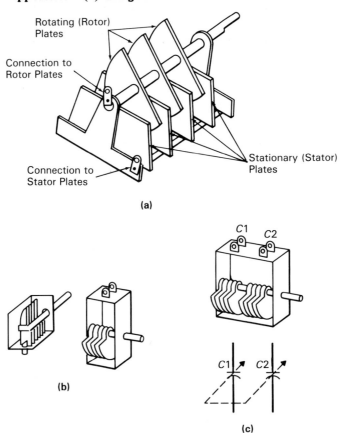

Capacitance and Capacitors

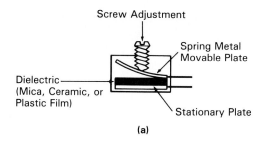

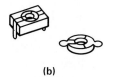

FIGURE 14-29 Adjustable Mica, Ceramic, or Plastic Capacitors. (a) Construction. (b) Physical Appearance.

(a)

(b)

(2) MICA, (3) CERAMIC, AND (4) PLASTIC (ADJUSTABLE)

Figure 14-29 illustrates some of the typical packages for mica, ceramic, or plastic film types of adjustable capacitors, which are also referred to as trimmers. The adjustable capacitor generally has one stationary plate and one spring metal moving plate. The screw forces the spring metal plate closer or farther away from the stationary plate, varying the distance between the plates and so changing capacitance. The two plates are insulated from one another by either mica, ceramic, or plastic film, and the advantages of each are the same as for fixed resistors.

These types of capacitors should only be adjusted with a plastic or nonmetallic alignment tool, because a metal screwdriver may affect the capacitance of the capacitor when nearby, making it very difficult to adjust for a specific value of capacitance.

14.9.3 THE ONE-FARAD CAPACITOR

Most capacitor values are normally measured in either microfarads or picofarads, which means that when fully charged they will supply either 1/1,000,000th or 1/1,000,000,000,000th of 1 ampere for 1 second. A 1 farad capacitor can supply one full ampere of current for 1 second of time.

The traditional picofarad and microfarad capacitors, which have been discussed, consist of two conductor plates separated by an insulator; when voltage is applied to the capacitor, current flows and a charge builds up within the capacitor and is stored. The amount of charge stored is dependent on the plate area, plate separation, and the dielectric constant of the insulator used. All three of these factors can be varied separately or in combination to create the capacitors that we have today.

The ceramic capacitor has become the most popular small value capacitor due to the high values of dielectric constants of 8000 or more that can be obtained from ceramic.

The electrolytic capacitor has become the most popular large value capacitor because of the extremely thin dielectric oxide layer that can be obtained.

Many techniques have been tried to extend the capacitance value up toward the farad; however, the conductive surface area is the main factor that needs to be increased in order to gain high values of capacitance, as seen in Figure 14-30(a). For example, if two pieces of aluminum the size of fingernails were held together, a fingernail thickness apart, a capacitance of approximately 1 pF would be created. This 1 pF capacitor would now have to be increased by a factor of 1,000,000,000,000 to obtain a capacitor of 1 farad.

To increase surface area, a process known as double etch has been used, as seen in Figure 14-30(b), where, instead of having a flat plate surface, as seen in Figure 14-30(a), large pits are etched onto the surface and then smaller bits are etched onto the larger pits to create a larger surface area. This method, however, still does not get us close to 1 F.

The 1 F capacitor uses a relatively new double-layer technique, the effect of which was first noticed more than a century ago by the German scientist Hermann von Helmholtz. Helmholtz discovered that a double layer of charge will build up on the surface between a solid and a liquid.

The double layer capacitor uses activated charcoal as the solid plates, and a liquid electrolyte between the two plates and an insulating separator allow a charge to build up within the capacitor due to the solid (charcoal) and liquid (electrolyte).

Activated charcoal is used as the plate because it has an almost infinite number of tiny particles making up its surface area, as seen in Figure 14-30(c), which yields a massive plate surface area. For example, 4 grams of activated charcoal would have an approximate surface area equivalent to a football field.

This 1 F capacitor package actually holds six 6 F capacitors connected in series,

FIGURE 14-30 (a) Traditional Plate Shape. (b) Double-etched Plate Area. (c) Activated Charcoal Surface Area. (d) One Inch Cube One Farad Capacitor.

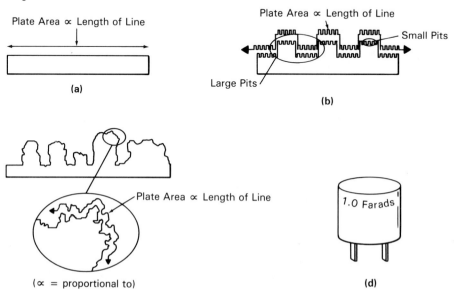

Capacitance and Capacitors

$$C_t \text{ (total series capacitance)} = \cfrac{1}{\cfrac{1}{6} + \cfrac{1}{6} + \cfrac{1}{6} + \cfrac{1}{6} + \cfrac{1}{6} + \cfrac{1}{6}}$$
$$= 1 \text{ F}$$

which yields an equivalent capacitance of 1 F. The reason for this is that the breakdown voltage of the double layer capacitor is approximately 1 V, and so by connecting six of these capacitors in series, the dielectric thickness is six times greater, increasing the breakdown voltage up to a more practical value (working dc voltage of 5 V). Due to the large value and the variations in activated charcoal, the tolerance of these capacitors can range from -20% to $+80\%$ of their value (0.8 to 1.8 F).

One major application for these capacitors is as a power backup for a computer's memory. If a computer has been used for a long time and a power failure occurs, all the information within the computer's memory will be lost. In this situation, the capacitor could be used as a backup source of power to supply power to the computer's memory (discharge) during power outages to prevent the loss of information. Figure 14-30(d) illustrates the 1 F capacitor.

SELF-TEST REVIEW (§ 14.9)

1. List the five types of fixed value capacitors.
2. Which fixed value capacitor is the most popular in applications where:
   a. Large values are required?
   b. Small values are required?
3. List the four basic types of variable capacitors.
4. Which variable capacitor types are best suited for:
   a. Large value variations?
   b. Small value variations?
5. Which capacitor type is said to be polarity conscious?

14.10
CODING OF CAPACITANCE VALUES

The capacitor value, tolerance, and voltage rating need to be shown in some way on the capacitor's exterior. Presently, two methods are used: (1) alphanumeric labels and (2) color coding.

14.10.1 ALPHANUMERIC LABELS

Manufacturers today most commonly use letters of the alphabet and numbers (alphanumerics) printed on either the disc or tubular body to indicate the capacitor's specifications, as illustrated in Figure 14-31.

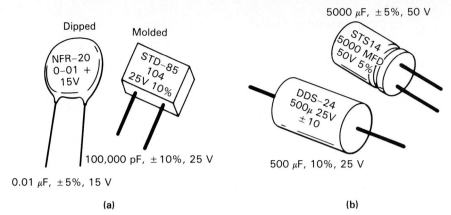

5000 μF, ±5%, 50 V

Dipped

NFR-20
0-01 +
15V

Molded

STD-85
104
25V 10%

STS14
5000 MFD
50V 5%

DDS-24
500μ 25V
± 10

100,000 pF, ±10%, 25 V

500 μF, 10%, 25 V

0.01 μF, ±5%, 15 V

(a)

(b)

FIGURE 14-31   **Alphanumeric Coding of Capacitors. (a) Disc Type. (b) Tubular.**

The tubular type, which tends to be larger in size, as seen in Figure 14-31(b), is the easier of the two since the information is basically uncoded. The value of capacitance and unit, typically the microfarad (μF or MF), tolerance figure (preceded by ± or followed by %), and voltage rating (followed by a V for voltage) are printed on all sizes of tubular cases. The remaining letters or numbers are merely manufacturer's codes for case size, series, and the like.

With disc type capacitors (dipped or molded), as seen in Figure 14-31(a), certain rules have to be applied when decoding the notations. Many capacitors of this type do not define the unit of capacitance; in this situation, try to locate a decimal point. If a decimal point exists, for example 0.01 or 0.001, then the value is in microfarads (10^{-6}). If no decimal point exists, for example 50 or 220, then the value is in picofarads (10^{-12}) and you must analyze the number in a little more detail.

EXAMPLE 14.9

What is the value of the following capacitor if it is labeled 50, 50 V, ±5?

SOLUTION

Since no decimal point is present, the unit is in picofarads:

$$50 \text{ pF}, \quad 50 \text{ V}, \quad 5\%$$

If no decimal point is present and three digits exist and the last digit is a zero, then the value is as stands and in picofarads. If the third digit is a number other than 0 (1 to 9), then it is a multiplier and describes the number of zeros to be added to the picofarad value.

EXAMPLE 14.10

If two capacitors are labeled with the following coded values, how should they be interpreted?
a. 220
b. 104

SOLUTION

Since no decimal point is present, they are both in picofarads.
a. If the last of the three digits is a zero, then the value is as it stands.

$$\underbrace{2 \quad 2 \quad 0}_{\text{three-digit value}}$$

b. If the third digit is a number from 1 to 9, then it is a multiplier.

$$\underbrace{1 \; 0}_{} \; \underbrace{4}_{}$$

two-digit value ⌐   └─── multipler

So
$$10 \times 10^4 = 100,000 \text{ pF}$$
or
$$100,000 \times 10^{-12} = 0.1 \times 10^{-6} = 0.1 \text{ } \mu\text{F}$$

The tolerance of the capacitor is sometimes clearly indicated, for example, ±5 or 10%; in other cases, a letter designation is used, such as:

$$F = \pm 1\%$$
$$G = \pm 2\%$$
$$J = \pm 5\%$$
$$K = \pm 10\%$$
$$M = \pm 20\%$$
$$Z = -20\%, +80\%$$

Unfortunately, there does not seem to be a standard among capacitor manufacturers, which can cause confusion when trying to determine the value of capacitance. Therefore, if you are not completely sure, you should always consult technical data or information sheets from the manufacturer.

14.10.2 COLOR CODING

Table 14-3 illustrates the capacitor color code, which is almost identical to the resistor color code except for certain tolerances. Although the alphanumeric labels are now commonplace, the color code is still being used by some manufacturers.

TABLE 14-3

Color	1st Digit	2nd Digit	Multiplier	Tolerance
Black		0	1	±20%
Brown	1	1	10^{-1}	±1%
Red	2	2	10^{-2}	±2%
Orange	3	3	10^{-3}	±3%
Yellow	4	4	10^{-4}	±4%
Green	5	5	10^{-5}	±5%
Blue	6	6	10^{-6}	
Violet	7	7	10^{-7}	
Gray	8	8	10^{-8}	
White	9	9	10^{-9}	±10%

SELF-TEST REVIEW (§ 14.10)

What are the following values of capacitance:

1. 470 ± 2
2. 0.47 ± 5

14.11
CAPACITIVE TIME CONSTANT

14.11.1 DC CHARGING

When a capacitor is connected across a dc voltage source, such as a battery or power supply, current will flow and the capacitor will charge up to a value of the dc source voltage, as seen in Figure 14-32(a) and (b). When the

FIGURE 14-32 (a) Capacitor Charging. (b) Capacitor Charged.

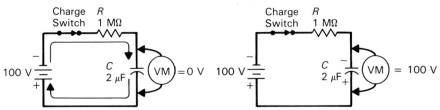

charge switch is first closed, there is no voltage across the capacitor at that instant and, therefore, a potential difference exists between the battery and capacitor. This causes current to flow and begin charging the capacitor.

Once the capacitor begins to charge, the voltage across the capacitor does not instantaneously rise to 100 V; it takes a certain amount of time before the capacitor voltage is equal to the battery voltage and, at that time, no potential difference exists between the voltage source and the capacitor. Consequently, no more current flows in the circuit as the capacitor has reached its full charge. The amount of time it takes for a capacitor to charge to the supplied voltage (100 V) is dependent on the circuit's resistance and capacitance value. If the circuit's resistance is increased, the opposition to current flow will be increased, and it will take the capacitor a longer period of time to obtain the same amount of charge because the circuit current available to charge the capacitor is less.

If the value of capacitance is increased, it again takes a longer time to charge to 100 V because a greater amount of charge is required to build up the voltage across the capacitor to 100 V.

The circuit's resistance (R) and capacitance (C) are the two factors that determine the charge time (τ). Mathematically, this can be stated as

$$\boxed{\tau = R \times C}$$

where τ = time constant (s)
R = resistance (Ω)
C = capacitance (F)

In this example, we are using a resistance of 1 MΩ and a capacitance of 2 μF, which means that the time constant is equal to

$$
\begin{aligned}
t &= R \times C \\
&= 2\ \mu F \times 1\ M\Omega \\
&= (2 \times 10^{-6}) \times (1 \times 10^{6}) \\
&= 2\ s
\end{aligned}
$$

Two seconds is the time, so what is the constant? The constant value that should be remembered throughout this discussion is 63.2.

Figure 14-33 illustrates the rise in voltage across the capacitor from 0 to a maximum of 100 V in five time constants ($5 \times 2\ s = 10\ s$). So where does 63.2 come into all this?

FIRST TIME CONSTANT   In $1RC$ seconds ($1 \times R \times C = 2\ s$), the capacitor will charge to 63.2% of the applied voltage (63.2 V).

SECOND TIME CONSTANT   In $2RC$ seconds ($2 \times R \times C = 4\ s$), the capacitor will charge to 63.2% of the remaining voltage. In the example, the capacitor will be charged to 63.2 V in one time constant, and therefore the voltage remaining is equal to 100 V $-$ 63.2 V $=$ 36.8 V. At the end of the second time constant, therefore, the capacitor will have charged to 63.2% of the remaining voltage (36.8 V) (63.2% $\times$ 36.8 V $=$ 23.3 V), which means it will have reached (63.2 + 23.3 = 86.5 V), 86.5% of the applied voltage (86.5 V).

THIRD TIME CONSTANT   In $3RC$ seconds (6 s), the capacitor will charge to 63.2% of the remaining voltage:

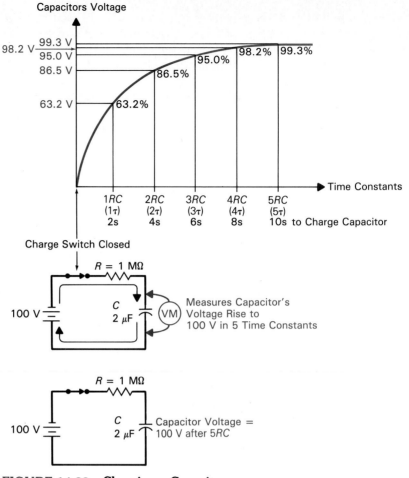

FIGURE 14-33 Charging a Capacitor.

$$\text{remaining voltage} = 100 \text{ V} - 86.5 \text{ V}$$
$$= 13.5 \text{ V}$$
$$63.2\% \text{ of } 13.5 \text{ V} = 8.532 \text{ V}$$

The capacitor will consequently have reached 63.2% (8.532 V) of the remaining voltage (13.5 V), which is equal to $86.5 + 8.532 = 95$ V, or 95% of the applied voltage.

FOURTH TIME CONSTANT   In $4RC$ seconds (8 s), the capacitor will have charged to 63.2% of the remaining voltage, reaching 98.2 V or 98.2% of the applied voltage.

FIFTH TIME CONSTANT   In $5RC$ seconds (10 s), the capacitor is considered to be fully charged since the capacitor will have reached 63.2% of the remaining voltage, 99.3 V or 99.3% of the applied voltage.

The voltage waveform produced by the capacitor acquiring a charge is known as an *exponential* waveform, and the voltage across the capacitor is said to rise exponentially.

Before the switch is closed and even at the instant the switch is closed, the capacitor is not charged, which means there is no capacitor voltage to

496   Capacitance and Capacitors

oppose the supply voltage and, therefore, a maximum current of V/R, 100 V/1 MΩ = 100 μA flows. This current begins to charge the capacitor, and a potential difference begins to build up across the plates of the capacitor, and this voltage opposes the supply voltage, causing a decrease in charging current. As the capacitor begins to charge, less of a potential difference exists between the supply voltage and capacitor voltage (across the resistor), and so the current begins to decrease.

To calculate the current at any time, we can use the formula

$$i = \frac{V_s - V_c}{R}$$

where i = instantaneous current
V_s = source voltage
V_c = capacitor voltage
R = resistance

For example, the current flowing in the circuit after 1 time constant will equal the source voltage, 100 V, minus the capacitor's voltage (voltage across R), which in 1 time constant will be 63.2% of the source voltage or 63.2 V, divided by the resistance.

$$
\begin{aligned}
i &= \frac{V_s - V_c}{R} \\
&= \frac{100 \text{ V} - 63.2 \text{ V}}{1 \text{ MΩ}} \\
&= 36.8 \text{ μA}
\end{aligned}
$$

As the charging continues, the potential difference across the plates builds up to the supply voltage and the current decays exponentially to zero in $5RC$ seconds (10 s), as seen in Figure 14-34.

The constant of 63.2 can be applied to the exponential fall of current from 100 μA to 0 in $5RC$ seconds.

When the switch was closed to commence charging of the capacitor, there was no charge on the capacitor; therefore, a maximum potential dif-

FIGURE 14-34   Charging a Capacitor.

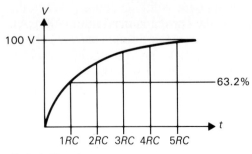

Switch Closed

Voltage across Capacitor
(Exponential Rise)

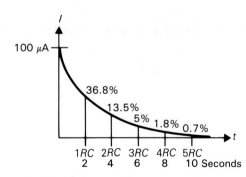

Switch Closed

Charge Current to Capacitor
(Exponential Fall)

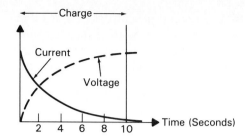

FIGURE 14-35 Voltage and Current Phase Difference.

ference existed between the battery and capacitor, causing a maximum current flow of 100 μA ($I = V/R$).

FIRST TIME CONSTANT   In 1RC seconds, the current will have exponentially decayed to 63.2% of its maximum (63.2% of 100 μA = 63.2 μA) to a value of 36.8 μA (100 μA − 63.2 μA). In the example of 2 μF and 1 MΩ, this occurs in 2 s.

SECOND TIME CONSTANT   In 2RC seconds (2 × R × C = 4 s), the current will decrease 63.2% of the remaining current, which is

$$63.2\% \text{ of } 36.8 \text{ μA} = 23.26 \text{ μA}$$

The current will drop 23.26 μA from 36.8 μA and reach 13.5 μA or 13.5%.

THIRD TIME CONSTANT   In 3RC seconds (6 s), the capacitor's charge current will decrease 63.2% of the remaining current (13.5 μA) to 5 μA or 5%.

FOURTH TIME CONSTANT   In 4RC seconds (8 s), the current will have decreased to 1.8 μA or 1.8%.

FIFTH TIME CONSTANT   In 5RC seconds (10 s), the charge current is now 0.7 μA or 0.7%. At this time, the charge current is assumed to be zero and the capacitor is now charged to a voltage equal to the applied voltage.

Studying the exponential rise of the voltage and the exponential decay of current in a capacitive circuit, you will notice an interesting relationship. In a pure resistive circuit, the current flow through a resistor would be in step with the voltage across that same resistor, in that an increased current would cause a corresponding increase in voltage drop across the resistor. Voltage and current are consequently said to be in step or in phase with one another.

With the capacitive circuit, the current flow in the circuit and voltage across the capacitor are not in step or in phase with one another. When the switch is closed to charge the capacitor, the current is maximum (100 μA), while the voltage across the capacitor is zero. After 5 time constants (10 s), the capacitor's voltage is now maximum (100 V) and the circuit current is zero, as seen in Figure 14-35. The circuit current flow is out of phase with the capacitor voltage, and this difference is referred to as a *phase shift*.

14.11.2 DC DISCHARGING A CAPACITOR

Figure 14-36 illustrates the circuit, voltage, and current waveforms that occur when a charged capacitor is discharged from 100 V to 0 V. The 2 μF capacitor,

498   **Capacitance and Capacitors**

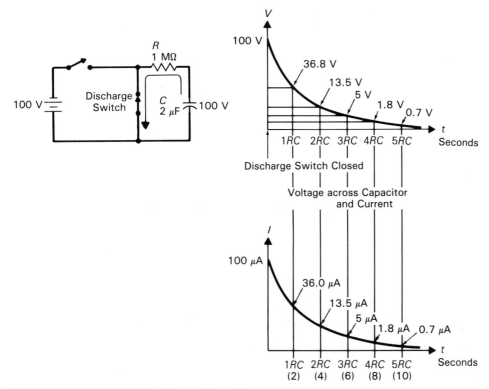

FIGURE 14-36 Discharging a Capacitor.

which was charged to 100 V in 10 s ($5RC$), is discharged from 100 to 0 V in the same amount of time.

Looking at the voltage curve, you can see that the voltage across the capacitor decreases exponentially, dropping 63.2% to 36.8 V in $1RC$ seconds, another 63.2% to 13.5 V in $2RC$ seconds, another 63.2% to 5 V in $3RC$ seconds, and so on, until zero.

The current flow within the circuit is dependent on the voltage in the circuit, which is across the 2 μF capacitor. As the voltage decreases, the current will also decrease by the same amount ($I\!\downarrow = V\!\downarrow/R$).

Discharge switch closed: $I = \dfrac{V}{R} = \dfrac{100\text{ V}}{1\text{ M}\Omega} = 100\ \mu A$ Max.

$1RC$ (2) seconds: $I = \dfrac{V}{R} = \dfrac{36.8\text{ V}}{1\text{ M}\Omega} = 36.8\ \mu A$

$2RC$ (4) seconds: $I = \dfrac{V}{R} = \dfrac{13.5\text{ V}}{1\text{ M}\Omega} = 13.5\ \mu A$

$3RC$ (6) seconds: $I = \dfrac{V}{R} = \dfrac{5\text{ V}}{1\text{ M}\Omega} = 5\ \mu A$

$4RC$ (8) seconds: $I = \dfrac{V}{R} = \dfrac{1.8\text{ V}}{1\text{ M}\Omega} = 1.8\ \mu A$

$5RC$ (10) seconds: $I = \dfrac{V}{R} = \dfrac{0.7\text{ V}}{1\text{ M}\Omega} = 0.7\ \mu A$ Zero

1. What is meant by the capacitor time constant?
2. In one time constant, a capacitor will have charged to what percentage of the applied voltage?
3. In one time constant, a capacitor will have discharged to what percentage of its full charge?
4. The charge or discharge of a capacitor follows a linear rate of change (true or false).

14.12
AC CHARGE AND DISCHARGE

Let's now return to our charged capacitor across a 100 V dc source, as seen in Figure 14-37(a). After a time equal to five time constants, the charge current will have charged the capacitor up to the supplied voltage of 100 V. After this, there will be no current flow because the capacitor's voltage is equal to the applied voltage.

If the 100 V source is reversed, as seen in Figure 14-37(b), the 100 V charge on the capacitor is now aiding the path of the battery instead of pushing or reacting against it. The discharge current will flow in the opposite direction to the charge current, until the capacitor voltage is equal to 0 V, at which time it will begin to charge in the reverse direction, as seen in Figure 14-37(c).

If the battery source is once more reversed, as seen in Figure 14-37(d), the charge current will discharge the capacitor to 0 V and then begin once more to charge it in the reverse direction.

The result of this procedure is that there is current flowing in the circuit at all times, except for the instant when the battery source polarity is reversed. The switching of the voltage source in this way is exactly the same as if we were applying an alternating voltage, as seen in Figure 14-37(f).

As we are already aware, current cannot flow through a capacitor due to the dielectric between the plates. However, when an alternating voltage is applied across a capacitor, the applied voltage reversal causes a charging and discharging of the capacitor, which makes it appear as though current is actually flowing through the capacitor. Current is not really flowing through the capacitor although it seems to be when an alternating voltage is applied. Any type of alternating current or in fact any fluctuating or changing dc current will appear to pass through the capacitor.

When a capacitor is conneced across a dc source, as seen in Figure 4-38(a), it charges and the voltage across the capacitor opposes the applied voltage and current is stopped; therefore, a capacitor creates an open circuit or block to dc after the charge current has ceased.

You can now understand why a capacitor is generally known as a short to ac, because the applied voltage is switching in polarity, and a block to

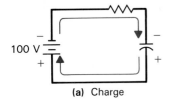

(a) Charge

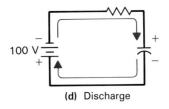

(b) Discharge

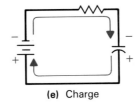

(c) Charge

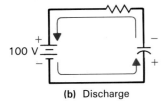

(d) Discharge

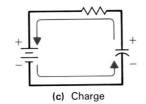

(e) Charge

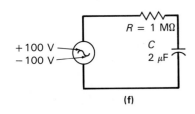

(f)

FIGURE 14-37   **AC Charge and Discharge.**

FIGURE 14-38   **(a) Capacitor Will Block DC. (b) Capacitor Will Pass AC.**

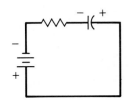

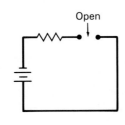

(a)

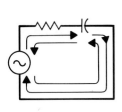

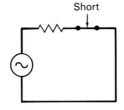

(b)

dc [Figure 14-38(b)]. This ability of a capacitor to block dc and pass ac is exploited and explained later in applications of capacitors.

SELF-TEST REVIEW (§ 14.12)

1. What differences occur when ac is applied to a capacitor rather than dc?
2. A capacitor's reaction to ac and dc accounts for why it is known as an ac short and dc block (true or false).

14.13
PHASE RELATIONSHIP BETWEEN AC CURRENT AND VOLTAGE

Thinking of the applied alternating voltage as a dc source that is being continually reversed sometimes makes it easier to understand how a capacitor reacts to ac. Whether the applied voltage is dc or ac, the rule holds true in that a 90° phase shift or difference exists between circuit current and capacitor voltage. The exact relationship between ac current and voltage in a capacitive circuit is illustrated in Figure 14-39.

This phase shift that exists between the circuit current and the capacitor voltage is normally expressed in degrees. At 0°, the capacitor is fully discharged (0 V), and the source is supplying a maximum circuit charge current. From 0° to 90°, the capacitor will charge toward a maximum positive value, and this increase in capacitor voltage will oppose the source voltage, whose circuit charging current will slowly decrease to 0 A. At 90°, the capacitor is fully charged, and the circuit current is at 0 A, and so the capacitor will return to the circuit the energy it consumed during the positive (+) charge cycle. This discharge circuit current is in the opposite direction to the positive

FIGURE 14-39

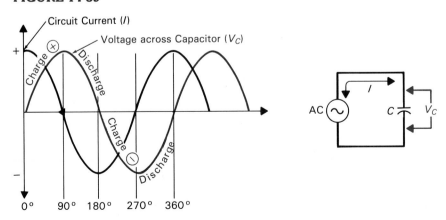

502   **Capacitance and Capacitors**

charge current. At 180°, the capacitor is fully discharged (0 V), and the source is supplying a maximum circuit charge current. From 180° to 270°, the capacitor will charge toward a maximum negative value, and this increase in capacitor voltage will oppose the source voltage, whose circuit charging current will slowly decrease to 0 A. At 270°, the capacitor is fully charged and the circuit current is at 0 A, and so the capacitor will discharge and return the energy it consumed during the negative ($-$) charge cycle. This discharge circuit current is in the opposite direction to the negative charge current.

Throughout this cycle, notice that the voltage across the capacitor follows the circuit current. This current, therefore, leads the voltage by 90°, and the 90° leading phase shift (current leading voltage) will only occur in a capacitive circuit.

SELF-TEST REVIEW (§ 14.13)

1. What is a voltage or current phase shift?
2. In a capacitive/resistive circuit, the current leads the voltage by exactly 90° (true or false).
3. In a purely resistive circuit, a 90° phase difference occurs between current and voltage (true or false).
4. Phase differences between voltage and current only occur when ac is applied (true or false).

SUMMARY

1. Capacitance is the ability of a circuit or device to store electrical charge, and the device or component specifically designed to have this capacity or capacitance is called a capacitor.
2. A capacitor has two or more parallel conductive plates that are separated by an insulating material known as a dielectric.
3. When a capacitor charges, the voltage across its plates exponentially rises to the supply voltage, while the charging current exponentially falls to zero.
4. When a capacitor discharges, the voltage across its plates and circuit current both exponentially fall to zero.
5. Just as a magnetic field is generated by current flow, an electric field is generated by a potential difference or voltage.
6. The energy of a capacitor is stored in the electric or electrostatic field within the dielectric.
7. A capacitor with the capacity of 1 farad (1 F) can store one coulomb of electrical charge when a voltage of 1 volt is applied across its plates.

8. The three factors determining the capacitance of a capacitor are:
   a. Plate area
   b. Distance between the plates
   c. Type of dielectric

9. When capacitors are connected in parallel, the total capacitance is equal to the sum of all the individual capacitances.

10. When capacitors are connected in series, the total capacitance is less than the value of the smallest value capacitor.

11. A fixed capacitor is a capacitor whose value remains constant and cannot be altered, while a variable capacitor is a capacitor whose value can be changed.

12. The fixed ceramic capacitor is the most popular small value capacitor, and the electrolytic is the most popular large value capacitor.

13. The electrolytic capacitor is polarized.

14. To increase the surface plate area and therefore capacitance value, a process known as double etch has been used.

15. The 1 F capacitor uses a relatively new double layer technique employing activated charcoal as the solid plates.

16. The capacitor value, tolerance, and voltage rating are generally indicated on the body of the capacitor by use of alphanumerics or color codes.

17. A capacitor will fully charge to the supply voltage and fully discharge to zero in 5 time constants, each time constant being equal to the product of capacitance and resistances.

18. A capacitor acts as an open or block to direct current (dc).

19. A capacitor will act as a short and pass alternating current (ac).

20. The voltage across a capacitor will lag the circuit current by 90°, or the circuit current will lead the capacitor voltage by 90°.

NEW TERMS

Activated charcoal	**Dielectric strength**	**Instantaneous current**
Adjustable capacitor	**Dielectric thickness**	**Mica capacitor**
Air capacitor	**Discharging**	**Microfarad**
Alignment	**Double etch**	**Paper capacitor**
Capacitance	**Electric field**	**Picofarad**
Capacitor	**Electric polarization**	**Plastic capacitor**
Capacitor breakdown	**Electrolytic capacitor**	**Plate area**
Capacitor leakage	**Electrostatic field**	**Plates**
Capacity	**Electrostatics**	**Static charge**
Ceramic capacitor	**Exponential fall**	**Tantalum capacitor**
Charge difference	**Exponential rise**	**Time constant**
Charging	**Farad**	**Total capacitance**
Condensers	**Fixed value capacitor**	**Trimmer capacitor**
Dielectric	**Ganged capacitor**	**Variable capacitor**
Dielectric constant	**Inductance**	**Volts per meter**

Capacitance and Capacitors

$$\text{Field strength (V/m)} = \frac{\text{charge difference (volts)}}{\text{distance between plates } (d) \text{ in meters}}$$

(V/m = volts per meter)

$$Q = C \times V$$

$$C = \frac{Q}{V}$$

$$V = \frac{Q}{C}$$

Q = change in coulombs
C = capacitance in farads
V = voltage in volts

$$C = \frac{(8.85 \times 10^{-12}) \times K \times A}{d}$$

C = capacitance in farads (F)
K = Dielectric constant
8.85×10^{-12} is a constant
d = distance between the plates in meters (m)

Capacitors in parallel:

$$C_t = C_1 + C_2 + C_3 + C_4 + \cdots$$

C_t = total capacitance in farads
C_1 = capacitance of C_1 in farads
C_2 = capacitance of C_2 in farads
C_3 = and so on

Capacitors in series:

Two capacitors

$$C_t = \frac{C_1 \times C_2}{C_1 + C_2}$$

More than two capacitors

$$C_t = \frac{1}{(1/C_1) + (1/C_2) + (1/C_3) + \cdots}$$

Voltage divider formula

$$V_{cx} = \frac{C_t}{C_x} \times V_t$$

V_{cx} = desired voltage
C_x = selected capacitor value
C_t = total capacitance
V_t = supplied voltage

$$t = R \times C$$

t = time constant in seconds
R = resistance in ohms
C = capacitance in farads

$$i = \frac{V_s - V_c}{R}$$

i = instantaneous current in amps
V_s = source voltage in volts
V_c = capacitor voltage in volts
R = resistance in ohms

REVIEW QUESTIONS

Multiple Choice Questions

1. Capacitors were originally referred to as:
   a. Vacuum tubes
   b. Condensers
   c. Inductors
   d. Suppressors

2. When a capacitor charges:
   a. The voltage across the plates rises exponentially
   b. The circuit current falls exponentially
   c. The capacitor charges to the source voltage in $5RC$ seconds
   d. All the above are true

3. A _____ field is generated by the flow of current and an _____ field is generated by voltage.
   a. Magnetic, electrostatic
   b. Electric, electrostatic
   c. Electric, magnetic
   d. All the above may be considered true

4. The strength of an electric field in a capacitor is proportional to the _____ and inversely proportional to the _____.
   a. Plate separation, charge
   b. Plate separation, potential plate difference
   c. Plate potential difference, plate separation
   d. Both (a) and (b) are true

5. The plates of a capacitor are generally made out of a
   a. Resistive material
   b. Semiconductor material

6. The energy of a capacitor is stored in:
   a. The magnetic field within the dielectric
   b. The magnetic field around the capacitor leads
   c. The electric field within the plates
   d. The electric field within the dielectric

7. What is the capacitance of a capacitor if it can store 24 coulombs of charge when 6 volts is applied across the plates?
   a. 2 μF
   b. 3 μF
   c. 4.7 μF
   d. None of the above

8. The capacitance of a capacitor is directly proportional to:
   a. The plate area
   b. The distance between the plates
   c. The constant of the dielectric used
   d. Both (a) and (c) are true
   e. Both (a) and (b) are true

9. The capacitance of a capacitor is inversely proportional to:
   a. The plate area
   b. The distance between the plates
   c. The dielectric used

c. Conductive material
d. Two of the above could be true

d. Both (a) and (c) are true

e. Both (a) and (b) are true

10. The breakdown voltage of a capacitor is determined by:

    a. The type of dielectric used

    b. The size of the capacitor plates

    c. The wire gauge of the connecting leads

    d. Both (a) and (b) are true

11. A capacitor should have a _____ leakage resistance to ensure a _____ leakage current.

    a. Large, small

    b. Large, medium

    c. Small, large

    d. Small, small

12. Total series capacitance of two capacitors is calculated by:

    a. Using the product over sum formula

    b. Using the voltage divider formula

    c. Using the series resistance formula on the capacitors

    d. Adding all the individual values

13. Total parallel capacitance is calculated by:

    a. Using the product over sum formula

    b. Using the voltage divider formula

    c. Using the parallel resistance formula on capacitors

    d. Adding all the individual values

14. The mica and ceramic fixed capacitors have:

    a. An arrangement of stacked plates

    b. A electrolyte substance between the plates

    c. An adjustable range

    d. All the above

15. An electrolytic capacitor:

    a. Is the most popular large-value capacitor

    b. Is polarized

    c. Can have either aluminum or tantalum plates

    d. All the above are true

16. Variable capacitors normally achieve a large variation in capacitance by varying _____, while adjustable trimmer capacitors only achieve a small capacitance range by varying _____.

    a. Dielectric constant, plate area

    b. Plate area, plate separation

    c. Plate separation, dielectric constant

    d. Plate separation, plate area

17. In one time constant, a capacitor will charge to _____ of the source voltage.

    a. 86.5%

    b. 63.2%

    c. 99.3%

    d. 98.2%

18. When ac is applied across a capacitor, a _____ phase shift exists between circuit current and capacitor voltage.

    a. 45°

    b. 60°

    c. 90°

    d. 63.2°

19. A capacitor consists of:

    a. Two insulated plates separated by a conductor

    b. Two conductive plates separated by a conductor

    c. Two conductive plates separated by an insulator

    d. Two conductive plates separated by conductive spacer

20. A 47 μF capacitor charged
to 6.3 V will have a stored
charge of:
a. 296.1 μC
b. 2.96×10^{-4} C

c. 0.296 mC
d. All the above are true
e. Both (a) and (b) are true

Essay Questions

21. Describe the construction and main parts of a capacitor. (14.1)
22. What happens to a capacitor during:
    a. Charge (14.2.1)
    b. Discharge (14.2.2)
23. Describe how electrostatics relates to capacitance and give the formula for electric field strength. (14.3)
24. Briefly explain the relationship between capacitance, charge, and voltage. (14.4)
25. Describe the three factors affecting the capacitance of a capacitor. (14.5)
26. Briefly describe:
    a. The term capacitor breakdown (14.6)
    b. The term capacitor leakage (14.7)
    c. Why the smallest value capacitor in a two capacitor series circuit has the largest voltage developed across it (14.8.2)
27. List the formula(s) used to calculate total capacitance when capacitors are connected in:
    a. Series (14.8.2)
    b. Parallel (14.8.1)
28. Describe the following types of capacitors: (14.9)
    Fixed
    a. Mica
    b. Ceramic
    c. Paper
    d. Plastic
    e. Electrolytic
    Variable
    a. Air
    b. Mica, ceramic, and plastic
29. Briefly explain the technique and materials used to create the 1 farad capacitor. (14.9.3)
30. Explain how the constant 63.2 is used in relation to the charge and discharge of a capacitor. (14.11)
31. How is the coding of a capacitor's value, tolerance, and voltage rating indicated on a capacitor? (14.10)
32. Describe the differences between dc and ac capacitor charging and discharging. (14.11, 14.12)
33. Would a mica dielectric concentrate on electrostatic field more or less than air; if so, by how much? (14.5)

Capacitance and Capacitors

34. Calculate the total capacitance of the circuits illustrated in Figure 14-40, and describe which capacitors are in parallel and which are in series. (14.8)

35. What advantages and disadvantages do tantalum electrolytic capacitors have over the aluminum type? (14.9.1)

36. Which of the fixed value capacitor types is polarity sensitive, and what does this mean? (14.9.1)

37. What is a ganged capacitor and what is its application? (14.9.2)

38. Describe one farad of capacitance. (14.4)

39. Briefly describe why series connected capacitors are treated like parallel-connected resistors, and why parallel connected capacitors are treated like series connected resistors when calculating total capacitance. (14.8)

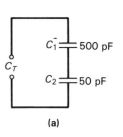

(a)

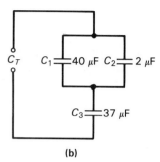

(b)

FIGURE 14-40

Practice Problems

40. If a 10 μF capacitor is charged to 10 V, how many coulombs of charge has it stored?

41. Calculate the electric field strength within the dielectric of a capacitor that is charged to 6 volts and the dielectric thickness is 32 micrometers (32-millionths of a meter).

42. If a 0.006-μF capacitor has stored 125×10^{-6} coulombs of charge, what potential difference would appear across the plates?

43. Calculate the capacitance of the capacitor that has the following parameter values: $A = 0.008$ m², $d = 0.00095$ m, the dielectric used is paper.

44. Calculate the total capacitance if the following are connected in:
    a. Parallel: 1.7 μF, 2.6 μF, 0.03 μF, 1200 pF
    b. Series: 1.6 μF, 1.4μF, 4 μF

45. If three capacitors of 0.025 μF, 0.04 μF, and 0.037 μF are connected in series across a 12 V source, what would be the voltage drop across each?

46. Give the value of the following alphanumeric capacitor value codes.
    a. 104   c. 0.01
    b. 125   d. 220

47. Give the value of the following color coded capacitor values.
    a. Yellow, violet, black
    b. Orange, yellow, violet

48. What would be the time constant of the following RC circuits:
    a. $R = 6$ kΩ, $C = 14$ μF
    b. $R = 12$ MΩ, $C = 1400$ pF
    c. $R = 170$ Ω, $C = 24$ μF
    d. $R = 140$ kΩ, $C = 0.007$ μF

49. If 10 V were applied across all the RC circuits in question 48, what would be the voltage across each capacitor after one time constant, and how much time would it take each capacitor to fully charge?

50. In one application, a capacitor is needed to store 25 μC of charge and will always have 125 V applied across its terminals. Calculate the capacitance value needed.

15

CAPACITIVE REACTANCE AND CAPACITOR APPLICATIONS

AFTER COMPLETING THIS CHAPTER, YOU WILL BE ABLE TO:

1. Define and explain capacitive reactance.
2. Describe the following terms as they relate to a series and parallel *RC* circuit:
   a. Impedance
   b. Phase angle
   c. Power and power factor
3. Explain some of the more common capacitor failures and how to use an ohmmeter and capacitance analyzer to test them.
4. Describe how the capacitor and resistor in combination can be used:
   a. To combine ac and dc.
   b. As an integrator.
   c. As a differentiator.
   d. As a filter.

INTRODUCTION

An RC circuit will react to current flow in a unique way. In this chapter, reactance and the characteristics of a series RC and parallel RC circuit will be examined, along with applications and capacitor testing.

15.1

CAPACITIVE REACTANCE

Resistance (R), by definition, is the opposition to current flow with the dissipation of energy and is measured in ohms. Capacitors oppose current flow like a resistor; however, a resistor dissipates energy, whereas a capacitor stores energy (charge) and then gives back its energy into the circuit (discharge). Because of this difference, a new term had to be used to describe the opposition offered by a capacitor. Capacitive reactance (X_C), by definition, is the opposition to current flow without the dissipation of energy and is also measured in ohms.

If capacitive reactance is basically opposition, then it is inversely proportional to the amount of current flow. If a large current is flowing within a circuit, then the opposition or reactance must be low. Conversely, a small current will be the result of a large opposition or reactance.

When a dc source is connected across a capacitor, current will flow only for a short period of time ($5RC$ seconds) to charge the capacitor. There is no further current flow. Consequently, the capacitive reactance to dc is infinite (maximum).

Alternating current is continuously reversing in polarity, resulting in the capacitor continuously charging and discharging. This means that charge and discharge currents are always flowing around the circuit. If we have an average value of current, then we must also have an average value of reactance or opposition.

Initially, when the capacitor's plates are uncharged, they will not oppose or react against the charging current and, therefore, maximum current will flow ($I\uparrow$) and the reactance will be very low ($X_C\downarrow$). As the capacitor charges, it will oppose or react against the charge current, which will decrease ($I\downarrow$), and so the reactance will increase ($X_C\uparrow$). The discharge current is also highest at the start of discharge ($I\uparrow$, $X_C\downarrow$) as the voltage of the charged capacitor is also high; but as the capacitor discharges, its voltage decreases and the discharge current will also decrease ($I\downarrow$, $X_C\uparrow$).

To summarize, at the start of a capacitor charge or discharge, the current is maximum and so the reactance is low. This value of current then begins to fall to zero and so the reactance increases.

If the applied alternating current is at a high frequency, as seen in Figure 15-1(a), it is switching polarity more rapidly than a lower frequency and there is very little time between the start of charge and discharge. As the charge and discharge currents are largest at the beginning of the charge and discharge of the capacitor, the reactance has very little time to build up and oppose the current, which is why the average current is a high value and the capacitive reactance is small at higher frequencies. With lower frequencies, as seen in Figure 15-1(b), the applied alternating current is switching at a slower rate and, therefore, the reactance, which is low at the beginning, has more time to build up and oppose the current.

Capacitive reactance is therefore inversely proportional to frequency:

$$\text{capacitive reactance } (X_C) \; \alpha \; \frac{1}{f\,(\text{frequency})}$$

Frequency, however, is not the only factor that determines capacitive reactance. Capacitive reactance is also inversely proportional to the value of capacitance. If a larger capacitor value is used a longer time is required to charge the capacitor ($t\uparrow = C\uparrow R$), which means that current will be flowing for a longer period of time and so the average current will be large ($I\uparrow$); consequently, the reactance must be small ($X_C\downarrow$). On the other hand, a small capacitance value will charge in a small amount of time ($t\downarrow = C\downarrow R$) and

FIGURE 15-1 Capacitive Reactance Is Inversely Proportional to Frequency.

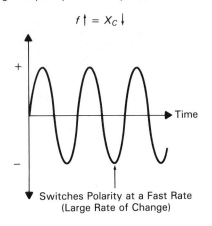

High Frequency = Low Capacitive Reactance

$f\uparrow = X_C\downarrow$

(a)

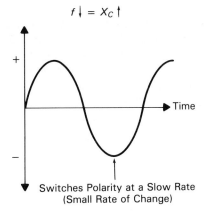

Low Frequency = High Reactance

$f\downarrow = X_C\uparrow$

(b)

the current is present for only a short period of time. The average current will therefore be small ($I\downarrow$), indicating a large reactance ($X_C\uparrow$).

Capacitive reactance (X_C) is inversely proportional to both frequency and capacitance and can be calculated by using the formula

$$X_C = \frac{1}{2\pi fC}$$

where X_C = capacitive reactance in ohms
f = frequency in hertz
C = capacitance in farads
2π = constant

SELF-TEST REVIEW (§ 15.1)

1. Define capacitive reactance.
2. State the formula for capacitive reactance.
3. Why is capacitive reactance inversely proportional to frequency and capacitance?
4. If $C = 4\ \mu F$, $F = 4$ kHz, calculate X_C?

EXAMPLE 15.1

Calculate the reactance of a 2 μF capacitor when a 10 kHz sine wave is applied.

SOLUTION

$$X_C = \frac{1}{2\pi fC}$$

$$= \frac{1}{2\times\pi\times 10\text{ kHz}\times 2\ \mu F} = 8\ \Omega$$

Calculator Sequence

Step	Keypad Entry	Display Response
1.	[2]	2.0
2.	[×]	
3.	[π]	3.1415927
4.	[×]	6.283185
5.	[1] [0] [EE] [3]	10E3
6.	[×]	6.2831.8
7.	[2] [EE] [6] [+/−]	2.−06
8.	[=]	0.1256637
9.	[1/x]	7.9577

Capacitive Reactance and Capacitor Applications

15.2
SERIES *RC* CIRCUIT

In a purely resistive circuit, as seen in Figure 15-2(a), the current flowing within the circuit and the voltage across the resistor are in phase with one another. In a purely capacitive circuit, as seen in Figure 15-2(b), the current flowing in the circuit leads the voltage across the capacitor by 90°.

   purely resistive: 0° phase shift
   purely capacitive: 90° phase shift

    If we connect a resistor and capacitor in series, as seen in Figure 15-3(a), we have probably the most commonly used electronic circuit, which has many applications. The voltage across the resistor (V_R) is always in phase with the circuit current (I), as can be seen in Figure 15-3(b), because maximum points and zero crossover points occur at the same time. The voltage across the capacitor (V_C) lags the circuit current by 90°.

    Since the capacitor and resistor are in series, the same current is supplied to both components, Kirchhoff's voltage law can be applied, which states that the sum of the voltage drops around a series circuit is equal to the voltage applied (V_S). By summing the series voltages, $V_C + V_R$, we end up

FIGURE 15-2 **Phase Relationships between *V* and *I*. (a) Resistive Circuit: Current and Voltage Are in Phase. (b) Capacitive Circuit: Current Leads Voltage by 90°.**

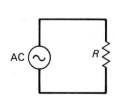
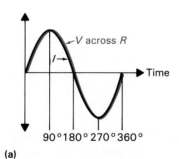

(a)

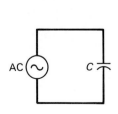
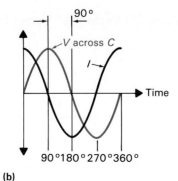

(b)

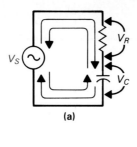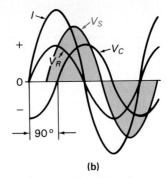

(a)

(b)

FIGURE 15-3 *RC* Series Circuit. (a) Circuit. (b) Waveforms.

with a resultant wave that represents the voltage applied (V_S); this is represented in Figure 15-3(b) by the shaded waveform. As expected, the circuit current leads the applied voltage, which is correct for a circuit with capacitance. However, with resistance and capacitance in the circuit, the phase shift is somewhere between 0° (resistive) and 90° (capacitive). In the example, the resistance of the resistor is equal to the reactance of the capacitor. Since $R = X_C$, the phase difference between circuit current and applied voltage is 45°. If the resistance (R) is larger than the reactance (X_C), the circuit will be mainly resistive and the phase shift will be between 0° and 45°, whereas, on the other hand, if the reactance (X_C) is larger than the resistance, the circuit will be mainly capacitive, and the phase shift will be between 45° and 90°.

15.2.1 VECTOR DIAGRAM

A vector (or phasor) is a quantity that has both magnitude and direction and is represented by a line terminated at the end by an arrowhead, as seen in Figure 15-4(a). A vector diagram is an arrangement of vectors to illustrate the magnitude and phase relationships between two or more quantities of the same frequency within an ac circuit. Figure 15-4(b) illustrates the basic parts of a vector diagram. The current (I) vector is at the 0° position and the size of the arrow represents the peak value of alternating current.

In Figure 15-5(a), (b), and (c), you can see the *RC* circuit, waveforms,

FIGURE 15-4 Vectors. (a) Vector. (b) Vector Diagram.

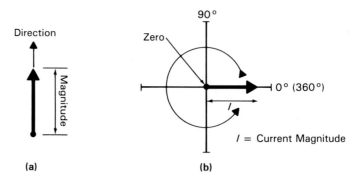

(a)

(b)

Capacitive Reactance and Capacitor Applications

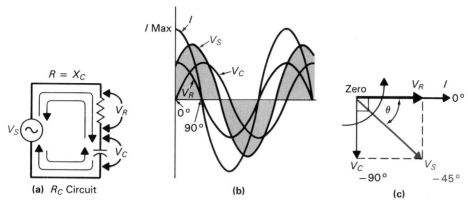

FIGURE 15-5 *RC* Series Circuit Analysis.

and vector diagram, respectively. In Figure 15-5(b), the current peak flowing in the series *RC* circuit occurs at 0° and will be used as a reference; therefore, the vector of current in Figure 15-5(c) is to the right in the 0° position. The voltage across the resistor (V_R) is in phase or coincident with the current (I), as seen in Figure 15-5(b), and the vector that represents the voltage across the resistor (V_R) overlaps or coincides with the I vector, at 0°.

The voltage across the capacitor (V_C) is, as seen in Figure 15-5(b), 90° out of phase (lagging) with the circuit's current, and so the V_C vector in Figure 15-5(c) is drawn at −90° (minus sign indicates lag) to the current vector, and the length of this vector represents the magnitude of the voltage across the capacitor. Since the ohmic value of the resistor (R) and the capacitor (X_C) are equal, the voltage drop across both components is the same. The V_R and V_C vectors are subsequently equal in length.

The source voltage (V_S) is, by Kirchhoff's voltage law, equal to the sum of the series voltage drops (V_C and V_R); however, since these voltages are not in phase with one another, we cannot simply add the two together (using vector addition). The source voltage (V_S) will be the sum of both V_C and V_R at a particular time. Studying the waveforms in Figure 15-5(b), you will notice that peak source voltage will occur at 45°. By vectorially adding the two voltages V_R and V_C in Figure 15-5(c), we obtain a resultant V_S vector that has both magnitude and phase. The angle theta (θ) formed between circuit current (I) and source voltage (V_S) will always be less than 90°, and in this example is equal to −45° because the voltage drops across R and C are equal due to R and X_C being of the same ohmic value.

If V_C and V_R are drawn to scale, then the peak source voltage (V_S) can be calculated by using the same scale; however, a mathematical rather than graphical method can be used to save the drafting time.

In Figure 15-6(a), we have taken the three voltages (V_R, V_C, and V_S) and formed a right angle triangle, as shown in Figure 15-6(b). Using the Pythagorean theorem for right angle triangles, which states that, if you take the square of a (V_R) and add it to the square of b (V_C), the square root of the result will equal c (the source voltage, V_S).

$$V_S = \sqrt{V_R^2 + V_C^2}$$

Series *RC* Circuit　　**517**

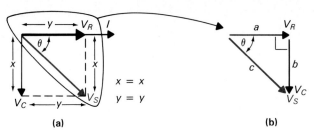

FIGURE 15-6

If two variables are known, then the other can be easily calculated by transposing the formula:

$$V_S = \sqrt{V_R^2 + V_C^2}$$

$$V_C = \sqrt{V_S^2 - V_R^2}$$

$$V_R = \sqrt{V_S^2 - V_C^2}$$

EXAMPLE 15.2

Calculate the source voltage applied across an *RC* series circuit if $V_R = 12$ V and $V_C = 8$ V.

SOLUTION

$$V_S = \sqrt{V_R^2 + V_C^2}$$
$$= \sqrt{12 \text{ V}^2 + 8 \text{ V}^2}$$
$$= \sqrt{144 + 64}$$
$$= 14.42 \text{ V}$$

Calculator Sequence

Step	Keypad Entry	Display Response
1.	① ②	12.0
2.	$\boxed{x^2}$	144.0
3.	$\boxed{+}$	
4.	⑧	8.0
5.	$\boxed{x^2}$	64.0
7.	$\boxed{=}$	208.0
8.	$\boxed{\sqrt{x}}$	14.42220

15.2.2 IMPEDANCE

Impedance (Z) is also measured in ohms and is the total circuit opposition to current flow. It is a combination of resistance (R) and reactance (X_C); however, in our capacitive and resistive circuit, a phase shift or difference exists, and just as V_C and V_R cannot be merely added together to obtain V_S, R and X_C cannot be simply summed to obtain Z.

If the current within a series circuit is constant (the same throughout the circuit), the resistance of a resistor (R) or reactance of a capacitor (X_C) will be directly proportional to the voltage across the resistor (V_R) or the capacitor (V_C).

$$V_R\!\updownarrow = I \times R\updownarrow, \qquad V_C\!\updownarrow = I \times X_C\!\updownarrow$$

A vector diagram can be drawn similarly to the voltage vector diagram to illustrate opposition, as seen in Figure 15-7(a). The current is used as a

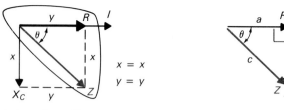

(a) Impedence Vector Diagram   (b)   **FIGURE 15-7**

reference (0°); the resistance vector (R) is in phase with the current vector (I), since V_R is always in phase with I. The capacitive reactance (X_C) vector is at $-90°$ to the resistance vector, due to the 90° phase shift between a resistor and capacitor. The lengths of the resistance vector (R) and capacitive reactance vector (X_C) are equal in this example. By vectorially adding R and X_C, we have a resulting impedance (Z) vector.

By using the three variables, which have again formed a right angle triangle [Figure 15-7(b)], we can apply the Pythagorean theorem to calculate the total opposition to current flow (Z), taking into account both R and X_C.

$$\boxed{Z = \sqrt{R^2 + X_C^2}}$$

$$R = \sqrt{Z^2 - X_C^2}$$

$$X_C = \sqrt{Z^2 - R^2}$$

EXAMPLE 15.3

Calculate the total impedance of a series RC circuit if $R = 27\ \Omega$, $C = 0.005\ \mu F$, and the source frequency $= 1$ kHz.

SOLUTION

The total opposition (Z) or impedance is equal to

$$Z = \sqrt{R^2 + X_C^2}$$

R is known, but X_C will need to be calculated.

$$X_C = \frac{1}{2\pi f C}$$

$$= \frac{1}{2 \times \pi \times 1\ \text{kHz} \times 0.005\ \mu F}$$

$$= 31.8\ \text{k}\Omega$$

Since $R = 27\ \Omega$ and $X_C = 31.8\ \text{k}\Omega$, then

$$Z = \sqrt{R^2 + X_C^2}$$

$$= \sqrt{27\ \Omega^2 + 31.8\ \text{k}\Omega^2}$$

$$= \sqrt{729 + 1 \times 10^9}$$

$$= 31.8\ \text{k}\Omega$$

Once impedance (total opposition) has been calculated, Ohm's law can be used to find total current.

$$I = \frac{V}{Z}, \quad Z = \frac{V}{I}, \quad V = I \times Z$$

By transposition, we can arrive at the usual combinations of Ohm's law.

15.2.3 PHASE ANGLE OR SHIFT (θ)

Referring back to the impedance vector diagram, you can see that we have used a simple example when R has always equaled X_C, and so the phase shift (θ) has always been $-45°$. If the resistance and reactance are different from one another, the phase shift (θ) will change, as seen in Figure 15-8. The phase shift, as can be seen, is dependent on the ratio of capacitive reactance to resistance (X_C/R).

In a purely resistive circuit, the total opposition (Z) is equal to the resistance of the resistor and so the phase shift (θ) is equal to 0° [Figure 15-9(a)].

In a purely capacitive circuit, the total opposition (Z) is equal to the capacitive reactance (X_C) of the capacitor, and so the phase shift (θ) is equal to $-90°$ [Figure 15-9(b)].

By the use of trigonometry (the science of triangles), we can derive a

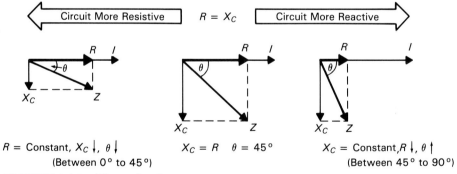

R = Constant, $X_C \downarrow$, $\theta \downarrow$ $X_C = R$ $\theta = 45°$ X_C = Constant, $R \downarrow$, $\theta \uparrow$
(Between 0° to 45°)   (Between 45° to 90°)

FIGURE 15-8   **Phase Angle.**

FIGURE 15-9   **(a) Purely Resistive Circuit. (b) Purely Capacitive Circuit.**

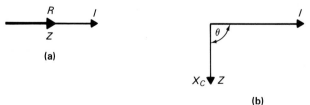

EXAMPLE 15.4

Calculate the phase shift or angle in two different
series RC circuits if:

circuit a. $V_R = 12$ V, $V_C = 8$ V

circuit b. $R = 27$ Ω, $X_C = 31.8$ kΩ

SOLUTION

$$\text{a. } \theta = \arctan \frac{V_C}{V_R}$$

$$= \arctan \frac{8 \text{ V}}{12 \text{ V}}$$

$$= 33.7°$$

$$\text{b. } \theta = \arctan \frac{X_C}{R}$$

$$= \arctan \frac{31.8 \text{ k}\Omega}{27 \text{ }\Omega}$$

$$= 89.95°$$

---

formula to calculate the degree of phase shift, since two quantities X_C and
R are known.

The phase angle, θ, is equal to

$$\boxed{\theta = \arctan \frac{X_C}{R}}$$

Since X_C/R is equal to V_C/V_R, the phase angle can also be calculated if V_R
and V_C are known.

$$\boxed{\theta = \arctan \frac{V_C}{V_R}}$$

This formula will determine by what angle V_R leads V_S.

(1) PURELY RESISTIVE CIRCUIT

In Figure 15-10, you can see the current, voltage, and power waveforms
generated by applying an ac voltage across a purely resistive circuit. The
applied voltage causes current to flow around the circuit, and the electrical
energy is converted into heat energy. This heat or power is dissipated and
lost and can be calculated by using the power formula.

$$p = V \times I$$

$$p = I^2 \times R$$

$$p = \frac{V^2}{R}$$

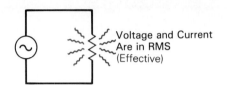

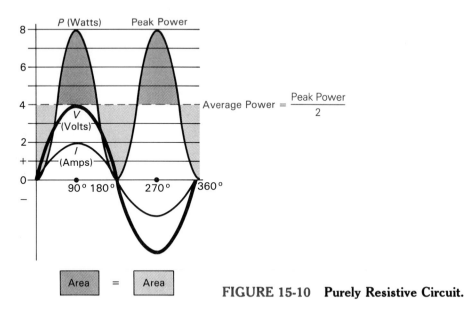

FIGURE 15-10 **Purely Resistive Circuit.**

Voltage and current are in phase with one another in a resistive circuit, and instantaneous power is calculated by multiplying voltage by current at every instant through 360° ($p = V \times I$).

The sinusoidal power waveform is totally positive, because a positive voltage multiplied by a positive current gives a positive value of power, and a negative voltage multiplied by a negative current will also produce a positive value of power. For these reasons, a resistor is said to generate a positive power waveform, which you may have noticed is twice the frequency of the voltage and current waveforms; two power cycles occur in the same time as one voltage and current cycle.

The power waveform has been split in half, and this line that exists between the maximum (8 W) and zero (0 W) points is the average value of power (4 W) that is being dissipated by the resistor.

(2) PURELY CAPACITIVE CIRCUIT

In Figure 15-11, you can see the current, voltage, and power waveforms generated by applying an ac voltage source across a purely capacitive circuit. As expected, the current leads the voltage by 90°, and the power wave is calculated by multiplying voltage by current, as before, at every instant through 360°. The resulting power curve is both positive and negative. During the positive alternation of the power curve, the capacitor is taking power as the capacitor charges. When the power alternation is negative, the capacitor is giving back the power it took as it discharges back into the circuit.

The average power dissipated is once again the value that exists between

522   Capacitive Reactance and Capacitor Applications

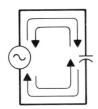

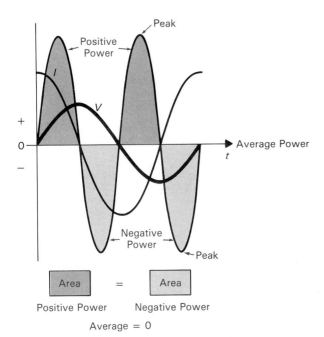

FIGURE 15-11 **Purely Capacitive Circuit.**

the maximum positive and maximum negative points and causes the area above this line to equal the area below. This average power level calculates out to be zero, which means that no power is dissipated in a purely capacitive circuit.

(3) RESISTIVE AND CAPACITIVE CIRCUIT

In Figure 15-12, you can see the current, voltage, and power waveforms generated by applying an ac voltage source across a series-connected *RC* circuit. The current leads the voltage by some phase angle less than 90°, and the power waveform is once again determined by the product of voltage and current. The negative alternation of the power cycle indicates that the capacitor is discharging and giving back the power that it consumed during the charge.

The positive alternation of the power cycle is much larger than the negative alternation because it is the combination of both the capacitor taking power during charge and the resistor consuming and dissipating power in the form of heat. The average power being dissipated by the resistor is now some positive value.

In conclusion, a resistive and capacitive circuit has a positive value of average power due to the heat being generated by the resistor.

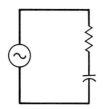

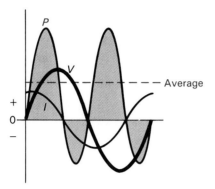

FIGURE 15-12 **Resistive and Capacitive Circuit.**

15.2.4 POWER FACTOR

In a purely resistive circuit, all the energy supplied to the resistor from the source is dissipated in the form of heat. This form of power is referred to as *resistive* power or *true* power.

In a purely capacitive circuit, all the energy supplied to the capacitor is stored from the source and then returned to the source, without energy loss. This form of power is referred to as *reactive* power or *imaginary* power.

When a circuit contains both capacitance and resistance, some of the supply is stored and returned by the capacitor and some of the energy is dissipated and lost by the resistor.

Figure 15-13(a) illustrates another vector diagram. Just as the voltage across a resistor is 90° out of phase with the voltage across a capacitor, and resistance is 90° out of phase with reactance, resistive power will be 90° out of phase with reactive power.

If we take the three variables from Figure 15-13(b) to form a right angle triangle as in Figure 15-13(c), we can vectorially add true power and reactive power to produce a resultant *apparent* power vector. Apparent power is the power that appears to be supplied to the load and includes both the true power dissipated by the resistance and the imaginary power delivered to the capacitor.

Applying the Pythagorean theorem, we can calculate apparent power by

$$P_A = \sqrt{P_R^2 + P_X^2}$$

P_A = apparent power

P_R = true power

P_X = reactive power

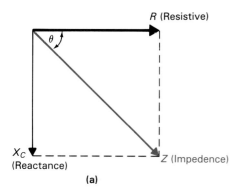

(a)

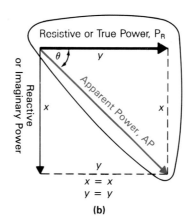

x = x
y = y

(b)

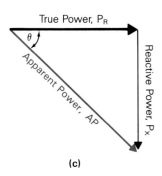

(c)

FIGURE 15-13   **Apparent Power.**

The power is a ratio of true power to apparent power and is therefore a measure of the loss in a circuit. It can be calculated by using the formula

$$PF = \frac{\text{true power (TP)}}{\text{apparent power (AP)}}$$

Figure 15-14 illustrates how the power factor can be anywhere between zero and one. With circuits that contain resistance and capacitance, the

FIGURE 15-14   **Resistive and Reactive Power Factor Comparison.**

Purely Resistive

In a resistive circuit the reactive power is zero, and therefore the true power = apparent power.

Purely Reactive

In a reactive circuit the resistive power (TP) is zero.

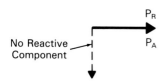

The power factor is therefore equal to:

$$PF = \frac{P_R}{P_A} = 1 \; \text{(maximum value)}$$

The power factor is therefore equal to:

$$PF = \frac{P_R}{P_A} = 0 \; \text{(minimum value)}$$

Series *RC* Circuit   **525**

power factor will be somewhere between zero (0 = reactive) and one (1 = resistive).

The power factor is a ratio of true power to apparent power. Since true power is determined by resistance and apparent power is dependent on impedance, as seen in Figure 15-13(a), the power factor can also be calculated by using the formula

$$PF = \frac{R}{Z}$$

As the ratio of true power to apparent power determines the angle θ, the power factor can also be determined by the angle. By the use of trigonometry, we can derive the formula

$$PF = \cos\theta$$

SELF-TEST REVIEW (§ 15.2)

1. What is the phase relationship between current and voltage in a series RC circuit?
2. What is a phasor diagram?
3. Define and state the formula for impedance.
4. What is the phase angle or shift in:
   a. A purely resistive circuit?
   b. A purely capacitive circuit?
   c. A series circuit consisting of R and C?

EXAMPLE 15.5

Calculate the following for a series RC circuit if $R = 2.2$ kΩ, $X_C = 3.3$ kΩ, and $V_S = 5$ V.
a. Z
b. I
c. θ
d. P_R
e. P_X
f. P_A
g. PF

SOLUTION

a. $Z = \sqrt{R^2 + X_C^2}$
$= \sqrt{2.2\ \text{k}\Omega^2 + 3.3\ \text{k}\Omega^2}$
$= 3.96$ kΩ

b. $I = \dfrac{V_S}{Z} = \dfrac{5\text{ V}}{3.96\text{ k}\Omega} = 1.26\text{ mA}$

c. $\theta = \arctan \dfrac{X_C}{R} = \arctan \dfrac{3.3\text{ k}\Omega}{2.2\text{ k}\Omega}$
   $= \arctan 1.5 = 56°$

d. True power $= I^2 \times R$
   $\qquad\qquad = 1.26\text{ mA}^2 \times 2.2\text{ k}\Omega$
   $\qquad\qquad = 3.49\text{ mW}$

e. Reactive power $= I^2 \times X_C$
   $\qquad\qquad\quad = 1.26\text{ mA}^2 \times 3.3\text{ k}\Omega$
   $\qquad\qquad\quad = 5.24 \times 10^{-3} \text{ or } 5.24\text{ mVAR}$

f. Apparent power $= \sqrt{TP^2 + RP^2}$
   $\qquad\qquad\quad = \sqrt{3.49\text{ mW}^2 + 5.24\text{ mW}^2}$
   $\qquad\qquad\quad = 6.29 \times 10^{-3} \text{ or } 6.29\text{ mVA}$

g. Power factor $= \dfrac{R}{Z} = \dfrac{2.2\text{ k}\Omega}{3.96\text{ k}\Omega} = 0.55$

   or

   $\qquad\quad = \dfrac{P_R}{P_A} = \dfrac{3.49\text{ mW}}{6.29\text{ mW}} = 0.55$

   or

   $\qquad\quad = \text{Cos } \theta = 0.55$

15.3
PARALLEL *RC* CIRCUIT

In Figure 15-15(a), you will see a parallel combination of a resistor and capacitor. The current through the resistor and capacitor is simply calculated by applying Ohm's law.

$$\text{resistor current } (I_R) = \dfrac{V_S}{R}$$

$$\text{capacitor current } (I_C) = \dfrac{V_S}{X_C}$$

Total current (I_t), however, is not as simply calculated. Resistor current (I_R) as expected is in phase with the applied voltage (V_S), as seen in the vector diagram Figure 15-15(b). Capacitor current will always lead the applied voltage by 90°, and, as the applied voltage is being used as our reference at 0° on the vector diagram, the capacitor current will have to be drawn at +90° in order to lead the applied voltage by 90°, since vector diagrams rotate in a counterclockwise direction.

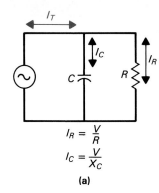

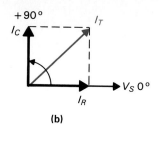

$$I_R = \frac{V}{R}$$

$$I_C = \frac{V}{X_C}$$

(a)

FIGURE 15-15 Parallel *RC* Circuits.

Total current is the vector sum of both the resistor and capacitor currents. Using the Pythagorean theorem, total current can be calculated by

$$I_t = \sqrt{I_R^2 + I_C^2}$$

The angle by which the total current (I_t) leads the source voltage (V_S) can be determined with the following formula:

$$\theta = \arctan \frac{R}{X_C}$$

Since the circuit is both capacitive and resistive, the total opposition or impedance of the parallel *RC* circuit can be calculated by

$$Z = \frac{V_S}{I_t}$$

The impedance of a parallel *RC* circuit is equal to the total voltage divided by the total current. Using basic algebra, this basic formula can be rearranged to express impedance in terms of reactance and resistance.

$$Z = \frac{R \times X_C}{\sqrt{R^2 + X_C^2}}$$

EXAMPLE 15.6

Calculate the following for a parallel *RC* circuit in which $R = 24\ \Omega$, $X_C = 14\ \Omega$ and $V_S = 10$ V.
a. I_R
b. I_C
c. I_T
d. Z
e. θ

Capacitive Reactance and Capacitor Applications

SOLUTION

a. $I_R = \dfrac{V_S}{R} = \dfrac{10\text{ V}}{24\ \Omega} = 416.66\text{ mA}$

b. $I_C = \dfrac{V_S}{X_C} = \dfrac{10\text{ V}}{14\ \Omega} = 714.28\text{ mA}$

c. $I_T = \sqrt{I_R^2 + I_C^2}$
$\qquad = \sqrt{416.66\text{ mA}^2 + 714.28\text{ mA}^2}$
$\qquad = \sqrt{0.173 + 0.510}$
$\qquad = 826.5\text{ mA}$

d. $Z = \dfrac{V_S}{I_t} = \dfrac{10\text{ V}}{826.5\text{ mA}} = 12\ \Omega$

 or

$\qquad = \dfrac{R \times X_C}{\sqrt{R^2 + X_C^2}} = \dfrac{24 \times 14}{\sqrt{24^2 + 14^2}}$
$\qquad = 12\ \Omega$

e. $\theta = \arctan\dfrac{R}{X_C} = \arctan\dfrac{24\ \Omega}{14\ \Omega} = 59.7°$

SELF-TEST REVIEW (§ 15.3)

1. What is the phase relationship between current and voltage in a parallel *RC* circuit?
2. Could a parallel *RC* circuit be called a voltage divider?
3. What formula is used to calculate (a) I_t ; (b) *Z*?
4. Will capacitor current lead or lag resistor current in a parallel *RC* circuit?

15.4
TESTING CAPACITORS

15.4.1 THE OHMMETER

A faulty capacitor may have one of three basic problems:

1. A short, which is easy to detect, and is caused by a contact from plate to plate.
2. An open, which is again quite easy to detect and is normally caused by one of the leads becoming disconnected from its respective plate.

Step 1: Discharge Capacitor

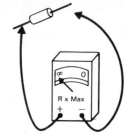

Step 2: Set Analog Ohmmeter to
High Ohms Range

Step 3: Meter Deflects Rapidly
to 0 Ohms Initially

Step 4: Meter Should Then Return to
Infinity as the Capacitor Charges

FIGURE 15-16 Testing a Capacitor of More Than 0.5 μF.

3. A leaky dielectric or capacitor breakdown, which is quite difficult to detect as it may only short at a certain voltage. This problem is usually caused by the deterioration of the dielectric, which starts displaying a much lower dielectric resistance than it was designed for. The capacitor with this type of problem is referred to as a *leaky* capacitor.

Capacitors of 0.5 μF and larger can be checked by using an analog ohmmeter in the procedure seen in Figure 15-16.

Steps

1. Ensure that the capacitor is discharged by shorting the leads together.
2. Set the ohmmeter to the highest ohms range scale. Zero the ohmmeter.
3. Connect the meter to the capacitor, observing the correct polarity if an electrolytic is being tested, and observe the meter pointer. The capacitor will initially be discharged and, therefore, maximum current will flow from the meter battery to the capacitor; maximum current means low resistance, which is why the meter's pointer deflects to the far right to indicate 0 ohms.
4. As the capacitor charges, it will cause current flow from the meter's battery to decrease, and consequently the meter needle will move toward the left side of the scale.

A good capacitor will cause the meter to react as just explained. A larger capacitance will cause the meter to move slowly to infinity (∞), as it will take a longer time to charge, while a smaller value of capacitance will charge at a much faster rate, causing the meter to deflect rapidly toward ∞. For this reason, the ohmmeter cannot be reliably used to check capacitors with values of less than 0.5 μF, becuse the capacitor charges up too quickly, and the meter does not have enough time to respond.

A shorted capacitor will cause the meter to deflect to zero ohms and remain in that position. An open capacitor will cause no meter deflection (infinite resistance) because there is no path for current to flow.

A leaky capacitor will deflect to the right, as normal. If the meter pointer returns almost all the way back to ∞, then only a small current is still flowing and the capacitor has a small dielectric leak. If the meter only comes back to halfway or a large distance away from infinity, a large amount of current is still flowing and the capacitor has a large dielectric leak (defect).

When using the analog ohmmeter to test capacitors, there are some other points that you should be aware of:

1. Electrolytics are noted for having a small yet noticeable amount of inherent leakage; therefore, do not expect the needle pointer to move all the way to the left (∞ ohms). Most electrolytic capacitors that are still functioning normally will show a resistance of 200 kΩ or more.

2. Some ohmmeters utilize internal batteries of up to 15 V, and so be careful not to exceed the voltage rating of the capacitor.

15.4.2 CAPACITANCE METER OR ANALYZER

The ohmmeter check tests the capacitor under a low-voltage condition. This may be adequate for some capacitor malfunctions; however, a problem that often occurs with capacitors is that they short or leak at a high voltage. The ohmmeter test is also adequate for capacitors of 0.5 μF or greater; however, a smaller capacitor cannot be tested because its charge time is too fast for the meter to respond. The ohmmeter cannot check for high-voltage failure, for small-value capacitance, or if the value of capacitance has changed through age or extreme thermal exposure.

A capacitance meter or analyzer, which is illustrated in Figure 15-17, can totally check all aspects of a capacitor in a range of values from approximately 1 pF to 20 farads. The tests that are generally carried out include:

(1) *Capacitor value change* [Figure 15-18(a)]: Capacitors will change their value over a period of time. Ceramic capacitors often change 10% to 15% within the first year as the ceramic material relaxes. Electrolytics change their value due to the electrolytic solution simply drying out. Some capacitors are simply labeled incorrectly by manufacturers or the technician cannot determine the correct value because of the labeling used. A value change accounts for approximately 25% of all defective capacitors.

(2) *Capacitor leakage* [Figure 15-18(b)]: Leakage occurs due to an imperfection of the dielectric. Although the dielectric's resistance is very high, a small amount of leakage current will flow between the plates. This resistance, which is between the plates and therefore effectively in parallel with the capacitor, can become too low and cause the circuit that the capacitor is in to malfunction. Most capacitance meters will perform a leakage test

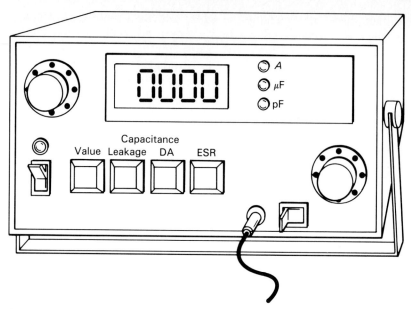

FIGURE 15-17 Capacitor Analyzer.

with operating potentials up to 650 V. Leakage accounts for approximately 40% of all defective capacitors.

(3) *Dielectric absorbtion* [Figure 15-18(c)]: This occurs mainly in electrolytics when they will take on a charge but will not fully discharge. This residual charge remains within the capacitor, similar to a small dc battery, and this changes the effective value of the capacitor once it is in circuit during operation. If this causes the value to change more than 15%, the capacitor should be rejected. Dielectric absorption accounts for approximately 25% of all defective capacitors.

(4) *Equivalent series resistance* (ESR) [Figure 15-18(d)]: Series resistance is found in the capacitor leads, lead to plate connection, and electrolyte

FIGURE 15-18 Problems with Capacitors. (a) Value Change. (b) Leakage. (c) Dielectric Absorption. (d) Equivalent Series Resistance.

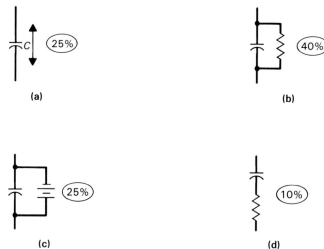

Capacitive Reactance and Capacitor Applications

(almost always occurs in electrolytics) and causes the effective circuit capacitance value to change. Equivalent series resistance accounts for 10% of all defective capacitors.

15.5
APPLICATIONS OF CAPACITORS

There are many applications of capacitors, some of which will be discussed now; others will be presented later and in your course of electronic studies.

15.5.1 COMBINING AC AND DC

Figure 15-19(a) and (b) shows how the capacitor can be used to combine ac and dc. The capacitor is large in value (electrolytic typically) and can be thought of as a very large bucket that will fill or charge to a dc voltage level. The ac voltage will charge and discharge the capacitor, which is similar to pouring in and pulling out (alternating) more and less water. The resulting waveform is a combination of ac and dc that varies above and below an average dc level. In this instance, the ac is said to be superimposed on a dc level.

15.5.2 *RC* INTEGRATOR

The term integrator is derived from a mathematical function in calculus. This combination of R and C, in some situations, displays this mathematical function. Figure 15-20 illustrates an integrator circuit that can be recognized by the series connection of R and C, but mainly from the fact that the output is taken across the capacitor.

If a 10 V square wave, is applied across the circuit, as seen in the waveforms in Figure 15-20(b), and the time constant of the *RC* combination calculates out to be 1 second, the capacitor will charge when the square

FIGURE 15-19 Superimposing AC on a DC Level.

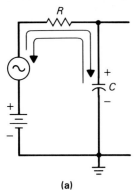

(a)

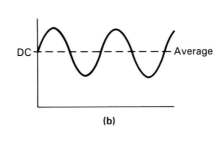

(b)

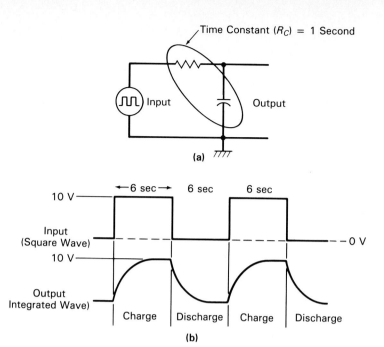

FIGURE 15-20 *RC* **Integrator. (a) Circuit. (b) Waveforms.**

wave input is positive toward the applied voltage (10 V) and reaches it in 5 time constants (5 seconds). Since the positive alternation of the square wave lasts for 6 seconds, the capacitor will be fully charged 1 second before the positive alternation ends, as seen in Figure 15-21.

When the positive half cycle of the square wave input ends after 6 seconds, the input falls to 0 V and the circuit is equivalent to that seen in Figure 15-22(a). The 10 V charged capacitor now has a path to discharge and in five time constants (5 seconds) is fully discharged, as seen in Figure 15-22(b).

If we now only allow the square wave to have 1 second during the positive and negative half cycles, the capacitor will not be able to charge fully toward 10 V. In fact, during the positive alternation of 1 second (1 time

FIGURE 15-21 **Integrator Response to Positive Step. (a) Equivalent Circuit. (b) Waveforms for Charge.**

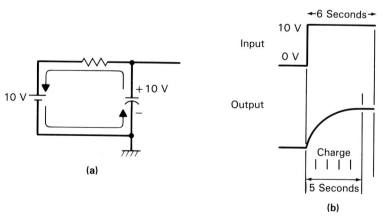

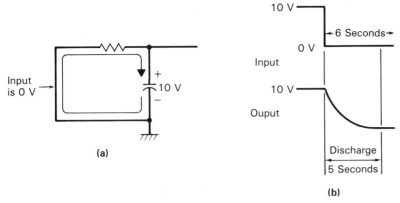

(a)

(b)

FIGURE 15-22   Integrator Response to Negative Step. (a) Equivalent Circuit. (b) Waveforms for Discharge.

constant), the capacitor will reach 63.2% of the applied voltage (6.32 V), and then during the 0 volt half-cycle, it will discharge to 63.2% of 6.32 V to 2.33 V, as seen by the waveform in Figure 15-23. The capacitor was initially discharged, and the output voltage will gradually build up and eventually level off to an average value of 5 V in about 5 time constants (5 seconds).

In summary, if the period or time of the square wave is decreased, or if the time constant is increased, the same effect results. The capacitor has

FIGURE 15-23   Integrator Response to Square Wave. (a) Circuit. (b) Waveforms.

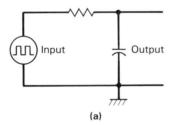

(a)

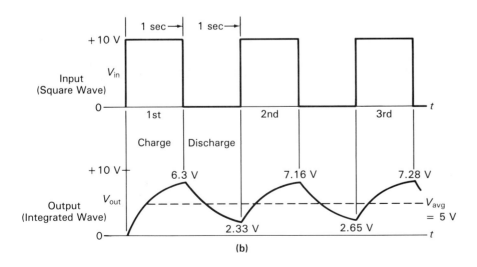

(b)

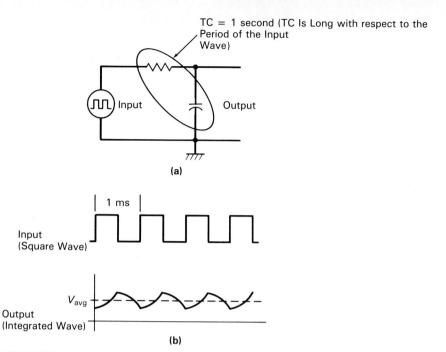

TC = 1 second (TC Is Long with respect to the Period of the Input Wave)

Input

Output

(a)

1 ms

Input
(Square Wave)

V_{avg}

Output
(Integrated Wave)

(b)

FIGURE 15-24   Integrator Response to Long Time Constant. (a) Circuit. (b) Waveforms.

less time to charge and discharge and reaches an average value of the input voltage, which for a square wave is half of its amplitude.

Figure 15-24 illustrates the circuit waveform produced when the time constant is long with respect to the period of the input waveform. If the time constant is further increased so that it is even longer with respect to the input waveform, the output will have an even smaller peak to peak variation.

15.5.3   *RC* DIFFERENTIATOR

Figure 15-25 illustrates the differentiator circuit, which is the integrator's opposite. In this case, the output is taken across the resistor instead of the capacitor, and the time constant is always short with respect to the input waveform.

The differentiator output waveform is taken across the resistor and is the result of the capacitor's charge and discharge. When the square wave swings positive, the equivalent circuit is that seen in Figure 15-26.

When the 10 V is initially applied (positive step of the square wave), all the voltage is across the resistor and, therefore, at the output, as the capacitor cannot charge instantly. As the capacitor begins to charge, more of the voltage is developed across the capacitor and less across the resistor. The voltage across the capacitor exponentially increases and reaches 10 V in 5 time constants, while the voltage across the resistor, and therefore the output, exponentially falls from its initial 10 V to 0 V in 5 time constants, at which time all the voltage is across the capacitor, then no voltage will be across the resistor.

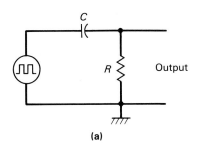

(a)

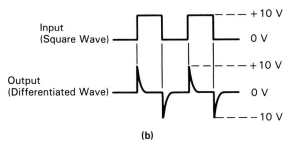

(b)

FIGURE 15-25   *RC* Differentiator. (a) Circuit. (b) Waveforms.

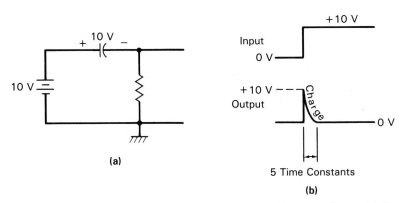

FIGURE 15-26   Differentiator's Response to Positive Step. (a) Equivalent Circuit. (b) Waveforms.

FIGURE 15-27   Differentiator's Response to a Negative Step. (a) Equivalent Circuit. (b) Waveforms.

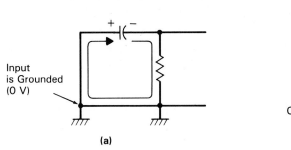

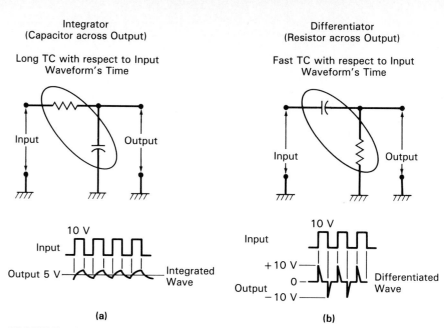

Integrator
(Capacitor across Output)

Long TC with respect to Input
Waveform's Time

Input

Output

10 V

Input

Output 5 V

Integrated
Wave

(a)

Differentiator
(Resistor across Output)

Fast TC with respect to Input
Waveform's Time

Input

Output

10 V

Input

+10 V
0
−10 V

Output

Differentiated
Wave

(b)

FIGURE 15-28   **Summary of Integrator and Differentiator.**

When the positive half-cycle of the square wave ends and the input falls to zero, as seen in Figure 15-27, the negative plate of the capacitor is now applied directly to the output. Since the capacitor cannot instantly discharge, the output drops suddenly down to − 10 V. This is the voltage across the resistor, and therefore the output. The capacitor now is in series with the resistor and therefore has a path through the resistor to discharge, which it does in 5 time constants to 0 V.

Figure 15-28(a) and (b) illustrates the integrator and differentiator circuits and waveforms, which are used extensively in many applications and equipment such as computers, robots, lasers, and communications.

15.5.4   *RC* FILTER

A filter is a circuit that allows certain frequencies to pass, but blocks other frequencies. In other words, it filters out the unwanted frequencies, but passes the wanted or selected ones.

There are two basic *RC* filters:

1. The low pass filter, which can be seen in Figure 15-29(a); as its name implies, it passes the low frequencies, but heavily attenuates the higher frequencies.

2. The high pass filter, which can be seen in Figure 15-29(b); as its name implies, it allows the high frequencies to pass, but heavily attenuates the lower frequencies.

With either the low or high pass filter, it is important to remember that capacitive reactance is inversely proportional to frequency ($X_c \propto 1/f$).

With the low pass filter seen in Figure 15-29(a), the output is connected

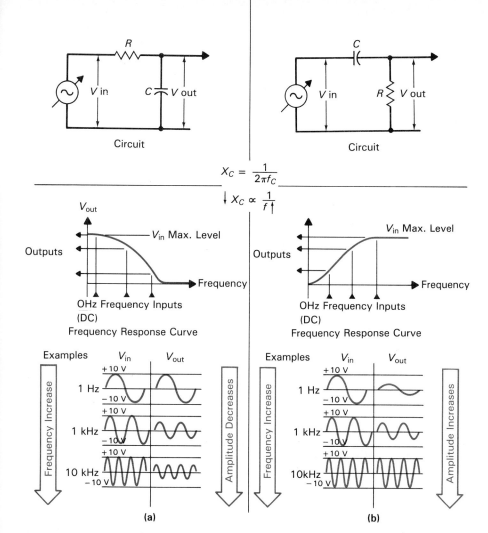

FIGURE 15-29 *RC* **Filters. (a) Low-pass Filter. (b) High-pass Filter.**

across the capacitor. As the frequency of the input increases, the amplitude of the output decreases. At dc (0 Hz) and low frequencies, the capacitive reactance is very large ($X_C\uparrow = 1/f\downarrow$) with respect to the resistor. All the input will appear across the capacitor, because the capacitor and resistor form a voltage divider, as seen in Figure 15-30. As with any voltage divider, the larger opposition to current flow will drop the largest voltage. Since the output voltage is determined by the voltage drop across the capacitor, almost all the input will appear across the capacitor and therefore be present at the output.

If the frequency of the input increases, the reactance of the capacitor will decrease ($X_C\downarrow = 1/f\uparrow$), and a larger amount of the signal will be dropped across the resistor. As frequency increases, the capacitor becomes more of a short circuit (lower reactance), and the output, which is across the capacitor, decreases, as seen in Figure 15-30(b).

Below the circuit of the low pass filter in Figure 15-29(a), you will see a graph known as the frequency response curve for the low pass filter. This curve illustrates that as the frequency of the input increases the voltage at the output will decrease.

Applications of Capacitors **539**

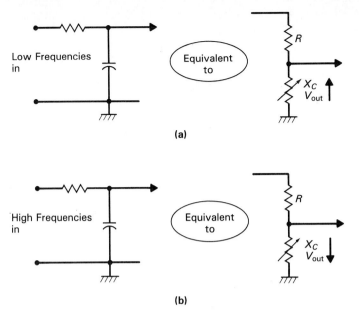

FIGURE 15-30 **Low-pass Filter. (a) Low-pass Filter, High X_C. (b) Low-pass Filter, Low X_C.**

With the high pass filter, seen in Figure 15-29(b), the capacitor and resistor have traded positions to show that the opposite effect for the low pass filter occurs

At low frequencies, the reactance will be high and almost all of the signal in will be dropped across the capacitor. Very little appears across the resistor and, consequently, the output, as seen in Figure 15-31(a). As the frequency of the input increases, the reactance of the capacitor decreases,

FIGURE 15-31 **High pass Filter. (a) High X_C. (b) Low X_C.**

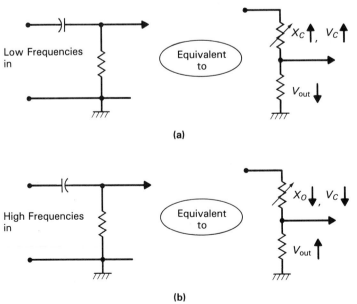

allowing more of the input signal to appear across the resistor and so appear at the output.

Below the circuit of the high pass filter in Figure 15-29(b), you will see the frequency response curve for the high pass filter. This curve illustrates that as the frequency of the input increases the voltage at the output increases.

SUMMARY

1. Capacitive reactance is the opposition to current flow without the dissipation of energy, is measured in ohms, and is inversely proportional to frequency and capacitance.
2. Impedance is also measured in ohms and is the total opposition (both reactive and resistive) to current flow.
3. The average power dissipated by a capacitor is zero.
4. The power stored by a resistor is referred to as resistive or true power.
5. The power consumed by the capacitor and then returned to the source is referred to as reactive or imaginary power.
6. Apparent power is the vectoral addition of true and imaginary power.
7. The power factor is a ratio of true power to apparent power and can be anywhere between zero and one.
8. A faulty capacitor may have one of three problems:
   a. A short between plates
   b. An open between leads and plates
   c. Capacitance leakage (leaky cap.) due to deterioration of the dielectric
9. electric
10. The ohmmeter can be used to check capacitors of 0.5 μF or more.
11. The capacitance analyzer, unlike the ohmmeter, can check for high voltage failure, small value capacitance, and value variation.
   Some of the applications of capacitors are:
   a. Combining ac and dc
   b. Integrating a square wave
   c. Differentiating a square wave
   d. Filtering

NEW TERMS

Apparent power	**High pass filter**	**Reactance**
Average power	**Imaginary power**	**Reactive power**
Capacitance meter	**Impedance**	**Resistive circuit**
Capacitive circuit	**Integrator**	**Series *RC* circuit**
Capacitive reactance	**Leaky capacitor**	**Superimposed**
Dielectric absorbtion	**Low pass filter**	**Theta**
Differentiator	**Parallel *RC* circuit**	**True power**
Equivalent series	**Positive power**	**Vector**
resistance	**Power factor**	**Vector diagram**
Filter	**Pythagorean theorem**	

NEW FORMULAS

$$X_C = \frac{1}{2\pi f C}$$

X_C = capacitive reactance in ohms
f = frequency in hertz
C = capacitance in farads

Series RC *Circuits*:

$$V_S = \sqrt{V_R^2 + V_C^2}$$

$V_C = \sqrt{V_S^2 - V_R^2}$

$V_R = \sqrt{V_S^2 - V_C^2}$

V_S = source voltage
V_R = voltage drop across resistor
V_C = voltage drop across capacitor

$$Z = \sqrt{R^2 + X_C^2}$$

$R = \sqrt{Z^2 - X_C^2}$

$X_C = \sqrt{Z^2 - R^2}$

Z = impedance in ohms
R = resistance in ohms
X_C = capacitive reactance in ohms

$$\theta = \arctan \frac{X_C}{R}$$

θ = phase angle or shift in degrees

$$\theta = \arctan \frac{V_C}{V_R}$$

$$P_A = \sqrt{P_R^2 + P_X^2}$$

P_A = apparent power
P_R = true power
P_X = reactive power

$$PF = \frac{P_R}{P_A} = \frac{R}{Z} = \cos\theta$$

PF = power factor

Parallel RC *Circuits*:

$$I_R = \frac{V}{R}$$

$$I_C = \frac{V}{X_C}$$

I_R = current through resistor
I_C = current through capacitor

$$I_t = \sqrt{I_R^2 + I_C^2}$$

I_t = total circuit current

$$Z = \frac{V}{I_t}$$

$$Z = \frac{R \times X_C}{\sqrt{R^2 + X_C^2}}$$

$$\theta = \arctan \frac{R}{X_C}$$

542   **Capacitive Reactance and Capacitor Applications**

Multiple Choice Questions

1. Capacitive reactance is inversely proportional to:
   a. Capacitance and resistance
   b. Frequency and capacitance
   c. Capacitance and impedance
   d. Both (a) and (c)

2. The impedance of an RC series circuit is equal to:
   a. The sum of R and X_C
   b. The square root of the sum of R^2 and X_C^2
   c. The square of the sum of R and X_C
   d. The sum of the square root of R and X_C

3. In a purely resistive circuit:
   a. The current flowing in the circuit leads the voltage across the capacitor by 90°
   b. The circuit current and resistor voltage are in phase with one another
   c. The current leads the voltage by 45°
   d. The current leads the voltage by a phase angle between 0 and 90°

4. In a purely capacitive circuit:
   a. The current flowing in the circuit leads the voltage across the capacitor by 90°
   b. The circuit current and resistor voltage are in phase with one another
   c. The current leads the voltage by 45°
   d. The current leads the voltage by a phase angle between 0 and 90°

5. In a circuit containing both capacitance and resistance:
   a. The current flowing in the circuit leads the voltage across the capacitor by 90°
   b. The circuit current and resistor voltage are in phase with one another
   c. The current leads the voltage by 45°
   d. Both (a) and (b)

6. In a series RC circuit, the source voltage is equal to:
   a. The sum of V_R and V_C
   b. The difference between V_R and V_C
   c. The vectoral sum of V_R and V_C
   d. The sum of V_R and V_C squared

7. As the source frequency is increased, the capacitive reactance will:
   a. Increase
   b. Decrease
   c. Be unaffected
   d. Increase, depending on harmonic content

8. The phase angle of a series RC circuit indicates by what angle V_S _____ V_R.
   a. Lags
   b. Leads
   c. Leads or lags
   d. None of the above

9. In a series RC circuit, the vector combination of R and X_C is the circuit's _____.
   a. Phase angle
   b. Apparent power
   c. Source voltage
   d. Impedance

10. In a parallel *RC* circuit, the total current is equal to:
    a. The sum of I_R and I_C
    b. The difference between I_R and I_C
    c. The vectoral sum of I_R and I_C
    d. The sum of I_R and I_C squared

11. _____ is the opposition offered by a capacitor to current flow without the dissipation of energy.
    a. Capacitive reactance
    b. Resistance
    c. Impedance
    d. Phase angle
    e. The power factor
    (*These choices also apply to Questions 12 and 13.*)

12. _____ is the total reactive and resistive circuit opposition to current flow.

13. _____ is the ratio of true (resistive) power to apparent (reactive) power and is therefore a measure of the loss in a circuit.

14. In a series *RC* circuit, the leading voltage will be measured across the:
    a. Resistor
    b. Capacitor
    c. Source
    d. Any of the choices are true

15. In a series *RC* circuit, the lagging voltage will be measured across the:
    a. Resistor
    b. Capacitor
    c. Source
    d. Any of the choices are true

Essay Questions _____

16. Give the formula and define the term capacitive reactance. (15.1)

17. In a series *RC* circuit, give the formulas for calculating: (15.2)
    a. V_S, when V_R and V_C are known
    b. Z, when R and X_C are known
    c. Z, when I and V are known
    d. θ, when X_C and R are known
    e. θ, when V_C and V_R are known
    f. Power factor, when R and Z are known
    g. Power factor, when P_R and P_A are known

18. In a parallel *RC* circuit, give the formulas for calculating: (15.3)
    a. I_R, when V and R are known
    b. I_C, when V and X_C are known
    c. I, when I_R and I_C are known
    d. Z, when V and I are known
    e. Z, when R and X_C are known

19. What is meant by long or short time constant, and do large or small values of *RC* produce a long or short time constant? (14.10)

20. Describe how the inverse relationship between frequency and capacitive reactance can be used for the application of filtering. (15.5.4)

21. Illustrate the voltage and current waveforms across a resistor and capacitor in series when a DC voltage is applied during charge and when the same resistor and capacitor are connected to discharge. (14.2 and 14.3)

22. Describe and illustrate how the capacitor can be used in the following applications:
    a. Combining AC and DC   (15.5.1)
    b. Integrating a square wave   (15.5.2)
    c. Differentiating a square wave   (15.5.3)
    d. Filtering high and low frequencies   (15.5.4)
23. Describe the difference between reactance, resistance, and impedance.   (15.2)
24. Sketch the phase relationships between: (15.2.3)
    a. V_R and I in a purely resistive circuit
    b. V_C and I in a purely capacitive circuit
25. What is positive power and negative power?

Practice Problems

26. Calculate the capacitive reactance of the following capacitor circuits with the following parameters.
    a. $f = 1$ kHz, $C = 2$ μF
    b. $f = 100$ Hz, $C = 0.01$ μF
    c. $f = 17.3$ MHz, $C = 47$ μF
27. In a series RC circuit, the voltage across the capacitor is 12 V and the voltage across the resistor is 6 V. Calculate the source voltage.
28. Calculate the impedance for the following series RC circuits.
    a. 2.7 MΩ, 3.7 μF, 20 kHz
    b. 350 Ω, 0.005 μF, 3 MHz
    c. $R = 8.6$ kΩ, $X_C = 2.4$ Ω
    d. $R = 4700$ Ω, $X_C = 2$ MΩ
29. In a parallel RC circuit with parameters of $V_S = 12$ V, $R = 4$ MΩ, and $X_C = 1.3$ kΩ, calculate:
    a. I_R
    b. I_C
    c. I_T
    d. Z
    e. θ
30. Calculate the total reactance in:
    a. A series circuit where $X_{C1} = 200$ Ω, $X_{C2} = 300$ Ω, $X_{C3} = 400$ Ω
    b. A parallel circuit where $X_{C1} = 3.3$ kΩ, $X_{C2} = 2.7$ kΩ
31. Calculate the capacitance needed to produce 10 kΩ of reactance at 20 kHz.
32. At what frequency will a 4.7 μF capacitor have a reactance of 2000 Ω?
33. A series RC circuit contains a resistance of 40 Ω and a capacitive reactance of 33 Ω across a 24 V source.
    a. Sketch the schematic diagram.
    b. Calculate Z, I, V_R, V_C, I_R, I_C, and θ.
34. A parallel RC circuit contains a resistance of 10 kΩ and a capacitive reactance of 5 kΩ across a 100 V source.
    a. Sketch the schematic diagram.
    b. Calculate I_R, I_C, I_T, Z, V_R, V_C, and θ.

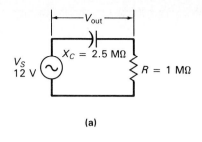
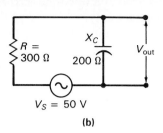

(a)

(b)

FIGURE 15-32

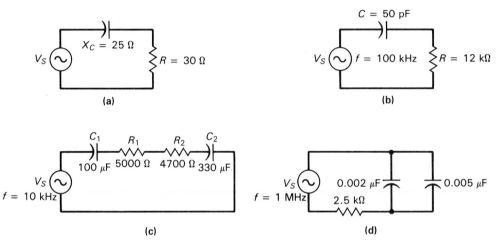

(a)

(b)

(c)

(d)

FIGURE 15-33

35. Calculate V_R and V_C for the circuits seen in Figure 15-32(a) and (b).

36. Calculate the impedance of the circuits shown in Figure 15-33.

37. In Figure 15-34, the output voltage, since it is taken across the capacitor, will _____ the voltage across the resistor by _____ degrees.

38. If the position of the capacitor and resistor in Figure 15-34 is reversed, the output voltage, since it is now taken across the resistor, will _____ the voltage across the capacitor by _____ degrees.

39. Calculate the resistive power, reactive power, apparent power, and power factor for the circuit seen in Figure 15-34; $V_{in} = 24$ V and $f = 35$ kHz.

40. Refer to Figure 15-35 and calculate the following:
  a. (Figure 15-35(a)) X_C, I, Z, I_R, θ, V_R, V_C.
  b. (Figure 15-35(b)) V_R, V_C, I_R, I_C, I_T, Z, θ.

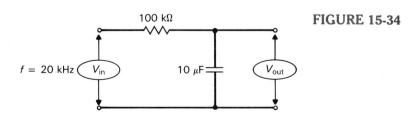

FIGURE 15-34

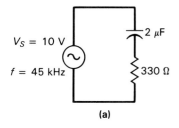

(a)

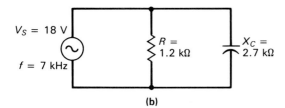

(b)

FIGURE 15-35

Troubleshooting Problems

41. Describe the three basic problems that normally occur with faulty capacitors.

42. Describe how to use the ohmmeter to check capacitors and also explain some of its limitations.

43. Describe the four basic tests performed by a capacitor meter or analyzer.

44. Which capacitor problem accounts for the largest percentage of defective capacitors. Explain exactly what is this malfunction.

45. If the needle of a meter goes to zero ohms and remains there, the capacitor is _____. If the capacitor is _____, however, no charging will occur and the meter will indicate infinite ohms.

16

INDUCTANCE AND INDUCTORS

AFTER COMPLETING THIS CHAPTER, YOU WILL BE ABLE TO:

1. Describe what is meant by self induction.
2. List and explain the factors affecting inductance.
3. Give the formula for inductance.
4. Identify inductors in series and parallel and understand how to calculate total inductance when inductors are in combination.
5. List and explain the fixed and variable types of inductors.
6. Explain the inductive time constant.
7. Give the formula for inductive reactance.
8. Describe all aspects relating to a series and parallel RL circuit.
9. State the three typical malfunctions of inductors and explain how they can be recognized.
10. Describe how inductors can be used for the following applications:
    a. RL integrator
    b. RL differentiator
    c. RL filter

INTRODUCTION

In previous chapters on electromagnetism, two important rules were discussed that relate to inductance. The first was that a magnetic field will build up around any current carrying conductor and, secondly, a voltage is induced into a conductor when it is subjected to a moving magnetic field. These two rules from the basis for a phenomenon known as self induction or inductance.

WHAT'S COOKING!

As a boy, George R. Stibitz was a natural experimenter intrigued by any electrical or electronic gadget. His parents were always anxious about these tinkerings, but wishing not to discourage his imagination and keeness to learn, his father, a professor of theology, often gave him devices to occupy him. On one occasion at the age of eight, his latest experiment in which he connected an electric motor to the electrical outlet, caused a circuit overload and almost sent the family home in Dayton, Ohio, up in smoke.

In 1937, Stibitz, a young mathematician working at Bell Telephone Laboratories (the research division of AT&T), had another one of his brainstorms and in his spare time he began building the prototype, which consisted of telephone-system components, batteries, and other devices all interconnected by a mass of wires. The machine he was putting together would be the first system able to achieve binary arithmetic in the United States; since it was assembled on the kitchen table, he called it the Model K.

After developing a more ambitious system in 1940 with veteran Bell engineer Samuel B. Williams, the new digital calculator could handle the complex number problems that were needed for the design of long-distance telephone networks. The system was installed in Bell Labs' Manhattan, New York, headquarters and drew a large amount of attention from the American Mathematical Society, who invited Stibitz to give a presentation on the machine at Dartmouth College in Hanover, New Hampshire. Stibitz later wrote, "With my usual genius for making things more difficult for myself and others, I suggested direct telegraph operation of the complex number calculator from Hanover, and this was decided upon." Both Stibitz and Williams worked tirelessly setting up the demonstration, and on September 11, 1940, Stibitz typed in complex number problems on a keyboard in Hanover and within a minute a correct answer came racing back from New York. This was the first demonstration of long-distance computing.

Three mathematicians who would themselves greatly influence computer science sat in the audience during this presentation. John von Neumann, Norbert Weiner, and John W. Mauchly in just a few years time would help invent ENIAC, the world's first large-scale electronic digital computer.

16.1
ELECTROMAGNETISM

Each electron has its own small magnetic field due to its velocity. When no current or electron flow exists in a circuit, as seen in Figure 16-1(a), the electrons within the conductor are misaligned (disorganized) and all the small magnetic fields around each electron will cancel with one another, causing no magnetic field to exist around the conductor.

When the switch is closed, as seen in Figure 16-1(b), the applied voltage causes the flow of electrons, and this organizes or aligns them in one direction. This causes all the small electron magnetic fields to aid one another, causing a combined magnetic field that exists around the conductor. Electromagnetism is the magnetic field that exists around a conductor when current is passed through it. If either the voltage increases or the resistance in the circuit decreases, the current will increase, and this increase in electron flow will cause a corresponding increase in the magnitude of the magnetic field.

Figure 16-2 illustrates how a conductor, when wound into a coil, de-

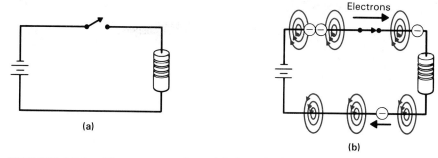

FIGURE 16-1   Electromagnetism. (a) Switch Open, No Current Flow, No Magnetic Field. (b) Switch Closed, Current Flow, Magnetic Field.

FIGURE 16-2   (a) DC Electromagnet. (b) AC Electromagnet.

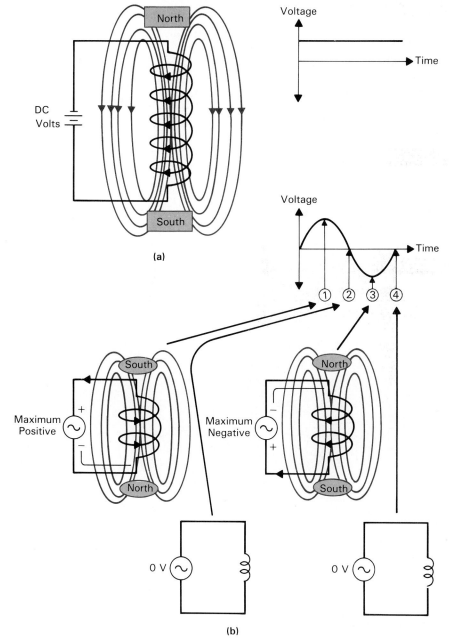

velops a magnetic field when direct and alternating current are passed through it. In Figure 16-2(a), direct current produces a constant magnetic field of a fixed polarity. In Figure 16-2(b), alternating current produces an alternating magnetic field that continually switches magnetic polarity.

16.2
ELECTROMAGNETIC INDUCTION

Current flow causes the generation of a magnetic field, but it is also possible to reverse the process and use a magnetic field to generate current. This

FIGURE 16-3 **Faraday Electromagnetic Induction Discoveries.**

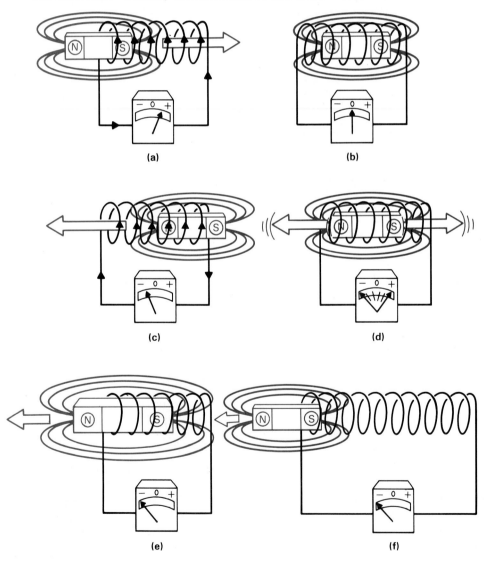

Inductance and Inductors

effect is known as *electromagnetic induction* and can occur in one of three ways:

1. If a conductor is moved so that it cuts a stationary magnetic field, a voltage will be induced into the conductor.
2. If a magnetic field is moved past a conductor so that the lines of flux cut the conductor, a voltage will be induced into the conductor.
3. If the magnetic field strength is changed about a conductor, a voltage will be induced into the conductor.

The key to inducing a voltage in the conductor is the movement of either the field or conductor relative to the other, which is stationary. This relative motion between the conductor and the magnetic field causes the electrons within the conductor to be moved in one direction or the other, depending on the direction of motion, and this electron flow constitutes current.

The magnitude of the induced voltage is dependent on three factors:

1. The speed at which the conductor passes through the magnetic field
2. The strength or flux density of the magnetic field
3. The length of the conductor in the magnetic field.

The direction of relative motion between the magnetic field and conductor determines the direction of current flow. This information pertaining to electromagnetic induction is summarized in Figure 16-3. Electromagnetism and electromagnetic induction are the two phenomena responsible for the property known as *inductance*.

SELF-TEST REVIEW (§ 16.1 and 16.2)

1. Define electromagnetism.
2. What is the difference between electromagnetism and electromagnetic induction?

16.3
SELF-INDUCTION

When current flows through a conductor or coil, as seen in Figure 16-4 (electromagnetism), the strength of the magnetic field rises in a short time from zero to maximum, expanding from the center of the conductor. The expanding magnetic lines of force have relative motion with respect to the stationary conductor, and so an induced voltage results (electromagnetic induction). The blooming magnetic field generated by the conductor is actually causing a voltage to be induced in the conductor that is generating the magnetic field. This effect of a current carrying coil of conductor inducing a voltage within itself is known as *self inductance*. This phenomenon was first discovered by Heinrich Lenz, who observed that the induced voltage

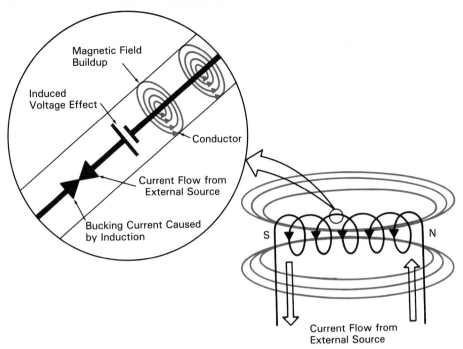

FIGURE 16-4 Self induction.

causes an induced (bucking) current to flow in the coil, which opposes the source current producing it.

As the current continues to rise toward its maximum value, the magnetic field further expands, and throughout this time of relative motion between field and conductor, an induced voltage will be present, which causes an induced current to oppose the change in the circuit current, as seen in Figure 16-5(a).

When the current reaches its maximum, the magnetic field, which is dependent on current, will also reach a maximum value and then no longer expand but remain stationary. When the current remains constant, no change will occur in the magnetic field and therefore no relative motion between conductor and magnetic field, resulting in no induced voltage or current to oppose circuit current, as seen in Figure 16-5(b). The coil has accepted electrical energy and is storing it in the form of a magnetic energy field, just as the capacitor stored electrical energy in the form of an electric field.

If the switch is put in position B, as seen in Figure 16-5(c), the current from the battery will be zero, and the magnetic field will collapse as it no longer has circuit current to support it. As the magnetic lines of force collapse, they cut the conducting coils, causing relative motion between the conductor and magnetic field. A voltage is induced in the coil, which will produce an induced current to flow in the same direction as the circuit current was flowing before the switch was opened. The coil is now converting the magnetic field energy into electrical energy and returning the original energy that it stored.

After a short period of time, the magnetic field will have totally collapsed, the induced voltage will be zero, and the induced current within the circuit will therefore also no longer be present.

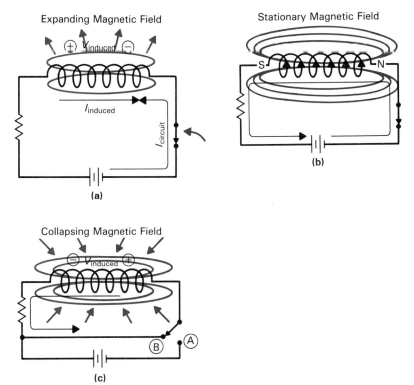

FIGURE 16-5 Self inductance. (a) Switch Closed. Increase in Circuit Current and, therefore, Increase in Magnetic Field Induces Voltage in Coil. This Results in an Induced Current to Oppose Circuit Current. (b) Constant Circuit Current. Stationary Magnetic Field and, therefore, No Induced Voltage or Current to Oppose Circuit Current. (c) Switch Opened. No Circuit Current and, therefore, Magnetic Field Collapses and Induces Voltage in Coil. This Results in an Induced Current in Same Direction as Circuit Current.

This induced voltage is called a counter or back emf. It opposes the applied emf (or battery voltage). The ability of a coil or conductor to induce or produce a counter emf within itself as a result of a change in current is called self inductance, or more commonly inductance (*L*). The unit of inductance is the henry (H), named in honor of Joseph Henry, an American physicist, for his experimentation within this area of science. The inductance of an inductor is 1 henry when a current change of 1 ampere per second causes an induced voltage of 1 volt. Inductance is, therefore, a measure of how much counter emf (induced voltage) can be generated by an inductor for a given amount of current change through that same inductor.

This counter emf or induced voltage can be calculated by the formula

$$V_{induced} = L \times \frac{\Delta_i}{\Delta_t}$$

where Δ_i = increment of change of current (*i*)

Δ_t = increment of change with respect to time (*t*)

L = inductance, in henries (*H*)

A larger inductance will create a larger induced voltage, and if the rate of change of current with respect to time is increased, then the induced voltage or counter emf will also increase.

EXAMPLE 16.1

What voltage is induced across an inductor of 4 henries when the current is changing at a rate of:
a. 1 amp/s?
b. 4 amps/s?

SOLUTION

a. $V_{ind} = L \times \dfrac{\Delta I}{\Delta t} = 4 \text{ H} \times 1 \text{ A/s} = 4 \text{ V}$

b. $V_{ind} = L \times \dfrac{\Delta I}{\Delta t} = 4 \text{ H} \times 4 \text{ A/s} = 16 \text{ V}$

The faster the coil current changes, the larger the induced voltage.

SELF-TEST REVIEW (§ 16.3)

1. Define self induction.
2. What is counter emf and how can it be calculated?
3. Calculate the voltage induced in a 2 mH inductor if the current is increasing at a rate of 4 kA/s.

16.4
THE INDUCTOR

An inductor is basically an electromagnet, as its construction and principle of operation are the same. We use the two different names because they have different applications. The purpose of the electromagnet or solenoid is to generate a magnetic field, while the purpose of an inductor or coil is to oppose any change of circuit current.

If a value of constant direct current is flowing in a circuit, as seen in Figure 16-6, a stationary magnetic field will be created by the inductor. If the current in the circuit is suddenly increased (by lowering circuit resistance), a voltage will be induced in the inductor. This induced voltage will attempt to oppose the applied source voltage from the battery, opposing the change in current and holding it at its previous constant level. Circuit

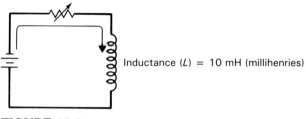

Inductance (L) = 10 mH (millihenries)

FIGURE 16-6

current will rise; however, as the counter emf is only present during the change from one current level to another. Once the new higher level of current has been reached and remains constant, there will no longer be a change. This lack of relative motion between field and conductor will no longer generate a counter emf, and so the current will remain at its new higher constant value.

This effect also happens in the opposite respect. If current decreases (by increasing circuit resistance), the magnetic lines of force will collapse because of the reduction of current and induce a voltage in the inductor, which will produce an induced current in the same direction as the circuit current. These two combine and tend to maintain the current at the higher previous constant level. Circuit current will fall, however, as the induced voltage and current are only present during the change (in this case the decrease from the higher current level to the lower); and once the new lower level of current has been reached and remains constant, the lack of change will no longer induce a voltage or current. So the current will then remain at its new lower constant value.

The inductor is, therefore, an electronic component that will oppose any changes in circuit current, and this ability or behavior is referred to as *inductance*.

SELF-TEST REVIEW (§ 16.4)

1. What is the difference between an electromagnet and an inductor?
2. The inductor will oppose any changes in circuit current (true or false).

16.5

FACTORS DETERMINING INDUCTANCE

The inductance of an inductor is determined by four factors:

1. Number of turns
2. Area of the coil

3. Length of the coil
4. Core material used within the coil

Let's now discuss these four factors in more detail, beginning with the number of turns.

16.5.1 NUMBER OF TURNS (*N*) (Figure 16-7)

If an inductor has a greater number of turns, the magnetic field produced by passing current through the coil will have more magnetic force than an inductor with fewer turns. A greater magnetic field will cause a larger counter emf, because more magnetic lines of flux will cut more coils of the conductor, producing a larger inductance value. Inductance (*L*) is therefore proportional to the number of turns (*N*):

$$L \propto N$$

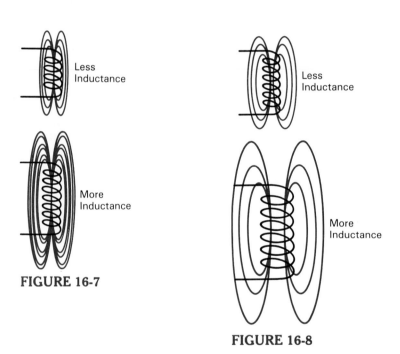

FIGURE 16-7

FIGURE 16-8

16.5.2 AREA OF COIL (*A*) (Figure 16-8)

If the area of the coil is increased for a given number of turns, more magnetic lines of force will be produced, and if the magnetic field is increased, then the inductance will also increase. Inductance is therefore proportional to the area of the coil (*A*):

$$L \propto A$$

16.5.3 LENGTH OF COIL (*l*) (Figure 16-9)

If, for example, four turns are spaced out (long length coil), the summation that occurs between all the individual coil magnetic fields will be small. On the other hand, if four turns are wound close to one another (small length coil), all the individual coil magnetic fields will easily interact and add together to produce a larger magnetic field and, therefore, greater inductance. Inductance is therefore inversely proportional to the length of the coil, in that a longer coil, for a given number of turns, produces a smaller inductance, and vice versa.

$$L \propto \frac{1}{l}$$

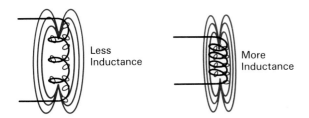

Less Inductance

More Inductance

FIGURE 16-9

16.5.4 CORE MATERIAL (μ) (Figure 16-10)

Most inductors have core materials such as nickel, cobalt, iron, steel, ferrite, or an alloy. These cores have magnetic properties that concentrate or intensify the magnetic field. Permeability is another factor that is proportional to inductance. The greater the permeability of the core material, the greater the inductance is.

$$L \propto \mu$$

16.5.5 FORMULA FOR INDUCTANCE

All the four factors described can be placed in a formula to calculate inductance.

$$\boxed{L = \frac{N^2 \times A \times \mu}{l}}$$

where L = inductance in henries
N = number of turns
A = cross-sectional area in square meters
μ = permeability
l = length of core in meters

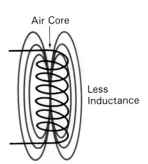

Air Core

Less Inductance

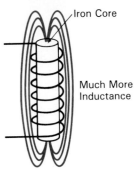

Iron Core

Much More Inductance

FIGURE 16-10

EXAMPLE 16.2

Refer to Figure 16-11(a) and (b) and calculate the inductance of each.

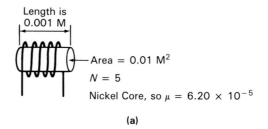

(a)

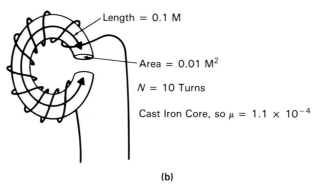

(b)

FIGURE 16-11

SOLUTION

a. $L = \dfrac{5^2 \times 0.01 \times 6.28 \times 10^{-5}}{0.001} = 15.7 \text{ mH}$

b. $L = \dfrac{10^2 \times 0.1 \times 1.1 \times 10^{-4}}{0.1} = 11 \text{ mH}$

SELF-TEST REVIEW (§ 16.5)

1. List the four factors that determine the inductance of an inductor.
2. State the formula for inductance.

16.6

INDUCTORS IN COMBINATION

Inductors oppose the change of current in a circuit and so are treated in a manner similar to resistors connected in combination. Two or more inductors in series merely extend the coil length and increase inductance. Inductors in parallel are treated in a manner similar to resistors, with the total inductance being less than that of the smallest inductor's value.

16.6.1 SERIES

When inductors are connected in series with one another, the total inductance is calculated by summing all the individual inductances.

EXAMPLE 16.3

Calculate the total inductance of the circuit shown in Figure 16-12.

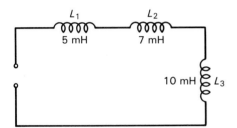

FIGURE 16-12   **Inductors in Series.**

SOLUTION

$$L_T = L_1 + L_2 + L_3$$
$$= 5\ \text{mH} + 7\ \text{mH} + 10\ \text{mH}$$
$$= 22\ \text{mH}$$

16.6.2 PARALLEL

When inductors are connected in parallel with one another, the general (two or more inductors) or product over sum (two inductors) formula can be used to find total inductance, which will always be less than the smallest inductor's value.

EXAMPLE 16-4

Determine L_T for the circuits in Figure 16-13(a) and (b).

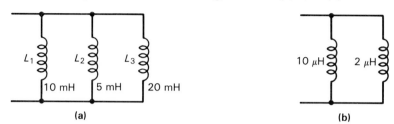

FIGURE 16-13 Inductors in Parallel.

SOLUTION

a. General formula:

$$L_T = \cfrac{1}{\left(\cfrac{1}{L_1}\right) + \left(\cfrac{1}{L_2}\right) + \left(\cfrac{1}{L_3}\right)}$$

$$= \cfrac{1}{\left(\cfrac{1}{10 \text{ mH}}\right) + \left(\cfrac{1}{5 \text{ mH}}\right) + \left(\cfrac{1}{20 \text{ mH}}\right)}$$

$$= 2.9 \text{ mH}$$

b. Product over sum:

$$L_T = \frac{L_1 \times L_2}{L_1 + L_2}$$

$$= \frac{10 \ \mu\text{H} \times 2 \ \mu\text{H}}{10 \ \mu\text{H} + 2 \ \mu\text{H}}$$

$$= \frac{20 \times 10^{-12} \text{H}}{12 \ \mu\text{H}} = 1.67 \ \mu\text{H}$$

SELF-TEST REVIEW (§ 16.6)

1. To calculate total inductance, inductors can be treated in the same manner as capacitors (true or false).
2. State the formula for calculating total inductance in:
   a. A series circuit
   b. A parallel circuit
3. Calculate the total circuit inductance if 4 mH and 2 mH are connected:
   a. In series
   b. In parallel

16.7
INDUCTOR TYPES

As with resistors and capacitors, inductors are basically divided into the two logical categories of fixed and variable, as seen by the symbols in Figure 16-14(a) and (b). Within these two categories, inductors are generally classified by the type of core material used.

With fixed inductors, the inductance value remains constant and cannot be altered. It is usually made of solid copper wire with an insulating enamel around the conductor, which is then wound into a coil. The three major types of fixed inductors on the market today are:

(a)

(b)

**FIGURE 16-14
(a) Fixed Inductor
Symbol. (b) Variable
Inductor Symbol.**

1. Air core
2. Iron core
3. Ferrite core

With variable inductors, the inductance value can be changed by adjusting the position of the movable core with respect to the stationary coil. There is basically only one type of variable inductor in wide use today, the ferrite core.

16.7.1 FIXED INDUCTORS

(1) AIR CORE

Figure 16-15 illustrates some typical air core fixed inductor types. The insulated copper wire can be wound on nonmagnetic materials such as plastic, ceramic, or Bakelite, which due to their nonmagnetic properties will have no effect on the inductance value. These materials act as a support or form for the inductor, whose wire size may be too small to support itself. If a

FIGURE 16-15 Air Core Fixed Inductors. (a) Air Core Inductor. (b) Construction.

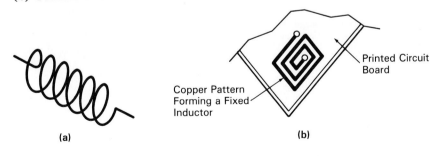

(a)

Copper Pattern Forming a Fixed Inductor

Printed Circuit Board

(b)

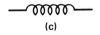
(c)

heavier or more rigid wire size is used, it can support itself and has no need for a nonmagnetic form.

Inductors of this type, which make use of a nonmagnetic form or do not have a form, are known as air core inductors. Air core inductors have the advantage that their inductance value does not vary with current, as do iron and ferrite core inductors. Their typical low values (below 10 μH) normally find application in high-frequency equipment, such as AM, FM, TV, and other communications transmitter and receiver circuits.

(2) IRON CORE

Figure 16-16 illustrates some typical iron core inductors. The name *choke* is used interchangeably with the word inductor, because choke basically describes an inductor's behavior. As we previously discovered, the inductor will oppose any change of current flow, whether it is in the form of pulsating dc or ac, because of its self inductance that generates a counter emf to limit or choke the flow of current.

An iron core has a higher permeability figure than air, which means it will concentrate the magnetic lines of force and, therefore, increase inductance.

$$L\uparrow = \frac{N^2 \times A \times \mu\uparrow}{l}$$

The E I inductor, seen in Figure 16-16(b), has two sections, one of which resembles the letter E and the other I. The conductor is wrapped around the center section of the E to form a coil. These inductors or chokes can be made with the highest inductance values, up to the hundreds of henrys and are used in dc and low frequency ac circuits.

(3) FERRITE CORE

Figure 16-17 illustrates two typical ferrite core inductors. Ferrite is a chemical compound, which is basically powdered iron oxide and ceramic. Although its permeability is less than iron, it has a higher permeability than air and can therefore obtain a higher inductance value than air for the same number of turns.

FIGURE 16-16 Iron Core Fixed Inductors. (a) Physical Appearance. (b) Construction. (c) Symbol.

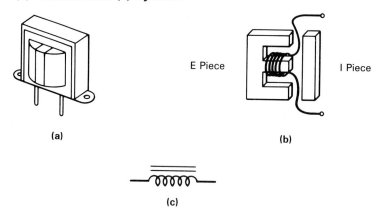

E Piece

I Piece

(a)

(b)

(c)

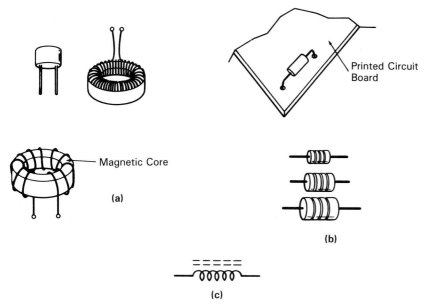

Printed Circuit Board

Magnetic Core

(a)

(b)

(c)

FIGURE 16-17 **Ferrite Core Fixed Inductors. (a) Toroidal Inductor.**
(b) Molded Inductor.

The toroidal inductor (doughnut shaped) illustrated in Figure 16-17(a) has an advantage over the cylindrical-shaped core in that the magnetic lines of force do not have to pass through air. With a cylindrical core, the magnetic lines of force emerge out of one end of the cylinder, propagate or travel through air, and then enter in the other end of the cylinder. The nonmagnetic air, which has a low permeability, will reduce the strength of the field and, therefore, the inductance value:

$$L\!\downarrow = \frac{N^2 \times A \times \mu\!\downarrow}{l}$$

If one could bend around either end of the cylinder to make a toroid, the magnetic lines of force emerging from one end of the inductor's core would pass through ferrite, which has a high permeability.

Toroidal type ferrite core inductors have a greater inductance than the cylindrical-type inductors; however, they are more expensive to manufacture. They are used in both low and high frequency applications.

The molded inductor, seen in Figure 16-17(b), consists of a spiral metal film deposited on a cylindrically shaped ferrite core. The entire assembly is encapsulated, and the available values range from 1.2 μH to 10 mH, with maximum currents of about 70 mA. Due to its small size, the value is often color coded onto the body instead of using alphanumeric or typographical labels, which can be used on all of the other larger sized inductors.

16.7.2 VARIABLE INDUCTORS

Figure 16-18 illustrates a typical ferrite core variable inductor. A screw adjustment moves a sliding ferrite core or slug farther in or out of the stationary coil. If the slug is all the way out, the permeability figure is low, because

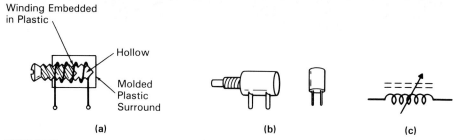

Winding Embedded
in Plastic

Hollow

Molded
Plastic
Surround

(a)

(b)

(c)

FIGURE 16-18   Ferrite Core Variable Inductors. (a) Construction.
(b) Physical Appearance. (c) Symbol.

an air core is being used ($L\downarrow \propto \mu\downarrow$). If the slug is adjusted and screwed into the coil, the air is replaced by the ferrite slug, which has a higher permeability and, therefore, inductance value ($L\uparrow \propto \mu\uparrow$).

Variable inductors should only be adjusted with a plastic or nonmetallic alignment tool. They are used extensively in radio circuits.

SELF-TEST REVIEW (§ 16.7)

1. List the three fixed inductor types.
2. Under which category would a fixed inductor with a nonmagnetic core come?
3. What is a ferrite compound?
4. What advantage does the toroid shaped inductor have?
5. What type of variable inductor is the only one in wide use today?
6. What factor is varied to change the value of the variable inductor?

16.8
INDUCTIVE TIME CONSTANT

Inductors will not have any effect on a constant value direct current (dc) source. If the dc is changing (pulsating), the inductor will oppose the change whether it is an increase or decrease in direct current, because a change in current causes the magnetic field to expand or contract, and in so doing it will cut the coil of the inductor and induce a voltage that will counter the applied emf.

16.8.1   DC CURRENT RISE

Figure 16.19(a) illustrates an inductor (L) connected across a dc source (battery) through a switch and series connected resistor. When the switch is closed, current will flow and the magnetic field will begin to expand around

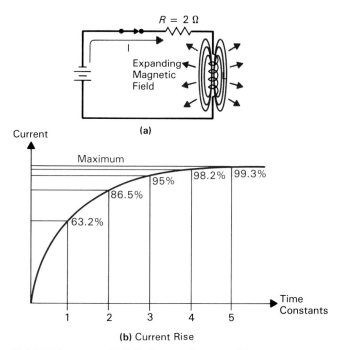

(a)

(b) Current Rise

FIGURE 16-19   DC Inductor Current Rise.

the inductor. This field cuts the coils of the inductor and induces a counter emf to oppose the rise in current. Current in an inductive circuit cannot rise instantly to its maximum value, which is determined by Ohm's law ($I = V/R$). Current will in fact take a time to rise to maximum, as graphed in Figure 16-19(b), due to the inductor's ability to oppose change.

It will actually take five time constants (5τ) for the current in an inductive circuit to reach maximum value. This time can be calculated by using the formula

$$\tau = \frac{L}{R} \text{ seconds}$$

The constant to remember is the same as before: 63.2%. In one time constant ($1 \times L/R$) the current in the LR circuit will have reached 63.2% of its maximum value. In two time constants ($2 \times L/R$), the current will have increased 63.2% of the remaining current, and so on through 5 time constants.

For example, if the maximum possible circuit current is 100 mA and an inductor of 4 H is connected in series with a resistor of 2 Ω, the current will increase as seen in Figure 16-20.

How quickly an inductor will allow the current to rise to its maximum value is proportional to the inductance and inversely proportional to the resistance. A larger inductance increases the strength of the magnetic field, and so the opposition or counter emf increases, and the longer it takes for current to rise to a maximum ($\tau\uparrow = L\uparrow/R$). If the circuit resistance is increased, the maximum current will be smaller, and a smaller maximum is reached more quickly than a higher.

Inductive Time Constant   **567**

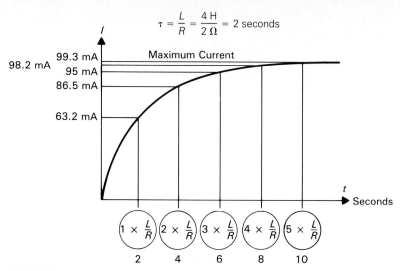

$$\tau = \frac{L}{R} = \frac{4\,H}{2\,\Omega} = 2 \text{ seconds}$$

FIGURE 16-20 Exponential Current Rise.

16.8.2 DC CURRENT FALL

When the inductor's dc source of current is removed, as seen in Figure 16-21(a), by placing the switch in position *B*, the magnetic field will collapse and cut the coils of the inductor, inducing a voltage and causing a current to flow in the same direction as the original source current. This current will exponentially decay, or fall from the maximum to zero level, in 5 time constants ($5 \times L/R = 5 \times 4/2 = 10$ seconds), as seen in Figure 16-21(b).

FIGURE 16-21 Exponential Current Fall.

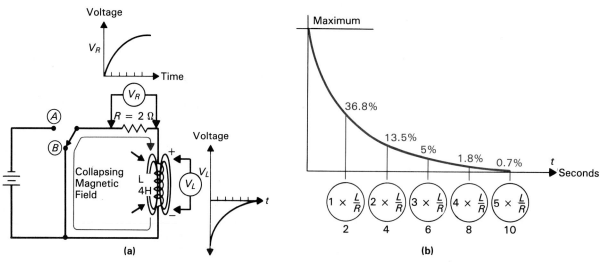

EXAMPLE 16.5

Calculate the circuit current at each of the five time constants if a 12 V dc source is connected across a series RL circuit, and $R = 60\ \Omega$ and $L = 24$ mH.

SOLUTION

$$\text{maximum current, } I_{max} = \frac{V_s}{R} = \frac{12\ V}{60\ \Omega} = 200\ mA$$

$$\text{time constant, } \tau = \frac{L}{R} = \frac{24\ mH}{60\ \Omega} = 400\ \mu s$$

At 1 time constant (400 μs after source voltage is applied), the current will be:

$$i = 63.2\% \text{ of } I_{max}$$
$$= 0.632 \times 200\ mA = 126.4\ mA$$

At 2 time constants (800 μs after source voltage is applied):

$$i = 86.5\% \text{ of } I_{max}$$
$$= 0.865 \times 200\ mA = 173\ mA$$

At 3 time constants (1200 μs or 1.2 ms):

$$i = 95\% \text{ of } I_{max}$$
$$= 0.95 \times 200\ mA = 190\ mA$$

At 4 time constants (1.6 ms):

$$i = 98.2\% \text{ of } I_{max}$$
$$= 0.982 \times 200\ mA = 196.4\ mA$$

At 5 time constants (2 ms):

$$i = 99.3\% \text{ of } I_{max}$$
$$= 0.993 \times 200\ mA = 198.6\ mA, \text{ approximately maximum (200 mA)}$$

16.8.3 AC RISE AND FALL

If an alternating (ac) voltage is applied across an inductor, as seen in Figure 16-22(a), the inductor will continuously oppose the alternating current because it is always changing.

There are three factors to consider in an inductive circuit: the applied voltage, the induced voltage, and the current, as seen in Figure 16-22(b).

The current in the circuit causes the magnetic field to expand and collapse and cut the conducting coils, resulting in an induced counter emf.

At points X and Y, the steepness of the current waveform indicates that the current will be changing at its maximum rate, and therefore the opposition or counter emf will also be maximum. When the current is at its maximum

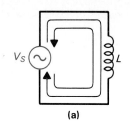

(a)

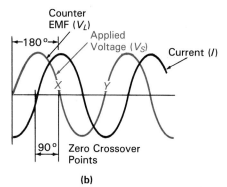

(b)

FIGURE 16-22

positive or negative value, it has a very small or no rate of change (flat peaks). Therefore, the opposition or counter emf should be very small or zero, as can be seen by the waveforms. The counter emf is, therefore, said to be 90° out of phase with the circuit current.

To summarize, we can say that the voltage across the inductor (V_L) or counter emf leads the circuit current by 90°.

SELF-TEST REVIEW (§ 16.8)

1. How does the inductive time constant relate to the capacitive time constant?
2. The greater the value of the inductor, the longer it would take for current to rise to a maximum (true or false).
3. A constant dc level is opposed continuously by an inductor (true or false).
4. What reaction does an inductor have to ac?

16.9
INDUCTIVE REACTANCE

Reactance is the opposition to current flow without the dissipation of energy, as opposed to resistance, which is the opposition to current flow with the dissipation of energy.

Inductive reactance (X_L) is the opposition to current flow offered by an inductor without the dissipation of energy. It is measured in ohms and can be calculated by using the formula:

$$\boxed{X_L = 2 \times \pi \times f \times L}$$

where X_L = inductive reactance in ohms
 f = frequency in hertz
 L = inductance in henries
 2π = 2π radians, 360° or 1 cycle

Inductive reactance is proportional to frequency ($X_L \propto f$) because a higher frequency (fast switching) will cause a greater amount of current change, and a greater change will generate a larger counter emf, which is an opposition or reactance against current flow. When 0 Hz is applied to a coil (dc), there exists no change and so the inductive reactance of an inductor to dc is zero ($X_L = 2\pi \times 0 \times L = 0$).

Inductive reactance is also proportional to inductance because a larger inductance will generate a greater magnetic field and subsequent counter emf, which is the opposition to current flow.

EXAMPLE 16.6

Calculate the current flowing in the circuit illustrated in Figure 16-23.

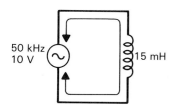

50 kHz
10 V

15 mH

FIGURE 16-23

SOLUTION

The current can be calculated by Ohm's law and is a function of the voltage and opposition, which in this case is inductive reactance.

$$I = \frac{V}{X_L}$$

However, we must first calculate X_L:

$$X_L = 2 \times \pi \times f \times L$$
$$= 6.28 \times 50 \text{ kHz} \times 15 \text{ mH}$$
$$= 4710 \ \Omega \quad \text{or} \quad 4.71 \text{ k}\Omega$$

Current is therefore equal to

$$I = \frac{V}{X_L} = \frac{10 \text{ V}}{4.71 \text{ k}\Omega} = 2.12 \text{ mA}$$

1. Define and state the formula for inductive reactance.
2. Why is inductive reactance proportional to frequency and the inductance value?
3. How does inductive reactance relate to Ohm's law?
4. Inductive reactance is measured in henries (true/false).

16.10
SERIES *RL* CIRCUIT

In a purely resistive circuit, as seen in Figure 16-24, the current flowing within the circuit and the voltage across the resistor are in phase with one

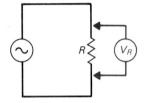

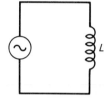

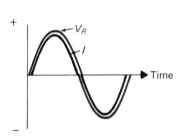

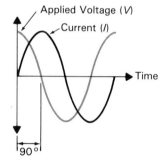

**FIGURE 16-24 Purely
Resistive Circuit; Current
and Voltage Are in Phase.**

**FIGURE 16-25 Purely
Inductive Circuit;
Current Lags Applied
Voltage by 90°.**

FIGURE 16-26 Series RL Circuit.

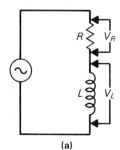

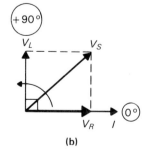

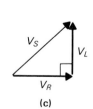

(a)

(b)

(c)

another. In a purely inductive circuit, as seen in Figure 16-25, the current will lag the applied voltage by 90°. If we connect a resistor and inductor in series, as seen in Figure 16-26(a), we will have the most common combination of R and L used in electronic equipment.

16.10.1 VOLTAGE

The voltage across the inductor and resistor can be calculated by using Ohm's law:

$$V_R = I \times R$$
$$V_L = I \times X_L$$

The vector diagram illustrates current (I) as the reference at 0° and, as expected, the voltage across the resistor (V_R) is in phase with the circuit current.

The voltage across the inductor (V_L) leads the circuit current and the voltage across the resistor (V_R) by 90°, or, the circuit current vector lags the voltage across the inductor by 90°.

As with capacitive vectors, we have to apply Kirchhoff's voltage law when calculating the value of applied or source voltage (V_S), which, due to the phase difference between V_R and V_L, is the vector sum of all the voltage drops.

By creating a right triangle from the three quantities, as seen in Figure 16-26(b), and applying the Pythagorean theorem, we arrive at a formula for source voltage.

$$V_S = \sqrt{V_R^2 + V_L^2}$$

As with any formula with three quantities, if two are known, the other can be calculated by simply rearranging the formula to:

$$V_S = \sqrt{V_R^2 + V_L^2}$$
$$V_R = \sqrt{V_S^2 - V_L^2}$$
$$V_L = \sqrt{V_S^2 - V_R^2}$$

16.10.2 IMPEDANCE (Z)

Impedance is the total opposition to current flow offered by a circuit with both resistance and reactance. It is measured in ohms and can be calculated by using Ohm's law:

$$Z = \frac{V}{I}$$

FIGURE 16-27 Impedance.

Just as a phase shift or difference exists between V_R and V_L and they cannot be added to find applied voltage, the same phase difference exists between R and X_L, so impedance or total opposition cannot be simply the sum of the two, as seen in Figure 16-27.

The impedance of a series *RL* circuit is equal to the square root of the sum of the squares of resistance and reactance, and, by rearrangement, X_L and R can also be calculated if the other two values are known:

$$Z = \sqrt{R^2 + X_L^2}$$
$$R = \sqrt{Z^2 - X_L^2}$$
$$X_L = \sqrt{Z^2 - R^2}$$

16.10.3 PHASE SHIFT

If a circuit is purely resistive, the phase shift (θ) is zero, and if a circuit is purely inductive, the phase shift is $+90°$. If the resistance and inductive reactance are equal, the phase shift will equal $+45°$, as seen in Figure 16-28.

The phase shift in an inductive and resistive circuit is the degrees of lead between the source voltage and current, and by looking at the examples in Figure 16-28, you can see that the phase angle is proportional to reactance and inversely proportional to resistance. Mathematically, it can be expressed as

$$\theta = \arctan \frac{X_L}{R}$$

FIGURE 16-28

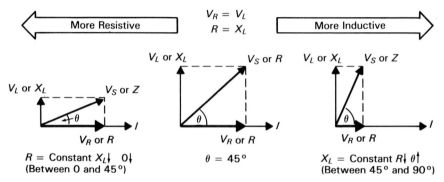

Inductance and Inductors

As the current is the same in both the inductor and resistor in a series circuit, the voltage drops across the inductor and resistor are directly proportional to reactance and resistance:

$$V_R \updownarrow \ = I \text{ (constant)} \times R \updownarrow , \qquad V_L \updownarrow \ = I \text{ (constant)} \times X_L \updownarrow$$

the phase shift can also be calculated by using the voltage drop across the inductor and resistor.

$$\theta = \arctan \frac{V_L}{V_R}$$

16.10.4 POWER

(1) PURELY RESISTIVE CIRCUIT

Figure 16-29 illustrates the current, voltage, and power waveforms produced when applying an ac voltage across a purely resistive circuit. Voltage and current are in phase, and power can be calculated by multiplying current by voltage ($P = V \times I$).

The sinusoidal power waveform is totally positive, as a positive voltage multiplied by a positive current produces a positive value of power, and a negative voltage multiplied by a negative current also produces a positive value of power. For this reason, the resistor is said to develop a positive

Voltage and Current
Are in RMS
(Effective) as Normal

FIGURE 16-29 Purely Resistive Circuit.

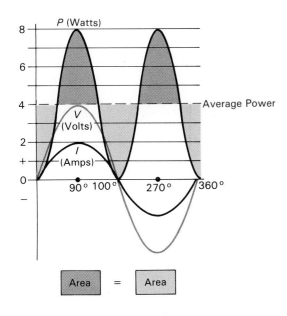

power waveform that is twice the frequency of the voltage or current waveform.

The average value of power dissipated by a purely resistive circuit is the halfway value between maximum and zero, in this example, 4 watts.

(2) PURELY INDUCTIVE CIRCUIT

The pure inductor, like the capacitor, is a reactive component, which means that it will consume power without the dissipation of energy. The capacitor holds its energy in an electric field, while the inductor consumes and holds its energy in a magnetic field and then releases it back into the circuit.

The power curve alternates equally above and below the zero line, as seen in Figure 16-30. During the first positive power half cycle, the circuit current is on the increase, to maximum (point *A*), the magnetic field is building up, and the inductor is storing electrical energy. When the circuit current is on the decline between *A* and *B*, the magnetic field begins to collapse and self-induction occurs and returns electrical energy back into the circuit. The power alternation is both positive when the inductor is consuming power and negative when the inductor is returning the power

FIGURE 16-30 Purely Inductive Circuit.

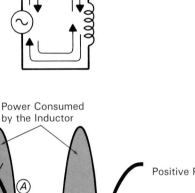

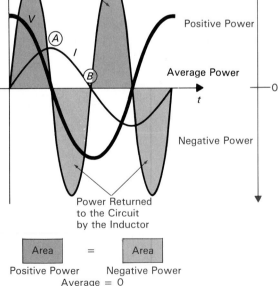

Inductance and Inductors

back into the circuit. As the positive and negative power alternations are equal but opposite, the average power dissipated is zero.

(3) RESISTIVE AND INDUCTIVE CIRCUIT

An inductor is different from a capacitor in that it has a small amount of resistance no matter how pure the inductor. For this reason, inductors will never have an average power figure of zero, because even the best inductor will have some value of inductance and resistance within it, as seen in Figure 16-31. The reason an inductor has resistance is that it is simply a piece of wire, and any piece of wire has a certain value of resistance, as well as inductance. This coil resistance should be, and normally is, very small and can usually be ignored; however, in some applications even this small resistance can prevent the correct operation of a circuit, so a value or term had to be created to specify the differences in the quality of inductor available.

The *quality factor* (Q) of an inductor is the ratio of the energy stored in the coil by its inductance to the energy dissipated in the coil by the resistance; therefore, the higher the Q, the better the coil is at storing energy rather than dissipating it:

$$\boxed{\text{quality } (Q) = \frac{\text{energy stored}}{\text{energy dissipated}}}$$

The energy stored is dependent on the inductive reactance (X_L) of the coil, and the energy dissipated is dependent on the resistance (R) of the coil. The quality factor of a coil or inductor can therefore also be calculated by using the formula

$$\boxed{Q = \frac{X_L}{R}}$$

Inductors, therefore, will never appear as pure inductance, but rather as an inductive and resistive (RL) circuit, and the resistance within the inductor will dissipate *true power*.

Figure 16-32 illustrates a circuit containing R and L and the power waveforms produced when $R = X_L$; the phase shift (θ) is equal to 45°.

The positive power alternation, which is above the zero line, is the combination of the power dissipated by the resistor and the power consumed by the inductor, while circuit current is on the rise. The negative power alternation is the power that was given back to the circuit by the inductor

FIGURE 16-31   Resistance within an Inductor.

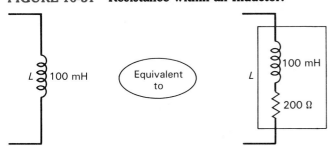

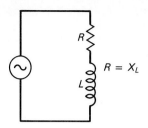

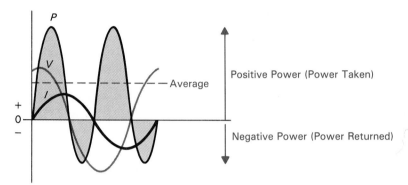

FIGURE 16-32 **Resisitive and Inductive Circuit.**

while the inductor's magnetic field was collapsing and returning the energy that was consumed.

(4) POWER FACTOR

When a circuit contains both inductance and resistance, some of the energy is consumed and then returned by the inductor (reactive or imaginary power),

---

EXAMPLE 16.7

Calculate the quality factor Q of a 22 mH coil connected across a 2 kHz, 10 V source if its internal coil resistance is 27 Ω.

SOLUTION

$$Q = \frac{X_L}{R}$$

The reactance of the coil is not known but can be calculated by the formula

$$X_L = 2\pi fL$$
$$= 2 \times \pi \times 2\ \text{kHz} \times 22\ \text{mH} = 276.5\ \Omega$$

Therefore,

$$Q = \frac{X_L}{R} = \frac{276.5\ \Omega}{27\ \Omega} = 10.24$$

---

and some of the energy is dissipated and lost by the resistor (resistive or true power).

Apparent power is the power that appears to be supplied to the load and is the vector sum of both the reactive and true power; it can be calculated by using the formula

$$P_A = \sqrt{P_R^2 + P_X^2}$$

where P_A = apparent power
P_R = true power
P_X = reactive power

The power factor is a ratio of the true power to the apparent power and is therefore a measure of the loss in a circuit.

$$PF = \frac{\text{true power } (P_R)}{\text{apparent power } (P_A)}$$

or

$$PF = \frac{R}{Z}$$

or

$$PF = \cos \theta$$

EXAMPLE 16.8

Calculate the following for a series RL circuit if $R = 40$ kΩ, $L = 450$ mH, $f = 20$ kHz, and $V_S = 6$ V.
a. X_L
b. Z
c. I
d. θ
e. Apparent power
f. PF

SOLUTION

a. $X_L = 2\pi f L = 2 \times \pi \times 20$ kHz $\times 450$ mH
    $= 56.5$ kΩ
b. $Z = \sqrt{R^2 + X_L^2}$
    $= \sqrt{40 \text{ k}\Omega^2 + 56.5 \text{ k}\Omega^2} = 69.23$ kΩ
c. $I = \dfrac{V_S}{Z} = \dfrac{6 \text{ V}}{69.23 \text{ k}\Omega} = 86.6$ μA
d. $\theta = \arctan \dfrac{X_L}{R} = \arctan \dfrac{56.5 \text{ k}\Omega}{40 \text{ k}\Omega} = 54.7°$

e. Apparent power $= \sqrt{\text{(true power)}^2 + \text{(reactive power)}^2}$

$$P_R = I^2 \times R = 86.6\ \mu A^2 \times 40\ k\Omega = 300\ \mu W$$
$$P_X = I^2 \times X_L = 86.6\ \mu A^2 \times 56.5\ k\Omega = 423.7\ \mu VAR$$
$$P_A = \sqrt{P_R^2 + P_X^2}$$
$$= \sqrt{300\ \mu W^2 + 423.7\ \mu W^2} = 519.2\ \mu VAR$$

f. $PF = \dfrac{P_R}{P_A} = \dfrac{300\ \mu W}{519.2\ \mu W} = 0.57$

$= \dfrac{R}{Z} = \dfrac{40\ k\Omega}{69.23\ k\Omega} = 0.57$

$= \text{Cos } \theta = \text{Cos } 54.7° = 0.57$

SELF-TEST REVIEW (§ 16.10)

1. In a purely inductive circuit, the current will lead the applied voltage by 90° (true or false).
2. Calculate the applied source voltage V_S in an RL circuit where V_R = 4 V and V_L = 2 V.
3. Define and state the formula for impedance when R and X_L are known.
4. If $R = X_L$, the phase shift will equal _____.
5. What is meant by positive power?
6. Define Q and state the formula when X_L and R are known.
7. What is the difference between true and reactive power?
8. State the power factor formula.

16.11
PARALLEL *RL* CIRCUIT

When R and L are connected in parallel across an ac source, as seen in Figure 16-33(a), the voltages across both components are equal because of the parallel connection. Total circuit current (I_T) is a combination of the resistive current (I_R) and inductive current (I_L), which are 90° out of phase with one another, as seen by the vector diagram in Figure 16-33(b). The total current is consequently the vector sum of both I_R and I_L.

EXAMPLE 16.9

Calculate the following for a parallel RL circuit if $R = 45\ \Omega$, $X_L = 1100$ Ω, $V_S = 24$ V.

a. I_R
b. I_L
c. I_T
d. Z
e. θ

SOLUTION

a. $I_R = \dfrac{V_S}{R} = \dfrac{24\text{ V}}{45\ \Omega} = 533.3\text{ mA}$

b. $I_L = \dfrac{V_S}{X_L} = \dfrac{24\text{ V}}{1100\ \Omega} = 21.8\text{ mA}$

c. $I_T = \sqrt{I_R^2 + I_L^2}$
$ = \sqrt{533.3\text{ mA}^2 + 21.8\text{ mA}^2} = 533.7\text{ mA}$

d. $Z = \dfrac{R \times X_L}{\sqrt{R^2 + X_L^2}} = \dfrac{45\ \Omega \times 1100\ \Omega}{\sqrt{45\ \Omega^2 + 1100\ \Omega^2}}$

$ = \dfrac{49.5\text{ k}\Omega}{1100.9\ \Omega} = 44.96\ \Omega$

e. $\theta = \arctan \dfrac{R}{X_L} = \arctan \dfrac{45\ \Omega}{1100\ \Omega} = 2.34°$

Therefore, it lags V_S by 2.34°.

FIGURE 16-33

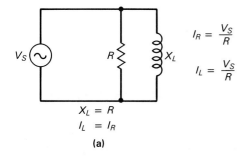

$I_R = \dfrac{V_S}{R}$

$I_L = \dfrac{V_S}{R}$

$X_L = R$
$I_L = I_R$

(a)

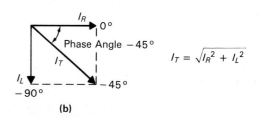

$I_T = \sqrt{I_R^2 + I_L^2}$

(b)

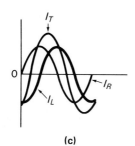

(c)

1. When R and L are connected in parallel with one another, the voltage across each will be different and dependent on the values of R and X_L (true/false).
2. State the formula for calculating total current (I_T) when I_R and I_L are known.

16.12
TESTING INDUCTORS

Basically, only three problems can occur with inductors:

1. An open
2. A complete short
3. A section short (value change)

16.12.1 OPEN [Figure 16-34(a)]

This problem can be isolated with an ohmmeter. Depending on the winding's resistance, the coil should be in the range of zero to a few hundred ohms. An open accounts for 75% of all defective inductors.

16.12.2 COMPLETE AND SECTION SHORT [Figure 16-34(b)]

A coil with one or more shorted turns or a complete short can be checked with an ohmmeter and thought to be perfectly good because of the normally

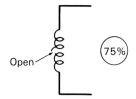

FIGURE 16-34   Defective Inductors.

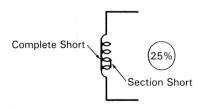

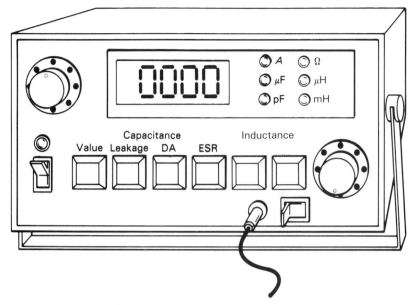

FIGURE 16-35 Capacitor and Inductor Analyzer.

low resistance of a coil, as it is just a piece of wire. But if it is placed in a circuit with a complete or section short present, it will not function effectively as an inductor, if at all. For these checks, an inductor analyzer needs to be used like the one seen in Figure 16-35, which can be used to check capacitance and inductance. Complete or section shorts account for 25% of all defective inductors.

SELF-TEST REVIEW (§ 16.12)

1. How could the following inductor malfunctions be recognized?
   a. An open
   b. A complete or section short
2. Which inductor malfunction accounts for almost 75% of all failures?

16.13
APPLICATIONS OF INDUCTORS

16.13.1 *RL* INTEGRATOR

In the *RL* integrator, the output is taken across the resistor, as seen in the circuit in Figure 16-36(a). The output is the same as the *RC* integrator's output. When the input rises from 0 to 10 V at the leading positive edge of the square wave input, the situation is as seen in Figure 16-37(a).

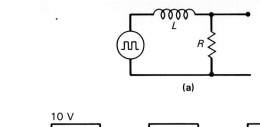

(a)

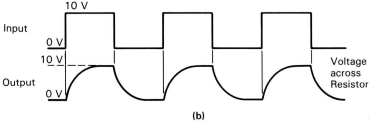

(b)

FIGURE 16-36 *RL* **Integrator.**

FIGURE 16-37

Expanding Magnetic Field, Self–Induction

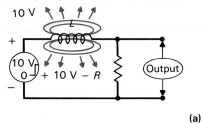

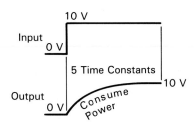

(a)

Stationary Magnetic Field, No Self–Induction

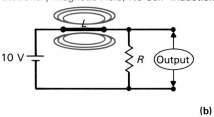

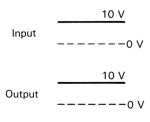

(b)

Collapsing Magnetic Field, Self–Induction

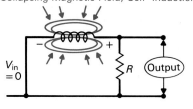

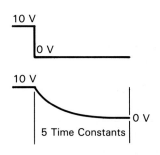

(c)

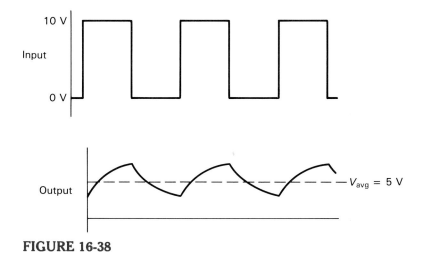

FIGURE 16-38

The inductor's 10 V counter emf opposes the sudden change from 0 to 10 V, and if 10 V is across the inductor, then 0 V must be across the resistor (Kirchhoff's voltage law), and therefore appearing at the output. After 5 time constants ($5 \times L/R$), the inductor's current in the circuit will have built up to maximum (V_{in}/R), and the inductor will be an equivalent short circuit, because no change and consequently no back emf exists, as seen in Figure 16-37(b). All the input voltage will now be across the resistor and, therefore, at the output.

When the square wave input drops to zero, the circuit is equivalent to that seen in Figure 16-37(c), and the collapsing magnetic field will cause an induced voltage within the conductor, which will cause current to flow within the circuit for 5 time constants, whereupon it will reach 0.

As with the *RC* integrator, if the period of the square wave is decreased or if the time constant is increased, the output will reach an average value of half the input square wave's amplitude, as seen in Figure 16-38.

16.3.2 *RL* DIFFERENTIATOR

In the *RL* differentiator, the output is taken across the inductor, as seen in Figure 16-39, and is the same as the *RC* differentiator's output. When the square wave input rises from 0 to 10 V, the inductor will generate a 10 V counter emf across it, as seen in Figure 16-40(a), and this 10 V will all appear across the output. As the circuit current exponentially rises, the voltage across the inductor, and therefore at the output, will fall exponentially to 0 V after 5 time constants.

When the square wave input falls from 10 to 0 V, the circuit is equivalent to that seen in Figure 16-40(b), and the collapsing magnetic field induces a counter emf. The sudden -10 V impulse at the output will decrease to 0 V as the circuit current decreases in 5 time constants.

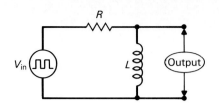

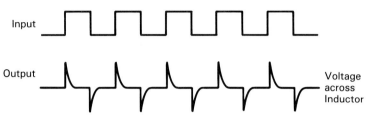

FIGURE 16-39 *RL* **Differentiator.**

FIGURE 16-40

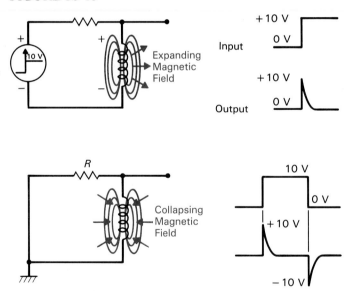

16.13.3 *RL* FILTER

The *RL* filter will achieve the same results as the *RC* filter in that it will pass some frequencies and block others, as seen in Figure 16-41. The inductive reactance of the coil and the resistance of the resistor form a voltage divider. Since inductive reactance is proportional to frequency ($X_L \propto f$), the inductor

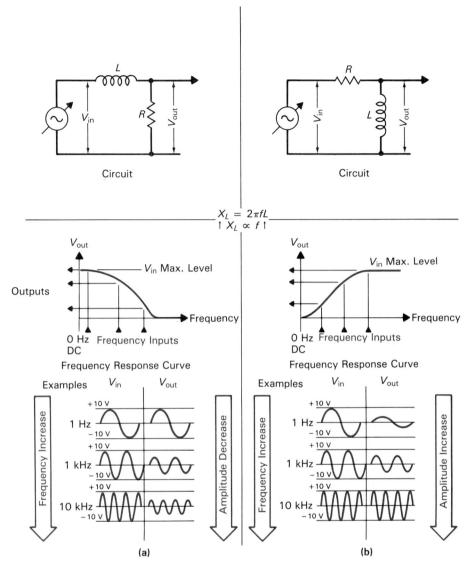

FIGURE 16-41   *RL* Filter. (a) Low Pass Filter. (b) High Pass Filter.

will drop less voltage at lower frequencies ($f\downarrow$, $X_L\downarrow$, $V_L\downarrow$) and more voltage at higher frequencies ($f\uparrow$, $X_L\uparrow$, $V_L\uparrow$).

With the low-pass filter seen in Figure 16-41(a) the output is developed across the resistor. If the frequency of the input is low, the inductive reactance will be low, and so almost all the input will be developed across the resistor and applied to the output. If the frequency of the input increases, the inductor's reactance will increase, resulting in almost all the input being dropped across the inductor and none across the resistor and, therefore, the output.

With the high pass filter, seen in Figure 16-41(b), the inductor and resistor have been placed in opposite positions, resulting in an effect opposite to the low pass filter.

Applications of Inductors   **587**

1. List three circuit applications of the inductor.
2. What are the waveform differences between an integrator and differentiator?
3. Will an *RL* filter act as a low pass filter or high pass filter, or can it be made to function as either?

SUMMARY

1. Each electron has its own very small magnetic field due to its velocity.
2. Electromagnetism is the magnetic field that exists around a conductor when current is passed through it.
3. Direct current produces a constant magnetic field of a fixed polarity.
4. Alternating current produces an alternating magnetic field that continually switches polarity.
5. Electromagnetic induction occurs when relative motion exists between a conductor and a magnetic field, causing an induced voltage and subsequent current flow.
6. Electromagnetism and electromagnetic induction are the two phenomena responsible for a property known as inductance.
7. Self inductance is the ability of a current carrying coil or conductor to induce a voltage within itself, which is why inductors oppose a change in current.
8. The induced voltage in an inductor is equal to the inductance times the rate of change of current and is called a counter or back emf.
9. An inductor is an electronic component that will oppose any changes in circuit current, and this ability or behavior is referred to as inductance.
10. The inductance of an inductor is determined by four factors:
    a. Number of turns
    b. Area of the coil
    c. Length of the coil
    d. Core material used within the coil
11. When inductors are connected in series with one another, the total inductance is calculated by adding all the individual inductances.
12. When inductors are connected in parallel with one another, the total inductance is less than the value of the smallest inductance.
13. The time constant of a series *RL* circuit is equal to the inductance divided by the resistance.
14. The voltage across an inductor leads the current by 90°.
15. Inductive reactance (X_L) is the opposition to current flow offered by an inductor and is measured in ohms.

16. An inductor acts as an open to a change (ac) and a short or low resistance to dc (X_L = the internal resistance of the inductor at 0 Hz).

17. Impedance is the total opposition to current flow offered by a circuit with both resistance and reactance and is measured in ohms.

18. Energy is stored by an inductor in its magnetic field, similarly to a capacitor storing energy in its electric field.

19. The *RL* integrator is a series circuit with the output taken across the resistor, while the *RL* differentiator is also a series circuit, but in this case the output is taken across the inductor.

20. Due to the proportional relationship between inductive reactance and frequency, the series *RL* circuit can also be used as a filter to pass either high or low frequencies.

NEW TERMS

Air core inductor
Choke
Ferrite core inductor
Form
Henry

Inductive reactance
Inductive time constant
Inductor
Iron core inductor
Q or quality factor

RL differentiator
RL filter
RL integrator
Self induction
Slug
Toroidal inductor

NEW FORMULAS

$$V_{ind} = L \times \frac{\Delta i}{\Delta t}$$

L = inductance in henries
Δi = increment of change of current
Δt = increment of change of time

$$L = \frac{N^2 \times A \times \mu}{l}$$

N = number of turns
A = cross-sectional area in square meters
μ = permeability
l = length of core in meters

Parallel RL Circuit:

$$I_R = \frac{V_S}{R}$$

$$I_L = \frac{V_S}{X_L}$$

$$I_T = \sqrt{I_R^2 + I_L^2}$$

$$Z = \frac{R \times X_L}{\sqrt{R^2 + X_L^2}}$$

$$\theta = \arctan \frac{R}{X_L}$$

$$L_T = \cfrac{1}{\left(\cfrac{1}{L_1}\right) + \left(\cfrac{1}{L_2}\right) + \left(\cfrac{1}{L_3}\right) + \cdots}$$

$$L_T = \frac{L_1 \times L_2}{L_1 + L_2}$$

Series RL Circuit:

$L_T = L_1 + L_2 + L_3 + \cdots \qquad L_T$ = total inductance in henries

$V_R = I \times R$

$V_L = I \times X_L$

$V_S = \sqrt{V_R^2 + V_L^2} \qquad V_R = \sqrt{V_S^2 - V_L^2} \qquad V_L = \sqrt{V_S^2 - V_R^2}$

$Z = \dfrac{V}{I} \qquad Z = \sqrt{R^2 + X_L^2} \qquad R = \sqrt{Z^2 - X_L^2} \qquad X_L = \sqrt{Z^2 - R^2}$

$\theta = \arctan \dfrac{X_L}{R} \qquad \theta = \arctan \dfrac{V_L}{V_R}$

Apparent power $= \sqrt{P_R^2 + P_X^2}$

$PF = \dfrac{\text{true power}}{\text{apparent power}} \qquad PF = \dfrac{R}{Z} \qquad PF = \cos \theta$

$$\boxed{t(\text{s}) = \frac{L}{R}}$$

t = inductive time constant in seconds

$$\boxed{X_L = 2\pi f L}$$

X_L = inductive reactance in ohms
f = frequency in hertz
L = inductance in henries

$$\boxed{Q = \frac{\text{energy stored}}{\text{energy dissipated}}}$$

Q = quality factor of an inductor

$$\boxed{Q = \frac{X_L}{R}}$$

P_R = true power in watts
P_X = reactive power in VAR
PF = power factor

REVIEW QUESTIONS

Multiple Choice Questions

1. Self inductance is a process by which a coil will induce a voltage within _____.

a. Another inductor
b. Two or more close proximity inductors
c. Itself
d. Both (a) and (b) are true

2. Mutual inductance is a process by which a coil will induce a voltage within _____.
   a. Another inductor
   b. Two or more close proximity inductors
   c. Itself
   d. Both (a) and (b) are true

3. The inductor stores electrical energy in the form of a _____ field, just as a capacitor stores electrical energy in the form of a _____ field.
   a. Electric, magnetic
   b. Magnetic, electric

4. The inductor is basically:
   a. An electromagnet
   b. A coil of wire
   c. A coil of conductor formed around a core material
   d. All the above are true
   e. None of the above are true

5. The inductance of an inductor is proportional to _____, and inversely proportional to _____.
   a. N, A, μ; l
   b. A, μ, l; N
   c. μ, l, N; A
   d. N, A, l; μ

6. The total inductance of a series circuit is:
   a. Less than the value of the smallest inductor
   b. Equal to the sum of all the inductance values
   c. Equal to the product over sum
   d. All the above are correct answers

7. The total inductance of a parallel circuit can be calculated by:
   a. Using the product-over-sum formula
   b. Using L divided by N for equal-value inductors.
   c. Using the general parallel resistance formula.
   d. All the above are true.

8. Air core fixed value inductors can use air or nonmagnetic forms, such as:
   a. Iron, cardboard
   b. Ceramic, copper
   c. Ceramic, cardboard
   d. Silicon, germanium

9. Ferrite is a chemical compound, which is basically powdered:
   a. Iron oxide and ceramic
   b. Iron and steel
   c. Mylar and iron
   d. Gauze and electrolyte

10. The ferrite core variable inductor varies inductance by changing:
   a. μ
   b. l
   c. N
   d. A

11. It will actually take _____ time constants for the current in an inductive circuit to reach a maximum value.
   a. 63.2
   b. 1
   c. 1.414
   d. 5

12. The time constant for a series inductive/resistive circuit is equal to:
   a. $L \times R$
   b. L/R
   c. V/R
   d. $2 \times \pi \times f \times L$

13. Inductive reactance (X_L) is proportional to:
   a. Time or period of the ac applied
   b. Frequency of the ac applied
   c. The stray capacitance that occurs due to the air acting as a dielectric between two turns of a coil
   d. The value of inductance
   e. Two of the above are true

14. In a series RL circuit, the source voltage (V_S) is equal to:
   a. The square root of the sum of V_R^2 and V_L^2
   b. The vector sum of V_R and V_L
   c. $I \times Z$
   d. Two of the above are partially true
   e. Answers (a), (b), and (c) are correct

15. In a purely resistive circuit, the phase shift is equal to _____, whereas in a purely inductive or capacitive circuit, the phase shift is _____ degrees.
   a. 45,0
   b. 90, 0
   c. 45, 90
   d. None of the above are true

16. With an RL integrator, the output is taken across the:
   a. Inductor
   b. Capacitor
   c. Resistor
   d. Transformer's secondary

17. With an RL differentiator, the output is taken across the:
   a. Inductor
   b. Capacitor
   c. Resistor
   d. Transformer's secondary

18. An inductor or choke between the input and output forms a:
   a. High pass filter
   b. Low pass filter

19. An inductor or choke connected to ground or in shunt forms a:
    a. Low pass filter
    b. High pass filter

20. Lenz's law states that when current is passed through a conductor a self induced voltage in a coil will:
    a. Aid the applied source voltage
    b. Aid the increasing current from the source
    c. Produce an opposing current
    d. Both (a) and (c) are true

21. When tested with an ohmmeter, an open coil would show:
    a. Zero resistance
    b. An infinite resistance
    c. A 100 200 Ω resistance
    d. Both (b) and (c) are correct

22. Inductive reactance:
    a. Increases with frequency
    b. Is proportional to inductance
    c. Reduces the amplitude of alternating current
    d. All the above

23. The current through an inductor _____ the voltage across the same inductor by _____.
    a. lags, 90°   c. leads, 90°
    b. lags, 45°   d. leads, 45°

24. The phasor combination of X_L and R is the circuit's:
    a. Reactance   c. Power factor
    b. Total resistance   d. Impedance

25. In a series RL circuit, where $V_R = 200$ mV and $V_L = 0.2$ V, $\theta =$:
    a. 45°   c. 0°
    b. 90°   d. 1°

Essay Questions ────────────────────────────────

26. Briefly describe the terms: (Introduction)
    a. Electromagnetism
    b. Electromagnetic induction

27. What is self induction and how does it relate to an inductor?   (16.3)

28. Give the formula for inductance, and explain the four factors that determine inductance.   (16.5)

29. List all the formulas for calculating total inductance when inductors are connected in: (16.6)
    a. Series
    b. Parallel

30. What is meant by a fixed and variable type inductor?   (16.7)

31. List the important factors of the: (16.7)
    a. Air, iron, and ferrite core fixed inductors
    b. Ferrite core variable inductor

32. Describe the current rise and fall through an inductor when: (16.8)
    a. DC is applied
    b. AC is applied

33. Define the following terms:
 a. Inductive reactance (16.9)
 b. Impedance (16.10.2)   **d.** Q factor (16.10.4)
 c. Phase shift (16.10.3)   **e.** Power factor (16.10.4)

34. With illustrations, describe how an inductor and resistor could be used for the following applications: (16.13)
 a. Integration
 b. Differentiation
 c. Filtering high and low frequencies

35. Explain briefly why an inductor acts as an open to an instantaneous change. Why does an inductor act like a short to dc? (16.9)

Practice Problems _____

36. Convert the following:
 a. 0.037 H to mH   **c.** 862 mH to H
 b. 1760 μH to mH   **d.** 0.256 mH to μH

37. Calculate the impedance (Z) of the following RL combinations:
 a. 22 MΩ, 25 μH, $f = 1$ MHz   **c.** 60 Ω, 0.05 H, $f = 1$ MHz
 b. 4 kΩ, 125 mH, $f = 100$ kHz

38. Calculate the voltage across a coil if:
 a. $d_i/d_t = 120$ mH/ms and $L = 2$ μH
 b. $d_i/d_t = 62$ μA/μs and $L = 463$ mH
 c. $d_i/d_t = 4$ A/s and $L = 25$ mH

39. Calculate the total inductance of the following series circuits:
 a. 75 μH, 61 μH, 50 mH   **b.** 8 mH, 4 mH, 22 mH

40. Calculate the total inductance of the following parallel circuits:
 a. 12 mH, 8 mH   **b.** 75 μH, 34 μH, 27 μH

41. Calculate the total inductance of the following series–parallel circuits:
 a. 12 mH in series with 4 mH, and both in parallel with 6 mH
 b. A two branch parallel arrangement made up of a 6 and 2 μH in series with one another, and an 8 and 4 μH in series with one another.
 c. Two parallel arrangements in series with one another, made up of a 1 and 2 μH in parallel and a 4 and 15 μH in parallel.

42. Determine the time constant of all the examples in question 41, and state how long it will take in each example for current to build up to maximum.

43. In a series RL circuit, if $V_L = 12$ V and $V_R = 6$ V, calculate:
 a. V_S
 b. I if $Z = 14$ kΩ   **d.** Q
 c. Phase angle   **e.** Power factor

44. What value of inductance is needed to produce 3.3 kΩ of reactance at 15 kHz?

45. At what frequency will a 330 μH inductor have a reactance of 27 kΩ?

46. Calculate the impedance of the circuits seen in Figure 16-42.

47. Referring to Figure 16-43, calculate the voltage across the inductor for all five time constants after the switch has been closed.

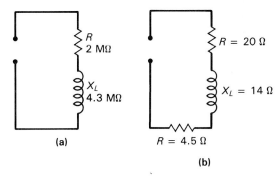

(a)

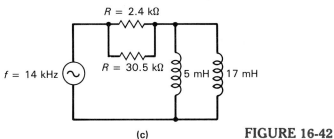

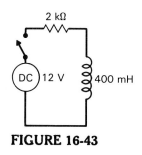

FIGURE 16-43

(b)

(c)   **FIGURE 16-42**

48. Referring to Figure 16-44, calculate:
 a. L_T
 b. X_L
 c. Z
 d. I_{R1}, I_{L1}, and I_{L2}
 e. θ
 f. True power, reactive power, and apparent power
 g. Power factor

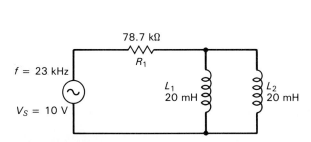

FIGURE 16-44

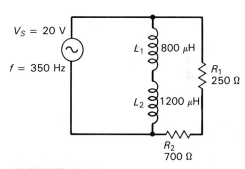

FIGURE 16-45

49. Referring to Figure 16-45, calculate
 a. R_T **f.** I_{RT}, I_{LT}
 b. L_T **g.** I_T
 c. X_L **h.** θ
 d. Z **i.** Apparent power
 e. V_{RT}, V_{LT} **j.** PF

Troubleshooting Problems

50. What problems can occur with inductors, and how can the ohmmeter be used to determine these problems?

17

TRANSFORMERS

AFTER COMPLETING THIS CHAPTER, YOU WILL BE ABLE TO:

1. Define mutual inductance and how it relates to transformers.
2. Describe the basic operation of a transformer.
3. Explain the differences between a loaded and unloaded transformer.
4. State what is meant by the coefficient of coupling.
5. List the three basic applications of transformers.
6. Describe how a transformer's turns ratio can be used to step up or step down voltage or current, or match impedances.
7. Describe the transformer dot convention as it relates to windings and phase.
8. List and explain the fixed and variable transformer types.
9. Describe the three basic transformer power losses.

INTRODUCTION

The transformer is an electrical device that by electromagnetic induction transforms electrical energy from one circuit to one or more other circuits. Step up transformers transform or convert a lower ac voltage to a higher ac voltage, while step down transformers reduce the ac input voltage to a lower ac output voltage.

WOOLEN MILL MAKES MINIS

At the age of 24 in 1950, Kenneth Olsen went to work at MIT's Digital Computer Laboratory as a research assistant. For the next seven years, Olsen worked on the SAGE (Semi Automatic Ground Environment) project, which was a new computer system designed to store the constantly changing data obtained from radars tracking long range enemy bombers capable of penetrating American air space. IBM won the contract to build the SAGE network and Olsen traveled to the plant in New York to supervise production.

After completing the SAGE project, Olsen was ready to move on to bigger and better things; however, he didn't want to go back to MIT and was not happy with the rigid environment at a large corporation like IBM.

In September of 1957, Olsen and fellow MIT colleague Harlan Anderson leased an old brick woolen mill in Maynard, Massachusetts, and after some cleaning and painting they began work in their new company, which they named Digital Equipment Corporation or DEC.

After three years, the company came out with their first computer, the PDP 1 (programmed data processor), and its ease of use, low cost, and small size made it an almost instant success. In 1965, a desk top PDP 8 was launched, and since in that era mini-skirts were popular, it was probably inevitable that the small machine was nicknamed a minicomputer. In 1969, the PDP 11 was unveiled and, like its predecessors, this number crunching data processor went on to be accepted in many applications, such as tracking the millions of telephone calls received on the 911 emergency line, directing welding machines in automobile plants, recording experimental results in laboratories, and processing all types of data in offices, banks, and department stores. Just 20 years after introducing low-cost computing in the 1960s, DEC was second only to IBM as a manufacturer of computers in the United States.

17.1
MUTUAL INDUCTANCE

The principle on which a transformer is based is an inductive effect known as *mutual inductance*.

Self inductance is the process by which a coil induces a voltage within itself, whereas mutual inductance is the process by which an inductor induces a voltage in another inductor.

Figure 17-1 illustrates two inductors that are magnetically linked, yet electrically isolated from one another. As the alternating current continually rises, falls, and then rises in the opposite direction, a magnetic field will build up, collapse, and then build up in the opposite direction.

If a second inductor, or secondary coil (L_2), is in close proximity with the first inductor or primary coil (L_1), which is producing the alternating magnetic field, a voltage will be induced into the nearby inductor, which causes current to flow in the secondary circuit through the load resistor. This phenomenon is known as mutual inductance or transformer action.

As with self inductance, mutual inductance is dependent on change. Direct current (dc) is a constant current and produces a constant or stationary magnetic field that does not change, as seen in Figure 17-2(a). Alternating current, however, is continually varying, and as the polarity of the magnetic

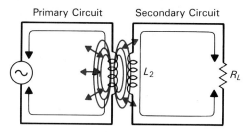

Primary Circuit   Secondary Circuit

L_2

R_L

Magnetic Field Alternately Builds up and Collapses   **FIGURE 17-1   Mutual Inductance.**

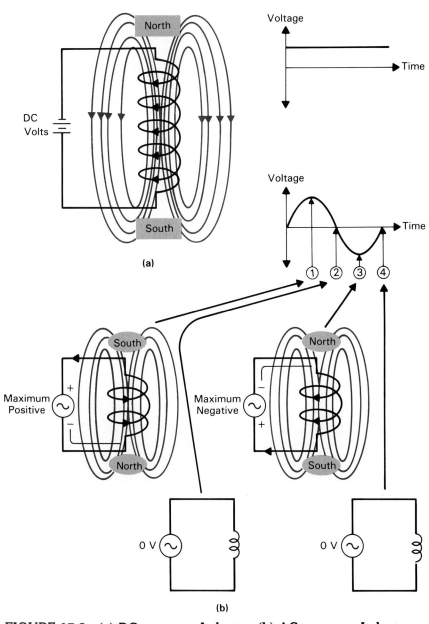

FIGURE 17-2   (a) DC across an Inductor. (b) AC across an Inductor.

field is dependent on the direction of current flow (left hand rule), the magnetic field will also be alternating in polarity, as seen in Figure 17-2(b); it is this continual building up and collapsing of the magnetic field that cuts the adjacent inductor's conducting coils and induces a voltage in the secondary circuit. Mutual induction is possible only with ac and cannot be achieved with dc due to the lack of change.

Self induction, is a measure of how much voltage an inductor can induce within itself. Mutual inductance is a measure of how much voltage is induced in the secondary coil due to the change in current in the primary coil.

SELF-TEST REVIEW (§ 17.1)

1. Define mutual inductance and how it differs from self-inductance.
2. Mutual induction is only possible with direct current flow (true or false).

17.2
BASIC TRANSFORMER

Figure 17-3(a) illustrates a basic transformer, which consists of two coils within close proximity to one another, to ensure that the second coil will be cut by the magnetic flux lines produced by the first coil, and thereby ensure mutual inductance. The ac voltage source is electrically connected (through

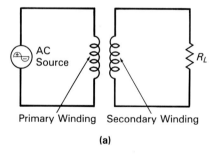

(a)

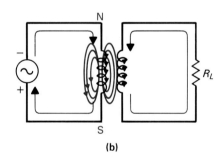

(b)

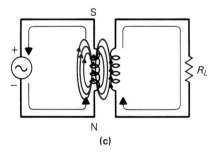

(c)

FIGURE 17-3

Transformers

wires) to the primary coil or winding, and the load is electrically connected to the secondary coil or winding.

In Figure 17-3(b), the ac voltage source has produced current flow in the primary circuit, as illustrated. This current flow produces a north pole at the top of the primary winding and, as the ac voltage input swings more negative, the current increase causes the magnetic field being developed by the primary winding to increase. This expanding magnetic field cuts the coils of the secondary winding and induces a voltage, and a subsequent current flows in the secondary circuit, which travels up through the load resistor. The ac voltage follows a sinusoidal pattern and moves from a maximum negative to zero and then begins to build up toward a maximum positive.

In Figure 17-3(c), the current flow in the primary circuit is in the opposite direction due to the ac voltage increase in the positive direction. As voltage increases, current increases, and the magnetic field expands and cuts the secondary winding, inducing a voltage and causing current to flow in the reverse direction down through the load resistor.

You may have noticed a few interesting points about the basic transformer just discussed.

1. As primary current increases, secondary current increases, and as primary current decreases, secondary current also decreases. It can, therefore, be said that the frequency of the alternating current in the secondary is the same as the frequency of the alternating current in the primary.

2. Although the two coils are electrically isolated from one another, energy can be transferred from primary to secondary, because the primary converts electrical energy into magnetic energy, and the secondary converts magnetic energy back into electrical energy.

SELF-TEST REVIEW (§ 17.2)

1. A transformer achieves electrical isolation (true/false).
2. The transformer converts electrical energy to magnetic, and then from magnetic back to electrical (true or false).
3. What are the names given to the windings of a transformer?

17.3
TRANSFORMER LOADING

Let's now carry our discussion of the basic transformer a little further and see what occurs when the transformer is not connected to a load, as seen in Figure 17-4.

Primary circuit current is determined by $I = V/Z$, where Z is the impedance of the primary coil (both its inductive reactance and resistance)

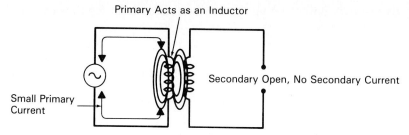

Primary Acts as an Inductor

Secondary Open, No Secondary Current

Small Primary Current

FIGURE 17-4 Unloaded Transformer.

and V is the applied voltage. Since no current can flow in the secondary, because an open in the circuit exists, the primary acts as a simple inductor, and the primary current is small due to the inductance of the primary winding. This small primary current lags the applied voltage due to the counter emf by approximately 90° because the coil is mainly inductive and has very little resistance.

When a load is connected across the secondary, as seen in Figure 17-5, a change in conditions occurs and the transformer acts differently. The important point that will be observed is that as we go from a no load to load condition the primary current will increase due to mutual inductance. Let's follow the steps one by one.

1. The ac applied voltage sets up an alternating magnetic field in the primary winding.
2. The continually changing flux of this primary field induces and produces a counter emf into the primary to oppose the applied voltage.
3. The primary's magnetic field also induces a voltage in the secondary winding, which causes current to flow in the secondary circuit through the load.
4. The current in the secondary winding produces another magnetic field that is opposite the field being produced by the primary.
5. This secondary magnetic field feeds back to the primary and induces a voltage that tends to cancel or weaken the counter emf that was set up in the primary by the primary current.
6. The primary's counter emf is therefore reduced, and so primary current can now increase.
7. This increase in primary current is caused by the secondary's magnetic field; consequently, the greater the secondary current, the stronger the secondary magnetic field, which causes a reduction in the primary's counter emf, and so a primary current increase.

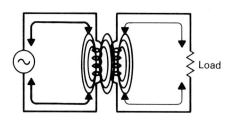

Load

FIGURE 17-5 Loaded Transformer.

602   Transformers

In summary, an increase in secondary current ($I_s\uparrow$) causes an increase in primary current ($I_p\uparrow$), and this effect in which the primary induces a voltage in the secondary (V_s) and the secondary induces a voltage into the primary (V_p) is known as *mutual inductance*.

SELF-TEST REVIEW (§ 17.3)

1. An increase in secondary current causes an increase in primary current (true/false).
2. The greater the secondary current, the greater the primary's counter emf (true/false).

17.4
COEFFICIENT OF COUPLING (*k*)

The voltage induced into the secondary winding is dependent on the mutual inductance between the primary and secondary, which is determined by how much of the magnetic flux produced by the primary actually cuts the secondary winding.

The coefficient of coupling (*k*) is a ratio of the number of magnetic lines of force that cut the secondary compared to the total number of magnetic flux lines being produced by the primary and is a figure between 0 and 1.

$$k = \frac{\text{flux linking secondary coil}}{\text{total flux produced by primary}}$$

If, for example, all the primary flux lines cut the secondary winding, the coefficient of coupling will equal 1. If only half of the total flux lines being produced cut the secondary winding, *k* will equal 0.5.

EXAMPLE 17-1

A primary coil is producing 65 μW (microwebers) of magnetic flux. Calculate the coefficient of coupling if 52 μW links with the secondary coil.

SOLUTION

$$k = \frac{52\ \mu\text{W}}{65\ \mu\text{W}} = 0.8$$

This means that 80% of the magnetic flux lines being generated by the primary oil are linking with the secondary coil.

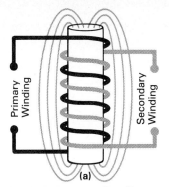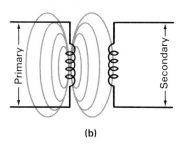

FIGURE 17-6 Coefficient of Coupling. (a) Wound on Same Core, Small Distance between Primary and Secondary, High *K*. (b) Winding Far Apart, Small *K*.

The coefficient of coupling depends on:

1. How close together the primary and secondary are to one another.
2. The type of core material used.

Figure 17-6 illustrates why the primary and secondary have to be in close proximity to one another to achieve a high coefficient of coupling.

SELF-TEST REVIEW (§ 17.4)

1. Define and state the formula for the coefficient of coupling.
2. If $k = 0.75$, then 75% of the total flux lines produced by the primary are cutting the coils of the secondary (true/false).
3. List the two factors that determine k.

17.5

TRANSFORMER RATIOS AND APPLICATIONS

Basically, transformers are used for one of three applications:

1. To step up (increase) or step down (decrease) current.
2. To step up or step down voltage
3. To match impedances.

In all three cases, the application can be achieved by something known as the turns ratio.

17.5.1 TURNS RATIO

The turns ratio is the ratio between the number of turns in the secondary winding (N_s) and the number of turns in the primary winding (N_p).

$$\boxed{\text{turns ratio} = \frac{N_s}{N_p}}$$

EXAMPLE 17.2

If the primary has 200 turns and the secondary has 600, what is the turns ratio (Figure 17-7)?

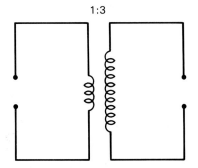

1:3

FIGURE 17-7

SOLUTION

$$\text{turns ratio} = \frac{N_s}{N_p}$$
$$= \frac{600}{200}$$
$$= \frac{3 \text{ (secondary)}}{1 \text{ (primary)}}$$
$$= 3$$

This simply means that there are three windings in the secondary to every one winding in the primary. Moving from a smaller number (1) to a larger number (3) means that we *stepped up* in value. Stepping up always results in a turns ratio figure greater than 1, in this case, 3.

EXAMPLE 17.3

If the primary has 120 turns and the secondary has 30 turns, what is the turns ratio (Figure 17-8)?

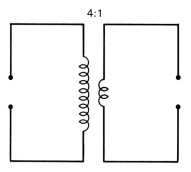

4:1

FIGURE 17-8

SOLUTION

$$\text{turns ratio} = \frac{N_s}{N_p}$$
$$= \frac{30}{120}$$
$$= \frac{1 \text{ (secondary)}}{4 \text{ (primary)}}$$
$$= 0.25$$

Said simply, there are four primary windings to every one secondary winding. Moving from a larger number (4) to a smaller number (1) means that we *stepped down* in value. Stepping down always results in a turns ratio figure of less than 1, in this case 0.25.

17.5.2 VOLTAGE RATIO

(1) STEP UP

If the secondary voltage (V_s) is greater than the primary voltage (V_p), the transformer is called a *step-up transformer* ($V_s > V_p$), as seen in Figure 17-

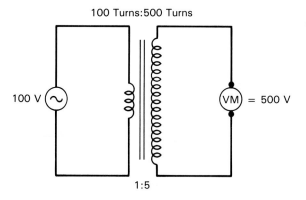

100 Turns:500 Turns

100 V

VM = 500 V

1:5

FIGURE 17-9 Step-up Transformer.

9. The voltage is stepped up or increased in much the same way as a generator voltage can be increased by increasing the number of turns.

If the ac primary voltage is 100 V and the turns ratio is a 1:5 step up, the secondary voltage will be 5 times that of the primary voltage, or 500 V, because the magnetic flux established by the primary cuts more turns in the secondary and, therefore, induces a larger voltage.

In this example, you can see that the ratio of the secondary voltage to the primary voltage is equal to the turns ratio; in other words

$$\boxed{\frac{V_s}{V_p} = \frac{N_s}{N_p}}$$

or

$$\frac{500}{100} = \frac{500}{100}$$

To calculate V_s, therefore, we can rearrange the formula and arrive at

$$\boxed{V_s = \frac{N_s}{N_p} \times V_p}$$

In our example, this is

$$V_s = \frac{500}{100} \times 100$$
$$= 500 \text{ V}$$

(2) STEP DOWN

If the secondary voltage (V_s) is smaller than the primary voltage (V_p), the transformer is called a *step down transformer* ($V_s < V_p$), as seen in Figure 17-10. The secondary voltage will be equal to

$$\boxed{V_s = \frac{N_s}{N_p} \times V_p}$$

$$\frac{10}{100} \times 1000 = 100 \text{ V}$$

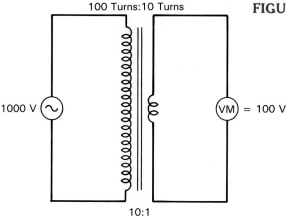

100 Turns:10 Turns

1000 V

(VM) = 100 V

10:1

FIGURE 17-10 Step-down Transformer.

EXAMPLE 17.4

Calculate the secondary voltage (V_s) if a $1:6$ step up transformer has 24 V ac applied to the primary.

SOLUTION

$$V_S = \frac{N_s}{N_p} \times V_p$$

$$= \frac{6}{1} \times V_p = 144 \text{ V}$$

The coupling coefficient (k) in this formula is always assumed to be 1, which for most iron core transformers is almost always the case. This means that all the primary magnetic flux is linking the secondary, and the secondary voltage is dependent on the number of secondary turns that are being cut by the primary magnetic flux.

The transformer can be used to transform the primary ac voltage into any other voltage, either up or down, by merely changing the transformer's turns ratio.

17.5.3 POWER AND CURRENT RATIO

The power in the secondary of the transformer is equal to the power in the primary ($P_p = P_s$). Power, as we know, is equal to $P = V \times I$, and if voltage is stepped up or down, the current automatically is stepped down or up, respectively, in the opposite direction to voltage to maintain the power constant.

For example, if the secondary voltage is stepped up ($V_s\uparrow$), the secondary current is stepped down ($I_s\downarrow$), so the output power is the same as the input power.

$$p_s = V_s\uparrow \times I_s\downarrow$$

This is an equal but opposite change. Therefore, $p_s = p_p$; and you cannot get more power out than you put in. The current ratio is therefore inversely proportional to the voltage ratio:

$$\boxed{\frac{V_s}{V_p} = \frac{I_p}{I_s}}$$

If the secondary voltage is stepped-up, the secondary current goes down:

$$\frac{V_s\uparrow}{V_p} = \frac{I_p}{I_s\downarrow}$$

If the secondary voltage is stepped-down, the secondary current goes up:

$$\frac{V_s\downarrow}{V_p} = \frac{I_p}{I_s\uparrow}$$

If the current ratio is inversely proportional to the voltage ratio, it is also inversely proportional to the turns ratio:

$$\frac{I_p}{I_s} = \frac{V_s}{V_p} = \frac{N_s}{N_p}$$

EXAMPLE 17.5

The step-up transformer in Figure 17-11 has a turns ratio of 5 to 1. Calculate
a. Secondary voltage (V_s)
b. Secondary current (I_s)
c. Primary power (P_p)
d. Secondary power (P_s)

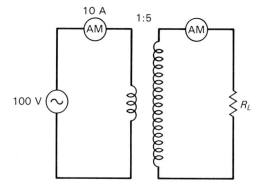

FIGURE 17-11

SOLUTION

The secondary has five times as many windings as the primary and, consequently, the voltage will be stepped up by 5 between primary and secondary. If the secondary voltage is going to be 5 times that of the primary, then the secondary current is going to decrease to $\frac{1}{5}$th of the primary.

a. $V_s = \dfrac{N_s}{N_p} \times V_p$

$\quad = \dfrac{5}{1} \times 100 \text{ V}$

$\quad = 500 \text{ V}$

b. $I_s = \dfrac{N_p}{N_s} \times I_p$

$\quad = \dfrac{1}{5} \times 10$

$\quad = 2 \text{ A}$

c. $P_p = V_p \times I_p = 100 \text{ V} \times 10 \text{ A} = 1000 \text{ VA}$

d. $P_s = V_s \times I_s = 500 \text{ V} \times 2 \text{ A} = 1000 \text{ VA}$

$\quad$ Therefore, $P_p = P_s$

By rearranging the current and turns ratio, we can arrive at a formula for secondary current, which is

$$I_s = \frac{N_p}{N_s} \times I_p$$

17.5.4 IMPEDANCE RATIO

The maximum power transfer theorem, which has been previously discussed and is summarized in Figure 17-12, states that maximum power is transferred from source (ac generator) to load (equipment) when the impedance of the load is equal to the internal impedance of the source. If these impedances are different, then a large amount of power could be wasted.

In most cases, it is required to transfer maximum power from a source that has an internal impedance (Z_i) that is not equal to the load impedance (Z_L). In this situation, a transformer can be inserted between the source and

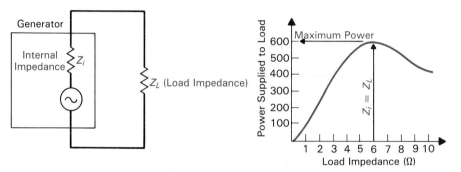

FIGURE 17-12 Maximum Power Transfer Theorem.

FIGURE 17-13 Impedance Matching.

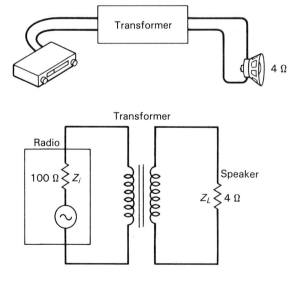

Transformers

the load to make the load impedance appear to equal the source's internal impedance.

For example, let's imagine that your car stereo system (source) has an internal impedance of 100 Ω and is driving a speaker (load) of 4 Ω impedance, as seen in Figure 17-13.

By choosing the correct turns ratio, the 4 Ω speaker can be made to appear as a 100 Ω load impedance, which will match the 100 Ω internal source impedance of the stereo system, resulting in maximum power transfer.

The turns ratio can be calculated by using the formula

$$\text{turns ratio} = \sqrt{\frac{Z_L}{Z_S}}$$

In our example, this will calculate out to be

$$\text{turns ratio} = \sqrt{\frac{Z_L}{Z_S}}$$
$$= \sqrt{\frac{4}{100}} = \frac{\sqrt{4}}{\sqrt{100}}$$
$$= \frac{2}{10} = \frac{1}{5}$$
$$= 0.2$$

If the turns ratio is less than 1, then a step down transformer is required. A turns ratio of 0.2 means a step down transformer is needed with a turns ratio of 5:1 (1/5 = 0.2).

EXAMPLE 17.6

Calculate the turns ratio needed to match the 22.2-Ω output impedance of an amplifier to two 16 Ω speaker connected in parallel.

SOLUTION

The total load impedance of two 16 Ω speakers in parallel will be

$$Z_L = \frac{\text{product}}{\text{sum}} = \frac{16 \times 16}{16 + 16} = 8 \ \Omega$$

The turns ratio will be

$$\text{turns ratio} = \sqrt{\frac{Z_L}{Z_S}}$$
$$= \sqrt{\frac{8 \ \Omega}{22.2 \ \Omega}}$$
$$= \sqrt{0.36} = 0.6$$

Therefore, a step down transformer is needed with a turns ratio of 1.67:1.

1. What is the turns ratio of a 402 turn primary and 1608 turn secondary, and is this transformer step up or step down?
2. State the formula for calculating the secondary voltage (V_s).
3. A transformer can, by adjusting the turns ratio, be made to step up both current and voltage between primary and secondary (true/false).
4. State the formula for calculating the secondary current (I_s).
5. What turns ratio is needed to match a 25-Ω source to a 75-Ω load?
6. Calculate V_s if $N_s = 200$, $N_p = 112$, and $V_p = 115$ V.

17.6
WINDINGS AND PHASE

The way in which the primary and secondary coils are wound around the core determines the polarity of the voltage induced into the secondary relative to the polarity of the primary. In Figure 17-14, you can see that, if the primary and secondary windings are both wound in a clockwise direction around the core, the voltage induced in the primary will be in phase with the voltage induced in the secondary. Both the input and output will also be in phase with one another if the primary and secondary are both wound in a counterclockwise direction.

In Figure 17-15, you will notice that in this case we have wound the primary and secondary in opposite directions, the primary being wound in a clockwise direction and the secondary in a counterclockwise direction. In this situation, the output ac sine wave voltage is 180° out of phase with respect to the input ac voltage.

In a schematic diagram, there has to be a way of indicating to the technician that the secondary voltage will be in phase or 180° out of phase

FIGURE 17-14 Primary and Secondary in Phase.

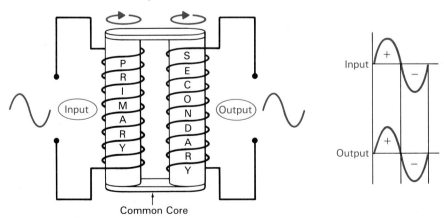

Common Core

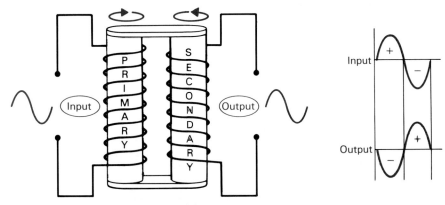

FIGURE 17-15   Primary and Secondary out of Phase.

with the input. The dot convention is a standard used with transformers and is illustrated in Figure 17-16(a) and (b).

A positive on the primary dot causes a positive on the secondary dot, and, similarly, since ac is applied, a negative on a primary dot will cause a negative on the secondary dot. In Figure 17-16(a), you can see that, when the top of the primary swings positive or negative, the top of the secondary will also follow suit and swing positive or negative, respectively, and you will now be able to determine that the transformer secondary voltage is in phase with the primary voltage.

In Figure 17-16(b), however, the dots are on top and bottom, which means that as the top of the primary winding swings positive, the bottom of the secondary will go positive, which means the top of the secondary will actually go negative, resulting in the secondary ac voltage being out of phase with the primary ac voltage.

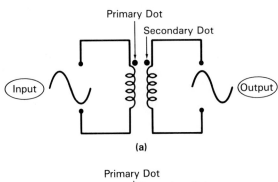

(a)

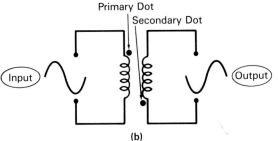

(b)

FIGURE 17-16   Dot Convention.
(a) In Phase Dots.
(b) Out of Phase Dots.

1. How does the winding of the primary and secondary affect the phase of the output with respect to the input?
2. Describe the transformer dot convention used on schematics.

17.7
TRANSFORMER TYPES

We will treat transformers the same as every other electronic component and classify them as either fixed or variable.

17.7.1 FIXED TRANSFORMERS

Fixed transformers have a turns ratio that cannot be varied. They are generally wound on a common core to ensure a high k, and can be classified by the type of core material used. The two types of fixed transformers are:

1. Air core [Figure 17-17(a)]
2. Iron or ferrite core [Figure 17-17(b)]

The air core transformers typically have a nonmagnetic core, such as ceramic or a cardboard hollow shell, and are used in high frequency applications. The electronic circuit symbol just shows the primary and secondary coils. The more common iron or ferrite core transformers concentrate the magnetic lines of force, resulting in improved transformer performance; they are symbolized by two lines running between the primary and secondary. The iron core transformer's lines are solid, while the ferrite core transformer's lines are dashed.

Figure 17-18 illustrates a few types of fixed transformers.

17.7.2 VARIABLE TRANSFORMERS

Variable transformers have a turns ratio that can be varied. They can be classified as follows:

1. Center tapped secondary
2. Multiple tapped secondary
3. Multiple winding
4. Single winding

614   Transformers

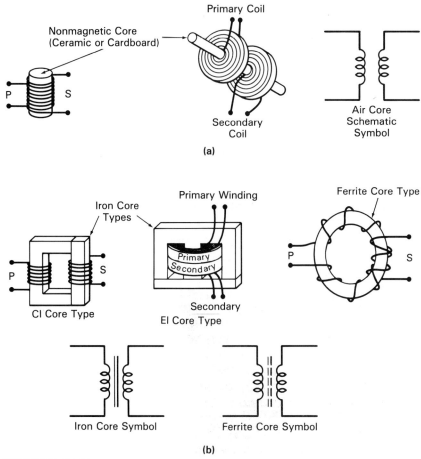

(a)

(b)

FIGURE 17-17 Fixed Transformers. (a) Air Core Transformers. (b) Iron or Ferrite Core Transformers.

Figure 17-19 illustrates the first two of these, the center tapped and the multiple tapped secondary types.

Transformers can often have tapped secondaries. A tapped winding will have a lead connected to one of the loops other than the two ends connections.

FIGURE 17-18 Physical Appearance of Some Transformer Types.

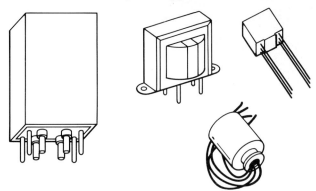

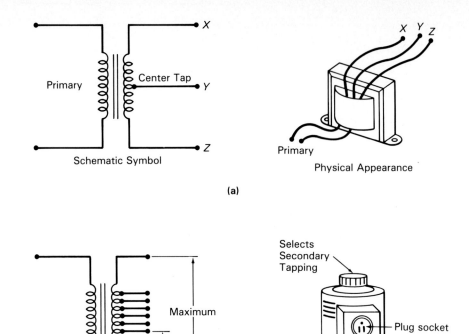

(a)

(b)

FIGURE 17-19   Secondary Tapped Transformers. (a) Center Tapped. (b) Multiple Tapped.

(1) CENTER-TAPPED SECONDARY

If the tapped lead is in the exact center of the secondary, the transformer is said to have a center-tapped secondary, as seen in Figure 17-20. With the center tapped transformer, the two secondary voltages are each half of the total secondary voltage. If we assume a 1:1 turns ratio and a 20 V ac primary voltage and therefore secondary voltage, each of the output voltages between either end of the secondary and the center tap will be 10 V waveforms, as seen in Figure 17-20. The two secondary outputs will be 180° out of phase

FIGURE 17-20   Center-tapped Secondary Transformer.

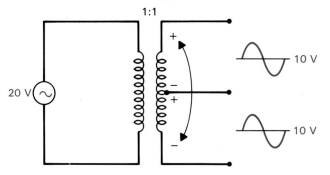

Secondary Winding

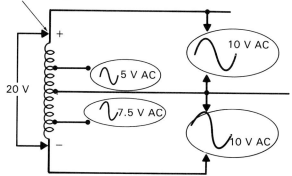

FIGURE 17-21 Multitapped Secondary Transformer.

with one another from the tap. When the top of the secondary swings positive (+), the bottom of the secondary will be negative, and vice versa.

(2) MULTIPLE-TAPPED SECONDARY

If several leads are attached to the secondary, the transformer is said to have a multiple tapped secondary. This multitapped transformer will tap or pick off different values of ac voltage, as seen in Figure 17-21.

(3) MULTIPLE WINDINGS

The simple transformer has only one primary and one secondary. In some applications, transformers can have multiple primaries and even multiple secondaries.

(a) Multiple Primaries The major application of this arrangement is to switch between two primary voltages and obtain the same secondary voltage. In Figure 17-22(a), the two primaries are connected in parallel, and so the primary only has an equivalent 100 turns; and since the secondary has 10 turns, this 10:1 step down ratio will result in a secondary voltage of

$$
\begin{aligned}
V_s &= \frac{N_s}{N_p} \times V_p \\
&= \frac{10}{100} \times 120 \\
&= 12 \text{ V}
\end{aligned}
$$

If the two primaries are connected in series, as seen in Figure 17-22(b), the now 200 turns primary will result in a secondary voltage of

$$
\begin{aligned}
V_s &= \frac{N_s}{N_p} \times V_p \\
&= \frac{10}{200} \times 240 \\
&= 12 \text{ V}
\end{aligned}
$$

Some portable electrical or electronic equipment, such as radios or shavers, have a switch that allows you to switch between 120 V ac (USA

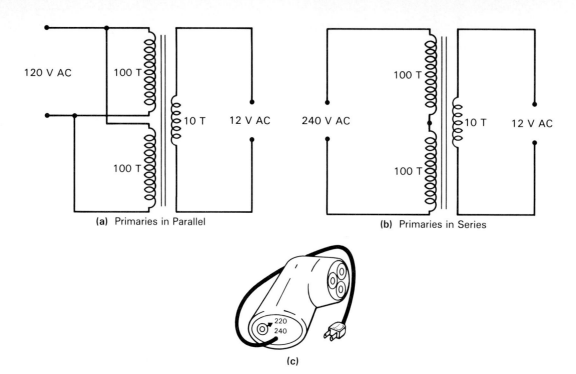

FIGURE 17-22 Multiple Primary Windings Transformer.

wall outlet voltage) and 240 V ac (European wall outlet voltage), as seen in Figure 17-22(c). By activating the switch, you can always obtain the correct voltage to operate the equipment, whether the wall socket is supplying 240 or 120 V.

(b) Multiple Secondaries In some applications, more than one secondary is wound onto a common primary and core. The advantage of this arrangement can be seen in Figure 17-23, where many larger and smaller voltages can be acquired from one primary voltage.

(4) SINGLE WINDING (AUTOTRANSFORMER)

The autotransformer is used in the automobile ignition system (switched dc pulse system) to raise or lower voltage up to about 20 kV for the spark plugs, as seen in Figure 17-24. Autotransformers are constructed by winding one continuous coil onto a core, which acts as both the primary and secondary. This yields three advantages for the single winding transformer:

1. They are smaller.
2. They are cheaper.
3. They are lighter than the normal separated primary/secondary transformer types.

Its disadvantage, however, is that no electrical isolation exists between primary and secondary. It is normally used as a step-up transformer for the automobile, or in televisions to obtain 20 kV. It can be arranged to step down voltage, as seen in Figure 17-24(b).

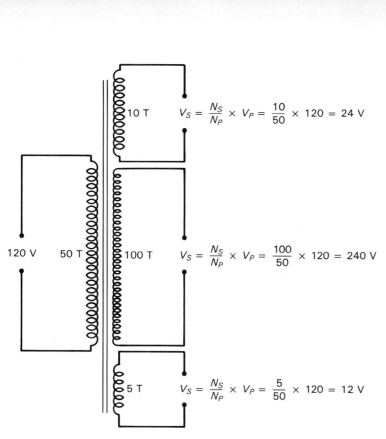

$$V_S = \frac{N_S}{N_P} \times V_P = \frac{10}{50} \times 120 = 24 \text{ V}$$

$$V_S = \frac{N_S}{N_P} \times V_P = \frac{100}{50} \times 120 = 240 \text{ V}$$

$$V_S = \frac{N_S}{N_P} \times V_P = \frac{5}{50} \times 120 = 12 \text{ V}$$

FIGURE 17-23 Multiple Secondary Windings Transformer.

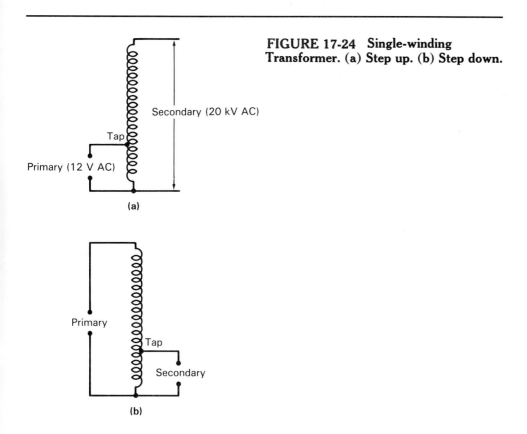

FIGURE 17-24 Single-winding Transformer. (a) Step up. (b) Step down.

Secondary (20 kV AC)

Tap

Primary (12 V AC)

(a)

Primary

Tap

Secondary

(b)

1. List the two types of fixed transformers.
2. List the four basic classifications for variable transformers.
3. The two outputs of a center tapped secondary transformer will always be 90° out of phase with one another (true/false).
4. In what application would a multiple primary transformer be used?
5. In what application would a single winding transformer be used?
6. What are the advantages of a single winding transformer?

17.8
TRANSFORMER RATINGS

A typical transformer rating could read 1 kVA, 500/100, 60 Hz. The 500 normally specifies the maximum primary voltage, the 100 normally specifies the maximum secondary voltage, and the 1 kVA is the apparent power rating. In this example, the maximum load current will equal

$$I_s = \frac{\text{apparent power}}{\text{secondary voltage } (V_S)} \qquad \left(\begin{array}{c} P = V \times I; \text{ therefore} \\ I = \dfrac{P}{V} \end{array} \right)$$

$$= \frac{1 \text{ kVA}}{100} = 10 \text{ A}$$

With this secondary voltage at 100 V and a maximum current of 10 A, the smallest load resistor that can be connected across the output of the secondary is

$$R_L = \frac{V_s}{I_s}$$

$$= \frac{100 \text{ V}}{10 \text{ A}}$$

$$= 10 \text{ } \Omega$$

Exceeding the rating of the transformer will cause overheating and even burning out of the windings.

EXAMPLE 17.7

Calculate the smallest value of load resistance that can be connected across a 3 kVA, 600/200, 60-Hz step down transformer.

SOLUTION

$$I = \frac{\text{apparent power}}{\text{secondary voltage}}$$

$$\frac{3 \text{ kVA}}{200 \text{ V}} = 15 \text{ A}$$

$$R_L = \frac{V_s}{I_s} = \frac{200 \text{ V}}{15 \text{ A}} = 13.3 \ \Omega$$

SELF-TEST REVIEW (§ 17.8)

1. What do each of the values mean when a transformer is rated as a 10 kVA, 200/100, 60 Hz?
2. If a 100 Ω resistor is connected across the secondary of a transformer that is to supply 1 kV and is rated at a maximum current of 8 A, will the transformer overheat and possibly burn out?

17.9
TRANSFORMER LOSSES

Internal losses cause the power delivered from the secondary winding of a transformer in reality to be less than the power fed into the primary winding.

(1) COPPER LOSSES

Due to ohmic resistance of the windings, heat ($I^2 \times R$) is dissipated in both the primary and secondary windings.

(2) IRON AND CORE LOSSES

(a) Hysteresis During each cycle, the core is taken through a cycle of magnetization; hence, energy is lost due to hysteresis and appears as heat in the core. This loss is proportional to frequency and is minimized by using a core of soft iron or an alloy such as stalloy or permalloy.

(b) Eddy-current Loss The continuously changing flux induces voltages in the conducting core, which results in local eddy currents in the core that combine to produce a large circulating current, as seen in Figure 17.25(a), that opposes the main flux and also generates heat. This loss, which is proportional to frequency, is minimized by using a laminated core, as seen in Figure 16-25(b).

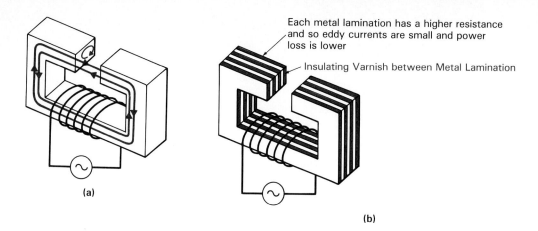

Each metal lamination has a higher resistance and so eddy currents are small and power loss is lower

Insulating Varnish between Metal Lamination

(a)

(b)

FIGURE 17-25   Eddy Currents. (a) Solid Core, Large Eddy Currents. (b) Laminated Core, Small Eddy Currents.

(3) MAGNETIC LEAKAGE

Not all the flux lines produced will cut the secondary winding. The result is that, for a given turns ratio, the secondary terminal voltage will be slightly lower than expected due to magnetic leakage.

SELF-TEST REVIEW (§ 17.9)

1. List the four types of transformer losses.
2. Why will a laminated transformer core reduce eddy current loss?

SUMMARY

1. Self inductance is the process by which a coil will induce a voltage within itself, whereas mutual inductance is the process by which an inductor will induce a voltage into another inductor.
2. The transformer is an electrical device that by electromagnetic induction transforms electrical energy from one circuit to one or more other circuits.
3. A step up transformer transforms or converts a lower ac voltage to a higher ac voltage; a step down transformer transforms a higher ac input voltage to a lower ac output voltage.
4. The basic transformer consists of two coils or inductors in close proximity to one another. The inductor or coil connected to the source is known as the primary winding. The coil connected across the load is referred to as the secondary winding.

5. The coefficient of coupling (k) is a ratio of the number of magnetic lines of force that cut the secondary to the total number of magnetic flux lines being produced by the primary, and is a figure between 0 and 1.

6. Transformers are used to:
   a. Step current up or step down current
   b. Step voltage up or step down voltage
   c. Match impedances

   All these applications are achieved by the turns ratio.

7. The turns ratio is the ratio between the number of secondary turns and the number of primary turns.

8. A step up transformer has a turns ratio greater than 1; a step down transformer's turns ratio is always less than 1.

9. If the secondary voltage is stepped up, the secondary current is stepped down, as primary power always equals secondary power; you cannot get more power out, than you put in.

10. To achieve maximum power transfer, the reflected load impedance must equal the source impedance; otherwise, an impedance mismatch will result in a large amount of power being wasted.

11. The dot convention on schematics is used to indicate the polarity of the voltage induced into the secondary relative to the polarity of the primary.

12. A fixed transformer has a turns ratio that cannot be varied, while a variable transformer has either a multiple tapped secondary or can have a single or multiple winding.

13. The single winding (autotransformer) transformer is smaller, cheaper, and lighter than most other transformers and is normally used as a step up transformer for the automobile or television set.

14. A laminated core transformer or inductor reduces the size of the eddy current within the core and consequently the amount of power lost in the form of heat.

NEW TERMS

Autotransformer	Fixed transformer	Secondary
Center-tapped secondary	Impedance ratio	Single-winding transformer
Coefficient of coupling	Magnetic leakage	Transformer
Copper losses	Multiple-tapped secondary	Transformer action
Core losses	Multiple winding	Turns ratio
Current ratio	Mutual inductance	Variable transformer
Dot convention	Power ratio	Voltage ratio
Eddy-current losses	Primary	
	Ratings	

NEW FORMULAS

$$\text{Coupling coefficient } (k) = \frac{\text{flux linking secondary coil}}{\text{total flux produced by primary}}$$

$$\text{Turns ratio} = \frac{N_s}{N_p}$$

N_s = number of turns in the secondary
N_p = number of turns in the primary

$$V_s = \frac{N_s}{N_p} \times V_p$$

V_s = secondary voltage
V_p = primary voltage

$$I_s = \frac{N_p}{N_s} \times I_p$$

I_s = secondary current
I_p = primary current

$$\text{Turns ratio} = \sqrt{\frac{Z_L}{Z_s}}$$

Z_L = load impedance
Z_s = source impedance

REVIEW QUESTIONS

Multiple Choice Questions

1. Transformer action is based on:
   a. Self inductance
   b. Air between the coils
   c. Mutual capacitance
   d. Mutual inductance

2. An increase in transformer secondary current will cause a(an) _____ in primary current.
   a. Decrease
   b. Increase

3. If 50% of the magnetic lines of force produced by the primary were to cut the secondary coil, the coefficient of coupling would be:
   a. 75
   b. 50
   c. 0.5
   d. 0.005

4. A step-up transformer will always have a turns ratio _____, while a step-down transformer has a turns ratio _____.
   a. $<1, >1$
   b. $>1, >1$
   c. $>1, <1$
   d. $<1, <1$

5. With an 80 V ac secondary voltage center tapped transformer, what would be the voltage at each output, and what would be the phase relationship between the two secondary voltages?
   a. 20 V, in phase with one another
   b. 30 V, 180° out of phase
   c. 40 V, in phase
   d. 40 V, 180° out of phase

6. One application of the autotransformer would be:
   a. To obtain the final anode high voltage supply for the cathode-ray tube in a television
   b. To obtain two outputs, 180° out of phase with one another
   c. To tap several different voltages from the secondary
   d. To obtain the same secondary voltage for different voltages

7. Alternating current can be used only with transformers because:
   a. It produces an alternating magnetic field.
   b. It produces a fixed magnetic field.
   c. Its magnetic field is greater than that of dc.
   d. Its rms is 0.707 of the peak.

8. Eddy current losses are reduced with laminated iron cores because:
   a. The air gap is always kept to a minimum.
   b. The resistance of iron is always low.
   c. The laminations are all insulated from one another.
   d. Current cannot flow in iron.

9. Assuming 100% efficiency, the output power, P_s, is always equal to:
   a. P_p
   b. $V_s \times I_s$
   c. $0.5 \times P_p$
   d. Both (a) and (c)
   e. Both (a) and (b)

10. If the primary winding of a transformer were open, the result would be:
    a. No flux linkage between primary and secondary
    b. No primary or secondary current
    c. No induced voltage in the secondary
    d. All the above

Essay Questions

11. What is a transformer?   (Introduction)
12. Describe mutual inductance and how it relates to transformers.   (17.1)
13. Why does loading the transformer's secondary circuit affect primary transformer current?   (17.3)
14. Describe what a coefficient of coupling figure of 0.9 means.   (17.4)
15. List the three basic applications of transformers, and then describe how each of these applications can be achieved by merely changing the transformer's turns ratio.   (17.5)
16. Briefly describe how the dot convention is used in schematics to describe phase.   (17.6)
17. Illustrate the schematic symbol and main points relating to the following transformers:

    a. Fixed air core (17.7.1)
    b. Fixed iron core (17.7.1)
    c. Fixed ferrite core (17.7.1)
    d. Center tapped secondary (17.7.2)
    e. Four output tapped secondary (17.7.2)
    f. Multiple primary and multiple secondary (17.7.2)
    g. Autotransformer (17.7.2)

18. Illustrate and explain the following transformer applications: (17.7)
    a. 120/240 V primary voltage to constant secondary voltage
    b. 20-kV secondary voltage using the autotransformer

19. Briefly describe what is meant by copper losses with transformers. (17.9)

20. Briefly describe the following iron and core losses and how they can be reduced: (17.9)
    a. Hysteresis
    b. Eddy current

21. Briefly describe the transformer loss known as magnetic leakage. (17.9)

22. What meter could be used to check a suspected open primary coil? (Chapter 16)

23. Could a transformer be considered a dc block? (17.1)

24. Would a step up voltage transformer step up or step down current? What would happen to secondary power? (17.5)

25. Briefly describe how transformers are used to reduce I^2R losses in ac power distribution. (Chapter 11)

Practice Problems _____

26. Calculate the turns ratio of the following transformers and state whether they are step up or step down:
    a. P = 12T, S = 24T
    b. P = 3T, S = 250T
    c. P = 24T, S = 5T
    d. P = 240T, S = 120T

27. Calculate the secondary ac voltage for all the examples in question 26 if the primary voltage equals 100 V.

28. Calculate the secondary ac current for all the examples in question 26 if the primary current equals 100 mA.

29. What turns ratio would be needed to match a source impedance of 24 Ω to a load impedance of 8 Ω?

30. What turns ratio would be needed to step:
    a. 120 V to 240 V
    b. 240 V to 720 V
    c. 30 V to 14 V
    d. 24 V to 6 V

31. For a 24 V, 12 turn primary, and 16, 2, 1, and 4 turn multiple secondary transformer, calculate each of the secondary voltages.

32. If a 2 to 1 step-down transformer has a primary input voltage of 120 V, 60 Hz and a 2 kVA rating, calculate the maximum secondary current and smallest load resistor that can be connected across the output.

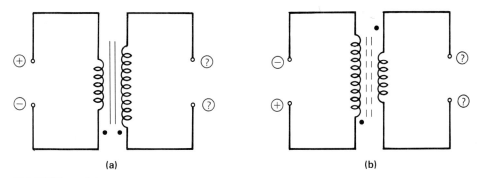

FIGURE 17-26

33. Indicate the polarity of the secondary voltages in Figure 17-26.
34. Referring to Figure 17-27, sketch the output, showing its polarity with respect to the input and its amplitude.

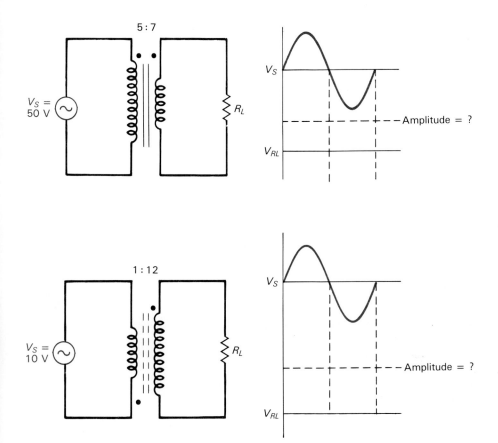

FIGURE 17-27

35. If a transformer is rated at 500 VA, 60 Hz and the primary voltage is 240 V ac and the secondary voltage is 600 V ac, calculate:
   a. Maximum load current
   b. Smallest value of R_L

18

RESISTIVE, INDUCTIVE, AND CAPACITIVE CIRCUITS (*RLC*)

AFTER COMPLETING THIS CHAPTER, YOU WILL BE ABLE TO:

1. Identify the difference between a series and parallel *RLC* circuit.
2. Explain the following as they relate to series *RLC* circuits:
   a. Impedance
   b. Current
   c. Voltage
   d. Power
   e. Resonance
3. Explain the following as they relate to parallel *RLC* circuits:
   a. Current
   b. Resonance
4. Identify and explain the following *RLC* circuit applications:
   a. Low pass filter
   b. High pass filter
   c. Band pass filter
   d. Band stop filter
5. Describe complex numbers in both rectangular and polar form.
6. Perform complex number arithmetic.
7. Describe how complex numbers apply to ac circuits containing series parallel *RLC* components.

INTRODUCTION

Up to this point we have discussed circuits containing resistance and capacitance (*RC*) and resistance and inductance (*RL*). Let's now combine all three properties of *R*, *L*, and *C* into first series and then parallel circuits to see their behavior and applications.

THE FAIRCHILDREN

On December 23, 1947, John Bardeen, Walter Brattain, and William Shockley first demonstrated how a semiconductor device, named the transistor, could be made to amplify. The device, however, had mysterious problems and was very unpredictable. Shockley continued his investigations, and in 1951 he presented the world with the first reliable junction transistor.

In 1956, all three shared the Nobel Prize in physics for their discovery, and much later, in 1972, Bardeen would win a rare second Nobel Prize for his research at the University of Illinois in the field of superconductivity.

Shockley left Bell Labs in 1955 to start his own semiconductor company near his home in Palo Alto and began recruiting personnel. He was, however, very selective, only hiring those who were bright, young, and talented. The company was a success, although many of the employees could not tolerate Shockley's eccentricities, such as posting everyone's salary and requiring that all employees rate one another. Two years later, eight of Shockley's most talented defected. The "traitorous eight," as Shockley called them, started their own company only a dozen blocks away, named Fairchild Semiconductor.

More than 50 companies would be founded by former Fairchild employees. One of the largest was started by Robert Noyce and two other colleagues from the group of eight Shockley defectors; they named their company Intel.

18.1
SERIES *RLC* CIRCUIT

Figure 18-1 begins our analysis of series *RLC* circuits by illustrating the current and voltage relationships. The circuit current is always the same throughout a series circuit and can therefore be used as a reference. Studying

FIGURE 18-1 Series *RLC* Circuit. (a) *RLC* Series Circuit Current: Current flow is always the same in all parts of a series circuit. (b) *RLC* Series Circuit Voltages: *I* is in phase with V_R by 90° and *I* leads V_C by 90°.

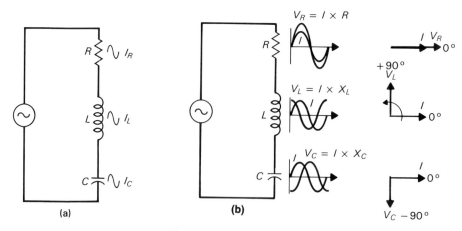

Resistive, Inductive, and Capacitive Circuits (*RLC*)

the waveforms and vector diagrams shown alongside the components, you can see that the voltage across a resistor is always in phase with the current, while the voltage across the inductor leads the current by 90° and the voltage across the capacitor lags the current by 90°.

Let's now analyze the impedance, current, voltage, and power distribution of this circuit in a little more detail.

18.1.1 IMPEDANCE

Impedance is the total opposition to current flow and is a combination of both reactance (X_L, X_C) and resistance (R). An example circuit is illustrated in Figure 18-2(a).

Capacitive reactance can be calculated by using the formula

$$X_C = \frac{1}{2\pi fC}$$

In the example,

$$X_C = \frac{1}{2\pi \times 60 \times 10\ \mu F} = 265.3\ \Omega$$

Inductive reactance is calculated by using the formula

$$X_L = 2\pi fL$$

FIGURE 18-2   Series Circuit Impedance.

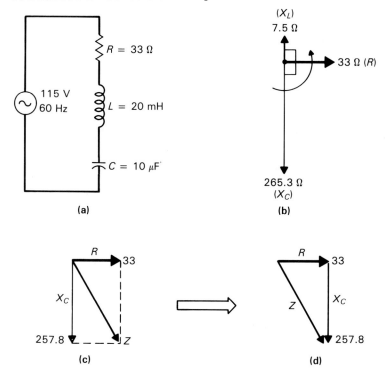

Series *RLC* Circuit   **631**

In the example,

$$X_L = 2\pi \times 60 \times 20 \text{ mH} = 7.5 \ \Omega$$

Resistance in the example is equal to $R = 33 \ \Omega$. Figure 18-2(b) illustrates these values of resistance and reactance in a vector diagram. In this vector diagram, you can see that X_L is drawn 90° ahead of R, and X_C is drawn 90° behind R; the capacitive and inductive reactances are 180° out of phase with one another and counteract to produce a vector diagram, as seen in Figure 18-2(c). The difference between X_L and X_C is equal to 257.8, and since X_C is greater than X_L, the resultant reactive vector is capacitive. Reactance, however, is not in phase with resistance, and impedance is the vector sum of the reactive (X) and resistance (R) vectors. The formula, based on the Pythagorean theorem, and illustrated in Figure 18-2(d), is

$$\boxed{Z = \sqrt{R^2 + X^2}}$$

In the example, therefore, the circuit impedence will be equal to

$$Z = \sqrt{R^2 + X^2} = \sqrt{33^2 + 257.8^2} = 260 \ \Omega$$

Since X is equal to the difference between (symbolized $\sim$) X_L and X_C $(X_L \sim X_C)$, the impedance formula can be modified slightly to incorporate the calculation to determine the difference between X_L and X_C.

$$\boxed{Z = \sqrt{R^2 + (X_L \sim X_C)^2}}$$

Using our example with this new formula, we arrive at the same value of impedance, and since the difference between X_L and X_C resulted in a capacitive vector, the circuit is said to act capacitively.

$$\begin{aligned} Z &= \sqrt{R^2 + (X_L \sim X_C)^2} \\ &= \sqrt{33^2 + (7.5 \sim 265.3)^2} \\ &= \sqrt{33^2 + 257.8^2} \\ &= 260 \ \Omega \end{aligned}$$

18.1.2 CURRENT

Once the total impedance of the circuit is known, Ohm's law can be applied to calculate the circuit current:

$$\boxed{I = \frac{V_s}{Z}}$$

In the example, circuit current is equal to

$$\begin{aligned} I &= \frac{V_s}{Z} \\ &= \frac{115 \text{ V}}{260 \ \Omega} \\ &= 0.44 \text{ A} \quad \text{or} \quad 440 \text{ mA} \end{aligned}$$

Resistive, Inductive, and Capacitive Circuits (*RLC*)

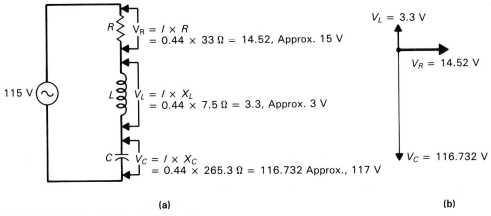

FIGURE 18-3 **Series Voltage Drops.**

18.1.3 VOLTAGE

Now that you know the value of current flowing in the series circuit, you can calculate the voltage drops across each component, as seen in Figure 18-3(a).

$$\boxed{\begin{aligned} V_R &= I \times R \\ V_L &= I \times X_L \\ V_C &= I \times X_C \end{aligned}}$$

The voltages in the example cannot be simply added together to obtain the applied voltage, because, like the resistances and reactances, they are not in phase, as seen in Figure 18-3(b).

A similar formula to impedance (Z) is used to calculate the applied voltage, which is, as we already know, 115 V in the example.

$$\boxed{V_s = \sqrt{V_R^2 + (V_L \sim V_C)^2}}$$
$$\begin{aligned} &= \sqrt{15^2 + (3 \sim 117)^2} \\ &= \sqrt{225 + 12{,}996} \\ &= \sqrt{13{,}221} \\ &= 115 \text{ V} \end{aligned}$$

18.1.4 POWER

The apparent power consumed by the circuit is calculated by

$$\boxed{\text{apparent power} = V_s \times I} = 115 \text{ V} \times 0.44 = 50.6 \text{ volt-amps}$$

while the true power is equal to

$$\boxed{\text{true power} = I^2 \times R} = 0.44^2 \times 33 \ \Omega = 6.4 \text{ watts}$$

The true or actual power dissipated by the resistor is as expected smaller than the apparent power that seems to be being used.

The power factor can be calculated, as usual, by

$$\boxed{PF = \text{Cos } \theta = \frac{R}{Z} = \frac{P_R}{P_A}} = 0.13$$

$$PF = 0 = \text{reactive circuit}$$
$$PF = 1 = \text{resistive circuit}$$

18.1.5 SUMMARY

As you can see, the analysis of a series *RLC* follows a fixed procedure, which is to calculate:

1. Inductive and capacitive reactance (X_L and X_C)
2. Circuit impedance (Z)

EXAMPLE 18.1

For a series circuit where $R = 10\ \Omega$, $L = 5$ mH, $C = 0.05\ \mu$F, and $V_S = 100$ V/2 kHz, calculate:
a. X_C e. V_R, V_C, and V_L
b. X_L f. Apparent power
c. Z g. True power
d. I h. Power factor

SOLUTION

a. $X_C = \dfrac{1}{2\pi f C} = 1.6$ kΩ

b. $X_L = 2\pi f L = 62.8\ \Omega$

c. $Z = \sqrt{R^2 + (X_L \sim X_C)^2}$
$\quad = \sqrt{(10\ \Omega)^2 + (1.6\text{ k}\Omega \sim 62.8\ \Omega)^2}$
$\quad = \sqrt{10\ \Omega^2 + 1.54\text{ k}^2}$
$\quad = 1.54$ *k*Ω (capacitive circuit due to high X_C)

d. $I = \dfrac{V_S}{Z} = \dfrac{100\text{ V}}{1.54\text{ k}\Omega} = 64.9$ mA

e. $V_R = I \times R = 64.9$ mA $\times$ 10 Ω = 0.65 V
$\quad V_C = I \times X_C = 64.9$ mA $\times$ 1.6 kΩ = 103.9 V
$\quad V_L = I \times X_L = 64.9$ mA $\times$ 62.8 Ω = 4.1 V

f. Apparent power = $V_S \times I = 100$ V $\times$ 64.9 mA = 6.49 VA

g. True power = $I^2 \times R = 64.9^2 \times 10\ \Omega = 42.17$ mW

h. PF $= \dfrac{R}{Z} = \dfrac{10\ \Omega}{1.5\text{ k}\Omega} = 0.006$ (reactive circuit)

634

3. Circuit current (I)
4. Component voltage drop (V_R, V_L, and V_C)
5. Power and power factor (PF)

18.1.6 RESONANCE

Resonance is a circuit condition that occurs when the inductive reactance (X_L) and the capacitive reactance (X_C) have been balanced. Figure 18-4 illustrates a parallel and series connected LC circuit. If a dc voltage is applied to the input of either circuit, the capacitor will act as an open (X_C = infinite Ω) and the inductor will act as a short (X_L = 0 Ω).

If a low frequency ac is now applied to the input, X_C will decrease from maximum, and X_L will increase from zero. As the ac frequency is further increased, the capacitive reactance will continue to fall ($X_C\!\downarrow \propto 1/f\!\uparrow$) and the inductive reactance to rise ($X_L\!\uparrow \propto f\!\uparrow$), as seen in Figure 18-5.

As the input ac frequency is further increased, a point will be reached where X_L will equal X_C, and this condition is known as *resonance*. The frequency at which $X_L = X_C$ in either a parallel or series LC circuit is known as the *resonant frequency* (f_0) and can be calculated by the following formula, which has been derived from the capacitive and inductive reactance formulas:

$$X_L = X_C$$

Therefore,

$$f_0 = \frac{1}{2\pi\sqrt{LC}}$$

(1) SERIES RESONANCE

Figure 18-6(a) illustrates a series RLC circuit at resonance ($X_L = X_C$). The ac input voltage causes current to flow around the circuit, and since all the components are connected in series, the same value of current (I_s) will flow through all the components. Since R, X_L, and X_C are all equal to 100 Ω and the current flow is the same throughout, the voltage dropped across each component will be equal, as is vectorially illustrated in Figure 18-6(b).

The voltage across the resistor is in phase with the series circuit current

FIGURE 18-4 **Resonance. (a) Parallel *LC* Circuit. (b) Series *LC* Circuit.**

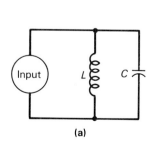

(a)

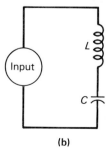

(b)

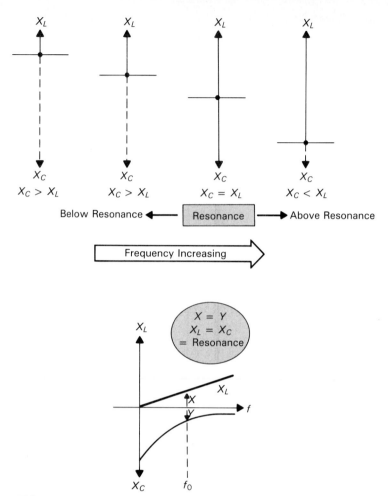

FIGURE 18-5 Frequency versus Reactance.

(I_s); however, since the voltage across the inductor (V_L) is 180° out of phase with the voltage across the capacitor (V_C), and both are equal to one another, V_L cancels V_C, when both are measured in series.

Three unusual characteristics occur when a circuit is at resonance, which do not occur at any other frequency.

(1) The first is that if V_L and V_C cancel, the voltage across L and C will measure 0 V on a voltmeter. Since there is effectively no voltage being dropped across these two components, all the voltage must be across the resistor $(V_R = 12 \text{ V})$. This is true; however, since the same current flows throughout the series circuit, a voltmeter will measure 12 V across C, 12 V across L, and 12 V across R, as seen in Figure 18-6(c). It now appears that the voltage drops around the series circuit (36 V) do not equal the voltage applied (12 V). This is not true as V_L *and* V_C cancel, because they are out of phase with one another, and so Kirchhoff's voltage law is still valid.

(2) The second unusual characteristic of resonance is that because the total opposition or impedance (Z) is equal to

$$Z = \sqrt{R^2 + (X_L \sim X_C)^2}$$

and the difference between X_L and X_C is 0 ($Z = \sqrt{R^2 + 0}$), the impedance of a series circuit at resonance is equal to the resistance value R ($Z = \sqrt{R^2} = R$). The applied ac voltage of 12 V is consequently forcing current to flow through this series RLC circuit. And since current is equal to $I_s = V/Z$ and $Z = R$, the circuit current at resonance is dependent on only the value of resistance. The capacitor and inductor are invisible and are seen by the source as simply a piece of conducting wire with no resistance, as illustrated in Figure 18-7. Since only resistance exists in the circuit, current (I_s) and voltage (V_s) are in phase with one another, and the power factor as expected for a purely resistive circuit will be equal to 1.

(3) To emphasize the third strange characteristic of series resonance, we will take another example, as seen in Figure 18-8. The circuit current in this example is equal to $I = V/R = 12 \text{ V}/10 = 1.2 \text{ A}$, since $Z = R$ at resonance. Since the same current flows throughout a series circuit, the voltage across each component can be calculated.

$$V_R = I \times R = 1.2 \text{ A} \times 10 = 12 \text{ V}$$

$$V_L = I \times X_L = 1.2 \text{ A} \times 100 = 120 \text{ V}$$

$$V_C = I \times X_C = 1.2 \text{ A} \times 100 = 120 \text{ V}$$

FIGURE 18-6   **Series Resonant Circuit.**

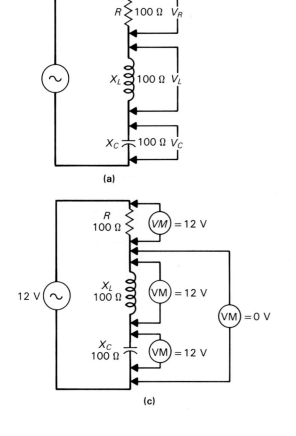

(a)

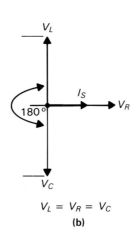

$$V_L = V_R = V_C$$

(b)

(c)

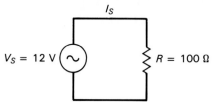

FIGURE 18-7 Impedance = *R* at Resonance.

FIGURE 18-8

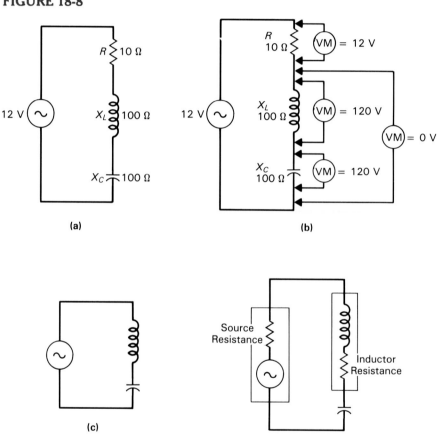

(a)

(b)

(c)

(d)

As V_L is 180° out of phase with V_C, the 120 V across the capacitor cancels with the 120 V across the inductor, resulting in 0 V across L and C, as seen in Figure 18-8(b). Since L and C have the ability to store energy, the voltage across them individually will appear larger than the applied voltage.

If the resistance in the circuit is removed completely, as seen in Figure 18-8(c), the circuit current, which is determined by the resistance only, will increase to a maximum ($I\uparrow = V/R\downarrow$) and consequently cause an infinitely high voltage across the inductor and capacitor ($V\uparrow = I\uparrow \times R$). In reality, the ac source will have some value of internal resistance, and the inductor, which is a long length of thin wire ($R\uparrow$), will have some value of resistance, as seen in Figure 18-8(d), which limits the series resonant circuit current.

In summary, we can say that in a series resonant circuit:

1. The inductor and capacitor electrically disappear due to their equal but opposite effect, resulting in 0 V drops across the series combination, and the circuit consequently seems purely resistive.

2. The current flow is large because the impedance of the circuit is low and equal to the series resistance (R), which has the source voltage developed across it.

3. The individual voltage drops across the inductor or capacitor can be larger than the source voltage if R is smaller than X_L and X_C.

(2) QUALITY FACTOR

As previously discussed under inductance, the Q factor is a ratio of inductive reactance to resistance and is used to express how efficiently an inductor will store rather than dissipate energy. In a series resonant circuit, the Q factor indicates the quality of the series resonant circuit, or is the ratio of the reactance to the resistance.

$$Q = \frac{X_L}{R}$$

or, since $X_L = X_C$,

$$Q = \frac{X_C}{R}$$

Another way to calculate the Q of a series resonant circuit is by using the formula

$$Q = \frac{V_L}{V_R} = \frac{V_C}{V_R} \quad \text{(at resonance only)}$$

or, since $V_R = V_s$

$$Q = \frac{V_L}{V_s} = \frac{V_C}{V_s} \quad \text{(at resonance only)}$$

If the Q and source voltage are known, the voltage across the inductor or capacitor can be found by transposition of the formula, as can be seen in the example in Figure 18-9.

The Q of a resonant circuit is almost entirely dependent on the inductor's

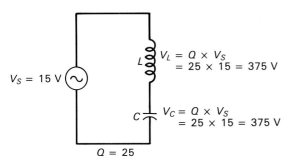

FIGURE 18-9 Quality Factor.

$V_S = 15\ V$

$V_L = Q \times V_S$
$\quad = 25 \times 15 = 375\ V$

$V_C = Q \times V_S$
$\quad = 25 \times 15 = 375\ V$

$Q = 25$

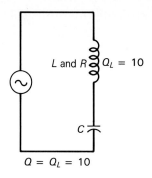

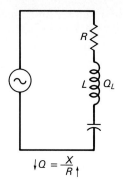

FIGURE 18-10 **Resistance within Inductor.**

coil resistance, because capacitors tend to have almost no resistance figure at all, only reactive, which makes them very efficient.

The inductor has a Q value of its own, and if only L and C are connected in series with one another, the Q of the series resonant circuit will be equal to the Q of the inductor, as seen in Figure 18-10. If the resistance is added in with L and C, the Q of the series resonant circuit will be less than that of the inductor's Q.

EXAMPLE 18.2

Calculate the resistance of the series resonant circuit illustrated in Figure 18-11.

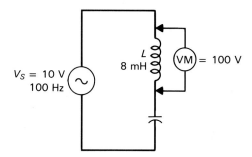

FIGURE 18-11

SOLUTION

$$Q = \frac{V_L}{V_s} = \frac{100 \text{ V}}{10 \text{ V}} = 10$$

Since $Q = X_L/R$, then $R = X_L/Q$; so if the inductive reactance can be found, then the R can be determined.

$$X_L = 2\pi \times f \times L$$
$$= 2 \times \pi \times 100 \times 8 \text{ mH}$$
$$= 5 \ \Omega$$

R will consequently equal

$$R = \frac{X_L}{Q}$$
$$= \frac{5}{10}$$
$$= 0.5 \ \Omega$$

---

(3) BANDWIDTH

A series resonant circuit is selective in that frequencies at resonance or slightly above or below will cause a larger current flow than frequencies well above or below the circuit's resonant frequency. The group or band of frequencies that causes the larger current flow is called the circuit's *bandwidth*.

Figure 18-12 illustrates a series resonant circuit and its bandwidth. The × marks on the curve illustrate where different frequencies were applied to the circuit and the resulting value of current measured in the circuit. The resulting curve produced is called a *frequency response curve*, as it illustrates the circuit's response to different frequencies. At resonance, $X_L = X_C$ and the two cancel, which is why maximum current flowed in the circuit (100 mA) when the resonant frequency (100 Hz) was applied.

The bandwidth includes the group or band of frequencies that cause 70.7% or more of the maximum current to flow within the series resonant circuit; in this example, frequencies from 90 to 110 Hz cause 70.7 mA or more, which is 70.7% of maximum (100 mA), to flow. The bandwidth in this example is equal to

$$\text{BW} = 110 - 90 = 20 \ \text{Hz}$$

(110 and 90 Hz are known as cutoff frequencies.)

Referring to the bandwidth curve in Figure 18-12(b), you may notice that 70.7% is also called the half-power points, although it does not exist halfway between 0 and maximum. This value of 70.7% is not the half-

FIGURE 18-12   Series Resonant Circuit Bandwidth. (a) Circuit. (b) Frequency Response Curve.

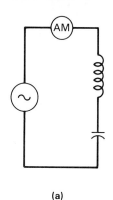
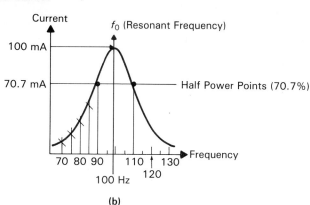

(a)

(b)

current point but is the half-power point, as we can prove with a simple example.

EXAMPLE 18.3

$R = 2 \text{ k}\Omega$ and $I = 100$ mA; therefore, power $= I^2 \times R \times 100$ mA$^2 \times 2$ k$\Omega = 20$ W. If the current is now reduced so that it is 70.7% of its original value, calculate the power dissipated.

SOLUTION

$P = I^2 \times R = 70.7$ mA$^2 \times 2$ k$\Omega = 10$ W.

In summary, the 70.7% current points are equal to the 50% or half-power points. A circuit's bandwidth is the band of frequencies that exists between the 70.7% current points or half-power points.

FIGURE 18-13   **Bandwidth of a Series Resonant Circuit.**

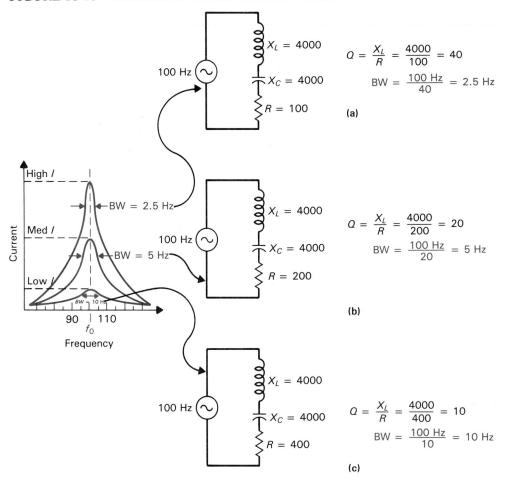

Resistive, Inductive, and Capacitive Circuits (*RLC*)

The bandwidth of a series resonant circuit can also be calculated by use of the formula

$$BW = \frac{f_0}{Q}$$

This formula states that the BW is proportional to the resonant frequency of the circuit and inversely proportional to the Q of the circuit.

Figure 18-13 illustrates three example response curves. In these three examples, the value of R is changed from 100 Ω to 200 Ω to 400 Ω. This does not vary the resonant frequency, but simply alters the Q and therefore the BW. The resistance value will determine the Q of the circuit, and since Q is inversely proportional to resistance, Q is proportional to current; consequently, a high value of Q will cause a high value of current.

In summary, the bandwidth of a series resonant circuit will increase as the Q of the circuit decreases (BW$\uparrow$ = $f_0/Q\downarrow$), and vice versa.

SELF-TEST REVIEW (§ 18.1)

1. List in order the procedure that should be followed to fully analyze a series *RLC* circuit.
2. State the formulas for calculating the following in relation to a series *RLC* circuit.
   a. Impedance
   b. Current
   c. Apparent power
   d. V_s
   e. True power
   f. V_R
   g. V_L
   h. V_C
3. Define resonance.
4. What is series resonance?
5. In a series resonant circuit, what are the three rather unusual circuit phenomena that take place?
6. How does Q relate to series resonance?
7. Define bandwidth.
8. Calculate BW if f_0 = 12 kHz and Q = 1000.

18.2
PARALLEL *RLC* CIRCUIT

Figure 18-14(a) and (b) depicts the analysis of parallel *RLC* circuits by illustrating the current and voltage relationships.

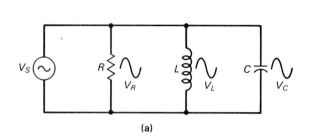 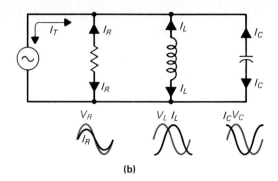

(a)   (b)

FIGURE 18-14   Parallel *RLC* Circuit. (a) *RLC* Parallel Circuit Voltage: Voltages across each component are all equal and in phase with one another in a parallel circuit. (b) *RLC* Parallel Circuit Currents: I_R is in phase with V_R, I_L lags V_L by 90°, and I_C leads V_C by 90°.

18.2.1   CURRENT

The total current (I_T) will be one of the first things to be calculated. An example circuit is illustrated in Figure 18-15(a), and the branch currents can be calculated by using the formulas

$$I_R = \frac{V}{R}$$

$$I_L = \frac{V}{X_L}$$

$$I_C = \frac{V}{X_C}$$

Figure 18-15(b) illustrates these branch currents vectorially, with I_R in phase with V_s, I_L lagging by 90°, and I_C leading I_R by 90°. The 180° phase difference between I_C and I_L results in a cancellation, as seen in Figure 18-15(c).

The total current (I_T) can be calculated by using the Pythagorean theorem on the right triangle, as illustrated in Figure 18-15(d).

$$I_T = \sqrt{I_R^2 + I_X^2}$$
$$= \sqrt{3.5^2 + 14.9^2}$$
$$= 15.3 \text{ A}$$

Since the total current (I_T) is known, the impedance of all three components in parallel can be calculated by the formula

$$Z = \frac{V}{I_T}$$
$$= \frac{115 \text{ V}}{15.3 \text{ A}}$$
$$= 7.5 \ \Omega$$

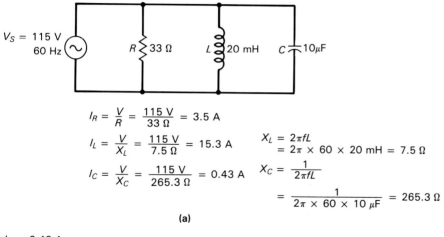

$$I_R = \frac{V}{R} = \frac{115\ V}{33\ \Omega} = 3.5\ A$$

$$I_L = \frac{V}{X_L} = \frac{115\ V}{7.5\ \Omega} = 15.3\ A \qquad \begin{aligned} X_L &= 2\pi fL \\ &= 2\pi \times 60 \times 20\ mH = 7.5\ \Omega \end{aligned}$$

$$I_C = \frac{V}{X_C} = \frac{115\ V}{265.3\ \Omega} = 0.43\ A \qquad X_C = \frac{1}{2\pi fL}$$

$$= \frac{1}{2\pi \times 60 \times 10\ \mu F} = 265.3\ \Omega$$

(a)

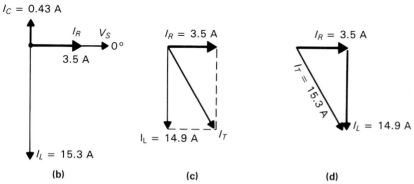

(b) (c) (d)

FIGURE 18-15 Example *RLC* Parallel Circuit.

18.2.2 PARALLEL RESONANCE

The parallel resonant circuit acts differently from the series resonant circuit, and these different characteristics need to be analyzed and discussed. Figure 18-16 illustrates a parallel resonant circuit. The inductive current could be calculated by using the formula

$$\boxed{I_L = \frac{V}{X_L}}$$

$$= \frac{10\ V}{1\ k\Omega}$$

$$= 10\ mA$$

The capacitive current could be calculated by using the formula

$$I_C = \frac{V_C}{X_C}$$

$$= \frac{10\ V}{1\ k\Omega}$$

$$= 10\ mA$$

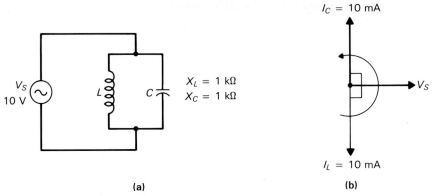

FIGURE 18-16 Parallel Resonant Circuit. (a) Circuit. (b) Vector Diagram.

Looking at the vector diagram in Figure 18-16(b), you can see that I_C leads the source voltage by 90° and I_L lags the source voltage by 90°, creating a 180° phase difference between I_C and I_L. This means that when 10 mA of current flows up through the inductor 10 mA of current will flow in the opposite direction down through the capacitor, as seen in Figure 18-17(a). During the opposite alternation, 10 mA will flow down through the inductor and 10 mA will travel up through the capacitor, as seen in Figure 18-17(b).

If 10 mA arrives into point X and 10 mA of current leaves point X, then no current can be flowing from the source (V_s) to the parallel LC circuit; the current is simply swinging or oscillating back and forth between the capacitor and inductor.

The source voltage (V_s) is needed initially to supply power to the LC circuit and start the oscillations; but once the oscillating process is in progress (assuming the ideal case), current is only flowing back and forth between

FIGURE 18-17

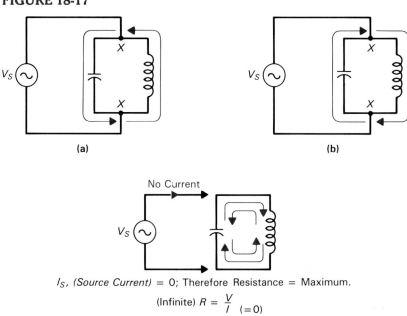

Resistive, Inductive, and Capacitive Circuits (*RLC*)

inductor and capacitor, and no current is flowing from the source. So the *LC* circuit appears as an infiinite impedance and the source can be disconnected, as seen in Figure 18-17(c).

(1) FLYWHEEL ACTION

Let's discuss in a little more detail this oscillating effect, which is called *flywheel action*. The name is derived from the fact that it resembles a mechanical flywheel, which, once started, will continually keep going until friction reduces the magnitude of the rotations to zero.

The electronic equivalent of the mechanical flywheel is a resonant parallel connected *LC* circuit. Figure 18-18(a) through (h) illustrates the continual energy transfer between capacitor and inductor, and vice versa. The direction of the circulating current reverses each half cycle at the frequency of resonance. Energy is stored in the capacitor in the form of an electric field between the plates on one half cycle, and then the capacitor discharges, supplying current to build up a magnetic field on the other half-cycle. The inductor stores its energy in the form of a magnetic field, which will collapse, supplying a current to charge the capacitor, which will then discharge, supplying a current back to the inductor, and so on. Due to the

FIGURE 18-18 Energy and Current Flow in an *LC* Parallel Circuit at Resonance.

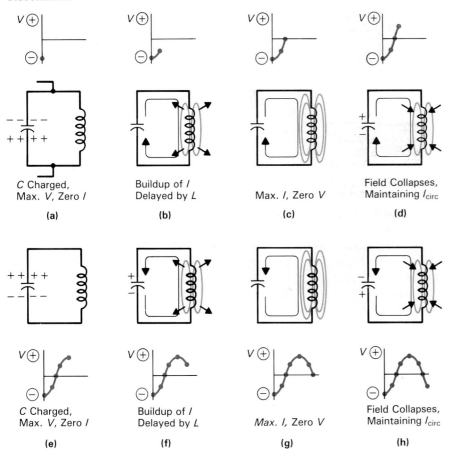

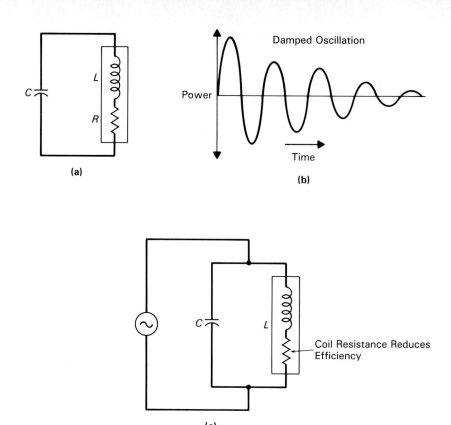

FIGURE 18-19 **Losses in Tanks.**

storing action of this circuit, it is sometimes related to the fluid analogy and referred to as a *tank circuit*.

(2) THE REALITY OF TANKS

A tank circuit, in the ideal condition, should oscillate indefinitely if no losses occur within the circuit. In reality, the resistance of the coil reduces that 100% efficiency, as does friction with the mechanical flywheel. This coil resistance is illustrated in Figure 18-19(a), and, unlike reactance, resistance is the opposition to current flow, with the dissipation of energy in the form of heat. As a small part of the energy is dissipated with each cycle, the oscillations will be reduced in size and eventually fall to zero, as seen in Figure 18-19(b).

If the ac source is reconnected to the tank, as seen in Figure 18-19(c), a small amount of current will flow from the source to the tank to top up the tank or replace the dissipated power. The higher the coil resistance is, the higher the loss and the larger the current flow from source to tank to replace the loss.

(3) QUALITY FACTOR

In the series resonant circuit, we are concerned with voltage drops since current remains the same throughout a series circuit, so

Resistive, Inductive, and Capacitive Circuits (*RLC*)

$$Q = \frac{V_C \text{ or } V_L}{V_s} \quad \text{(at resonance only)}$$

In a parallel resonant circuit, we are concerned with circuit currents rather than voltage, and so

$$Q = \frac{I_{tank}}{I_s} \quad \text{(at resonance only)}$$

The quality factor, Q, can also be expressed as the ratio between reactance and resistance:

$$Q = \frac{X_L}{R} \quad \text{(at any frequency)}$$

Another formula, which is the most frequently used when discussing and using parallel resonant circuits, is

$$Q = \frac{Z_{tank}}{X_L} \quad \text{(at resonance only)}$$

This formula states that the Q of the tank is proportional to the tank impedance. A higher tank impedance results in a smaller current flow from source to tank. This assures less power is dissipated, and that means a higher-quality tank.

Of all the three Q formulas for parallel resonant circuits, $Q = I_{tank}/I_s$, $Q = X_L/R$, or $Q = Z_{tank}/X_L$, the latter is the easiest to use as both X_L and the tank impedance can easily be determined in most cases where C, L, and R internal for the inductor are known.

(4) BANDWIDTH

Figure 18-20 illustrates a parallel resonant circuit and two typical response curves. These response curves summarize what we have previously de-

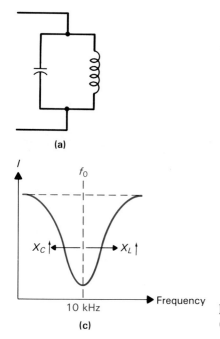

(a)

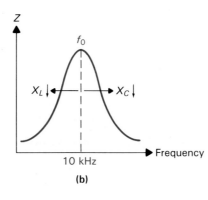

(b)

(c)

FIGURE 18-20 Parallel Resonant Circuit Bandwidth.

scribed, in that a parallel resonant circuit has maximum impedance, [Figure 18-20(b)] and minimum current [Figure 18-20(c)] at resonance. The current versus frequency response curve in Figure 18-20(c) is the complete opposite to the series resonant response curve. At frequencies below resonance (<10 kHz), X_L is low and X_C is high, and the inductor offers a low reactance, producing a high current path and low impedance. On the other hand, at frequencies above resonance (>10 kHz), the capacitor displays a low reactance, producing a high current path and low impedance. The parallel resonant circuit is like the series resonant circuit in that it responds to a band of frequencies close to its resonant frequency.

The bandwidth (BW) can be calculated by use of the formula

$$\boxed{\text{BW} = \frac{f_0}{Q}}$$

EXAMPLE 18.4

Calculate the bandwidth of the circuit illustrated in Figure 18-21.

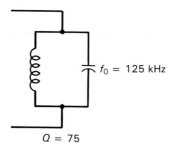

$f_0 = 125$ kHz

$Q = 75$

FIGURE 18-21

SOLUTION

$$\begin{aligned}
\text{BW} &= \frac{f_0}{Q} \\
&= \frac{125 \text{ kHz}}{75} \\
&= 1.7 \text{ kHz} \qquad (124.15 \text{ kHz to } 125.85 \text{ kHz})
\end{aligned}$$

---

(5) SELECTIVITY

Selectivity, by definition, is the ability of a tuned circuit to respond to a desired frequency and ignore all others. Parallel resonant LC circuits are sometimes too selective, as the Q is too large, producing too narrow a bandwidth, as seen in Figure 18-22(a) (BW↓ = f_0/Q↑).

In this situation, because of the very narrow response curve, a high resistance value can be placed in parallel with the LC circuit to provide an alternate path for line current. This process is known as loading or damping the tank and will cause an increase in line current and decrease in Q

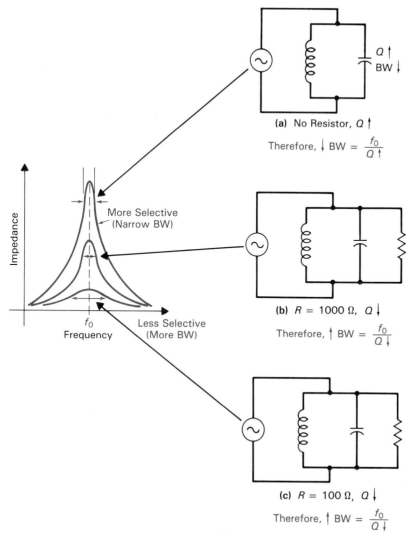

(a) No Resistor, $Q \uparrow$

Therefore, $\downarrow$ BW $= \dfrac{f_0}{Q \uparrow}$

(b) $R = 1000 \; \Omega$, $Q \downarrow$

Therefore, $\uparrow$ BW $= \dfrac{f_0}{Q \downarrow}$

(c) $R = 100 \; \Omega$, $Q \downarrow$

Therefore, $\uparrow$ BW $= \dfrac{f_0}{Q \downarrow}$

FIGURE 18-22   Varying Bandwidth by Loading the Tank Circuit.

$(Q\downarrow = I_{tank}/I_{line}\uparrow)$. The decrease in q will cause a corresponding increase in BW (BW$\uparrow = f_0/Q\downarrow$), as seen by the examples in Figure 18-22, which illustrates a 1000 Ω loading resistor [Figure 18-22(b)] and a 100 Ω loading resistor [Figure 18-22(c)].

In summary, a parallel resonant circuit can be made less selective with a broader bandwidth if a resistor is added in parallel, providing an increase in current and a decrease in impedance, which widens the bandwidth.

SELF-TEST REVIEW (§ 18.2)

1. State the formulas for calculating the following in relation to a parallel *RLC* circuit:
    a. I_R
    b. I_T

    c. I_C
    d. I_L
2. What are the differences between a series and parallel resonant circuit?
3. Describe flywheel action.
4. Calculate the Q of a tank if $X_L = 50 \ \Omega$ and $R = 25 \ \Omega$.
5. When calculating bandwidth for a parallel resonant circuit, can the series resonant bandwidth formula be used?
6. What is selectivity?

18.3

APPLICATIONS OF *RLC* CIRCUITS

In the previous chapters, you saw how *RC* and *RL* filter circuits are used as low pass or high pass filters to pass some frequencies and block others. There are basically four types of filter:

1. Low pass filter, which passes frequencies below a cutoff frequency.
2. High pass filter, which passes frequencies above a cutoff frequency
3. Band pass filter, which passes a band of frequencies
4. Band stop filter, which stops a band of frequencies

18.3.1 LOW-PASS FILTER

Figure 18-23(a) illustrates how an inductor and capacitor can be connected to act as a low pass filter. At low frequencies, X_L has a small value compared to the load resistor (R_L), so nearly all the low frequency input is developed and appears at the output across R_L. Since X_C is high at low frequencies, nearly all the current passes through R_L rather than C.

At high frequencies, X_L increases and drops more of the applied input across the inductor rather than the load. The capacitive reactance, X_C, aids

FIGURE 18-23 **Low-pass Filter. (a) Circuit. (b) Frequency Response.**

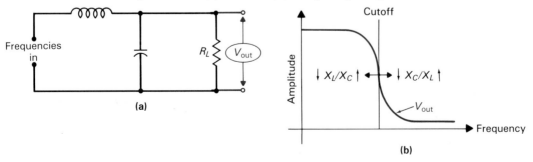

this blocking of high frequency effect by decreasing its reactance and providing an alternative path for current to flow.

Since the inductor basically blocks alternating current and the capacitor shunts alternating current, the net result is to prevent high frequency signals from reaching the load. The way in which this low pass filter responds to frequencies is graphically illustrated in Figure 18-23(b).

18.3.2 HIGH-PASS FILTER

Figure 18-24(a) illustrates how an inductor and capacitor can be connected to act as a high pass filter. At high frequencies, the reactance of the capacitor (X_C) is low, while the reactance of the inductor (X_L) is high, so all the high frequencies are easily passed by the capacitor and blocked by the inductor, so they all are routed through to the output and load.

At low frequencies, the reverse condition exists, resulting in a low X_L and high X_C. The capacitor drops nearly all the input, and the inductor shunts the signal current away from the output load.

18.3.3 BAND-PASS FILTER

Figure 18-25(a) illustrates a series resonant band pass filter, and Figure 18-25(b) shows a parallel resonant band pass filter. Figure 18-25(c) shows the frequency response curve produced by the band pass filter. At resonance, the series resonant LC circuit has a very low impedance and will consequently pass the resonant frequency to the load with very little drop across the L and C components.

Below resonance, X_C is high, and the capacitor drops a large amount of the input signal; above resonance, X_L is high and the inductor drops most of the input frequency voltage. This circuit will therefore pass a band of

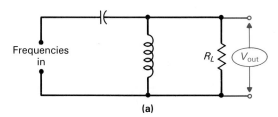

(a)

FIGURE 18-24 **High pass Filter.**
(a) Circuit.
(b) Frequency Response.

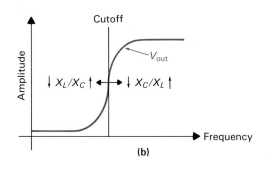

(b)

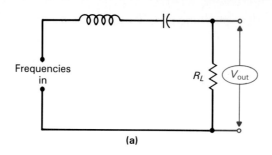

(a)

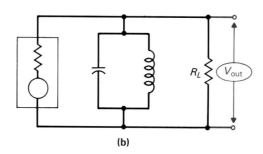

(b)

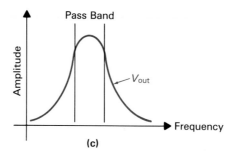

(c)

FIGURE 18-25   Band-pass Filter. (a) Series Resonant Band-pass Filter. (b) Parallel Resonant Band-pass Filter. (c) Frequency Response.

frequencies centered around the resonant frequency of the series LC circuit and block all other frequencies above and below this resonant frequency.

Figure 18-25(b) illustrates how a parallel resonant LC circuit can be used to provide a band-pass response. The series resonant circuit was placed in series with the output, whereas the parallel resonant circuit will have to be placed in parallel with the output to provide the same results. At resonance, the parallel resonant circuit or tank has a high impedance, so very little current will be shunted away from the output; it will be passed on to the output, and almost all the input will appear at the output across the load.

Above resonance, X_C is small, so most of the input is shunted away from the output by the capacitor; below resonance, X_L is small, and the shunting action occurs again, but this time through the inductor.

Figure 18-26 illustrates how a transformer can be used to replace the inductor to produce a band-pass filter. At resonance, maximum flywheel current flows within the parallel circuit made up of the capacitor and the primary of the transformer (L), which is known as a *tuned* transformer. With maximum flywheel current, there will be a maximum magnetic field, which means that there will be maximum power transfer between primary and

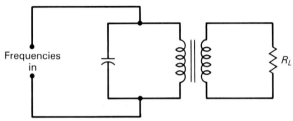

FIGURE 18-26   Parallel Resonant Band Pass Circuit Using a Transformer.

secondary. So nearly all the input will be coupled to the output (coupling coefficient $k = 1$) and appear across the load at and around a small band of frequencies centered on resonance.

Above and below resonance, current within the parallel resonant circuit will be smaller. So the power transfer ability will be less, effectively keeping the frequencies outside of the pass band from appearing at the output.

18.3.4 BAND STOP FILTER

Figure 18-27(a) illustrates a series resonant and Figure 18-27(b) a parallel resonant band stop filter. Figure 18-27(c) shows the frequency response curve produced by a band stop filter. The band stop filter operates in completely the opposite way as a band pass filter in that it blocks or attenuates a band of frequencies centered on the resonant frequency of the *LC* circuit.

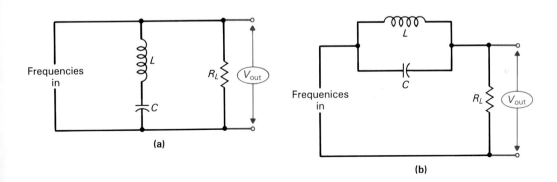

(a)

(b)

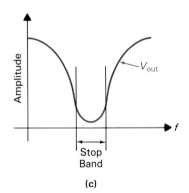

(c)

FIGURE 18-27   Band Stop Filter. (a) Series Resonant Band Stop Filter. (b) Parallel Resonant Band Stop Filter. (c) Frequency Response.

FIGURE 18-28 **Tuning in of Station by Use of Band Pass Filter.**

In the series resonant circuit in Figure 18-27(a), the *LC* impedance is very low at and around resonance, so these frequencies are rejected or shunted away from the output. Above and below resonance, the series circuit has a very high impedance, which results in almost no shunting of the signal away from the output.

In the parallel resonant circuit, in Figure 18-27(b), the *LC* circuit is in series with the load and output. At resonance, the impedance of a parallel resonant circuit will be very high, and the band of frequencies centered around resonance will be blocked. Above and below resonance, the impedance of the tank is very low, so nearly all the input is developed across the output.

Filters are necessary in applications such as televison or radio, where we need to tune in (select or pass) one frequency that contains the information we desire, yet block all the millions of other frequencies that are also carrying information, as seen in Figure 18-28.

SELF-TEST REVIEW (§ 18.3)

1. Of the four types of filters, which:
   a. Would utilize the inductor as a shunt?
   b. Would utilize the capacitor as a shunt?
   c. Would use a series resonant circuit as a shunt?
   d. Would use a parallel resonant circuit as a shunt?
2. In what applications can filters be found?

18.4
COMPLEX NUMBERS

After reading this section, you will realize that there is really nothing complex about complex numbers. The complex number system allows us to determine the *magnitude* and *phase angle* of electrical quantities by adding, subtracting, multiplying, and dividing phasor quantities, and is an invaluable tool in ac circuit analysis.

FIGURE 18-29 Real Number Line.

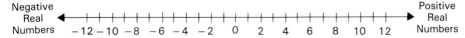

18.4.1 THE REAL NUMBER LINE

Real numbers can be represented on a horizontal line, known as the real number line, as in Figure 18-29. Referring to this line, you can see that positive numbers exist to the right of the point corresponding to zero, while negative numbers exist to the left. This representation satisfied most mathematicians for a short time, as they could indicate numbers such as 2 or 5 as a point on the line. Numbers corresponding to the $\sqrt{9}$ could also be represented, as three points to the right of zero ($\sqrt{9} = +3$). However, a problem was reached if they wished to indicate a point corresponding to $\sqrt{-9}$. The $\sqrt{-9}$ is not $+3$ [since $(+3) \times (+3) = +9$], and it is not -3 [since $(-3) \times (-3) = +9$]. So it was eventually realized that the square root of a negative number could not be indicated on the real number line, as it is not a real number.

**FIGURE 18-30
Imaginary Number
Line.**

18.4.2 THE IMAGINARY NUMBER LINE

Mathematicians decided to call the square root of a negative number, such as $\sqrt{-4}$ or $\sqrt{-9}$, imaginary numbers, which are not fictitious or imaginary, but simply a particular type of number.

Just as real numbers can be represented on a real number line, imaginary numbers can be represented on an imaginary number line, as seen in Figure 18-30. The imaginary number line is vertical to distinguish it from the real number line, and when working with electrical quantities a $\pm j$ prefix, known as the *j operator*, is used for values that appear on the imaginary number line.

18.4.3 THE COMPLEX PLANE

A complex number is the combination of a real and imaginary number and is represented on a two-dimensional plane called the complex plane, as in

Complex Numbers

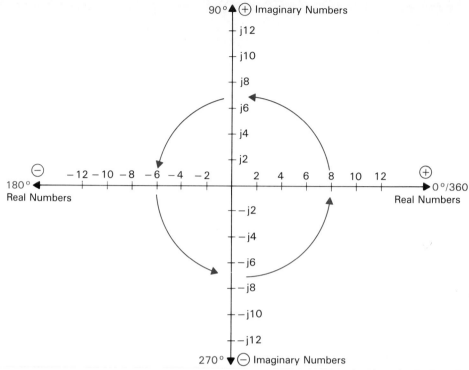

FIGURE 18-31   The Complex Plane.

Figure 18-31. Generally, the real number appears first, followed by the imaginary number. Here are some examples of complex numbers.

Real Numbers		Imaginary Numbers
3	+	$j4$
−2	+	$j4$
−3	−	$j2$

Complex numbers, therefore, are merely terms that need to be added as phasors, and all you have to do basically is draw a vector representing the real number and then draw another vector representing the imaginary number.

EXAMPLE 18.5

Find the points in the complex plane in Figure 18-32 that correspond to the following complex numbers.

$$W = 3 + j4$$
$$X = 5 - j7$$
$$Y = -4 + j6$$
$$Z = -3 - j5$$

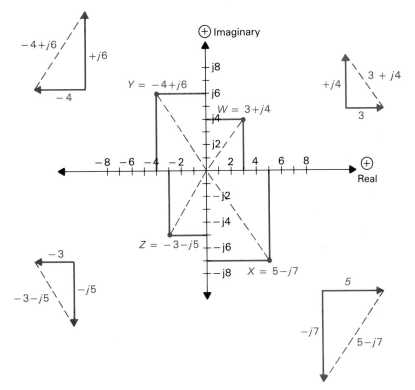

FIGURE 18-32

SOLUTION

By first locating the point corresponding to the real number on the horizontal line and then plotting it against the imaginary number on the vertical line, the points can be determined as shown in Figure 18-32.

A number like $3 + j4$ specifies two phasors in rectangular coordinates, so this system is the rectangular representation of a complex number. There are several other ways to describe a complex number, one of which is the polar representation of a complex number, which will be discussed next.

18.4.4 POLAR COMPLEX NUMBERS

Phasors can also be expressed in polar form, as seen in Figure 18-33, which compares rectangular and polar notation. With the rectangular notation in Figure 18-33(a), the horizontal coordinate is the real part and the vertical coordinate is the imaginary part of the complex number. With the polar notation in Figure 18-33(b), the magnitude of the phasor (a) and the angle relative to the positive real axis (measured in a counterclockwise direction) are stated.

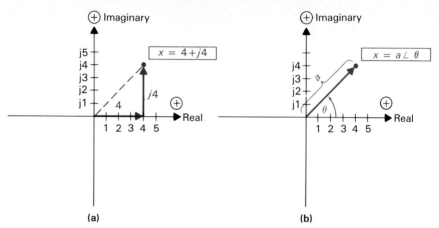

FIGURE 18-33 **Representing Phasors. (a) Rectangular Notation. (b) Polar Notation.**

EXAMPLE 18.6

Sketch the following polar numbers:
a. $5\angle 60°$
b. $3\angle 220°$

FIGURE 18-34

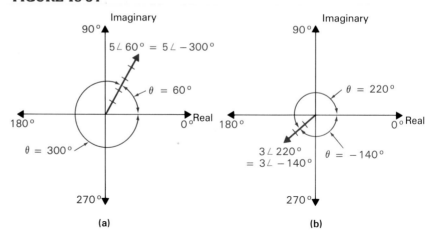

SOLUTION

As you can see in Figure 18-34, an equivalent negative angle, which is calculated by subtracting the given positive angle from 360°, can also be used.

18.4.5 RECTANGULAR/POLAR CONVERSIONS

Many scientific calculators have a feature that allows you to convert rectangular numbers to polar numbers, and vice versa. These conversions are based on the Pythagorean Theorem and trigonometric functions.

Resistive, Inductive, and Capacitive Circuits (*RLC*)

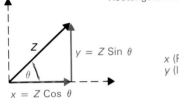

Polar Notation

Rectangular Notation

x (Real number) $= Z \cos \theta$
y (Imaginary number) $= Z \sin \theta$

FIGURE 18-35 **Polar-to-Rectangular Conversion.**

(1) POLAR-TO-RECTANGULAR CONVERSION

The polar notation states the magnitude and angle, as seen in Figure 18-35(a). The following examples will show how this conversion can be achieved.

EXAMPLE 18.7

Convert the following polar numbers to rectangular form:
a. $5\angle 30°$
b. $18\angle -35°$
c. $44\angle 220°$

SOLUTION

a. Real number $= 5 \cos 30° = 4.33$
   Imaginary number $= 5 \sin 30° = 2.5$
   Polar number, $5\angle 30°$ = rectangular number, $4.33 + j2.5$
b. Real number $= 18 \cos (-35°) = 14.74$
   Imaginary number $= 18 \sin (-35°) = -10.32$
   Polar number, $18\angle -35°$ = rectangular number, $14.74 - j10.32$
c. Real number $= 44 \cos 220° = -33.7$
   Imaginary number $= 44 \sin 220° = -28.3$
   Polar number, $44 \angle 220°$ = rectangular number, $-33.7 - j28.3$

(2) RECTANGULAR-TO-POLAR CONVERSION

The rectangular notation states the horizontal (imaginary) and vertical (real) sides of a triangle, as seen in Figure 18-36. The following examples will show how the conversion can be achieved.

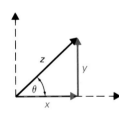

Polar Notation

Rectangular Notation

FIGURE 18-36 **Rectangular-to-Polar Conversion.**

Magnitude, $Z = \sqrt{x^2 + y^2}$

Angle, $\theta = \arctan \left(\frac{y}{x}\right)$

EXAMPLE 18.8

Convert the following rectangular numbers to polar form.
a. $4 + j3$
b. $16 - j14$

SOLUTION

a. Magnitude $= \sqrt{4^2 + 3^2} = 5$
   Angle $= \arctan(3/4) = 36.9°$
   Rectangular number, $4 + j3 =$ polar number, $5\angle 36.9°$
b. Magnitude $= \sqrt{16^2 + (-14)^2} = 21.3$
   Angle $= \arctan(-14/16) = -41.2°$
   Rectangular number, $16 - j14 =$ polar number, $21.3\angle -41.2°$

18.4.6 COMPLEX NUMBER ARITHMETIC

Since a phase difference exists between real and imaginary (j) numbers, certain rules should be applied when adding, subtracting, multiplying, or dividing complex numbers.

(1) ADDITION

The sum of two complex numbers is equal to the sum of their real and imaginary parts.

EXAMPLE 18.9

Add the following complex numbers:
a. $(3 + j4) + (2 + j5)$
b. $(4 + j5) + (2 - j3)$

SOLUTION

a. $(3 + j4) + (2 + j5) = (3 + 2) + (j4 + j5) = 5 + j9$
b. $(4 + j5) + (2 - j3) = (4 + 2) + (j5 - j3) = 6 + j2$

(2) SUBTRACTION

The difference of two complex numbers is equal to the difference between the separate real and imaginary parts.

EXAMPLE 18.10

Subtract the following complex numbers:
a. $(4 + j3) - (2 + j2)$
b. $(12 + j6) - (6 - j3)$

SOLUTION

a. $(4 + j3) - (2 + j2) = (4 - 2) + j(3 - 2) = 2 + j1$
b. $(12 + j6) - (6 - j3) = (12 - 6) + j[6 - (-3)] = 6 + j9$

---

(3) MULTIPLICATION

Multiplication of two complex numbers is more easily achieved if they are in polar form. The simple rule to remember is to multiply the magnitudes and then add the angles algebraically.

EXAMPLE 18.11

Multiply the following complex numbers:
a. $5\angle 35° \times 7\angle 70°$
b. $4\angle 53° \times 12\angle -44°$

SOLUTION

a. Multiple the magnitudes: $5 \times 7 = 35$.
   Algebraically add the angles: $\angle(35° + 70°) = \angle 105° = 35\angle 105°$.
b. Multiply the magnitudes: $4 \times 12 = 48$.
   Algebraically add the angles: $\angle[53° + (-44°)] = \angle 9° = 48\angle 9°$

---

(4) DIVISION

Division is also more easily carried out in polar form. The rule to remember is to divide the magnitudes, and then subtract the denominator angle from the numerator angle.

EXAMPLE 18.12

Divide the following complex numbers:

a. $60\angle 30°$ by $30\angle 15°$
b. $100\angle 20°$ by $5\angle -7°$

SOLUTION

a. Divide the magnitudes: $60/30 = 2$.
   Subtract the denominator angle from the numerator angle: $\angle(30° - 15°)$
   $= \angle 15° = 2\angle 15°$.
b. Divide the magnitudes: $100/5 = 20$.
   Subtract the angles: $\angle[20° - (-7°)] = \angle 27° = 20\angle 27°$.

18.4.7 HOW COMPLEX NUMBERS APPLY TO AC CIRCUITS

Complex numbers find an excellent application in ac circuits due to all the phase differences that occur between different electrical quantities, such as X_L, X_C, R, and Z as seen in Figure 18-37. The positive real number line, at an angle of 0°, is used for resistance, which in this example is 3 Ω.

FIGURE 18-37   Applying Complex Numbers to Series AC Circuits. (a) Impedance Phasors. (b) Series Circuit. (c) Voltage and Current Phasors.

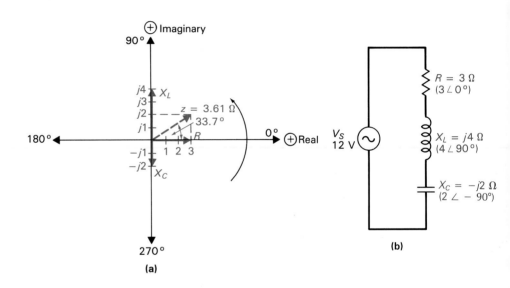

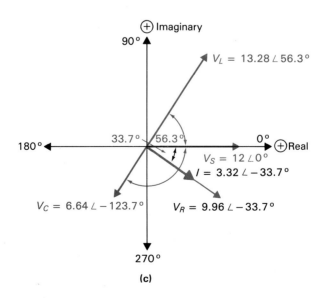

On the positive imaginary number line, at an angle of 90° $(+j)$, inductive reactance (X_L) is represented, which in this example is $j4\ \Omega$ $(X_L = 4\ \Omega)$. The voltage drop across an inductor (V_L) is proportional to its inductive reactance (X_L), and both are represented on the $+j$ imaginary number line, since the voltage drop across an inductor will always lead the current (which in a series circuit is always in phase with the resistance) by 90°.

On the negative imaginary number line, at an angle of $-90°$ or 270° $(-j)$, capacitive reactance (X_C) is represented, which in this example is $-j2$ $(X_C = 2\ \Omega)$. The voltage drop across a capacitor (V_C) is proportional to its capacitive reactance (X_C), and both are represented on the $-j$ imaginary number line since the voltage drop across a capacitor always lags current (charge and discharge) by $-90°$.

(1) SERIES AC CIRCUITS

Referring to Figure 18-37(a) and (b) once again, we can calculate total impedance by simply adding the phasors.

(a) Z_T **(Rectangular)** Total series impedance is equal to the sum of all the resistances and reactances:

$$Z = \sqrt{R^2 + (X_L \sim X_C)^2}$$
$$Z = R + jX_L - jX_C$$

(b) Z_T **(Polar)** The total series impedance can be converted from rectangular to polar form:

$$\text{Magnitude} = \sqrt{3^2 + 2^2} = 3.61\ \Omega$$

$$\text{Angle} = \arctan\left(\frac{2}{3}\right) = 33.7°$$

$$= 3.61\angle 33.7°$$

(c) **Current** Once the magnitude of Z_T is known (3.61 Ω), I can be calculated. The source voltage of 12 V is a real positive number (0°) and is therefore represented as $12\angle 0°$. Current is equal to

$$I = \frac{V_S}{Z_T} = \frac{12\angle 0°}{3.61\angle 33.7°}$$

Polar division   Divide the magnitudes: $\dfrac{12}{3.61} = 3.32$ A

Subtract the angles: $0° - 33.7° = -33.7°$
$I = 3.32\angle -33.7°$

(d) **Phase Angle** The circuit current has an angle of $-33.7°$, which means that it lags V_T. This negative phase angle is expected since this series circuit is inductive $(X_L > X_C)$, in which case current should lag voltage by some phase angle. This phase angle is less than 45° because the net reactance is less than the circuit resistance, and is shown in Figure 18-37(c).

(e) **Voltage Drops** The component voltage drops are calculated with the following formulas and shown in Figure 18-37(c).

$V_R = I \times R = (3.32\angle - 33.7°) \times (3\angle 0°)$
Multiply the magnitudes: $3.32 \times 3 = 9.96$ V
Algebraically add the angles: $\angle(-33.7° + 0°) = -33.7°$
$V_R = 9.96\angle - 33.7°$

$V_L = I \times X_L = (3.32\angle - 33.7°) \times (4\angle 90°)$
$\times$ Magnitudes: $3.32 \times 4 = 13.28$ V
$+$ Angles: $\angle(-33.7 + 90°) = 56.3°$
$V_L = 13.28\angle 56.3°$

$V_C = I \times X_C = (3.32\angle - 33.7°) \times (2\angle - 90°)$
$\times$ Magnitudes: $3.32 \times 2 = 6.64$ V
$+$ Angles: $\angle(-33.7 + -90°) = -123.7°$
$V_C = 6.64\angle - 123.7°$

(f) Phase Relationships As seen in Figure 18-37(c), the voltage across an inductor leads the circuit current by $+90°$, while the voltage across a capacitor lags the circuit current by $-90°$. The source voltage acts as the zero reference phase and leads the circuit current and the voltage across the resistor (V_R) by $33.7°$.

(g) Source Voltage Although the source voltage is known, it can be checked to verify all the previous calculations, since the sum of all the individual voltage drops should equal the source voltage.

	Polar		Rectangular
$V_L =$	$13.28\angle 56.3$	$=$	$7.37 + j11.05$
$V_R =$	$9.96\angle - 33.7°$	$=$	$8.29 - j5.53$
$V_C =$	$6.64\angle - 123.7°$	$=$	$-3.68 - j5.52$
			$11.98 + j0$

(2) SERIES–PARALLEL AC CIRCUITS

Figure 18-38(a) illustrates a series–parallel ac circuit containing an R_L branch, an R_C branch, and an RLC branch.

(a) Impedance of Each Branch The three branches will each have a value of impedance that will be equal to:

	Rectangular		Polar
$Z_1 =$	$10 + j5$	$=$	$11.2\angle 26.6°$ Ω
$Z_2 =$	$25 - j15$	$=$	$29.2\angle - 31.0°$ Ω

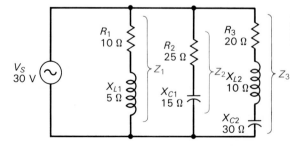

FIGURE 18-38 **Applying Complex Numbers to Series–Parallel AC.**

Resistive, Inductive, and Capacitive Circuits (*RLC*)

The third branch is capacitive since the difference between $-j30$ (X_{C2}) and $+j10$ (X_{L2}) is $-j20$.

$$Z_3 = 20 - j20 = 28.3\angle -45° \ \Omega$$

(b) Branch Currents The three branch currents, I_1, I_2, and I_3, are calculated by dividing the source voltage (V_S) by the individual branch impedances.

$$I_1 = \frac{V_S}{Z_1} = \frac{30\angle 0°}{11.2\angle 26.6°}$$

Divide magnitudes: $30 \div 11.2 = 2.68$
Subtract angles: $\angle(0° - 26.6°) = -26.6°$
$I_1 = 2.68\angle -26.6° = 2.4 - j1.2$ A

$$I_2 = \frac{V_S}{Z_2} = \frac{30\angle 0°}{29.2\angle -31°} = 1.03\angle +31° = 0.88 + j0.5 \text{ A}$$

$$I_3 = \frac{V_S}{Z_3} = \frac{30\angle 0°}{28.3\angle -45°} = 1.06\angle +45° = 0.75 + j0.7 \text{ A}$$

(c) Total current

$$\begin{aligned} I_T &= I_1 + I_2 + I_3 \\ &= (2.4 - j1.2) + (0.88 + j0.5) + (0.75 + j0.7) \\ &= (2.4 + 0.88 + 0.75) + [-j1.2 + (+j0.5) + (+j0.7)] \\ &= 4.03 \text{ A} \end{aligned}$$

In polar form, this will equal $4.03\angle 0°$ A.

(d) Total Impedance

$$Z_T = \frac{V_S}{I_T} = \frac{30\angle 0°}{4.03\angle 0°} = 7.77\angle 0° \ \Omega$$

Polar	Rectangular
$7.77\angle 0°$	$= 5.5 - j3.3$

The complex ac circuit seen in Figure 18-38 is therefore equivalent to a 7.77 Ω resistor in series with no reactance.

SELF-TEST REVIEW (§ 18.4)

1. In complex numbers, resistance is a/an _____ term and reactance is a/an _____ term. (imaginary/real)
2. Convert the following rectangular number to polar form: $5 + j6$.
3. Convert the following polar number to rectangular form: $33\angle 25°$.
4. What is a complex number?

SUMMARY

1. The analysis of series *RLC* circuits follows a fixed procedure, which is:
   a. Calculate inductive and capacitive reactance (X_L and X_C).
   b. Calculate circuit impedance (Z).
   c. Calculate circuit current (I).
   d. Calculate the component voltage drops (V_R, V_L, and V_C).
2. Resonance is a circuit condition that occurs when the inductive reactance (X_L) and the capacitive reactance (X_C) are balanced or equal.
3. In a series resonant *RLC* circuit, the impedance is purely resistive and the current is maximum at resonance.
4. The frequency of resonance is inversely proportional to the inductance and capacitance values in the circuit, which means low values of L and C will produce a high resonant frequency, and vice versa.
5. In a series resonant circuit, the reactive voltages V_C and V_L cancel one another due to their equal but opposite values, resulting in a 0 V drop across the series combination.
6. The Q of a circuit is a ratio of reactance to resistance or reactive power (stored) to resistive (lost) power and indicates the quality of the circuit.
7. The bandwidth of a series resonant circuit includes the band of frequencies that causes half-power or 70.7% or greater of the maximum current to flow within the circuit.
8. The smaller the bandwidth, the higher the Q and therefore the better the selectivity.
9. A tuned circuit is a common term used to describe a resonant circuit.
10. With a parallel resonant circuit, the impedance is maximum (purely resistive), the source current is minimum at resonance, and the circulating current is maximum at resonance.
11. A low pass filter passes frequencies below the cutoff frequency, while a high pass filter passes frequencies above its cutoff frequency.
12. A band pass filter passes a band of frequencies, whereas a band stop filter stops a band of frequencies.
13. A complex number represents the magnitude and phase angle of electrical quantities.
14. The horizontal line on the complex plane is called the real number line or axis.
15. The imaginary number line is vertical, and values are prefixed by a $\pm j$ to distinguish them from real numbers.
16. Both the rectangular and polar number forms will specify the magnitude and phase angle of a phasor in the two-dimensional complex plane.
17. A rectangular number consists of a real and an imaginary (j) number.
18. A polar number consists of a magnitude and an angle.
19. The sum of two complex numbers is equal to the sum of their real parts and the sum of their imaginary parts.
20. The subtraction of two complex numbers is equal to the difference between the separate real and imaginary parts.

Resistive, Inductive, and Capacitive Circuits (*RLC*)

21. Multiplication of two complex numbers is more easily achieved if they are in polar form. The result is obtained by multiplying the magnitudes and then algebraically adding the angles.

22. Division of two complex numbers is also more easily achieved when the complex numbers are in polar form. The result is obtained by first dividing the magnitudes, and then subtracting the denominator angle from the numerator angle.

23. Complex numbers find an excellent application in ac circuits due to all the phase differences that occur among the different electrical quantities.

NEW TERMS

Band pass filter
Band stop filter
Bandwidth
Complex numbers
Complex plane
Flywheel action
Frequency response curve

Imaginary number
Parallel resonance
Polar number
Real number
Rectangular number
Resonance

Resonant frequency
Selectivity
Series resonance
Tank
Tuned transformer

NEW FORMULAS

Series *RLC* Circuits

$$X_C = \frac{1}{2\pi f C}$$

$$X_L = 2\pi f L$$

$$Z = \sqrt{R^2 + (X_L \sim X_C)^2}$$

$$I = \frac{V_S}{Z}$$

$$V_R = I \times R \qquad V_L = I \times X_L \qquad V_C = I \times X_C$$

$$V_S = \sqrt{V_R^2 + (V_L \sim V_C)^2}$$

Resonance $(X_L = X_C)$

$$f_0 = \frac{1}{2\pi\sqrt{LC}}, \quad f_0 = \text{resonant frequency}$$

$$Z = R$$

$$Q = \frac{X_L}{R} = \frac{X_C}{R} = \frac{V_L}{V_R} = \frac{V_C}{V_R} = \frac{V_L}{V_S} = \frac{V_C}{V_S}$$

$$BW = \frac{f_0}{Q}, \quad BW = \text{bandwidth}$$

Power

Apparent power $= V_S \times I$ (volt-amps)

True power $= I^2 \times R$ (watts)

$$PF = \text{Cos } \theta = \frac{R}{Z} = \frac{TP}{AP}$$

Parallel *RLC* Circuits

$$I_R = \frac{V}{R} \qquad I_L = \frac{V}{X_L} \qquad I_C = \frac{V}{X_C}$$

$$I_T = \sqrt{I_R^2 + I_X^2}$$

$$Z = \frac{V}{I_T}$$

$$Q = \frac{I_{tank}}{I_S} = \frac{Z_{tank}}{X_L}$$

Complex Numbers
Polar-to-rectangular: Real number = magnitude × Cos θ
Imaginary number = magnitude × Sin θ
Rectangular-to-polar: Magnitude = $\sqrt{\text{(real number)}^2 + \text{(imaginary number)}^2}$

$$\text{Angle} = \arctan \left(\frac{\text{imaginary number}}{\text{real number}} \right)$$

Addition = sum of the real and sum of the imaginary parts (rectangular form)

Subtraction = difference between the real and the difference between imaginary parts (rectangular form)

Multiplication = multiply the magnitudes and then add the angles algebraically (polar form)

Division = divide the magnitudes and then subtract the denominator angle from the numerator angle (polar form)

REVIEW QUESTIONS

Multiple Choice Questions

1. Capacitive reactance is _____ to frequency and capacitance, while inductive reactance is _____ to frequency and inductance.
   a. Proportional, inversely proportional
   b. Inversely proportional, proportional
   c. Proportional, proportional
   d. Inversely proportional, inversely proportional

2. Resonance is a circuit condition that occurs when:
   a. V_L equals V_C
   b. X_L equals X_C
   c. L equals C
   d. Both (a) and (c)
   e. Both (a) and (b)

3. As frequency is increased, X_L will _____, while X_C will _____.
   a. Decrease, increase
   b. Increase, decrease
   c. Remain the same, decrease
   d. Increase, remain the same

4. In an *RLC* series resonant circuit, with $R = 500 \, \Omega$ and $X_L = 250 \, \Omega$, what would be the value of X_C?
   a. $2 \, \Omega$
   b. $125 \, \Omega$
   c. $250 \, \Omega$
   d. $500 \, \Omega$

5. At resonance, the voltage drop across both a series connected inductor and capacitor will equal:
   a. 70.7 V
   b. 50% of the source
   c. 10 V
   d. Zero

670   Resistive, Inductive, and Capacitive Circuits (*RLC*)

6. In a series resonant circuit the current flow is _____, as the impedance is _____ and equal to _____.
   a. Large, small, R
   b. Small, large, X
   c. Large, small, X
   d. Small, large, R

7. A circuit's bandwidth includes a group or band of frequencies that cause _____ or more of the maximum current, or more than _____ of the maximum power to appear at the output.
   a. 110, 90
   b. 50%, 70.7%
   c. 70.7%, 50%
   d. Both (a) and (c)

8. The bandwidth of a circuit is proportional to the:
   a. Frequency of resonance
   b. Q of the tank
   c. Tank current
   d. Two of the above are true

9. Series or parallel resonant circuits can be used to create:
   a. Low pass filters
   b. Low pass and high pass filters
   c. Band pass and band stop filters
   d. All the above

10. Flywheel action occus in:
    a. A tank circuit
    b. In a parallel LC circuit
    c. A series LC circuit
    d. Two of the above are true
    e. None of the above

11. $25\angle 39°$ is an example of a complex number in:
    a. Polar form
    b. Rectangular form
    c. Algebraic form
    d. None of the above

12. $3 + j10$ is an example of a complex number in:
    a. Polar form
    b. Rectangular form
    c. Algebraic form
    d. None of the above

13. Which complex number form is usually more convenient for addition and subtraction?
    a. Rectangular
    b. Polar

14. Which complex number form is usually more convenient for multiplication and division?
    a. Rectangular
    b. Polar

15. In complex numbers, resistance is a real term, while reactance is a/an _____.

a. j term

b. Imaginary term

c. Value appearing on the vertical axis

d. All the above

Essay Questions

16. Illustrate with phasors and describe the current and voltage relationships in a series *RLC* circuit. (18.1)

17. Describe the four-step procedure for the analysis of a series *RLC* circuit. (18.1.5)

18. Define resonance and give the formula for calculating the frequency of resonance. (18.1.6)

19. Describe the three unusual characteristics of a circuit that is at resonance.

20. Define the following:
    a. Flywheel action
    b. Quality figure
    c. Bandwidth
    d. Selectivity

21. Describe what is meant by a frequency response curve. (18.1.6)

22. Illustrate with phasors and describe the current and voltage relationships in a parallel *RLC* circuit. (18.2)

23. Describe the differences between a series and parallel resonant circuit. (18.1.6 and 18.2.2)

24. Explain how loading a tank affects bandwidth and selectivity. (18.2.2)

25. Illustrate the circuit and explain the operation of the following, with their corresponding response curves. (18.3)
    a. Low pass filter
    b. High pass filter
    c. Band pass filter
    d. Band stop filter

26. Describe why capacitive reactance is written as $-jX_C$ and inductive reactance is written as jX_L. (18.4.7)

27. How are capacitive and inductive reactances written in polar form? (18.4.4)

28. List the rules used to perform complex number: (18.4.6)
    a. Addition (rectangular)
    b. Subtraction (rectangular)
    c. Multiplication (polar)
    d. Division (polar)

29. Describe briefly how the real number and imaginary number lines are used for ac circuit analysis and what electrical phasors are represented at 0°, 90°, and −90°. (18.4.7)

30. Referring to Figure 18-37, describe why the series circuit current (I) is not in phase with the source voltage (V_s). (18.4.7)

Practice Problems

31. Calculate the values of capacitive and inductive reactance for the following when connected across a 60 Hz source.
    a. 0.02 μF
    b. 18 μF
    c. 360 pF
    d. 2700 nF
    e. 4 mH
    f. 8.18 H
    g. 150 mH
    h. 2 H

32. If a 1.2 kΩ resistor, a 4 mH inductor and 8 μF capacitor are connected in series across a 120 V/60 Hz source, calculate:

    a. X_C
    b. X_L
    c. Z
    d. I
    e. V_R
    f. V_L
    g. V_C
    h. Apparent power
    i. True power
    j. Resonant frequency
    k. Circuit quality factor
    l. Bandwidth

33. If a 270 Ω resistor, a 150 mH inductor and 20 μF capacitor are all connected in parallel with one another across a 120 V/60 Hz source, calculate:
    a. X_L
    b. X_C
    c. I_R
    d. I_L
    e. I_C
    f. I_T
    g. Z
    h. Q factor
    i. Resonant frequency
    j. Bandwidth

34. Calculate the impedance of a series circuit if $R = 750\ \Omega$, $X_L = 25\ \Omega$, and $X_C = 160\ \Omega$.

35. Calculate the impedance of a parallel circuit with the same values as those of question 34.

36. State the following series circuit impedances in rectangular and polar form:
    a. $R = 33\ \Omega, X_C = 24\ \Omega$
    b. $R = 47\ \Omega, X_L = 17\ \Omega$

37. Convert the following impedances to rectangular form:
    a. $25\angle 37°$
    b. $19\angle -20°$
    c. $114\angle -114°$
    d. $59\angle 99°$

38. Convert the following impedances to polar form:
    a. $-14 + j14$
    b. $27 + j17$
    c. $-33 - j18$
    d. $7 + j4$

39. Add the following complex numbers:
    a. $(4 + j3) + (3 + j2)$
    b. $(100 - j50) + (12 + j9)$

40. Perform the following mathematical operations:
    a. $(35\angle -24°) \times (13\angle 50°)$
    b. $(100 - j25) - (25 + j5)$
    c. $(98\angle 80°) \div (40\angle 17°)$

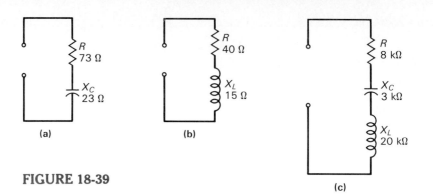

FIGURE 18-39

41. State the impedances of the circuits seen in Figure 18-39 in rectangular and polar form. What is Z_T in ohms and its phase angle?

42. Calculate the impedances of both circuits shown in Figure 18-40 in polar form. Then combine the two impedances as if the circuits were parallel connected, using the product over sum method. Express the combined impedance in polar form.

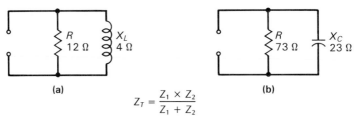

(a) $Z_T = \dfrac{Z_1 \times Z_2}{Z_1 + Z_2}$ (b)

FIGURE 18-40

43. Referring to Figure 18-41, calculate:
    a. Z_T (rectangular and polar)
    b. Circuit current and phase angle
    c. Voltage drops
    d. V_C, V_L, and V_R phase relationships

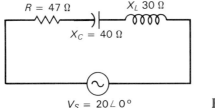

$R = 47\ \Omega$ $X_L\ 30\ \Omega$

$X_C = 40\ \Omega$

$V_S = 20\angle 0°$ **FIGURE 18-41**

44. Sketch an impedance and voltage phasor diagram for the circuit in Figure 18-41.

45. Referring to Figure 18-42, calculate:
    a. Impedance of the two branches
    b. Branch currents
    c. Total current
    d. Total impedance

674

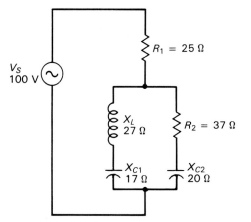

V_S
100 V

$R_1 = 25\ \Omega$

X_L
27 Ω

$R_2 = 37\ \Omega$

X_{C1}
17 Ω

X_{C2}
20 Ω

FIGURE 18-42

46. Sketch an impedance and current phasor diagram for the circuit in Figure 18-42.
47. Referring to Figure 18-41, verify that the sum of all the individual voltage drops is equal to the total voltage.
48. Referring to Figure 18-42, verify that the sum of all the branch currents is equal to the total current.
49. Determine whether questions 43 and 45 would be easier to answer with or without the use of complex numbers.
50. Is the circuit in Figure 18-42 more inductive or capacitive?

APPENDIXES

Absorption The loss or dissipation of energy as it travels through a medium. For example, radio waves lose some of their electromagnetic energy as they travel through the atmosphere.

AC Abbreviation for alternating current.

Accelerate To go faster.

AC coupling A circuit or component that couples or passes the ac signal, yet blocks any dc level.

AC/DC If indicated on a piece of equipment, it means that the equipment will operate from either an ac or dc supply.

AC generator A device that transforms or converts a mechanical input into an ac electrical power output.

Acoustic Relating to sound or the science of sound.

AC power supply A power supply that delivers one or more sources of ac voltage.

Activate To put ac voltage to work or make active.

Active component A component that amplifies a signal or achieves some sort of gain between input and output.

Active equipment Equipment that will transmit and receive, as opposed to passive equipment, which only receives.

AC voltage An alternating voltage.

ADC An abbreviation for analog-to-digital converter.

Adjustable resistor A resistor whose value can be changed.

Admittance (symbolized Y) A measure of how easily ac will flow through a circuit. It is equal to the reciprocal of impedance and is measured in siemens.

Aerial Another term used for antenna.

AF Abbreviation for audio frequency.

AFC Abbreviation for automatic frequency control.

AGC Abbreviation for automatic gain control.

Alkaline cell also called an alkaline manganese cell, it is a primary cell that delivers more current than the carbon–zinc cell.

Aligning tool Small, nonconductive non-magnetic screwdriver used to adjust receiver or transmitter tuned circuits.

Alligator clip A spring clip normally found on the end of test leads and used to make temporary connections.

Alphanumeric Having numerals, symbols, and letters of the alphabet.

Alpha wave Also called alpha rhythm, it is a brain wave between 9 to 14 Hz.

Alternating current An electric current that rises from zero to a maximum in one direction, falls to zero, and then rises to a maximum in the opposite direction, and then repeats another cycle. The positive and negative alternations being equal.

Alternation One ac cycle consists of a positive and negative alternation.

Alternator Another term used to describe an ac generator.

AM Abbreviation for amplitude modulation.

Ambient temperature The temperature of the air surrounding the components.

Ammeter A meter placed in the path of current flow to measure the amount.

Ampere Abbreviated amp, it is the unit of electric current.

Ampere-hour If you multiply the amount of current, in amps, that can be supplied for a given time frame in hours, a value of ampere-hours is obtained.

Ampere-hour meter A meter that measures the amount of current drawn in a unit of time (hours).

Ampere-turn This is the unit of magnetomotive force.

Ampere-turn per meter The base unit of magnetic field strength.

Amplification The process of making bigger or increasing the voltage, current, thus increasing the power of a signal.

Amplifier A circuit or device that achieves amplification.

Amplitude The magnitude or size an alternation varies from zero.

Amplitude distortion The changing of a wave shape so that it does not match its original form.

Analog data Information that is represented in a continuously varying form, as opposed to digital data, which have two distinct and discrete values.

Anode A positive electrode or terminal.

Antenna This device converts an electrical wave into an electromagnetic wave that radiates away from the antenna.

Antenna transmission line A system of conductors connecting the transmitter or receiver to the antenna.

Arc A discharge of electricity through a gas. For example, lightning is a discharge of static electricity buildup through air.

Armature The rotating or moving component of a magnetic circuit.

Atom The smallest particle of an element.

Atomic number The number of positive charges or protons in the nucleus of an atom.

Attenuation The loss or decrease in energy of a signal.

Audio Relating to all the frequencies that can be heard by the human ear; from the Latin word meaning ''I hear.''

Audio frequency A frequency that can be detected by a human ear. Approximately between 20 Hz to 20 kHz.

Autotransformer A single-winding tapped transformer.

Average value This mean value is found when the area of a wave above a line is equal to the area of the wave below the line.

Avionics The field of aviation electronics.

AWG Abbreviation for American wire gauge.

Axial lead A component that has its connecting leads extending from either end of its body.

Balanced bridge A condition that occurs when a bridge is adjusted to produce a zero output.

Band-pass filter A filter circuit that passes a group or band of frequencies between a lower and upper cutoff frequency, while heavily attenuating any other frequency outside this band.

Bandwidth The width of the group or band of frequencies between the half-power points.

Barometer A meter used to measure atmospheric pressure.

Barretter A temperature-sensitive device having a positive temperature coefficient; that is, as temperature increases, resistance increases.

Battery A dc voltage source containing two or more cells that converts chemical energy into electrical energy.

Battery charger A piece of electrical equipment that converts ac input power to a pulsating dc output, which is then used to charge a battery.

Baud The unit of signaling speed describing the number of elements per second.

Binary A number system having only two levels and used in digital electronics.

Bolometer A device whose resistance changes when heated.

BW Abbreviation for bandwidth.

Bypass capacitor A capacitor that is connected to provide a low impedence path for ac.

Byte A group of 8 binary digits or bits.

Cable A group of two or more insulated wires.

CAD Abbreviation for computer-aided design.

Calculator A device that can come in a pocket (battery operated) or desktop (ac power) size and is used to achieve arithmetic operations.

Capacitance (C) Measured in farads, it is the ability of a capacitor to store an electrical charge.

Capacitance meter An instrument used to measure the capacitance of a capacitor or circuit.

Capacitive reactance (X_C) Measured in ohms, it is the ability of a capacitor to oppose current flow without the dissipation of energy.

Capacitor A device that stores electric energy in the form of an electric field that exists within a dielectric (insulator) between two conducting plates each connected to a connecting lead. This device was originally called a condensor.

Capacitor-input filter A filter in which a capacitor is connected to shunt away low-frequency signals such as ripple from a power supply.

Capacitor microphone This microphone contains a stationary and movable plate separated by air. The arriving sound waves cause movements in the movable plate, which affects the distance between the plates and therefore the capacitance, generating a varying AF electrical wave. Also called an electrostatic microphone.

Capacitor speaker Also called the electrostatic speaker, its operation depends on mechanical forces generated by an electrostatic field.

Capacity The output current capabilities of a device or circuit.

Carbon-film resistor A thin carbon film deposited on a ceramic form to create resistance.

Carbon microphone A microphone whose operation depends on the variation in the resistance of carbon granules.

Carbon resistor A fixed resistor consisting of carbon particles mixed with a binder, which is molded and then baked. Also called a composition resistor.

Cardiac pacemaker A device used to control the frequency or rhythm of the heart by stimulating it through electrodes.

Carrier A wave that has one of its characteristics varied by an information signal. The carrier will then transport this information between two points.

Cassette A thin, flat, rectangular device containing a length of magnetic tape that can be recorded onto or played back.

Cathode A term used to describe a negative electrode or terminal.

Cathode-ray tube (CRT) This vacuum tube device is used to display data in a visual form by use of electron beams.

Cell A single unit having only two plates that convert chemical energy into a dc electrical voltage.

Celsius-temperature scale A scale that defines the freezing point of water as 0°C and the boiling point as 100°C (named after Anders Celsius).

Center tap A midway connection between the two ends of a winding or resistor.

Ceramic capacitor A capacitor in which the dielectric material used between the plates is ceramic.

Cermet A ceramic–metal mixture used in film resistors.

Charge The quantity of electrical energy stored in a battery or capacitor.

Charging current The current that flows to charge a capacitor or battery when a voltage is applied.

Chassis A metal box or frame in which components, boards, and units are mounted.

Chassis ground A connection to the metal box or frame that houses the components and associated circuitry.

Choke An inductor used to impede the flow of alternating or pulsating current.

Circuit The interconnection of many components to provide an electrical path between two or more points.

Circuit breaker A reusable fuse. This device will open a current-carrying path without damaging itself once the current value exceeds its maximum current rating.

Clock Generally, a square waveform used for the synchronization and timing of several circuits.

Closed circuit A circuit having a complete path for current to flow.

Coaxial cable A transmission line in which a center signal-carrying conductor is completely covered by a solid dielectric, another conductor, and then an outer insulating sleeve.

Coefficient This is related to the ratio of change under certain conditions.

Coercive force (H) The magnetizing force needed to reduce the residual magnetism within a material to zero.

Coil A number of turns of wire wound around a core to produce magnetic flux (an electromagnet) or to react to a changing magnetic flux (an inductor).

Color code A set of colors used to indicate a component's value, tolerance, and rating.

Common A connection or point shared by two or more components or circuits.

Communication The transmission of information between two points.

Component (1) A device or part in a circuit or piece of equipment. (2) In vector diagrams, it can mean a part of a wave, voltage, or current.

Computer A piece of equipment used to process information or data.

Conductance (G) A measure of how well a circuit or path conducts or passes current. It is equal to the reciprocal of resistance.

Conduction Use of a material to pass electricity or heat.

Conductivity The reciprocal of resistivity; it is the ability of a material to conduct current.

Conductor A length of wire whose properties are such that it will carry an electric current.

Connection When two or more devices are attached together so that a path exists between them, there is said to be a connection between the two.

Connector A conductive device that makes a connection between two points.

Constant A fixed value.

Constant current A current that remains at a fixed, unvarying level despite variations in load resistance.

Contact The current-carrying part of a switch, relay, or connector.

Continuity Occurs when a complete path for electric current exists.

Continuity test A resistance test to determine whether a path for electric current exists.

Core A magnetic material within a coil used to intensify or concentrate the magnetic field.

Coulomb The unit of electric charge. One coulomb equals 6.24×10^{18} electrons.

Counter electromotive force Abbreviated counter emf, or sometimes called back emf, it is the voltage generated in an inductor due to an alternating or pulsating current and is always of opposite polarity to that of the applied voltage.

CRT Abbreviation for cathode ray tube.

Current (I) Measured in amperes or amps, it is the flow of electrons through a conductor.

Current-limiting resistor A resistor that is inserted into the path of current flow to limit the amount of current to some desired level.

Cutoff frequency The frequency at which the gain of the circuit falls below 0.707 of the maximum current or half-power (-3 dB).

Cycle When a repeating wave rises from zero to some positive value, back to zero, and then to a maximum negative value before returning back to zero, it is said to have completed one cycle. The number of cycles occurring in one second is the frequency measured in hertz (cycles per second).

DAC Abbreviation for digital-to-analog converter.

Damping The reduction in the magnitude of oscillation due to absorption.

D'Arsonval movement When a direct current is passed through a small, lightweight, movable coil, the magnetic field produced interacts with a fixed permanent magnetic field and rotates, moving an attached pointer positioned over a scale.

DC Abbreviation for direct current.

DC block A component used to prevent the passage of direct current, normally a capacitor.

DC generator A device used to convert a mechanical energy input into a dc electrical output.

Dead short A short circuit having almost no resistance.

Decibel Abbreviated dB, it is a unit used to indicate the loss ($-$dB) or gain ($+$dB) of voltage or power between two points.

Design engineer An engineer responsible for the design of a product for a specific application.

Device A component or part.

Diagnostic Related to the detection and isolation of a problem or malfunction.

Dielectric The insulating material between two (*di*) plates in which the *electric* field exists.

Digital Relating to devices or circuits that have outputs of only two distinct levels or steps, for example, on–off, 0–1, open–closed, and so on.

Digital computer A piece of electronic equipment used to process digital data.

Digital data Data represented in digital form.

Digital multimeter A multimeter used to measure amps, volts, and ohms and indicate the results on a digital readout display.

Digital display A display in which either light-emitting diodes (LEDs) or liquid crystal diodes (LCDs) are used and each of the segments can be either turned on or off.

DIP Abbreviation for dual in-line package.

Direct current (dc) Current flow in only one direction.

Direct voltage (dc voltage) A voltage that causes electrons to flow in only one direction.

Discharge The release of energy from either a battery or capacitor.

Disconnect The breaking or opening of an electric circuit.

Discrete component An individual circuit component that has been manufactured independently and placed in its own package.

Display A visual presentation of information or data.

Display unit A unit designed to display digital data.

Dissipation The release of electrical energy in the form of heat.

Distributed capacitance Also known as self-capacitance, it is any capacitance other than that within a capacitor, for example, the capacitance between the coil of an inductor or between two conductors (lead capacitance).

Distributed inductance Any inductance other than that within an inductor, for example, the inductance of a length of wire (line inductance).

Double-pole, double-throw switch (DPDT) A switch having two movable contacts (double pole) that can be thrown or positioned in one of two positions (double throw).

Double-pole, single-throw switch (DPST) A switch having two movable contacts (double pole) that can be thrown in only one position (single throw) to either make a connection, or opened to disconnect both poles.

Dropping resistor A resistor whose value has been chosen to drop or develop a given voltage across it.

Dry cell A dc voltage-generating chemical cell using a non-liquid (paste) type of electrolyte.

Dual in-line package A package that has two (dual) sets or lines of connecting pins.

Dual-trace oscilloscope An oscilloscope that can simultaneously display two signals.

Dynamic Related to conditions or parameters that change.

E-core A laminated form, in the shape of the letter E, onto which inductors and transformers are wound.

Eddy currents Small currents induced in a conducting core due to the variations in alternating magnetic flux.

Efficiency A ratio of output power to input power, normally expressed in percent.

Electrical equipment Equipment designed to manage or control the flow of power.

Electric charge Electric energy stored on the surface of a material.

Electric current Electron movement or motion.

Electric field Also called a voltage field, it is a field or force that exists in the space between two different potentials or voltages.

Electrician A person invovled with the design, repair, and assembly of electrical equipment.

Electricity The science that states that certain particles possess a force field, which with electrons is negative and with protons is positive. Electricity can be divided into two groups: static and dynamic. Static electricity deals with charges at rest, while dynamic electricity deals with charges in motion.

Electric motor A device that converts an electrical energy input into a mechanical energy output.

Electroacoustic transducer A device that achieves an energy transfer from electric to acoustic (sound), and vice versa. Examples include a microphone and a loudspeaker.

Electroluminescence The transmission or conversion of electrical energy into light energy.

Electrolyte An electrically conducting liquid (wet) or paste (dry).

Electrolytic capacitor A capacitor having an electrolyte between the two plates; due to chemical action, a very thin layer of oxide is deposited on only the positive plate, which accounts for why this type of capacitor is polarized.

Electromagnetic communication The use of an electromagnetic wave to pass information between two points. Also called *wireless* communication.

Electromagnetic energy Radiant electromagnetic energy, such as radio and light waves.

Electromagnetic field A field having both an electric (voltage) and magnetic (current) field.

Electromagnetic spectrum A list or diagram showing the entire range of electromagnetic radiation.

Electromagnetic wave A wave that consists of both an electric and magnetic variation.

Electromagnetism Relates to the magnetic field generated around a conductor when current is passed through it.

Electromechanical transducer A device that transforms electrical energy into mechanical energy (motor), and vice versa (generator).

Electromotive force The force that causes the motion of electrons due to a potential difference between two points.

Electron The smallest subatomic particle of negative charge that orbits the nucleus of the atom.

Electronic equipment Equipment designed to control the flow of information.

Electronics The science related to the behavior of electrons in devices.

Electrostatic Related to static electric charge.

Electrostatic field The force field produced by static electrical charges.

Element 104 different chemical substances or elements exist on Earth and can be categorized as either being a gas, solid, or liquid.

Energy The capacity to do work.

Engineer A person who designs and develops materials to achieve desired results.

Equipment A term used to describe electrical or electronic units.

Equivalent resistance (R_{eq}) The total resistance of all the individual resistances in a circuit.

Facsimile An electronic process whereby pictures or images are scanned and the graphical information is converted into electrical signals that can be reproduced locally or transmitted to a remote point, where a likeness or facsimile of the original can be produced.

Fahrenheit temperature scale A temperature scale that indicates the freezing point of water at 32°F and boiling point at 212°F.

Farad (F) The unit of capacitance.

FCC Abbreviation for Federal Communications Commission.

Ferrite bead A ferrite composition in the form of a bead.

Ferrite core A ferrite core normally shaped like a doughnut.

Ferrites A compound composed of iron oxide, a metallic oxide, and ceramic. The metal oxides include zinc, nickel, cobalt, or iron.

Fiber optics A laser's light output carries information that is conveyed between two points by thin glass optical fibers.

Filament A thin thread of, for example, carbon or tungsten, which when heated by the passage of electric current will emit light.

Filter A network composed of resistor, capacitor, and inductors used to pass certain frequencies yet block others through heavy attenuation.

Fission Atomic or nuclear fission is the process of splitting the nucleus of heavy elements such as uranium and plutonium into two parts, which results in large releases of radioactivity and heat (fission or division of a nuclei).

Fixed component A component whose value or characteristics cannot be varied or changed.

Floating ground A ground potential that is not tied or in reference to Earth.

Flow-soldering A flow or wave soldering technique is used in large-scale electronic assembly to solder all the connections on a printed circuit board by moving the board over a wave or flowing bath of molten solder.

Fluorescent lamp A gas-filled glass tube that when bombarded by the flow of electric current causes the gas to ionize and then release light.

Flywheel effect The sustaining effect of oscillation in an *LC* circuit due to the charging and discharging of the capacitor and the expansion and contraction of the magnetic field around the inductor.

Force A physical action capable of moving a body or modifying its movement.

Free electrons Electrons that are not in any orbit around a nucleus.

Frequency The rate or recurrences of a periodic wave normally within a unit of one second, measured in hertz (cycles/second).

Frequency meter A meter used to measure the frequency or cycles per second of a periodic wave.

Frequency response An indication of how well a device or circuit responds to the different frequencies applied to it.

Frictional electricity The generation of electric charges by rubbing one material against another.

Full-scale deflection (FSD) The deflection of a meter's pointer to the farthest position on the scale.

Full-wave rectifier A rectifier that makes use of the full ac wave (in both the positive and negative cycle) when converting ac to dc.

Function generator A signal generator that can function as a sine, square, rectangular, triangular, or sawtooth waveform generator.

Fundamental frequency This sine wave is always the lowest frequency and largest amplitude component of any waveform shape and is used as a reference.

Fuse This circuit- or equipment-protecting device consists of a short, thin piece of wire that melts and breaks the current path if the current exceeds a rated damaging level.

Fuse holder A housing used to support a fuse with two connections.

Fusion Atomic or nuclear fusion is the process of melting the nuclei of two light atoms together to create a heavier nucleus, which results in a large release of energy (fusion or combining of nuclei).

Gain An increase in power from one point to another. Normally expressed in decibels.

Gamma rays High-frequency electromagnetic radiation from radioactive particles.

Ganged The mechanical coupling of two or more capacitors, switches, potentiometers, or any other components so that the activation of one control will operate all.

Gas Any aeriform or completely elastic fluid, which is not a solid or liquid. All gases are produced by the heating of a liquid beyond its boiling point.

Geiger counter A device used to detect nuclear particles.

Generator A device used to convert a mechanical energy input into an electrical energy output.

Giga The prefix for one billion (10^9).

Greenwich mean time (GMT) Also known as universal time, it is a standard based on the Earth's rotation with respect to the sun's position. The solar time at the meridian of Greenwich, England, which is at zero longitude.

Ground A connection to earth, literally, as a metal spike is driven in the earth and is used as a voltage reference point.

Hardware Electrical, electronic, mechanical, or magnetic devices or components. The physical equipment.

Harmonic A sine wave that is smaller in amplitude and some multiple of the fundamental frequency.

Henry (H) The unit of inductance.

Hertz (Hz) The unit of frequency. One hertz is equal to one cycle per second.

High fidelity (hi-fi) Sound reproduction equipment that is as near to the original sound as possible.

High-pass filter A network or circuit designed to pass any frequencies above a critical or cutoff frequency and reject or heavily attenuate all frequencies below.

High Q Q is an abbreviation for quality and generally relates to inductors that have a high value of inductance and very little coil resistance.

High tension Lethal voltage in the kilovolt range and above.

High-voltage probe An accessory to the voltmeter that has added multiplier resistors within the probe to divide up the large potential being measured by the probe.

***H*-lines** Invisible lines of magnetic flux.

Hologram A three-dimensional picture created with a laser.

Holography The science dealing with three-dimensional optical recording.

Horizontal Parallel to the horizon or perpendicular to the force of gravity.

Horizontally polarized wave An electromagnetic wave that has the electric field lying in the horizontal plane.

Hybrid circuit A circuit that combines two technologies (passive and active or discrete and integrated components) onto one microelectronic circuit. Passive components are generally made by thin-film techniques, while active components are made utilizing semiconductor techniques. Intergrated circuits can be mounted on the microelectronic circuit and connected to discrete components also on the small postage-stamp-size boards.

Hydroelectric The generation of electric power by the use of water in motion.

Hysteresis The amount that the magnetization of a material lags the magnetizing force due to molecular friction.

IC Abbreviation for integrated circuit.

Impedance (Z) Measured in ohms, it is the total opposition a circuit offers to current flow (reactive and resistive).

Impedance matching The matching of the source impedance to the load impedance causes maximum power to be transferred.

Incandescence The state of a material when it is heated to such a high temperature that it emits light.

Induced current The current that flows due to an induced voltage.

Induced voltage The voltage generated in a conductor when it is moved through a magnetic field.

Inductance The property of a circuit or component to oppose any change in current as the magnetic field produced by the change in current causes an induced countercurrent to oppose the original change.

Inductive circuit A circuit that has a greater inductive reactance figure than capacitive reactance figure.

Inductive reactance (X_L) Measured in ohms, it is the opposition to alternating or pulsating current flow without the dissipation of energy.

Inductor A length of conductor used to introduce inductance into a circuit.

Infinite Having no limits.

Infinity An amount larger than any number can indicate.

Information Data or meaningful signals.

Infrared Electromagnetic heat radiation whose frequencies are above the microwave frequency band and below red in the visible band.

Inhibit To stop an action or block data from passing.

In phase Two or more waves of the same frequency whose maximum positive and negative peaks occur at the same time.

Input impedance The impedance seen when looking into the input terminals of a device.

Insulated When a nonconductive material is used to isolate conducting materials from one another.

Insulating material A material that will in nearly all cases prevent the flow of current due to its chemical composition.

Insulation resistance The resistance of the insulating material. The greater the insulation resistance, the better the insulator.

Insulator A material that has few electrons per atom and those electrons are close to the nucleus and cannot be easily removed.

Integrated When two or more components are combined (into a circuit) and then incorporated in one package or housing.

Integrator A device that approximates and whose output is proportional to an integral of the input signal.

Intermittent Occurring at random intervals of time. An intermittent component, circuit, or equipment problem is undesirable and difficult to troubleshoot, as the problem needs to occur before isolation of the fault can begin.

Internal resistance No voltage source is 100% efficient in that not all the energy in is converted to electrical energy out; some is wasted in the form of heat dissipation. The internal series resistance of a voltage source represents this inefficiency.

Ion An atom that has an equal number of protons ($+$ charge) and electrons ($-$ charge) is considered a neutral atom. If more electrons exist in orbit around the atom it is considered a negative ion, whereas if less electrons are in orbit around the atom it is referred to as a positive ion.

Jack A socket or connector into which a plug may be inserted.

Junction A contact or connection between two or more wires or cables.

Kilovolt-ampere 1000 volts at 1 amp.

Kilowatt-hour 1000 watts for 1 hour.

Lag The difference in time between two waves of the same frequency expressed in degrees.

Laminated core A core made up of sheets of magnetic material insulated from one another by an oxide or varnish.

Lamp A device that produces light.

Laser A device that produces a very narrow, intense beam of light. The name is an acronym for ''light amplification by stimulated emission of radiation.''

Lead–acid cell A cell made up of lead plates immersed in a sulfuric acid electrolyte.

Leakage A small, undesirable flow of current through an insulator or dielectric.

LED Abbreviation for light-emitting diode.

Lie detector A piece of electronic equipment, also called a polygraph, that determines whether a person is telling the truth by looking for dramatic changes in blood pressure, body temperature, breathing rate, heart rate, and skin moisture in response to certain questions.

Light Electromagnetic radiation in a band of frequencies that can be received by the human eye.

Limiter A circuit or device that prevents some portion of its input from reaching the output.

Linear A relationship between input and output in which the output varies in direct proportion to the input.

Live A term used to describe a circuit or piece of equipment that is on and has current flow within it.

Load A source drives a load, and whatever component, circuit, or piece of equipment is connected to the source can be called a load and will have a certain load resistance, which will consequently determine the load current.

Load impedance The total reactive and resistive opposition of a load.

Loading effect A large load resistance will cause a small load current to flow, and so the loading down of the source or loading effect will be small (light load), whereas a small load resistance will cause a large load current to flow from the source, which will load down the source (heavy load).

Logic A science dealing with the principles and applications of gates and switches.

Loss Term used to describe a decrease in power.

Low-pass filter A network or circuit designed to pass any frequencies below a critical or cutoff frequency and reject or heavily attenuate all frequencies above.

Magnet A body that can be used to attract or repel magnetic materials.

Magnetic circuit breaker A circuit breaker that is tripped or activated by use of an electromagnet.

Magnetic coil A spiral of a conductor, which is called an electromagnet.

Magnetic core A material that exists in the center of the magnetic coil to either support the windings (nonmagnetic-material) or intensify the magnetic flux (magnetic-material).

Magnetic field The magnetic lines of force traveling from the north to the south pole of a magnet.

Magnetic poles The points of a magnet from which magnetic lines of force leave (north pole) and arrive (south pole).

Magnetism The property of some materials to attract and repel others.

Magnetomotive force The force that produces a magnetic field.

Main frame Large computers that initially were only affordable to medium and large businesses who had the space for them. Minicomputers came after main frames and were affordable to any business, and now we have microcomputers which are easily affordable to anyone.

Matched impedance A condition that occurs when the source impedance is equal to the load impedance, resulting in maximum power being transferred.

Matching The connection of two components or circuits so that maximum energy is transferred or coupled between the two.

Measurement Determining the presence and magnitude of variables.

Medical electronics The branch of electronics involved with therapeutic or diagnostics in medicine.

Mercury cell A primary cell that has mercuric-oxide cathode, a zinc anode, and a potassium-hydroxide electrolyte.

Meter (1) Any electrical or electronic measuring device. (2) In the metric system, it is a unit of length equal to 39.37 inches or 3.28 feet.

Mica capacitor A fixed capacitor that uses mica as the dielectric between its plates.

Microphone This electroacoustic transducer responds and converts a sound wave input into an equivalent electrical wave out.

Microwave A term used to describe a band of very small wavelength radio waves within the UHF, SHF, and EHF bands.

Mismatch A term used to describe a difference between the source impedance and load impedance, which will prevent maximum power transfer.

Modulation The process whereby an information signal is used to modify some characteristic of another higher-frequency wave known as a carrier.

Molecule The smallest particle of a compound that still retains its chemical characteristics.

Moving-coil microphone A microphone that makes use of a moving coil between a fixed magnetic field. Also called a dynamic microphone.

Moving-coil pickup This dynamic phonograph pickup uses a coil between a fixed magnetic field, which is moved back and forth by the needle or stylus.

Moving-coil speaker This dynamic speaker uses a coil placed between a fixed magnetic field and converts the electrical wave input into sound waves.

Multimeter A piece of electronic test equipment that can perform multiple tasks in that it can be used to measure voltage, current, or resistance.

Mutual inductance The ability of one inductor's magnetic lines of force to link with another inductor.

Navigation equipment Electronic equipment designed to aid in the direction of aircraft and ships to their destination.

Negative (neg.) (1) Some value less than zero. (2) A terminal that has an excess of electrons.

Negative ion An atom that has more than the normal neutral amount of electrons.

Negative temperature coefficient An effect that describes that if temperature increases, resistance or capacitance will decrease.

Neon bulb A glass envelope filled with neon gas, which when ionized by an applied voltage will glow red.

Network A combination and interconnection of components, circuits, or systems.

Neutral When an object is neither positive nor negative.

Neutron A subatomic particle residing within the nucleus and having no electrical charge.

Nickel–cadmium cell The most popular secondary cell; it uses a nickel-oxide positive electrode and cadmium negative electrode.

Node A junction or branch point.

Noise Unwanted electromagnetic radiation within an electrical or mechanical system.

Normally closed (N.C.) A designation which states that the contacts of a switch or relay are connected normally; however, when activated, these contacts will open.

Normally open (N.O.) A designation which states that the contacts of a switch or relay are normally not connected; however, when activated, these contacts will close.

North pole The pole of a magnet out of which magnetic lines of force are assumed to originate.

Nuclear energy The atomic energy or power released in a nuclear reaction when either a neutron is used to split an atom into smaller atoms (fission) or when two smaller nuclei are joined together (fusion).

Nuclear reactor A unit that maintains a continuous self-supporting nuclear reaction (fission).

Nucleus The core of an atom; it contains both positive (protons) and neutral (neutrons) subatomic particles.

Octave An interval between two sounds whose fundamental frequencies differ by a ratio of 2 to 1.

Ohm The unit of resistance, symbolized by the Greek letter omega (Ω).

Ohmmeter A measurement device used to measure electric resistance.

Ohm's law A relationship between the three electrical phenomena of voltage, current, and resistance, which states that the current flow within a circuit is directly proportional to the voltage applied across the circuit and inversely proportional to its resistance.

Ohms per volt A value that indicates the sensitivity of a voltmeter. The higher the ohms per volt rating, the more sensitive the meter.

Open circuit A break in the path of current flow.

Operational amplifier A special type of high-gain amplifier; also called an op amp.

Oscillate The continual repetition of or passing through a cycle.

Oscillator An electronic circuit that converts dc to a continuous alternating current out.

Oscilloscope An instrument used to view signal amplitude, frequency, and shape at different points throughout a circuit.

Out of phase When the maximum and minimum points of two or more waveforms do not occur at the same time.

Output The terminals at which a component, circuit, or piece of equipment delivers current, voltage, or power.

Output impedance The impedance measured across the output terminals of a device without the load connected.

Output power The amount of power a component, circuit, or system can deliver to its load.

Overload This situation occurs when the load is greater than the component, circuit, or system was designated to handle (load resistance too small, load current too high), resulting in waveform distortion and/or overheating.

Overload protection A protective device such as a fuse or circuit breaker that automatically disconnects or opens a current path when it exceeds an excessive value.

Paper capacitor A fixed capacitor using oiled or waxed paper as a dielectric.

Parallel Also called shunt; it is a circuit having two or more paths for current flow.

Parallel resonant circuit A circuit having an inductor and capacitor in parallel with one another, offering a high impedance at the frequency of resonance.

Passband A band or range of frequencies that will be passed by a filter.

Passive component A component that does not amplify a signal, such as a resistor or capacitor.

Passive system A system that emits no energy; in other words it only receives; it does not transmit and consequently reveal its position.

Peak The maximum or highest-amplitude level.

Peak to peak The difference between the maximum positive and maximum negative values.

Period The time taken to complete one complete cycle of a periodic or repeating waveform.

Permanent magnet A magnet, normally made of hardened steel, that retains its magnetism indefinitely.

Permeability a measure of how much better a material is as a path for magnetic lines of force with respect to air, which has a permeability of 1 (symbolized by the Greek letter mu, μ).

Permanence The magnetic equivalent of electrical conductance and consequently equal to the reciprocal of reluctance, just as conductance is equal to the reciprocal of resistance.

Phase The angular relationship betwen two waves, normally between current and voltage in an ac circuit.

Phase angle The phase difference between two waves, normally expressed in degrees.

Phase shift The change in phase of a waveform between two points, given in degrees of lead or lag.

Phonograph A piece of equipment used to reproduce sound.

Phosphor A luminescent material applied to the inner surface of a cathode ray tube that when bombarded with electrons will emit light.

Photoconductive cell A material whose resistance decreases or conductance increases when light strikes it.

Photodetector A component used to detect or sense light.

Photometer A meter used to measure light intensity.

Photon A discrete portion of electromagnetic energy. A small packet of light.

Photovoltaic cell A component, commonly called a solar cell, used to convert light energy into electric energy (voltage).

Pi A value representing the ratio between the circumference and diameter of a circle and equal to approximately 3.142 (symbolized by the Greek letter π).

Piezoelectric crystal A crystal material that will generate a voltage when mechanical pressure is applied and conversely will undergo mechanical stress when subjected to a voltage.

Pitch A term used to describe the inflection or frequency scale of sounds. When the pitch is increased by one octave, twice the original frequency will be the result.

Plate A conductive electrode in either a capacitor or battery.

Plug A movable connector that is normally inserted into a socket.

Polarity A term used to describe positive and negative charges.

Positive A point that attracts electrons, as opposed to negative, which supplies electrons.

Positive ion An atom that has lost one or more of its electrons and therefore has more protons than electrons, resulting in a net positive charge.

Potential difference (pd) A voltage difference between two points, which will cause current to flow in a closed circuit.

Potential energy Energy that has the potential to do work because of its position relative to others.

Potentiometer A three-lead variable resistor that through mechanical turning of a shaft can be used to produce a variable voltage or potential.

Power The amount of energy converted by a component or circuit in a unit of time, normally seconds. It is measured in units of watts (joules/second).

Power dissipation The amount of heat energy generated by a device in one second when current flows through it.

Power factor A ratio of actual power to apparent power.

Power loss A ratio of power absorbed to power delivered.

Power supply A piece of electrical equipment used to deliver either ac or dc voltage.

Prefix A name used to designate a factor or multiplier.

Primary The first winding of a transformer that is connected to the source, as opposed to the secondary that is connected to the load.

Primary cell A cell that produces electrical energy out through an internal electrochemical action; once discharged, it cannot be reused.

Printed circuit board (PCB) An insulating board that has conductive tracks printed onto the board to make the circuit.

Propogation The traveling of electromagnetic, electrical, or sound waves through a medium.

Propogation time The time it takes for a wave to travel between two points.

Proton A subatomic particle within the nucleus that has a positive charge.

Pulse The rise and fall of some quantity for a period of time.

Pulse fall time The time it takes for a pulse to decrease from 90% to 10% of its maximum value.

Pulse rise time The time it takes for a pulse to increase from 10% to 90% of its maximum value.

Q The quality factor of an inductor or capacitor; it is the ratio of a component's reactance (energy stored) to its effective series resistance (energy dissipated).

Radar An acronym for "radio detection and ranging"; it is a system that measures the distance and direction of objects.

Radioastronomy A branch of astronomy that studies the radio waves generated by celestial bodies and uses these emissions to obtain more information about them.

Radio broadcast The transmission of music, voice, and other information on radio carrier waves that can be received by the general public.

Radiocommunication A term used to describe the communication of information between two or more points by use of radio or electromagnetic waves.

RC Abbreviation for resistance–capacitance.

RC circuit A circuit containing both a resistor and capacitor.

RC time constant In one time constant, which is equal to the product of resistance and capacitance in seconds, a capacitor will have charged or discharged 63.2% of the maximum applied voltage.

Reactance (X) The opposition to current flow without the dissipation of energy.

Receiver A unit or piece of equipment used for the reception of information.

Rectangular wave Also known as a pulse wave; it is a repeating wave that only alternates between two levels or values and remains at one of these values for a small amount of time relative to the other.

Rectification The process that converts alternating current (ac) into direct current (dc).

Rectifier A device that achieves rectification.

Reed relay A relay that consists of two thin magnetic strips within a glass envelope with a coil wrapped around the envelope so that when it is energized the relay's contacts or strips will snap together, making a connection between the two leads attached to each of the reed strips.

Relay An electromechanical device that opens or closes contacts when a current is passed through a coil.

Reluctance The resistance to the flow of magnetic lines of force.

Remanence The amount a material remains magnetized after the magnetizing force has been removed.

Residual magnetism The magnetism remaining in the core of an electromagnet after the coil current has been removed.

Resistance Symbolized R and measured in ohms (Ω), it is the opposition to current flow with the dissipation of energy in the form of heat.

Resistivity A measure of a material's resistance to current flow.

Resistor A component made of a material that opposes the flow of current and therefore has some value of resistance.

Resistor color code A coding system of colored stripes on a resistor that indicates the resistors-value and tolerance.

Resonance A circuit condition that occurs when the inductive reactance (X_L) is equal to the capacitive reactance (X_C).

Resonant circuit A circuit containing an inductor and capacitor tuned to resonate at a certain frequency.

Resonant frequency A frequency at which a circuit or object will produce a maximum amplitude output.

RF An abbreviation for radio frequency.

Rheostat A two-terminal variable resistor.

Rise time The time it takes a positive edge of a pulse to rise from 10% to 90% of its high value.

RMS Abbreviation for root mean square.

RMS value The rms value of an ac voltage, current, or power waveform is equal to 0.707 times the peak value. The rms value is the effective or dc value equivalent of the ac wave.

Rotary switch An electromechanical device that has a rotating shaft connected to one terminal that is capable of making or breaking a connection.

Sawtooth wave A repeating waveform that rises from zero to a maximum value linearly and then falls to zero and repeats.

Scale A set of markings used for measurement.

Schematic diagram An illustration of the electrical or electronic scheme of a circuit, with all the components represented by their respective symbols.

Scientific notation Numbers are entered and displayed in terms of a power of 10. For example:

Number	Scientific Notation
7642	7.642×10^3
64,000,000	64×10^6
0.0012	1.2×10^{-3}
0.000096	96×10^{-6}

Secondary The output wording of a transformer that is connected across the load.

Selectivity A characteristic of a circuit to discriminate between the wanted signal and the unwanted signal.

Semiconductor A conductor with a resistivity figure between that of an insulator and a metal.

Series circuit A circuit in which the components are connected end to end so that current has only one path to follow throughout the circuit.

Series–parallel network A network or circuit that contains components that are connected in both series and parallel.

Series resonance A condition that occurs when the inductive and capacitive reactances are equal and both components are connected in series with one another, and the impedance is minimum.

Seven-segment display A component that normally has eight LEDs, seven of which are mounted into segments or bars that make up the number 13, and the eighth LED is used as a decimal point.

Shield A metal grounded cover that is used to protect a wire, component, or piece of equipment from stray magnetic and/or electric fields.

Short circuit Also called a short; it is a low-resistance connection between two points in a circuit, typically causing a large amount of current flow.

Shorted out A term used to describe a component that has either internally malfunctioned, resulting in a low-resistance path through the component, or a component that has been bypassed by a low-resistance path.

Signal A conveyor of information.

Signal-to-noise ratio A ratio of the magnitude of the signal to the magnitude of the noise, normally expressed in decibels.

Signal voltage The rms or effective voltage value of a signal.

Silicon (Si) A nonmetallic element (atomic number 14) used in pure form as a semiconductor.

Silicon transistor A transistor using silicon as the semiconducting material.

Silver (Ag) A precious metal that does not easily corrode and is more conductive than copper.

Silvered mica capacitor A mica capacitor with silver deposited directly onto the mica sheets, instead of using conducting metal foil.

Silver solder A solder composed of silver, copper, and zinc with a melting point lower than silver but higher than the standard lead–tin solder.

Simplex Communication in only one direction at a time, for example, facsimile and television.

Simulcast The broadcasting of a program simultaneously in two different forms, for example, a program on both AM and FM.

Sine The sine of an angle of a right-angle triangle is equal to the opposite side divided by the hypotenuse.

Sine wave A wave whose amplitude is the sine of a linear function of time. It is drawn on a graph that plots amplitude against time or radial degrees relative to the angular rotation of an alternator.

Single in-line package (SIP) A package containing several electronic components (generally resistors) with a single row of external connecting pins.

Single-pole, double-throw (SPDT) A three-terminal switch or relay in which one terminal can be thrown in one of two positions.

Single-pole, single-throw (SPST) A two-terminal switch or relay that can either open or close one circuit.

Single sideband (SSB) An AM radio communication technique in which the transmitter suppresses one sideband and the carrier and therefore only transmits a single sideband.

Single-throw switch A switch containing only one set of contacts, which can be either opened or closed.

Sink A device, such as a load, that consumes power.

Sintering The process of bonding either a metal or powder by cold-pressing it into a desired shape and then heating to form a strong, cohesive body.

Sinusoidal Varying in proportion to the sine of an angle or time function; for example, alternating current (ac) is sinusoidal.

SIP Abbreviation for single in-line package.

Skin effect The tendency of high-frequency (rf) currents to flow near the surface layer of a conductor.

Slide switch A switch having a sliding bar, button, or knob.

Slow-acting relay A slow-operating relay that when energized may not pull up the armature for several seconds.

Slow-blow fuse A fuse that can withstand a heavy current (up to ten times its rated value) for a small period of time without blowing.

Snap switch A switch containing a spring under tension or compression that causes the contacts to come together suddenly when activated.

SNR Abbreviation for signal-to-noise ratio.

Soft magnetic material A ferromagnetic material that is easily demagnetized.

Software The program of instructions that directs the operation of a computer.

Solar cell A photovoltaic cell that converts light into electric energy. They are especially useful as a power source for space vehicles.

Solder A metallic alloy that is used to join two metal surfaces.

Soldering The process of joining two metallic surfaces to make an electrical contact by melting solder (usually tin and lead) across them.

Soldering gun A soldering tool having a trigger switch and pistol shape that at its tip has a fast-heating resistive element for soldering.

Soldering iron A soldering tool having an internal heating element that is used for soldering.

Solenoid A coil and movable iron core that when energized by an alternating or direct current will pull the core into a central position.

Solid conductor A conductor having a single solid wire, as opposed to strands.

Solid state Pertaining to circuits and devices that use solid semiconductors such as silicon. Solid-state electronics devices have a solid material between their input and output pins (transistors, diodes), whereas vacuum tube electronics uses tubes, which have a vacuum between input and output.

Sonar An acronym for "sound navigation and ranging." A system using sound waves to determine a target's direction and distance.

Sonic Pertaining to the speed of sound waves.

Sound wave A traveling wave propogated in an elastic medium that travels at a speed of approximately 1133 ft/s.

Source (S) A device that supplies the signal power or electric energy to a load.

Source impedance The impedance a source presents to a load.

South pole The pole of a magnet into which magnetic lines of force are assumed to enter.

Spark A momentary discharge of electric energy due to the breakdown of air or some other dielectric material separating two terminals.

SPDT Abbreviation for single-pole, double-throw.

Speaker Also called a loudspeaker; it is an electroacoustic transducer that converts an electrical wave input into a mechanical sound wave (acoustic) output into the air.

Spectrum The frequency spectrum displays all the frequencies and their application.

Spectrum analyzer An instrument that can display the frequency domain of a waveform, plotting amplitude against frequency of the signals present.

Speed of light A physical constant equal to 186,282.397 miles/s, 2.997925×10^8 m/s, 161,870 nautical miles/s, or 328 yards/μs.

Speed of sound The speed at which a sound wave travels through a medium. In air it is equal to about 1133 ft/s or 335 m/s, while in water it is equal to approximately 4800 ft/s or 1463 m/s. Also known as sonic speed.

Speedup capacitor A capacitor connected in a circuit to speed up an action due to its inherent behavior.

SPST Abbreviation for single-pole, single-throw.

Square wave A wave that alternates between two fixed values for an equal amount of time.

Standard An exact value used as a basis for comparison or calibration.

Static A crackling noise heard on radio receivers caused by electric storms or electric devices in the vicinity.

Static electricity Stationary electricity.

Stator The stationary part of some rotating device.

Statute mile A distance unit equal to 5280 ft or 1.61 km.

Step-up transformer A transformer in which the ac voltage induced in the secondary is greater (due to more secondary windings) than the ac voltage applied to the primary.

Stereo sound A sound system in which the sound is delivered through at least two channels and loudspeakers arranged to give the listener a replica of the original performance.

Stranded conductor A conductor composed by a group of twisted wires.

Stray capacitance The undesirable capacitance that exists between two conductors such as two leads or a lead and a metal chassis.

Subassembly Components contained in a unit for convenience in assembling or servicing the equipment.

Subatomic Particles such as electrons, protons, and neutrons that are smaller than atoms.

Superconductor A metal such as lead or niobium that, when cooled to within a few degrees of absolute zero, can conduct current with no electrical resistance.

Superheterodyne receiver A radio frequency receiver that converts all rf inputs to a common intermediate frequency (IF) before demodulation.

Superhigh frequency (SHF) A frequency band between 3 and 30 GHz, so designated by the Federal Communications Commission (FCC).

Superposition theorem A theorem designed to simplify networks containing two or more sources. It states: In a network containing more than one source, the current at any point is equal to the algebraic sum of the currents produced by each source acting separately.

Supersonic Faster than the speed or velocity of sound (Mach 1).

Supply voltage The voltage produced by a power source or supply.

SW Abbreviation for shortwave.

Sweep generator A test instrument designed to generate a radio-frequency voltage that continually and automatically varies in frequency within a selected frequency range.

Swing The amount a frequency or amplitude varies.

Switch A manual, mechanical, or electrical device used for making or breaking an electric circuit.

Switching transistor A transistor designed to switch either on or off.

Synchronization Also called sync; it is the precise matching or keeping in step of two waves or functions.

Synchronous Two or more circuits or devices in step or in phase.

Sync pulse or signal A pulse waveform generated to synchronize two processes.

System A combination or linking of several parts or pieces of equipment to perform a particular function.

Tachometer An instrument that produces an output voltage that indicates the angular speed of the input in revolutions per minute.

Tank circuit A circuit made up of a coil and capacitor that is capable of storing electric energy.

Tantalum capacitor An electrolytic capacitor having a tantalum foil anode.

Tap An electrical connection to some point, other than the ends, on the element of a resistor or coil.

Technician An expert in troubleshooting circuit and system malfunctions. Along with a thorough knowledge of all test equipment and how to use it to diagnose problems, the technician is also familiar with how to repair or replace faulty components. Technicians basically translate theory into action.

Telegraphy The communication between two points by sending a series of coded current pulses either through wires or by radio.

Telemetry The transmission of instrument reading to a remote location either through wires or by radio waves.

Telephone An apparatus designed to convert sound waves into electrical waves, which are then sent to and reproduced at a distant point.

Telephone line The wires existing between subscribers and central stations.

Telephony A telecommunications system involving the transmission of speech information, therefore allowing two or more persons to converse verbally.

Teletypewriter An electric typewriter that like a teleprinter can produce coded signals corresponding to the keys pressed or print characters corresponding to the coded signals received.

Television (TV) A system that converts both audio and visual information into corresponding electric signals that are then transmitted through wires or by radio to a receiver, which reproduces the original information.

Telex Teletypewriter exchange service.

Temperature coefficient of frequency The rate frequency changes with temperature.

Temperature coefficient of resistance The rate resistance changes with temperature.

Tera (T) A prefix that represents 10^{12}.

Terminal A connecting point for making electric connections.

Tesla (T) The SI unit of magnetic flux density (1 tesla = 1 Wb/m^2).

Test A sequence of operations designed to verify the correct operation or malfunction of a system.

Thermal relay A relay activated by a heating element.

Thermistor A temperature-sensitive semiconductor that has a negative temperature coefficient of resistance (as temperature increases, resistance decreases).

Thermocouple A temperature transducer consisting of two dissimilar metals welded together at one end to form a junction that when heated will generate a voltage.

Thermostat A temperature-sensitive device that opens or closes a circuit.

Thevenin's theorem A theorem that replaces any complex network with a single voltage source in series with a single resistance. It states: Any network of resistors can be replaced with an equivalent voltage source (V_{Th}) and an equivalent series resistance (R_{Th}).

Thick-film capacitor A capacitor consisting of two thick-film layers of conductive film separated by a deposited thick-layer dielectric film.

Thick-film resistor A fixed-value resistor consisting of a thick-film resistive element made from metal particles and glass powder.

Thin-film capacitor A capacitor in which both the electrodes and the dielectric are deposited in layers on a substrate.

Three-phase supply An ac supply that consists of three ac voltages that are 120° out of phase with one another.

Threshold The minimum point at which an effect is produced or indicated.

Time constant The time needed for either a voltage or current to rise to 63.2% of the maximum or fall to 36.8% of the initial value. The time constant of an *RC* circuit is equal to the product of *R* and *C*, while the time constant of an *RL* circuit is equal to the inductance divided by the resistance.

Time-division multiplex (TDM) The transmission of two or more signals on the same path but at different times.

Toggle switch A spring-loaded switch that is put in one of two positions, either on or off.

Tolerance A permissible deviation from a specified value, normally expressed as a percentage.

TO package A cylindrical, metal can type of package for some semiconductor components.

Toroidal coil A coil wound on a doughnut-shaped core.

Torque A moving force.

Transducer Any device that converts energy from one form to another.

Transformer A device consisting of two or more coils that are used to couple electric energy from one circuit to another, yet maintain electrical isolation between the two.

Transistor (TRANSfer resISTOR) A semiconductor device having three main electrodes called the emitter, base, and collector that can be made to either amplify or rectify.

Transmission The sending of information.

Transmission line A conducting line used to couple signal energy between two points.

Transmitter The equipment used to achieve transmission.

Trigger A pulse used to initiate a circuit action.

Trimmer A small-value variable resistor, capacitor, or inductor.

Tuned circuit A circuit that can have its components' values varied so that the circuit responds to one selected frequency yet heavily attenuates all other frequencies.

Turns ratio The ratio of the number of turns in the secondary winding to the number of turns in the primary winding of a transformer.

Two phase Two repeating waveforms having a phase difference of 90°.

UHF Abbreviation for ultrahigh frequency.

Ultrasonic Signals that are just above the range of human hearing of approximately 20 kHz.

Uncharged Having a normal number of electrons and therefore no electrical charge.

VA Abbreviation for volt-ampere.

Vacuum tube An electron tube evacuated to such a degree that its electrical characteristics are essentially unaffected by the presence of residual gas or vapor. Eventually replaced by the transistor for amplification and rectification.

Valence shell The outermost shell formed by electrons.

Variable A quantity that can be altered or controlled to assume a number of distinct values.

Variable capacitor A capacitor whose capacitance can be changed by varying the effective area of the plates or the distance between the plates.

Variable resistor See rheostat and potentiometer.

VCR Abbreviation for video cassette recorder.

Vector A quantity that has both magnitude and direction. They are normally represented as a line, the length of which indicates magnitude and the orientation of which, due to the arrowhead on one end, indicates direction.

Vector diagram An arrangement of vectors showing the phase relationships between two or more ac quantities of the same frequency.

Very high frequency (VHF) An electromagnetic frequency band from 30 to 300 MHz as set by the FCC.

Very low frequency (VLF) A frequency band from 3 to 30 kHz as set by the FCC.

Video Relating to any picture or visual information, from the Latin word meaning "I see."

Voice coil A coil attached to the diaphragm of a moving coil speaker, which is moved through an air gap between the pole pieces of a permanent magnet.

Voice synthesizer A synthesizer that can simulate speech in any language by stringing together phonemes.

Volt (V) The unit of voltage, potential difference, or electromotive force. One volt is the force needed to produce 1 ampere of current in a circuit containing 1 ohm of resistance.

Voltage (V or E) The term used to designate electrical pressure or the force that causes current to flow.

Voltage divider A fixed or variable series resistor network that is connected across a voltage to obtain a desired fraction of the total voltage.

Voltage drop The voltage or difference in potential developed across a component or conductor due to the loss of electric pressure as a result of current flow.

Voltage gain Also called voltage amplification, it is equal to the difference between the output voltage level and the input signal voltage level. This value is normally expressed in decibels, which are equal to 20 times the logarithm of the ratio of the output voltage to the input voltage.

Voltage rating The maximum voltage a component can safely withstand without breaking down.

Voltaic cell A primary cell having two unlike metal electrodes immersed in a solution that chemically interacts with the plates to produce an emf.

Volt-ampere (VA) The unit of apparent power in an ac circuit containing reactance. Apparent power is equal to the product of voltage and current.

Voltmeter An instrument designed to measure the voltage or potential difference. Its scale can be graduated in kilovolts, volts, or millivolts.

Volume The magnitude or power level of a complex audio frequency (AF) wave, expressed in volume units (VU).

VOM Abbreviation for volt-ohm-milliammeter.

VRMS Abbreviation for volts root-mean-square.

W Abbreviation for watt.

Wall outlet A spring-contact outlet mounted on the wall to which a portable appliance is connected to obtain electric power.

Watt (W) The unit of electric power required to do work at a rate of 1 joule/second. One watt of power is expended when 1 ampere direct current flows through a resistance of 1 ohm. In an ac circuit, the true power is effective volts multiplied by effective amperes, multiplied by the power factor.

Wattage rating The maximum power a device can safely handle continuously.

Watt-hour (Wh) The unit of electrical work, equal to a power of 1 watt being absorbed continuously for 1 hour.

Wave An electric, electromagnetic, acoustic, mechanical, or other form whose physical activity rises and falls or advances and retreats periodically as it travels through some medium.

Waveform The shape of a wave.

Waveguide A rectangular or circular metal pipe used to guide electromagnetic waves at microwave frequencies.

Wavelength (λ) The distance between two points of corresponding phase and is equal to waveform velocity or speed divided by frequency.

Weber (Wb) The unit of magnetic flux. One weber is the amount of flux that when linked with a single turn of wire for an interval of 1 second, will induce an electromotive force of 1 V.

Wet cell A cell using a liquid electrolyte.

Winding One or more turns of a conductor wound to form a coil.

Wire A single solid or stranded group of conductors having a low resistance to current flow.

Wire gauge The American wire gauge (AWG) is a system of numerical designations of wire sizes, with the first being 0000 (the largest size) and then going to 000, 00, 0, 1, 2, 3, and so on up to the smallest sizes of 40 and above.

Wireless A term describing radio communication that requires no wires between the two communicating points.

Wirewound resistor A resistor in which the resistive element is a length of high-resistance wire or ribbon, usually Nichrome, wound onto an insulating form.

Wire-wrapping A method of prototyping in which solderless connections are made by wrapping wire around a rectangular terminal.

Woofer A large loudspeaker designed primarily to reproduce low audio-frequency signals at large power levels.

Work Work is done anytime energy is transformed from one type to another, and the amount of work done is dependent on the amount of energy transformed.

X Symbol for reactance.

X **axis** The horizontal axis.

Y Symbol for admittance.

Y **axis** The vertical axis.

Z **axis** An axis perpendicular to both the *X* and *Y* axes.

AC Alternating current
AC/DC AC or DC
A/D Analog to digital
ADC Analog-to-digital converter
AF Audio frequency
AFC Automatic frequency control
AGC Automatic gain control
AM Amplitude modulation
AM/FM AM or FM
AVC Automatic volume control
AWG American wire gauge
BW Bandwidth
C Capacitance; capacitor
CAD Computer-aided design
CAM Computer-aided manufacturing
CATV Cable TV
CB Citizen's band
CPU Central processing unit
CRT Cathode ray tube
D/A Digital to analog
DC Direct current
DIP Dual in-line package
DMM Digital multimeter
DPDT Double pole, double throw
DPST Double pole, single throw
DVM Digital voltmeter
ECG Electrocardiogram
EHF Extremely high frequency
EHV Extra high voltage
ELF Extremely low frequency
EMF Electromotive force
EMI Electromagnetic interference
EW Electronic warfare

FET Field-effect transistor
FM Frequency modulation
4PDT Four pole, double throw
4PST Four pole, single throw
FSD Full-scale deflection
G Gravitational force
HF High frequency
IC Integrated circuit
IF Intermediate frequency
I/O Input/output
IR Infrared
JFET Junction FET
L Coil; inductance
LC Inductance–capacitance
LCD Liquid crystal display
LDR Light-dependent resistor
LED Light-emitting diode
LF Low frequency
LO Local oscillator
LSI Large-scale integration
MF Medium frequency
MOS Metal oxide semiconductor
MPU Microprocessor unit (μP)
N Negative
NC Normally closed
NO Normally open
NPN Negative–positive–negative
P Positive; peak
PA Public address
PCB Printed circuit board
PLL Phase-locked loop
PM Phase modulation; permanent magnet
PNP Positive–negative–positive

P-P Peak to peak
PRF Pulse repetition frequency
PRT Pulse repetition time
R Resistance
RAM Random-access memory
RC Resistance–capacitance
RF Radio frequency
RFI Radio frequency interference
RLC Resistance–inductance–capacitance
RMS Root mean square
ROM Read-only memory
SCR Silicon-controlled rectifier
SHF Superhigh frequency
SIP Single in-line package
SNR Signal-to-noise ratio
SPDT Single pole, double throw

SPST Single pole, single throw
SSB Single sideband
SW Shortwave
SWR Standing-wave ratio
TC Time constant; temperature coefficient
TR Transmit–receive
TTL Transistor–transistor logic
TV Television
UHF Ultrahigh frequency
UHV Ultrahigh voltage
UV Ultraviolet
VCO Voltage-controlled oscillator
VHF Very high frequency
VLF Very low frequency
VOM Volt-ohm-milliammeter

APPENDIX C   Safety

Safety precautions should always be your first priority when working on electronic equipment, as there is always the possibility of receiving an electric shock. A shock is a sudden, uncontrollable reaction as current passes through your body and causes your muscles to contract and a certain amount of pain. Figure C-1 lists the physiological effects of different amounts of current, and as you can see, even a current as small as 10 mA can be fatal.

Any shock is dangerous, since even the mildest could surprise you and cause an involuntary action that could injure yourself or someone else. For example, a muscular spasm could throw you against a sharp object or move your arm to a point of higher voltage.

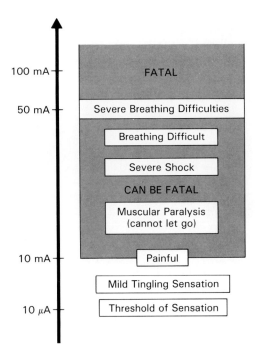

FIGURE C-1   **Physiological Effects of Electric Current.**

C.1
BODY RESISTANCE

Voltage is the force that causes current to flow through a circuit. The amount of current is dependent on the circuit's resistance and the value of voltage ($I = V/R$). The greatest danger, therefore, is from high voltage points (more than 25 V). The resistance of your body also determines current, since a low resistance will cause a more dangerous high current. The human body has a resistance of about 10,000 to 50,000 Ω, depending on how good a contact you make with the "live" (power present) conductor. Skin resistance is generally quite high and will subsequently oppose current flow; however, your resistance is lowered if your skin is wet due to perspiration or if your skin has a cut or an abrasion.

EXAMPLE

Calculate the current through a body resistance of 10 kΩ if it came in contact with 100 V. How much power is the body dissipating?

SOLUTION

$$I = \frac{V}{R} = \frac{100 \text{ V}}{10 \text{ k}\Omega} = 10 \text{ mA}$$
$$P = V \times I = 100 \text{ V} \times 10 \text{ mA} = 1 \text{ watt}$$

This current will be painful and possibly fatal.

Power is a point to consider. In the previous example, the body resistance was dissipating 1 watt. A power source may not be able to supply a fatal high power, and its output voltage will be pulled down due to the excessive load current through the body resistance. The current then is reduced to an amount that the power source can deliver.

C.2
PRECAUTIONS TO BE USED WHEN TROUBLESHOOTING

The following procedures and precautions have been acquired from experienced technicians and should all be applied whenever possible.

1. Except when absolutely necessary, do not work on electronic circuits or equipment when power is on.

2. When troubleshooting inside equipment, people tend to lean one hand on the chassis and hold the test lead or probe in the other hand. If the probing hand comes in contact with a high voltage, current will flow from one hand, through your chest (heart and lungs), and finally through the other hand to the chassis ground. To avoid this dangerous situation, always place the free hand in a pocket or behind your back while testing a piece of equipment with power on.

3. When troubleshooting equipment with power on, try to insulate yourself by not standing on a damp floor or leaning against any metal object and by wearing rubber soled shoes or standing on a rubber mat.

4. Electrolytic and other large-value capacitors can hold a dangerous voltage charge even after the equipment has been turned off, and left for several hours. Be sure always to check if these capacitors are fully discharged by shorting their two terminals with a pair of *insulated* pliers or *insulated* test lead.

5. All tools should be well insulated. If not, it is not only dangerous to you, but probes that are not well insulated right down to the tip can cause short circuits between two points, resulting in additional circuit and equipment problems.

6. Always switch off the equipment and remove the power (since some equipment has power present even when off) before replacing any components. When removing components that get hot, such as resistors, allow enough time for them to cool down after the equipment has been turned off.

7. Necklaces, rings, and wristwatches have a low resistance and should be removed when working on equipment.

8. Inspect the equipment before working on it, and if it is in poor condition (frayed, cracked, or burnt power cords, chipped plugs), turn off the equipment and replace these hazards.

9. Make sure someone is present who can render assistance in the event of an emergency.

10. Make it a point to know the location of the power off switch.

11. Cathode ray tubes are highly evacuated and should therefore be handled with extreme care. If broken, the relativey high external pressure will cause an implosion (burst inward), which will result in the inward metal parts and glass fragment being violently expelled outward.

C.3

FIRST AID, TREATMENT, RESUSCITATION

C.3.1 INJURY TO PERSONS

Injury may be caused in a number of ways:

(1) SHOCK

Electric shock is the effect produced on the body and in particular on the nerves by an electric current passing through it. Its magnitude depends on the strength of the current, which, in turn, depends on the voltage. Its effect varies according to the ohmic resistance of the body, which varies in different persons, and also according to the parts of the body between which the current flows (contact in the cardiac region can be particularly dangerous). It also depends on the current flow and on the surface resistance of the skin, which is much reduced when the skin is wet and is reduced to zero if the skin is penetrated.

Shock can be felt from voltages as low as 15 V, and at 20 to 25 V most people would experience pain. At about this voltage or a little higher, the victim may find himself unable to let go of the conductor and may suffer burning. It is believed that death can, in appropriate conditions, be caused by voltages as low as 70, but generally the danger below 120 V ac is believed to be small (although not entirely negligible). Most serious and fatal accidents occur at the industrial 200 to 240 V ac and from 25 to 30 milliamps and over.

Injury can also be caused by a minor shock, not serious in itself, but which has the effect of contracting the muscles sufficiently to result in a fall or some other reaction.

(2) BURNS

Burns can be caused by the passage through the body of a heavy current if the body is in contact with a conductor, or by direct contact with an electrically heated surface. Burns can also be caused by the intense heat generated by the arcing from a short circuit. All cases of burns require immediate medical attention.

(3) EXPLOSION

An explosion can be caused by the ignition of flammable gases by a spark from an electric contact. In all cases where a flammable or ignitable atmosphere or vapor is present, special care is necessary.

(4) EYE INJURIES

These can be caused by exposure to the strong ultraviolet rays of an electric arc. In these cases, the eyes may become inflamed and painful after a lapse of several hours, and there may be temporary loss of sight. Although very painful, the condition usually passes off within 24 hours. Lasers are also dangerous to the eyes due to their intense concentrated beam.

Precautions to protect the eyes must always be taken by wearing protective goggles when clipping leads or soldering.

Permanent injury to the eyes can arise from the energy propagated by microwave equipment. No one should look along a wave guide when it is on or examine a highly directional radiator at close distances.

(5) BODY INJURIES FROM MICROWAVE AND RADIO-FREQUENCY EQUIPMENT

The energy in microwave and radio-frequency equipment can damage the body, especially those parts with a low blood supply. The eyes are particularly vulnerable. The highest energy level to which operators should be subject is 1.0 mW/cm^2 and intensities exceeding 10 mW/cm^2 should always be avoided.

C.3.2 RESUSCITATION

You should familiarize yourself with the various methods of artificial respiration by contacting your local Red Cross for complete instruction.

The mouth-to-mouth method of artificial respiration is the most effective of the resuscitation techniques. It is comparatively simple and produces the best and quickest results when correctly applied.

MOUTH-TO-MOUTH RESUSCITATION METHOD

It is essential to commence artificial respiration without delay. *Do not touch the victim with your bare hands until the circuit is broken.* If this is not possible, *protect yourself* with dry insulating material and pull the victim clear of the conductor.

STEP 1. Lay the patient on his back and, if on a slope, have the stomach slightly lower than the chest.

STEP 2. Make a brief inspection of the mouth and throat to ensure they are clear of obvious obstructions.

STEP 3. Give the patient's head the maximum backward tilt so that the chin is prominent and the neck stretched to give a clear airway, as seen in Figure C-2(a).

FIGURE C-2 **Mouth-to-Mouth Resuscitation.**

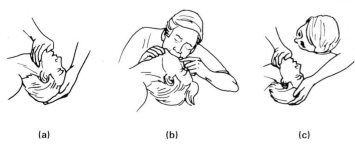

(a) (b) (c)

First Aid, Treatment, Resuscitation

STEP 4.   Sealing off the patients nose with your thumb and finger, open your mouth wide and make an airtight seal over the open patient's mouth and then blow, as seen in Figure C-2(b).

STEP 5.   After exhaling, turn your head to watch for chest movement, while inhaling deeply in readiness for blowing again, as seen in Figure C-2(c).

STEP 6.   If the chest does not rise, check that the patient's mouth and throat are free of obstruction and that the head is tilted back as far as possible, and then blow again.

APPENDIX D Electronic Schematic Symbols

RESISTORS

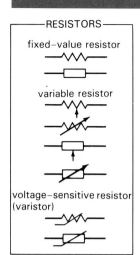

fixed-value resistor

variable resistor

voltage-sensitive resistor
(varistor)

SOURCES

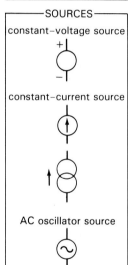

constant-voltage source

constant-current source

AC oscillator source

BATTERIES

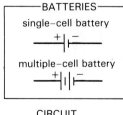

single-cell battery

multiple-cell battery

CIRCUIT PROTECTORS

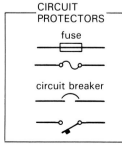

fuse

circuit breaker

CAPACITORS

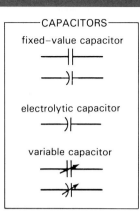

fixed-value capacitor

electrolytic capacitor

variable capacitor

CRYSTALS

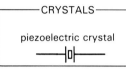

piezoelectric crystal

LAMPS

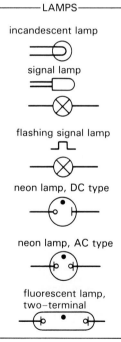

incandescent lamp

signal lamp

flashing signal lamp

neon lamp, DC type

neon lamp, AC type

fluorescent lamp,
two-terminal

GROUND

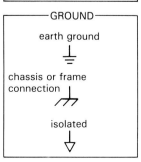

earth ground

chassis or frame
connection

isolated

INDUCTORS

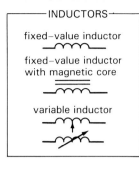

fixed-value inductor

fixed-value inductor
with magnetic core

variable inductor

AUDIO DEVICES

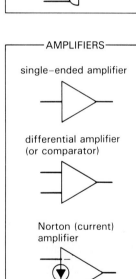

loudspeaker

Microphone

AMPLIFIERS

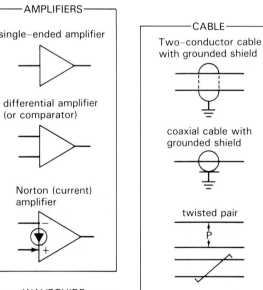

single-ended amplifier

differential amplifier
(or comparator)

Norton (current)
amplifier

WAVEGUIDE

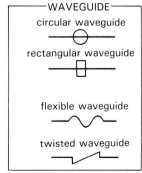

circular waveguide

rectangular waveguide

flexible waveguide

twisted waveguide

TRANSFORMERS

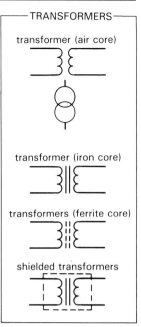

transformer (air core)

transformer (iron core)

transformers (ferrite core)

shielded transformers

CABLE

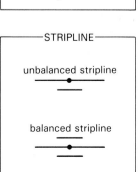

Two-conductor cable
with grounded shield

coaxial cable with
grounded shield

twisted pair

STRIPLINE

unbalanced stripline

balanced stripline

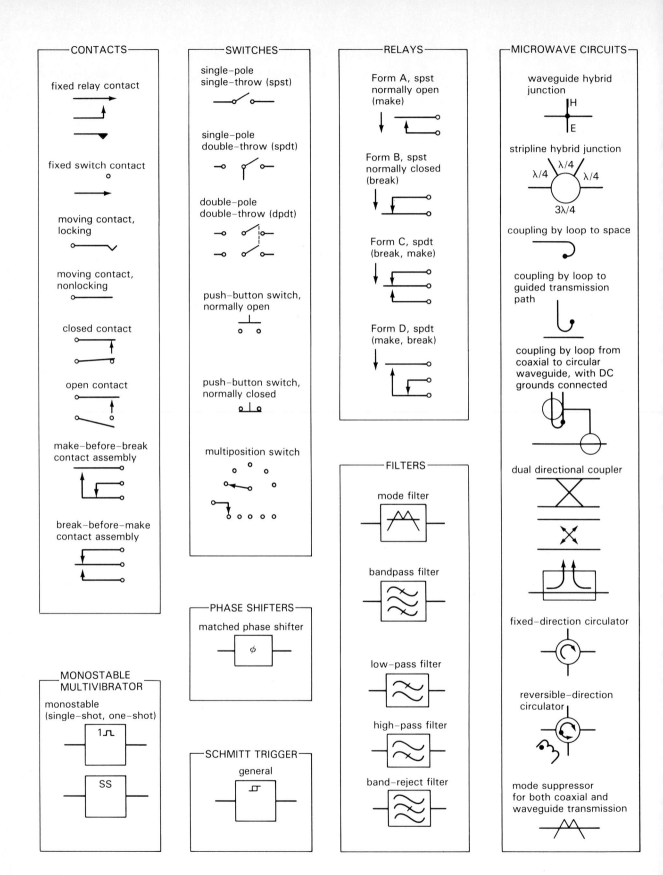

Appendix D   Electronic Schematic Symbols

DIODES

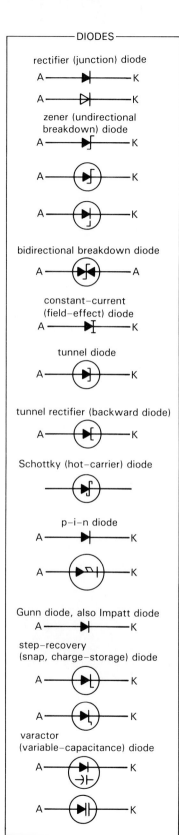

rectifier (junction) diode

A ———▷|——— K

A ———◁|——— K

zener (undirectional breakdown) diode

A ———▷|——— K

A K

A ———▷|——— K

bidirectional breakdown diode

A ———◁▷|——— A

constant-current (field-effect) diode

A ———▷|——— K

tunnel diode

A K

tunnel rectifier (backward diode)

A K

Schottky (hot-carrier) diode

A K

p-i-n diode

A ———▷|——— K

A K

Gunn diode, also Impatt diode

A ———▶|——— K

step-recovery (snap, charge-storage) diode

A K

A K

varactor (variable-capacitance) diode

A K

A K

TRANSISTORS

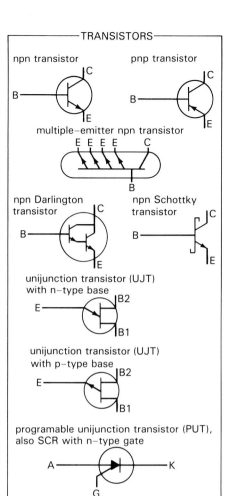

npn transistor

pnp transistor

multiple-emitter npn transistor

E E E E C

B

npn Darlington transistor

npn Schottky transistor

unijunction transistor (UJT) with n-type base

E B2

B1

unijunction transistor (UJT) with p-type base

E B2

B1

programable unijunction transistor (PUT), also SCR with n-type gate

A K

G

THYRISTORS

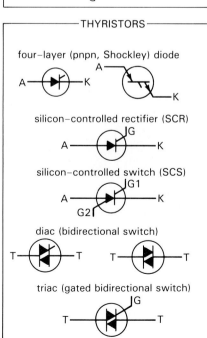

four-layer (pnpn, Shockley) diode

A K   A K

silicon-controlled rectifier (SCR)

A G K

silicon-controlled switch (SCS)

A G1 K

G2

diac (bidirectional switch)

T T   T T

triac (gated bidirectional switch)

G

T T

FIELD-EFFECT TRANSISTORS (FETs)

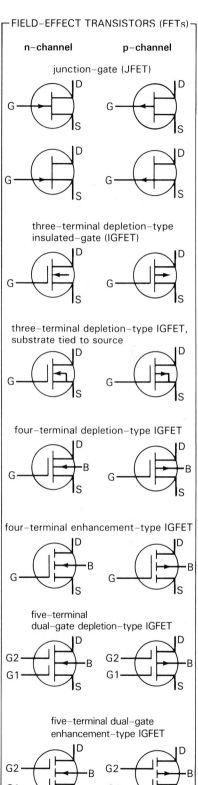

n-channel   **p-channel**

junction-gate (JFET)

three-terminal depletion-type insulated-gate (IGFET)

three-terminal depletion-type IGFET, substrate tied to source

four-terminal depletion-type IGFET

four-terminal enhancement-type IGFET

five-terminal dual-gate depletion-type IGFET

five-terminal dual-gate enhancement-type IGFET

Electronic Schematic Symbols

OPTOELECTRONIC DIODES

light–emitting diode (LED)

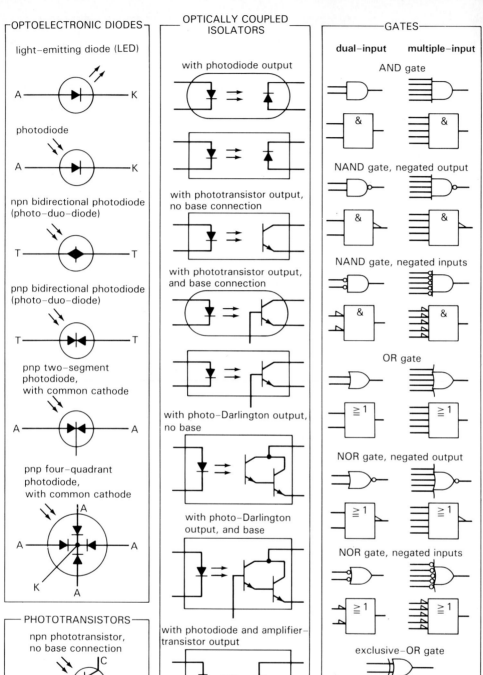

photodiode

npn bidirectional photodiode
(photo–duo–diode)

pnp bidirectional photodiode
(photo–duo–diode)

pnp two–segment
photodiode,
with common cathode

pnp four–quadrant
photodiode,
with common cathode

PHOTOTRANSISTORS

npn phototransistor,
no base connection

npn phototransistor,
with base connection

OPTICALLY COUPLED ISOLATORS

with photodiode output

with phototransistor output,
no base connection

with phototransistor output,
and base connection

with photo–Darlington output,
no base

with photo–Darlington
output, and base

with photodiode and amplifier–
transistor output

with NAND–gate–photo–
detector output

GATES

dual–input   multiple–input

AND gate

NAND gate, negated output

NAND gate, negated inputs

OR gate

NOR gate, negated output

NOR gate, negated inputs

exclusive–OR gate

inverter gate

FLIP–FLOPS

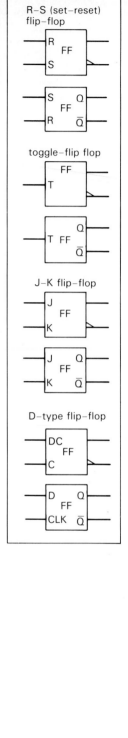

R–S (set–reset)
flip–flop

toggle–flip flop

J–K flip–flop

D–type flip–flop

APPENDIX E   Conversions, Constants, Units and Prefixes

GREEK ALPHABET

Name	Capital	Lowercase	Designates
Alpha	A	α	Angles, coefficients, attenuation constant, absorption factor, area.
Beta	B	β	Angles, coefficients, phase constant.
Gamma	Γ	γ	Specific quantity, angles, electrical conductivity, propagation constant, complex propagation constant (cap).
Delta	Δ	δ	Density, angles, increment or decrement (cap or small), determinant (cap), permittivity (cap).
Epsilon	E	ϵ	Dielectric constant, permittivity, base of natural (Napierian) logarithms, electric intensity.
Zeta	Z	ζ	Coordinate, coefficients.
Eta	H	η	Intrinsic impedance, efficiency, surface charge density, hysteresis, coordinates.
Theta	Θ	θ	Phase angle, reluctance.
Iota	I	ι	Unit vector.
Kappa	K	κ	Susceptibility, coupling coefficient.
Lambda	Λ	λ	Wavelength, attenuation constant, permeance (cap).
Mu	M	μ	Prefix *micro-*, permeability, amplification factor.
Nu	N	ν	Reluctivity, frequency.
Xi	Ξ	ξ	Coordinates.
Omicron	O	o	——
Pi	Π	π	3.1416 (circumference divided by diameter).
Rho	P	ρ	Resistivity, volume charge density, coordinates.
Sigma	Σ	σ	Surface charge density, complex propagation constant, electrical conductivity, leakage coefficient, sign of summation (cap).
Tau	T	τ	Time constant, volume resistivity, time-phase displacement, transmission factor, density.
Upsilon	Υ	υ	——
Phi	Φ	ϕ	Magnetic flux, angles, scalar potential (cap).
Chi	X	χ	Electric susceptibility, angles.
Psi	Ψ	ψ	Dielectric flux, phase difference, coordinates, angles.
Omega	Ω	ω	Angular velocity ($2\pi f$), resistance in ohms (cap), solid angles (cap).

SYMBOLS AND UNITS

Quantity	Symbol	Unit	Symbol
Charge	Q	coulomb	C
Current	I	ampere	A
Voltage, potential difference	V	volt	V
Electromotive force	δ	volt	V
Resistance	R	ohm	Ω
Conductance	G	mho (siemens)	A/V, or mho (S)
Reactance	X	ohm	Ω
Susceptance	B	mho	A/V, or mho
Impedance	Z	ohm	Ω
Admittance	Y	mho	A/V, or mho
Capacitance	C	farad	F
Inductance	L	henry	H
Energy, work	W	joule	J
Power	P	watt	W
Resistivity	ρ	ohm-meter	Ωm
Conductivity	σ	mho per meter	mho/m
Electric displacement	D	coulomb per square meter	C/m^2
Electric field strength	E	volt per meter	V/m
Permittivity (absolute)	ϵ	farad per meter	F/m
Relative permittivity	ϵ_r	(numeric)	
Magnetic flux	Φ	weber	Wb
Magnetomotive force	$\mathscr{F}$	ampere (ampere-turn)	A
Reluctance	$\mathscr{R}$	ampere per weber	A/Wb
Permeance	$\mathscr{P}$	weber per ampere	Wb/A
Magnetic flux density	B	tesla	T
Magnetic field strength	H	ampere per meter	A/m
Permeability (absolute)	μ	henry per meter	H/m
Relative permeability	μ_r	(numeric)	
Length	l	meter	m
Mass	m	kilogram	kg
Time	t	second	s
Frequency	f	hertz	Hz
Angular frequency	ω	radian per second	rad/s
Force	F	newton	N
Pressure	p	newton per square meter	N/m^2
Temperature (absolute)	T	Kelvin	K
Temperature (International)	t	degree Celsius	°C

718

SI SYMBOLS AND UNITS

Unit	Symbol	Unit	Symbol	Unit	Symbol
ampere	A	kelvin	K	milliliter	mL
ampere-hour	Ah	kiloampere	kA	millimeter	mm
ampere per meter	A/m	kilobaud	kBd	milliohm	mΩ
angstrom	Å	kilobit	kb	millirad	mrd
attoampere	aA	kiloelectronvolt	keV	milliradian	mrad
bar	bar	kilogauss	kG	milliroentgen	mR
barn	b	kilogram	kg	millisecond	ms
baud	Bd	kilohertz	kHz	millisiemens	mS
bel	B	kilohm	kΩ	millivolt	mV
bit	b	kilometer	km	milliwatt	mW
bit per second	b/s	kilometer per hour	km/h	minute (time)	min
calorie	cal	kilovolt	kV	nanoampere	nA
candela	cd	kilovoltampere	kVA	nanofarad	nF
candela per square	cd/m^2	kilowatt	kW	nanohenry	nH
meter		kilowatthour	kWh	nanometer	nm
centimeter	cm	lambert	L	nanosecond	ns
circular mil	cmil	liter	L	nanovolt	nV
coulomb	C	lumen	lm	nanowatt	nW
cubic centimeter	cm^3	lumen per square foot	lm/ft^2	nautical mile	nmi
cubic foot per minute	ft^3/min	lumen per square	lm/m^2	neper	Np
cubic meter	m^3	meter		oersted	Oe
cubic meter per second	m^3/s	lumen per watt	lm/W	ohm	Ω
curie	Ci	lumen second	lm·s	ohm per volt	Ω/V
decibel	dB	lux	lx	picoampere	pA
degree (plane angle)	°	maxwell	Mx	picofarad	pF
degree Celsius	°C	megabar	Mbar	picosecond	ps
degree Fahrenheit	°F	megabit	Mb	picowatt	pW
degree Rankine	°R	megaelectronvolt	MeV	pound	lb
dyne	dyn	megahertz	MHz	rad	rd
electronvolt	eV	megampere	MA	radian	rad
erg	erg	megavolt	MV	revolution per minute	r/min
farad	F	megawatt	MW	revolution per second	r/s
femtoampere	fA	megawatt-hour	MWh	roentgen	R
femtovolt	fV	megohm	MΩ	second (time)	s
femtowatt	fW	meter	m	siemens	S
foot	ft	microampere	μA	square foot	ft^2
foot per minute	ft/min	microbar	μbar	square inch	in.^{2}
foot per second	ft/s	microfarad	μF	square meter	m^2
gauss	G	microhenry	μH	square mile	mi^2
gigabit	Gb	microhm	$\mu\Omega$	square yard	yd^2
gigaelectronvolt	GeV	micrometer	μm	steradian	sr
gigahertz	GHz	microsecond	μs	teraelectronvolt	TeV
gigawatt	GW	microsiemens	μS	terahertz	THz
gigohm	GΩ	microvolt	μV	teraohm	TΩ
gilbert	Gb	microwatt	μW	tesla	T
gram	g	mil	mil	var	var
henry	H	mile	mi	volt	V
hertz	Hz	mile per hour	mi/h	volt-ampere	VA
horsepower	hp	milliampere	mA	watt	W
hour	h	millibar	mbar	watt-hour	Wh
inch	in.	milligauss	mG	watt per steradian	W/sr
inch per second	in./s	milligram	mg	weber	Wb
joule	J	millihenry	mH	yard	yd

METRIC CONVERSION CHART
(APPROXIMATIONS)

When You Know	Multiply by	To Find
Length		
millimeters (mm)	0.04	inches (in.)
centimeters (cm)	0.4	inches (in.)
meters (m)	3.3	feet (ft)
meters (m)	1.1	yards (yd)
kilometers (km)	0.6	miles (mi)
Area		
sq. centimeters (cm²)	0.16	sq. inches (in.²)
sq. meters (m²)	1.2	sq. yards (yd²)
sq. kilometers (km²)	0.4	sq. miles (mi²)
hectares (ha) (10 000 m²)	2.5	acres
Mass (weight)		
grams (g)	0.035	ounces (oz)
kilograms (kg)	2.2	pounds (lb)
tonnes (1 000 kg) (t)	1.1	short tons
Volume		
milliliters (mL)	0.03	fluid ounces (fl oz)
liters (L)	2.1	pints (pt)
liters (L)	1.06	quarts (qt)
liters (L)	0.26	gallons (gal)
cubic meters (m³)	35	cubic feet (ft³)
cubic meters (m³)	1.3	cubic yards (yd³)
Temperature (exact)		
Celsius (°C)	9/5 (°C) + 32	Fahrenheit (°F)
Temperature (exact) to Metric		
Fahrenheit (°F)	5/9 (°F − 32)	Celsius (°C)
Length		
inches (in.)	*2.5	centimeters (cm)
feet (ft)	30	centimeters (cm)
yards (yd)	0.9	meters (m)
miles (mi)	1.6	kilometers (km)

* in. = 2.54 cm exactly

When You Know	Multiply by	To Find
Area		
sq. inches (in.^{2})	6.5	sq. centimeters (cm^2)
sq. feet (ft^2)	0.09	sq. meters (m^2)
sq. yards (yd^2)	0.8	sq. meters (m^2)
sq. miles (mi^2)	2.6	sq. kilometers (km^2)
acres	0.4	hectares (ha)
Mass (weight)		
ounces (oz)	28	grams (g)
pounds (lb)	0.45	kilograms (kg)
short tons (2 000 lb)	0.9	tonnes (t)
Volume		
teaspoons (tsp)	5	milliliters (mL)
tablespoons (tbsp)	15	milliliters (mL)
fluid ounces (fl oz)	30	milliliters (mL)
cups (c)	0.24	liters (L)
pints (pt)	0.47	liters (L)
quarts (qt)	0.95	liters (L)
gallons (gal)	3.8	liters (L)
cubic feet (ft^3)	0.03	cubic meters (m^3)
cubic yards (yd^3)	0.76	cubic meters (m^3)

Multiplication Factor	Prefix	Symbol
1 000 000 000 000 000 000 = 10^{18}	exa	E
1 000 000 000 000 000 = 10^{15}	peta	P
1 000 000 000 000 = 10^{12}	tera	T
1 000 000 000 = 10^9	giga	G
1 000 000 = 10^6	mega	M
1 000 = 10^3	kilo	k
100 = 10^2	hecto	h
10 = 10	deka	da
0.1 = 10^{-1}	deci	d
0.01 = 10^{-2}	centi	c
0.001 = 10^{-3}	milli	m
0.000 001 = 10^{-6}	micro	μ
0.000 000 001 = 10^{-9}	nano	n
0.000 000 000 001 = 10^{-12}	pico	p
0.000 000 000 000 001 = 10^{-15}	femto	f
0.000 000 000 000 000 001 = 10^{-18}	atto	a

LETTER SYMBOLS FROM PIONEERS IN SCIENCE

Unit	Symbol	Name
ampere	A	Ampere
bel	B	Bell
coulomb	C	Coulomb
curie	Ci	Curie
farad	F	Faraday
gauss	G	Gauss
gilbert	Gb	Gilbert
henry	H	Henry
hertz	Hz	Hertz
joule	J	Joule
kelvin	K	Kelvin
lambert	L	Lambert
maxwell	Mx	Maxwell
neper	Np	Neper
oersted	Oe	Oersted
ohm	Ω	Ohm
roentgen	R	Roentgen
siemens	S	Siemens
tesla	T	Tesla
volt	V	Volta
watt	W	Watt
weber	Wb	Weber

CONVERSION FACTORS AND CONSTANTS

$$\pi = 3.1416 \qquad 2\pi = 6.2832$$
$$\pi^2 = 9.8696 \qquad (2\pi)^2 = 39.478$$
$$e = 2.7183 \qquad \sqrt{2} = 1.4142$$
$$\sqrt{3} = 1.7321 \qquad \log_\pi = 0.4972$$

1 meter = 39.37 inches = 3.28 feet
1 kilometer = 0.621 mile (about $\frac{3}{5}$ mile)
1 inch = 2.54 centimeters
1 kilogram = 2.2 pounds
1 liter = 1.06 quarts
1 ounce = 28.35 grams
1 horsepower = 746 watts

PHYSICAL CONSTANTS

Constant	Symbol	Rounded Value
Electronic charge	e	1.602×10^{-19} C
Speed of light in vacuum	c	2.9979×10^8 m/s
Permittivity of vacuum, electric constant	ϵ_0, Γ_e	8.8542×10^{-12} F/m
Permeability of vacuum, magnetic constant	μ_0, Γ_m	$4\pi \times 10^{-7}$ H/m
Planck constant	h	6.63×10^{-34} J·s
Boltzmann constant	k	1.38×10^{-23} J/K
Faraday constant	F	9.649×10^4 C/mol
Proton gyromagnetic ratio	γ	2.6752×10^8 rad/sT
Standard gravitational acceleration	g_n	9.80665 m/s^2
Normal atmospheric pressure	atm	101 325 N/m^2

Conversions, Constants, Units and Prefixes

APPENDIX F   Formulas

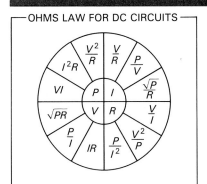

OHMS LAW FOR DC CIRCUITS

OHM'S LAW FOR AC CIRCUITS

Series	**Parallel**
$Z = \sqrt{R^2 + (X_L - X_C)^2}$	$Z = \dfrac{V_A}{I_t}$
$V = IZ$	$V = IZ$
$I = \dfrac{V}{Z}$	$I_t = \sqrt{I_R^2 + (I_L - I_C)^2}$
$AP = I^2Z$	$AP = V_A I_t$
$TP = I_R^2 R$	$TP = I_R^2 R$
$PF = \dfrac{R}{Z}$	$PF = \dfrac{TP}{AP}$
$V_R = IR$	$V_R = I_R R$
$V_L = IX_L$	$V_L = I_L X_L$
$V_C = IX_C$	$V_C = I_C X_C$
$I_R = \dfrac{V_R}{R}$	$I_R = \dfrac{V}{R}$
$I_L = \dfrac{V_L}{X_L}$	$I_L = \dfrac{V}{X_L}$
$I_C = \dfrac{V_C}{X_C}$	$I_C = \dfrac{V}{X_C}$
$V_A = \sqrt{V_R^2 + (V_L - V_C)^2}$	$V_A = I_t Z_t$

RESISTORS IN SERIES

$$R_{total} = R_1 + R_2 + R_3 + \cdots$$

TWO RESISTORS IN PARALLEL

$$R_t = \frac{R_1 R_2}{R_1 + R_2} \qquad R_1 = \frac{R_t R_2}{R_2 - R_t}$$

EQUAL RESISTORS IN PARALLEL

$$R_{total} = \frac{R}{n}, \text{ where } n \text{ is the number of resistors}$$

RESISTORS IN PARALLEL, GENERAL FORMULA

$$R_{total} = \frac{1}{\dfrac{1}{R_1} + \dfrac{1}{R_2} + \dfrac{1}{R_3} + \cdots}$$

SINUSOIDAL VOLTAGES AND CURRENTS

effective value $= 0.707 \times$ peak value
average value $= 0.637 \times$ peak value
peak value $= 1.414 \times$ effective value
effective value $= 1.11 \times$ average value
peak value $= 1.57 \times$ average value
average value $= 0.9 \times$ effective value

CONDUCTANCE, SUSCEPTANCE, AND ADMITTANCE

$$G = \frac{1}{R} \text{ (for DC circuit)}$$

$$G = \frac{R}{R^2 + X^2} \text{ (for AC circuit)}$$

$$B = \frac{1}{X} \text{ (when resistance is 0)}$$

$$B = \frac{X}{R^2 + X^2}$$

$$Y = \frac{1}{Z} = \frac{1}{\sqrt{R^2 + X^2}}$$

REACTANCE FORMULAS

$$X_C = \frac{1}{2\pi f C} \qquad C = \frac{1}{2\pi f X_C}$$

$$X_L = 2\pi f L \qquad L = \frac{X_L}{2\pi f}$$

RESONANT FREQUENCY FORMULAS

$$f = \frac{1}{2\pi\sqrt{LC}} \quad \text{or} \quad f = \frac{159.2^*}{\sqrt{LC}} \qquad f = \frac{1}{2\pi RC}$$

$$L = \frac{1}{4\pi^2 f^2 C} \quad \text{or} \quad L = \frac{25{,}330^*}{f^2 C} \qquad f = \frac{R}{2\pi L}$$

$$C = \frac{1}{4\pi^2 f^2 L} \quad \text{or} \quad C = \frac{25{,}330^*}{f^2 L}$$

* Where in the second formula f is in kHz and L and C are in microunits.

IMPEDANCE FORMULAS

$$Z = \sqrt{R^2 + (X_L - X_C)^2} \text{ (for series circuit)}$$

$$Z = \frac{RX}{\sqrt{R^2 + X^2}} \text{ (for } R \text{ and } X \text{ in parallel)} \qquad Z = \frac{V}{I}$$

POWER FACTOR

$PF = \cos\theta$ where θ is the angle of lead or lag

$$PF = \frac{\text{true power}}{\text{apparent power}} = \frac{P}{VI}$$

$$PF = \frac{R}{Z}$$

Q

$$Q = \frac{X_L}{R} \text{ or } \frac{X_C}{R}$$

TRANSFORMER RELATIONSHIPS

$$\frac{N_p}{N_s} = \frac{V_p}{V_s} = \frac{I_s}{I_p} = \sqrt{\frac{Z_p}{Z_s}} \qquad P_p = P_s$$

$$\% \text{ eff.} = \frac{P_{out}}{P_{in}} \times 100$$

EFFICIENCY (FOR ANY DEVICE)

$$\text{Eff.} = \frac{\text{output}}{\text{input}}$$

DECIBEL FORMULAS

When impedances are equal,

$$dB = 10 \log\frac{P_1}{P_2} = 20 \log\frac{V_1}{V_2} = 20 \log\frac{I_1}{I_2}$$

When impedances are unequal,

$$dB = 10 \log\frac{P_1}{P_2} = 20 \log\frac{V_1 \sqrt{Z_2}}{V_2 \sqrt{Z_1}} = 20 \log\frac{I_1 \sqrt{Z_1}}{I_2 \sqrt{Z_2}}$$

$$dB = 10 \log\frac{P_1}{P_2} = 20 \log\frac{V_1}{V_2} + 10 \log\frac{Z_2}{Z_1}$$

$$= 20 \log\frac{I_1}{I_2} + 10 \log\frac{Z_1}{Z_2}$$

DECIBEL TABLE

dB	Power Ratio	Voltage or Current Ratio	dB	Power Ratio	Voltage or Current Ratio
0	1.00	1.00	10	10.0	3.2
0.5	1.12	1.06	15	31.6	5.6
1.0	1.26	1.12	20	100	10
1.5	1.41	1.19	25	316	18
2.0	1.58	1.26	30	1,000	32
3.0	2.00	1.41	40	10,000	100
4.0	2.51	1.58	50	10^5	316
5.0	3.16	1.78	60	10^6	1,000
6.0	3.98	2.00	70	10^7	3,162
7.0	5.01	2.24	80	10^8	10,000
8.0	6.31	2.51	90	10^9	31,620
9.0	7.94	2.82	100	10^{10}	10

FREQUENCY AND WAVELENGTH

$$f \text{ kHz} = \frac{3 \times 10^5}{\lambda \text{ meter}} \qquad \lambda \text{ meter} = \frac{3 \times 10^5}{f \text{ kHz}}$$

$$f \text{ MHz} = \frac{3 \times 10^4}{\lambda \text{ centimeter}} \qquad \lambda \text{ cm} = \frac{3 \times 10^4}{f \text{ MHz}}$$

$$f \text{ MHz} = \frac{984}{\lambda \text{ feet}} \qquad \lambda \text{ feet} = \frac{984}{f \text{ MHz}}$$

$$\lambda \text{ meter} = \frac{3 \times 10^8}{f \text{ Hz}}$$

$$\lambda \text{ cm} = \frac{3 \times 10^{10}}{f \text{ Hz}}$$

SERIES CIRCUITS

$$V_t = V_1 + V_2 + V_3 + V_n$$
$$I_t = I_1 = I_2 = I_3 = I_n$$
$$R_t = R_1 + R_2 + R_3 + R_n$$
$$P_t = P_1 + P_2 + P_3 + P_n$$
$$L_t = L_1 + L_2 + L_3 + L_n$$
$$C_t = \frac{C_1 \times C_2}{C_1 + C_2} \quad \text{(two capacitors)}$$
$$C_t = \frac{1}{\dfrac{1}{C_1} + \dfrac{1}{C_2} + \dfrac{1}{C_3} + \dfrac{1}{C_n}} \quad \text{(two or more capacitors)}$$
$$C_n = \frac{C}{N} \quad \text{(two or more capacitors of equal value)}$$

TIME CONSTANT

$$\text{TC} = RC$$
$$\text{TC} = \frac{L}{R}$$

POWER

$$P_t = I_t^2 \times R_t$$
$$P_t = I_t \times V_a$$
$$P_t = IV \times \text{Cos} \angle \theta$$
$$P_a = I_tV_t = I_t^2Z_t$$
$$P_X = I^2X_L$$
$$P_X = I^2X_C$$
$$\text{PF} = \frac{P_t}{P_a} = \text{Cos} \angle \theta$$

BANDWIDTH

$$\text{BW} = \frac{F_0}{Q}$$
$$\text{BW} = f_2 - f_1$$

TRIGONOMETRIC FORMULAS

$$\sin A = \frac{a}{c} = \frac{\text{OPPOSITE SIDE}}{\text{HYPOTENUSE}}$$
$$\cos A = \frac{b}{c} = \frac{\text{ADJACENT SIDE}}{\text{HYPOTENUSE}}$$
$$\tan A = \frac{a}{b} = \frac{\text{OPPOSITE SIDE}}{\text{ADJACENT SIDE}}$$
$$\operatorname{cosec} A = \frac{c}{a} = \frac{1}{\sin A}$$
$$\sec A = \frac{c}{b} = \frac{1}{\cos A}$$
$$\cot A = \frac{b}{a} = \frac{1}{\tan A}$$
$$c^2 = a^2 + b^2$$

PARALLEL CIRCUITS

$$V_t = V_{\text{branch 1}} = V_{\text{branch 2}} = V_n$$
$$R_t = \frac{R_1 \times R_2}{R_1 + R_2} \quad \text{(two resistors)}$$
$$R_t = \frac{1}{\dfrac{1}{R_1} + \dfrac{1}{R_2} + \dfrac{1}{R_3} + \dfrac{1}{R_n}} \quad \text{(two or more resistors)}$$
$$R_t = \frac{R}{N} \quad \text{(two or more resistors of equal value)}$$
$$I_t = I_1 + I_2 + I_3 + I_n$$
$$P_t = P_1 + P_2 + P_3 + P_n$$
$$L_t = \frac{L_1 \times L_2}{L_1 + L_2} \quad \text{(two inductors)}$$
$$L_t = \frac{1}{\dfrac{1}{L_1} + \dfrac{1}{L_2} + \dfrac{1}{L_3} + \dfrac{1}{L_n}} \quad \text{(two or more inductors)}$$
$$L_t = \frac{L}{N} \quad \text{(two or more inductors of equal value)}$$
$$C_t = C_1 + C_2 + C_3 + C_n$$

TIME DURATION OF ONE CYCLE

100 kHz —	10 μsec
200 kHz —	5 μsec
250 kHz —	4 μsec
1 MHz—	1 μsec
4 MHz—	0.25 μsec
10 MHz—	0.1 μsec

$$f = \frac{1}{t}$$
$$t = \frac{1}{f}$$

APPENDIX G   Codes

NUMERATION

Arabic	Roman	Greek	Binary Numbers	Arabic	Roman	Greek	Binary Numbers
1	I	α'	1	50	L	ν'	110010
2	II	β'	10	60	LX	ξ	111100
3	III	γ'	11	70	LXX	ο'	1000110
4	IV	δ'	100	80	LXXX	π'	1010000
5	V	ε'	101	90	XC	ϙ'	1011010
6	VI	ϛ'	110	99	XCIX	ϙθ'	1100011
7	VII	ζ'	111	100	C	ρ'	1100100
8	VIII	η'	1000	111	CXI	ρια'	1101111
9	IX	θ'	1001	150	CL	ρν'	10010110
10	X	ι'	1010	200	CC	σ'	11001000
11	XI	ια'	1011	300	CCC	τ'	100101100
12	XII	ιβ'	1100	400	CD	υ'	110010000
13	XIII	ιγ'	1101	500	D	φ'	111110100
14	XIV	ιδ'	1110	600	DC	χ'	1001011000
15	XV	ιε'	1111	700	DCC	ψ'	1010111100
16	XVI	ιϛ'	10000	800	DCCC	ω'	1100100000
17	XVII	ιζ	10001	900	CM	ϡ'	1110000100
18	XVIII	ιη'	10010	1,000	M	‚α	1111101000
19	XIX	ιθ'	10011	1,111	MCXI	‚αρια'	10001010111
20	XX	κ'	10100	1,500	MD	‚αφ'	10111011100
21	XXI	κα'	10101	2,000	MM	‚β	11111010000
22	XXII	κβ'	10110	3,000	MM̲M	‚γ	101110111000
23	XXIII	κγ'	10111	4,000	M̲V	‚δ	111110100000
24	XXIV	κδ'	11000	5,000	V̅	‚ε	1001110001000
25	XXV	κε'	11001	6,000	V̅M	‚ϛ	1011101110000
29	XXIX	κθ'	11101	10,000	X̅	‚ι	10011100010000
30	XXX	λ'	11110	20,000	X̅X	‚κ	100111000100000
40	XL	μ'	101000	100,000	C̅	‚ρ	11000011010100000

Decimal Number System			Binary Number System						
100 Place	10 Place	1 Place	64 Place	32 Place	16 Place	8 Place	4 Place	2 Place	1 Place
		0	0	0	0	0	0	0	0
		1	0	0	0	0	0	0	1
		2	0	0	0	0	0	1	0
		3	0	0	0	0	0	1	1
		4	0	0	0	0	1	0	0
		5	0	0	0	0	1	0	1
		6	0	0	0	0	1	1	0
		7	0	0	0	0	1	1	1
		8	0	0	0	1	0	0	0
		9	0	0	0	1	0	0	1
	1	0	0	0	0	1	0	1	0
	1	1	0	0	0	1	0	1	1
	1	2	0	0	0	1	1	0	0
	1	3	0	0	0	1	1	0	1
	1	4	0	0	0	1	1	1	0
	1	5	0	0	0	1	1	1	1
	1	6	0	0	1	0	0	0	0
	1	7	0	0	1	0	0	0	1
	1	8	0	0	1	0	0	1	0
	1	9	0	0	1	0	0	1	1
	2	0	0	0	1	0	1	0	0
	2	1	0	0	1	0	1	0	1
	2	2	0	0	1	0	1	1	0
	2	3	0	0	1	0	1	1	1
	2	4	0	0	1	1	0	0	0
	2	5	0	0	1	1	0	0	1
	2	6	0	0	1	1	0	1	0
	2	7	0	0	1	1	0	1	1
	2	8	0	0	1	1	1	0	0
	2	9	0	0	1	1	1	0	1
	3	0	0	0	1	1	1	1	0
	3	1	0	0	1	1	1	1	1
	3	2	0	1	0	0	0	0	0
	3	3	0	1	0	0	0	0	1
	3	4	0	1	0	0	0	1	0
	3	5	0	1	0	0	0	1	1
	3	6	0	1	0	0	1	0	0
	3	7	0	1	0	0	1	0	1
	3	8	0	1	0	0	1	1	0
	3	9	0	1	0	0	1	1	1
	4	0	0	1	0	1	0	0	0
	4	1	0	1	0	1	0	0	1
	4	2	0	1	0	1	0	1	0
	4	3	0	1	0	1	0	1	1
	4	4	0	1	0	1	1	0	0
	4	5	0	1	0	1	1	0	1
	4	6	0	1	0	1	1	1	0
	4	7	0	1	0	1	1	1	1
	4	8	0	1	1	0	0	0	0
	4	9	0	1	1	0	0	0	1
	5	0	0	1	1	0	0	1	0
	5	1	0	1	1	0	0	1	1
	5	2	0	1	1	0	1	0	0
	5	3	0	1	1	0	1	0	1
	5	4	0	1	1	0	1	1	0
	5	5	0	1	1	0	1	1	1
	5	6	0	1	1	1	0	0	0
	5	7	0	1	1	1	0	0	1
	5	8	0	1	1	1	0	1	0
	5	9	0	1	1	1	0	1	1
	6	0	0	1	1	1	1	0	0
	6	1	0	1	1	1	1	0	1
	6	2	0	1	1	1	1	1	0
	6	3	0	1	1	1	1	1	1

Decimal Number System			Binary Number System						
100 Place	10 Place	1 Place	64 Place	32 Place	16 Place	8 Place	4 Place	2 Place	1 Place
	6	4	1	0	0	0	0	0	0
	6	5	1	0	0	0	0	0	1
	6	6	1	0	0	0	0	1	0
	6	7	1	0	0	0	0	1	1
	6	8	1	0	0	0	1	0	0
	6	9	1	0	0	0	1	0	1
	7	0	1	0	0	0	1	1	0
	7	1	1	0	0	0	1	1	1
	7	2	1	0	0	1	0	0	0
	7	3	1	0	0	1	0	0	1
	7	4	1	0	0	1	0	1	0
	7	5	1	0	0	1	0	1	1
	7	6	1	0	0	1	1	0	0
	7	7	1	0	0	1	1	0	1
	7	8	1	0	0	1	1	1	0
	7	9	1	0	0	1	1	1	1
	8	0	1	0	1	0	0	0	0
	8	1	1	0	1	0	0	0	1
	8	2	1	0	1	0	0	1	0
	8	3	1	0	1	0	0	1	1
	8	4	1	0	1	0	1	0	0
	8	5	1	0	1	0	1	0	1
	8	6	1	0	1	0	1	1	0
	8	7	1	0	1	0	1	1	1
	8	8	1	0	1	1	0	0	0
	8	9	1	0	1	1	0	0	1
	9	0	1	0	1	1	0	1	0
	9	1	1	0	1	1	0	1	1
	9	2	1	0	1	1	1	0	0
	9	3	1	0	1	1	1	0	1
	9	4	1	0	1	1	1	1	0
	9	5	1	0	1	1	1	1	1
	9	6	1	1	0	0	0	0	0
	9	7	1	1	0	0	0	0	1
	9	8	1	1	0	0	0	1	0
	9	9	1	1	0	0	0	1	1
1	0	0	1	1	0	0	1	0	0
1	0	1	1	1	0	0	1	0	1
1	0	2	1	1	0	0	1	1	0
1	0	3	1	1	0	0	1	1	1
1	0	4	1	1	0	1	0	0	0
1	0	5	1	1	0	1	0	0	1
1	0	6	1	1	0	1	0	1	0
1	0	7	1	1	0	1	0	1	1
1	0	8	1	1	0	1	1	0	0
1	0	9	1	1	0	1	1	0	1
1	1	0	1	1	0	1	1	1	0
1	1	1	1	1	0	1	1	1	1
1	1	2	1	1	1	0	0	0	0
1	1	3	1	1	1	0	0	0	1
1	1	4	1	1	1	0	0	1	0
1	1	5	1	1	1	0	0	1	1
1	1	6	1	1	1	0	1	0	0
1	1	7	1	1	1	0	1	0	1
1	1	8	1	1	1	0	1	1	0
1	1	9	1	1	1	0	1	1	1
1	2	0	1	1	1	1	0	0	0
1	2	1	1	1	1	1	0	0	1
1	2	2	1	1	1	1	0	1	0
1	2	3	1	1	1	1	0	1	1
1	2	4	1	1	1	1	1	0	0
1	2	5	1	1	1	1	1	0	1
1	2	6	1	1	1	1	1	1	0
1	2	7	1	1	1	1	1	1	1

The American Standard Code for Information Interchange (ASCII)

Char	b7	b6	b5	b4	b3	b2	b1
•	0	1	0	0	0	0	0
!	0	1	0	0	0	0	1
"	0	1	0	0	0	1	0
#	0	1	0	0	0	1	1
$	0	1	0	0	1	0	0
%	0	1	0	0	1	0	1
&	0	1	0	0	1	1	0
'	0	1	0	0	1	1	1
(	0	1	0	1	0	0	0
)	0	1	0	1	0	0	1
*	0	1	0	1	0	1	0
+	0	1	0	1	0	1	1
,	0	1	0	1	1	0	0
−	0	1	0	1	1	0	1
.	0	1	0	1	1	1	0
/	0	1	0	1	1	1	1
0	0	1	1	0	0	0	0
1	0	1	1	0	0	0	1
2	0	1	1	0	0	1	0
3	0	1	1	0	0	1	1
4	0	1	1	0	1	0	0
5	0	1	1	0	1	0	1
6	0	1	1	0	1	1	0
7	0	1	1	0	1	1	1
8	0	1	1	1	0	0	0
9	0	1	1	1	0	0	1
:	0	1	1	1	0	1	0
;	0	1	1	1	0	1	1
<	0	1	1	1	1	0	0
=	0	1	1	1	1	0	1
>	0	1	1	1	1	1	0
?	0	1	1	1	1	1	1
@	1	0	0	0	0	0	0
A	1	0	0	0	0	0	1
B	1	0	0	0	0	1	0
C	1	0	0	0	0	1	1
D	1	0	0	0	1	0	0
E	1	0	0	0	1	0	1
F	1	0	0	0	1	1	0
G	1	0	0	0	1	1	1
H	1	0	0	1	0	0	0
I	1	0	0	1	0	0	1
J	1	0	0	1	0	1	0
K	1	0	0	1	0	1	1
L	1	0	0	1	1	0	0
M	1	0	0	1	1	0	1
N	1	0	0	1	1	1	0
O	1	0	0	1	1	1	1

Char	b7	b6	b5	b4	b3	b2	b1
P	1	0	1	0	0	0	0
Q	1	0	1	0	0	0	1
R	1	0	1	0	0	1	0
S	1	0	1	0	0	1	1
T	1	0	1	0	1	0	0
U	1	0	1	0	1	0	1
V	1	0	1	0	1	1	0
W	1	0	1	0	1	1	1
X	1	0	1	1	0	0	0
Y	1	0	1	1	0	0	1
Z	1	0	1	1	0	1	0
[	1	0	1	1	0	1	1
\	1	0	1	1	1	0	0
]	1	0	1	1	1	0	1
∧	1	0	1	1	1	1	0
—	1	0	1	1	1	1	1
'	1	1	0	0	0	0	0
a	1	1	0	0	0	0	1
b	1	1	0	0	0	1	0
c	1	1	0	0	0	1	1
d	1	1	0	0	1	0	0
e	1	1	0	0	1	0	1
f	1	1	0	0	1	1	0
g	1	1	0	0	1	1	1
h	1	1	0	1	0	0	0
i	1	1	0	1	0	0	1
j	1	1	0	1	0	1	0
k	1	1	0	1	0	1	1
l	1	1	0	1	1	0	0
m	1	1	0	1	1	0	1
n	1	1	0	1	1	1	0
o	1	1	0	1	1	1	1
p	1	1	1	0	0	0	0
q	1	1	1	0	0	0	1
r	1	1	1	0	0	1	0
s	1	1	1	0	0	1	1
t	1	1	1	0	1	0	0
u	1	1	1	0	1	0	1
v	1	1	1	0	1	1	0
w	1	1	1	0	1	1	1
x	1	1	1	1	0	0	0
y	1	1	1	1	0	0	1
z	1	1	1	1	0	1	0
{	1	1	1	1	0	1	1
|	1	1	1	1	1	0	0
}	1	1	1	1	1	0	1
~	1	1	1	1	1	1	0
DEL	1	1	1	1	1	1	1

INTERNATIONAL MORSE CODE

A · −	N − ·	Á · − − · −	8 − − − · ·		
B − · · ·	O − − −	Ä · − · −	9 − − − − ·		
C − · − ·	P · − − ·	É · · − · ·	0 − − − − −		
D − · ·	Q − − · −	Ñ − − · − −	, (comma) − − · · − −		
E ·	R · − ·	Ö − − − ·	. · − · − · −		
F · · − ·	S · · ·	Ü · · − −	? · · − − · ·		
G − − ·	T −	1 · − − − −	; − · − · − ·		
H · · · ·	U · · −	2 · · − − −	: − − − · · ·		
I · ·	V · · · −	3 · · · − −	' (apostrophe) · − − − − ·		
J · − − −	W · − −	4 · · · · −	- (hyphen) − · · · · −		
K − · −	X − · · −	5 · · · · ·	/ − · · − ·		
L · − · ·	Y − · − −	6 − · · · ·	parenthesis − · − − · −		
M − −	Z − − · ·	7 − − · · ·	underline · · − − · −		

728

APPENDIX H Equipment Troubleshooting Procedure

The first step in troubleshooting is to determine what function does not operate or what that function is doing wrong. Then, by carrying out a few tests in the area that is responsible for that function, you can isolate the problem to a specific printed circuit board or to a particular circuit on that board. The following is a six-step equipment troubleshooting procedure designed to logically and systematically isolate a problem.

Name:	Date:	
Equipment Description:	Start Time:	Stop time:

Step 1: Symptom Recognition

From the presentation and all other visual indications, list what functions do not operate or what these functions are doing wrong.

_____

_____

_____

_____

> **SYMPTOM RECOGNITION**
> Tools: Sight, sounds, smell, controls, knowledge.

Step 2: List Probable Faulty Units

Referring to the overall equipment block diagram, list the probable block/board that could cause the malfunction listed in step 1.

_____

_____

_____

_____

> **LIST PROBABLE FAULTY UNITS**
> Tools: Block interconnect diagram, thought controls, notes.

Step 3: Localizing Faulty Block/Board

In the following table, list the test point and data utilized to isolate the malfunctioning block.

Test Point	Normal Reading	Actual Reading

> **LOCALIZING FAULTY BLOCK/UNIT**
> Tools: Sight, sound, smell, test equip., knowledge, thought, block and circuit diagrams.

Step 4: Localizing Faulty Circuit in PCB

In the following table, list the test points and data utilized to isolate the faulty circuit within the printed circuit board.

Test Point	Normal Reading	Actual Reading

LOCALIZING FAULTY CIRCUIT IN PCB Tools: Sight, sound, smell, test equip., knowledge, thought, circuit diagrams, theory of operation.

Step 5: Localizing Faulty Component

In the following table, list the test points and data utilized to isolate the faulty component within the circuit.

Test Point	Normal Reading	Actual Reading

LOCALIZING FAULTY COMPONENT Tools: Sight, sound, smell, test equip., knowledge, thought, circuit diagrams, theory of operation.

FAULTY COMPONENT IS:

Step 6: Analysis of Failure

Explain why this component caused the symptoms listed in step 1 and how the problem was resolved.

_____

_____

_____

_____

APPENDIX I Protoboards

The solderless prototyping board (protoboard) or breadboard is designed to accommodate the many experiments described in this text's associated lab manual. This protoboard will hold and interconnect resistors, capacitors, inductors, and many other components, as well as provide electrical power. Figure I-1 shows an experimental circuit wired up on a protoboard.

Figure I-2 shows the top view of a basic protoboard. The individual sections snap together to form as large an area as needed. The two main

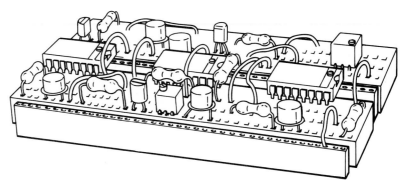

FIGURE I-1 Experimenting with the Protoboard.

FIGURE I-2 Internal Construction of Solderless Protoboard.

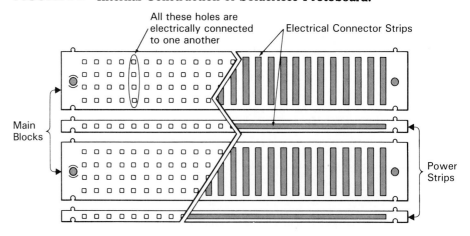

All these holes are electrically connected to one another

Electrical Connector Strips

Main Blocks

Power Strips

blocks in this figure have, as you can see by the cross section on the right side, electrical connector strips running vertically. These conductive strips make a connection between the four holes, and so a component lead could be inserted into one of the holes. Then a hookup wire can come out of one of the other vertical holes and the two will be electrically connected. The narrow blocks have an electrical connector strip running from end to end, as seen in the cross section. They are usually connected to a power supply, for example, positive voltage on one and ground on the other, so that these voltages can be tapped off to supply power to the circuit at any point along the strip.

Other protoboards may vary slightly as far as layout, but you should be able to determine the pattern of conductive strips by making a few checks with an ohmmeter.

Problem solving is an integral part of your electronics study, for it is only through the solutions of problems that decisions are tested and verified.

J.1
CALCULATOR SELECTION

It is very important that your "full-function" scientific or engineering calculator have the capabilities necessary for your particular application. For this reason, Table J-1 lists all the functions that your calculator should have, with some blank columns so that you can check off these functions against several different types.

Caculators use a keypad to enter the data to be processed and a display to show the result. There are two systems of data entry:

1. *Algebraic entry system* (AES): Here the basic arithmetic operations ($+$, $-$, $\div$, $\times$, and so on) are placed between the two numbers; for example, $8 + 4$.
2. *Reverse Polish notation* (RPN): This system is based on the work of Jan Lukasiewicz, a Polish mathematician. With this system, the previous example would be entered as 8 enter, 4 enter, and then $+$.

The programmable calculator is another feature you may want to consider, and once this tool is mastered you will find it invaluable in applications where you have to repeatedly solve a function for many different values.

Before you purchase, talk to your instructor and other students, and shop around to find out what works best for you.

TABLE J-1

Keypad Symbols	Function	Description
+	Add	Simple arithmetic operations
−	Subtract	
×	Multiply	
÷	Divide	
CHS or +/−	Change sign	
y^x or x^y	Exponential	Raise number to power
$1/x$	Reciprocal	Calculate the reciprocal
$\sqrt{x}$	Square root	Calculate square root
x^2	Square x	Square a number
FIX	Fix point notation	Display and rounding
SCI	Scientific notation	
ENG	Engineering notation	
STO	Store	Memory
RCL	Recall	
x▸y or EXC	Exchange x and y	
EE or EEX	Enter exponent	
log	Common logarithm	Logarithmic and exponential functions
10^x or INV LOG	Common antilog	
ln x	Natural logarithm	
e^x or INV LN	Natural antilogarithm	
SIN	Sine	Trigonometric functions
COS	Cosine	
TAN	Tangent	
SIN⁻¹ or INV SIN	Arc sine	Inverse trigonometric functions
COS⁻¹ or INV COS	Arc cosine	
TAN⁻¹ or INV TAN	Arc tangent	
DEG or DRG	Degree	Angular mode selection
RAD	Radian	
→P	Rectangular to polar	Polar/rectangular coordinate conversion
→R	Polar to rectangular	

Decibels (dB) were originally based on the response of the human ear to sound waves. The decibel is a measure of the smallest "volume change" an ear can detect; this is illustrated in Figure K-1. Referring to the graph, you can see that a sound level of 1 mW can hardly be heard (a whisper) but is used as a reference of 0 dB. The power has to be increased to 1.3 mW before the ear can detect that a change has occurred in the sound level. This point is labeled 1 dB since it is the smallest detectable change.

   If the sound level is further increased, the next change in power level is detected at 1.6 mW, and this is labeled the 2 dB point. As the power level

FIGURE K-1

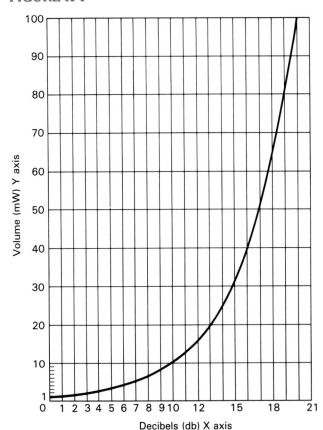

Whisper = 0 dB
Conversation = 45 dB
Industrial Equipment = 90 dB
Very Loud Music = 110 dB
Threshold of Pain = 130 dB

Decibels X	Y Volume
0	1.0
1	1.3
2	1.6
3	2.0
4	2.5
5	3.2
6	4.0
7	5.0
8	6.3
9	7.9
10	10.0
11	12.5
12	15.8
13	20.0
14	25.1
15	31.6
16	39.8
17	50.1
18	63.1
19	79.4
20	100.0

is further increased and the points plotted, you can see that, to notice a change in sound level, 1 dB, requires more and more power. For example, a power increase from 10 to 20 mW is a 3-dB change (10 to 13 dB), while a change from 50 to 100 mW is also detected as a 3-dB change.

K.1

WATTS TO DECIBELS

Decibels are therefore a measure of change. To calculate the amount of change, two power levels need to be known: a starting power, which will

TABLE K-1
COMMON POWER RATIOS

Gain (+dB)	Loss (−dB)
+3 dB = ×2	−3 dB = ÷2 or ½
+6 dB = ×4	−6 dB = ÷4 or ¼
+9 dB = ×8	−9 dB = ÷8
+12 dB = ×16	−12 dB = ÷16
+15 dB = ×32	−15 dB = ÷32
+10 dB = ×10	−10 dB = ÷10
+20 dB = ×100	−20 dB = ÷100
+30 dB = ×1000	−30 dB = ÷1,000
+40 dB = ×10,000	−40 dB = ÷10,000
+50 dB = ×100,000	−50 dB = ÷100,000

be called P_{in}, and an ending power, which will be called P_{out}. The following formula can be used to calculate the amount of change.

$$dB = 10 \times \log \frac{P_{out}}{P_{in}}$$

The final decibel value will be an indication of the amount of change between input and output, not a measure of the actual output power. The value could come out as a +dB, indicating a gain in power, or a −dB figure, signifying a loss in power between input and output.

Table K-1 lists the common power ratios.

EXAMPLE K.1

Calculate the power gain of an amplifier that has an input of 10 mW and an output of 1 watt.

SOLUTION

$$dB = 10 \times \log \frac{P_{out}}{P_{in}}$$

$$= 10 \times \log \frac{1000 \text{ mW}}{10 \text{ mW}} = +20 \text{ dB}$$

An amplifier gain of 20 dB does not mean that the output is 20 times greater than the input; we already know that the output is 100 times greater than the input (10 mW to 1000 mW). It simply means that a 20-dB gain has occurred between input and output, and 20 dB equates to ×100.

EXAMPLE K.2

A cable has an input of 1.5 kW and an output of 1.2 kW. Calculate the cable loss.

SOLUTION

$$dB = 10 \times \log \frac{P_{out}}{P_{in}}$$

$$= 10 \times \log \frac{1.2 \text{ kW}}{1.5 \text{ kW}}$$

$$= -1 \text{ dB}$$

In this example, the cable has a negative gain or loss of −1 dB between input and output.

Up to now, two power levels have had to be known in order to calculate the amount of gain or loss. The power at a point, however, can be expressed in decibels with respect to a reference. Telephone companies have established a standard of 1 mW as a reference power, and all other power levels are

---

EXAMPLE K.3

Calculate the gain of the following radio receiver.

$$\text{Antenna gain} = +7 \text{ dB}$$
$$\text{Coaxial cable} = -2 \text{ dB}$$
$$\text{Preamplifier} = +45 \text{ dB}$$
$$\text{Amplifier} = +15 \text{ dB}$$
$$\text{Speaker} = -5 \text{ dB}$$

SOLUTION

Total gain in decibels is +60 dB or a gain of 1,000,000 between the antenna input and the speaker output.

---

expressed with respect to the reference. Power levels below 1 mW are indicated as $-$dBm, and positive power levels above 1 mW are written as $+$dBm.

K.2

DECIBELS TO WATTS

By transposition of the original power formula, we can obtain the following:

$$dB = 10 \log \frac{P_{out}}{P_{in}}$$

$$P_{out} = P_{in} \text{ antilog } \frac{dB}{10}$$

$$P_{in} = \frac{P_{out}}{\text{antilog } (dB/10)}$$

EXAMPLE K.4

Calculate the output power of a 35-dB amplifier if the input is 4 mW.

SOLUTION

$$P_{out} = P_{in} \text{ antilog } \frac{dB}{10}$$
$$= 4 \text{ mW} \times \text{ antilog } \frac{35}{10} = 12.65 \text{ watts}$$

EXAMPLE K.5

Calculate the input power needed for a 30-dB amplifier to produce a 2-watt output.

SOLUTION

$$P_{in} = \frac{P_{out}}{\text{antilog } (dB/10)}$$
$$= \frac{2W}{\text{antilog } (30/10)} = 2 \text{ mW}$$

K.3

VOLTAGE OR CURRENT GAIN IN DECIBELS

A voltage or current gain or loss can also be expressed in decibels. The equation for voltage can be derived by substituting P_{out} and P_{in} in the power formula with V^2/R to obtain:

$$dB = 20 \log \frac{V_{out}}{V_{in}}$$

Similarly, a current formula can be derived:

$$dB = 20 \log \frac{I_{out}}{I_{in}}$$

Table K-2 makes a comparison between the common decibel values for power and voltage ratios. Looking at this table you will notice that

$$dB = 10 \log \left(\frac{P_{out}}{P_{in}} \right)$$

$$dB = 20 \log \frac{I_{out}}{I_{in}}$$

$$dB = 20 \log \frac{V_{out}}{V_{in}}$$

The corresponding voltage or current decibel values are twice those of the same power ratio. For example, double the power is 3 dB, whereas double the voltage or current is 6 dB. Also, 100 times the power is 20 dB, whereas 100 times the voltage or current is 40 dB. This is because the voltage and current decibel formulas have a multiplying factor of 20 rather than 10.

In summary, the decibel is a measure of change between input and output, and knowing the decibel figure for an amplifier or radio receiver, for example, helps give the user a better idea on the performance of these circuits or devices.

TABLE K-2
COMMON POWER AND VOLTAGE RATIOS

Decibels		Power Ratio	Voltage/Current Ratio
Gain:	40	10,000	100
	20	100	10
	6	4	2
	3	2	1.4
	1	1.2	1.1
Loss:	−1	1/1.2	1/1.1
	−3	1/2	1/1.4
	−6	1/4	1/2
	−20	1/100	1/10
	−40	1/10,000	1/100

APPENDIX L The Frequency Spectrum

The term "frequency" describes the number of alternations occurring in one second. Direct current (DC) is a steady or constant current that does not alternate and is therefore listed as zero cycles per second or 0 Hz. This appendix illustrates in detail the entire range of frequencies from the lowest subaudible frequency to the highest cosmic rays, along with their applications.

The following page is an overall summary of the complete range of frequencies or spectrum, as it is normally called. The subsequent pages cover each band or section of frequencies in a lot more detail and list the different frequency applications.

This and all of the other appendixes will be useful as references throughout your course of electronic study.

OVERVIEW OF FREQUENCY SPECTRUM

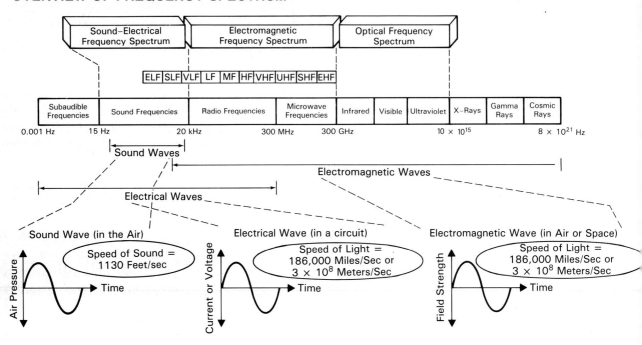

0 Hz	16 →16 kHz	16 → 30 kHz	10 kHz —→ 300 GHz

Direct Current (DC)
  Motors, Relays,
  Supply Voltages
	Audio Frequencies (AC)
	  Motors, Amplifiers,
	  Music Equipment,
	  Speakers, Microphones,
	  Oscillators
		Sound (Ultrasonic)
		  Frequencies
		  Sonar, Music,
		  Speech
			Electromagnetic (Radio)
			  Frequencies
			  Voice Communications
			  Television
			  Navigation
			  Medical, Scientific, and Military

300 GHz→ 4 × 10^14	4 × 10^14 → 7.69 × 10^14	4 × 10^14 → 6 × 10^16	9.375 × 10^15 → 3 × 10^19

Infrared (IR)
  Heating, Photography,
  Sensing, Military
	Visible
	  Color, Photography,
	  Movies, TV
		Ultraviolet (UV)
		  Sterilizing,
		  Medical
			X–Rays
			  Medical, Gauge
			  Thickness, Inspection

3 × 10^19 → 5 × 10^20	5 × 10^20 —→

Gamma Rays
  Deeper Penetrating Than
  X–Rays, Detection of
  Radiation
	Cosmic Rays
	  Present in Outer
	  Space

SOUND SPECTRUM

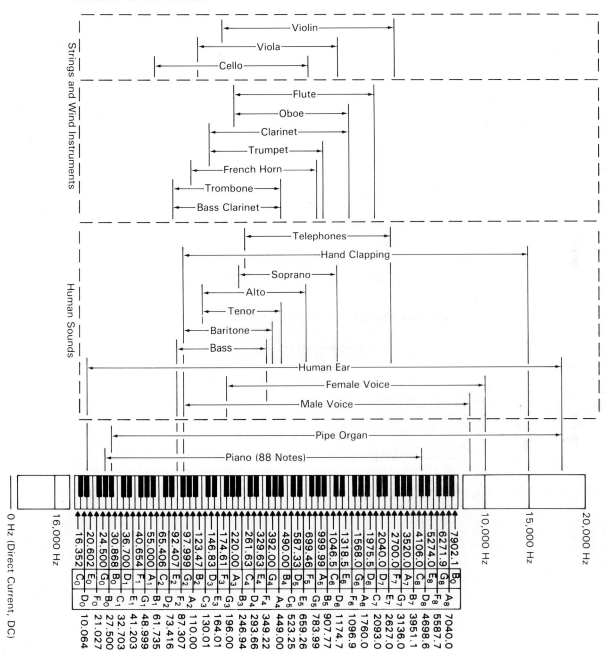

VLF (Very Low-Frequency Band)

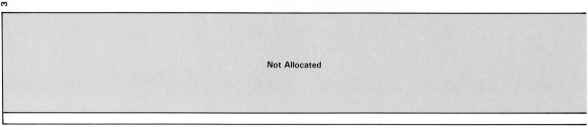

Not Allocated

3 kHz

Not Allocated

10

14

Not Allocated	Omega Long-Range Radionavigation	International Fixed Public
	Radiolocation	Ship to Shore Maritime Mobile

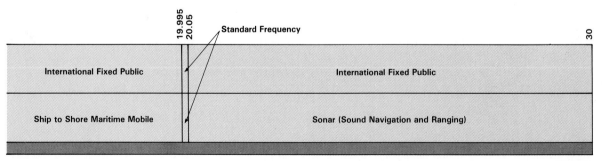

19.995
20.05

Standard Frequency

30

International Fixed Public	International Fixed Public
Ship to Shore Maritime Mobile	Sonar (Sound Navigation and Ranging)

30 kHz

Appendix L The Frequency Spectrum

LF (Low-Frequency Band)

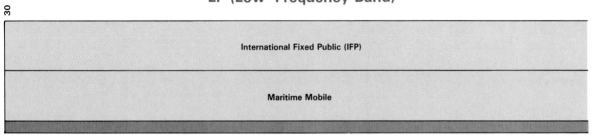

30

International Fixed Public (IFP)
Maritime Mobile

30 kHz

59 **61** **70** **90**

International Fixed Public (IFP)	Standard Frequency	(IFP)	(IFP)	Loran C. Radionavigation
			Decca Maritime Mobile Navigation (British)	
Maritime Mobile		**MARITIME MOBILE**	Radiolocation (Radar)	

110 **130** **160**

Loran C. Radionavigation	(IFP)	(IFP)	(IFP)
	Maritime Mobile	Maritime Mobile (MM)	Maritime Mobile (MM)
	Radar (Radio Detection and Ranging)		

200 **285** **300**

| (IFP) | Aeronautical Mobile Communications | Maritime Radio-navigation |
| Maritime Mobile (MM) | Maritime Radionavigation | Aeronautical Radio-navigation |

300 kHz

MF (Medium-Frequency Band)

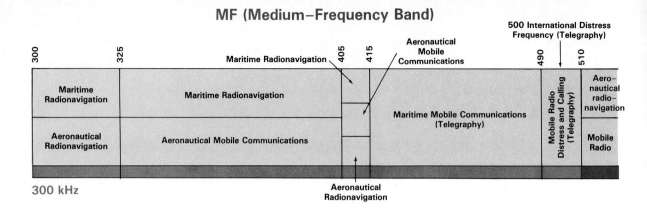

300 kHz

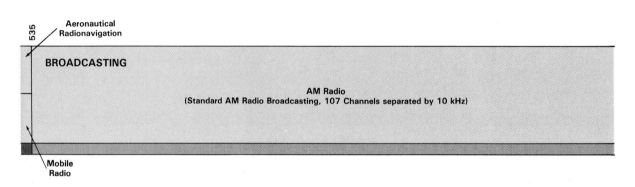

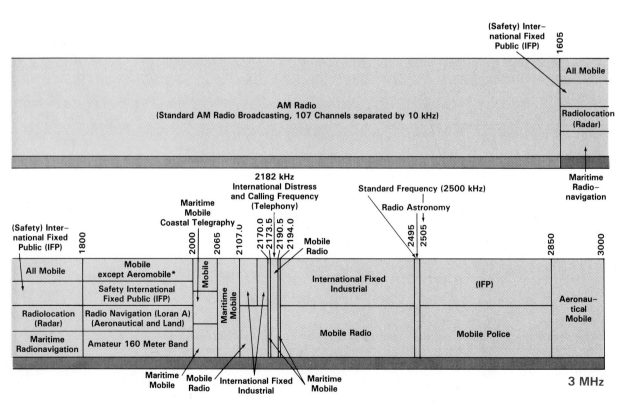

3 MHz

HF (High-Frequency Band)

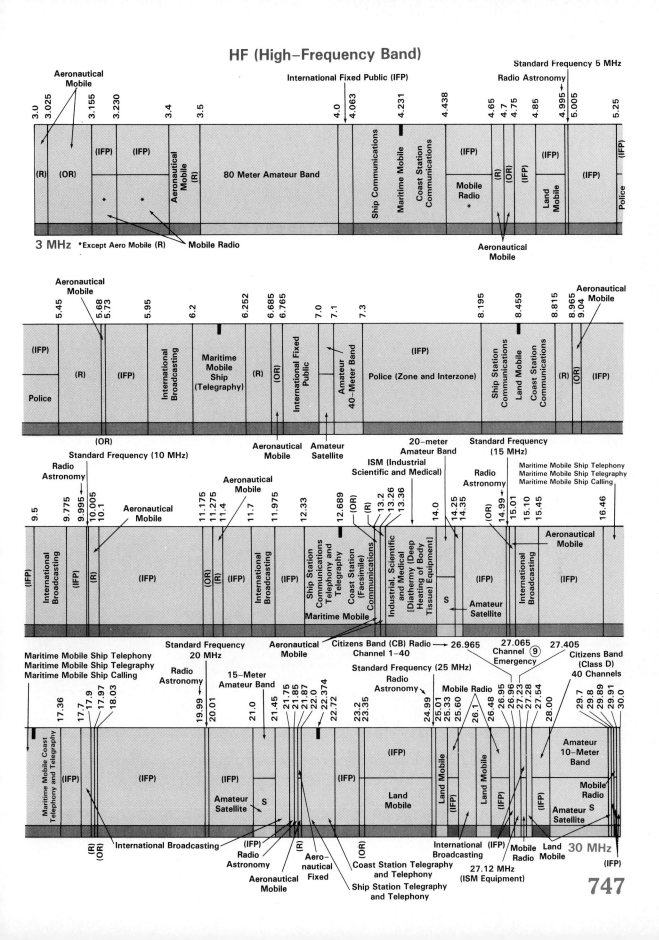

747

VHF (Very High–Frequency Band)

30 MHz

Frequency markers (MHz): 30.0, 30.56, 32.0, 33.0, 34.0, 35.0, 36.0, 37.0, 38.0, 39.0, 40.0, 40.68 ISM, 42.0, 42.95, 46.6, 47.0, (Mobile), 49.6, 50.0

- Mobile Radio (IFP)
- Land Mobile
- Mobile Radio (IFP)
- Land Transportation Public Safety Industrial
- Mobile Radio (IFP)
- Industrial Domestic Public Safety
- Mobile Radio (IFP)
- Public Safety Industrial
- Mobile and Radio Astronomy (IFP)
- Public Safety
- Mobile Radio (IFP)
- Public Safety
- Industrial Domestic Public Land Transportation
- (IFP)
- Industrial Public Safety
- 6-Meter Band Amateur
- (IFP)

Frequency markers (MHz): 54.0, 60.0, 66.0, 72.0, 73.0, 74.6, 75.0, 75.4, 76.0, 82.0, 88.0

International Fixed and Citizen's Band Class C

- 6 Meter Band Amateur
- ② ③ ④ TV Channels 2–4 (VHF)
- (IFP)
- Radio Astronomy
- (IFP)
- ⑤ ⑥ TV Channels 5–6 (VHF)
- FM Radio 100 Channels with 200 kHz Spacing

Frequency markers (MHz): 108.0, 117.975, 121.5 Emergency Frequency, 121.9625, 123.0875, 123.5875, 128.8125, 132.0125, 136.0, 137.0, 138.0, 144.0, 146.0, 148.0, 149.9, 150.05, 150.8, 156.25, 157.0375, 157.1875, 157.45, 161.575, 161.625, 161.775, 162.0125

- Aeronautical Mobile
- Aeronautical Marker Beacons
- Mobile and Space Telecommand
- 2-Meter Amateur Band
- Maritime Mobile
- FM Radio 100 Channels with 200 kHz Spacing
- VOR (VHF Omnirange) Omnidirectional Radio Range
- (R)
- Private Aircraft
- S
- (R)
- (R)
- (R)
- Mobile Radio
- (IFP)
- Space Research
- S
- (IFP)
- Land Transportation Public Safety Industrial Domestic Public
- Land Transportation Public Safety, Industrial, and Domestic Public
- Mobile and Remote Pickup (IFP)
- Flying Schools
- Meteorological Satellite
- Radio Astronomy
- Space Research
- Amateur Satellite
- Satellite Radionavigation
- (IFP)
- Land Mobile
- Maritime Mobile

Frequency markers (MHz): 173.2, 173.4, 174.0, 180.0, 186.0, 192.0, 198.0, 204.0, 210.0, 216.0, 220.0, 225.0, 240.0, 243.0, 300

- Telemetry
- Amateur 1.25-Meter Band
- Aircraft Survival Frequency (Emergency)
- Mobile Radio
- Mobile and Remote Pickup (IFP)
- ⑦ ⑧ ⑨ ⑩ ⑪ ⑫ ⑬ TV Channels 7–13 (VHF)
- (IFP)
- Radio-location
- Amateur
- (IFP)
- Mobile Radio
- (IFP)
- Mobile Radio
- Mobile Satellite
- (IFP)
- Land Mobile
- Mobile Radio
- Radio-location

300 MHz

UHF (Ultra High–Frequency Band)

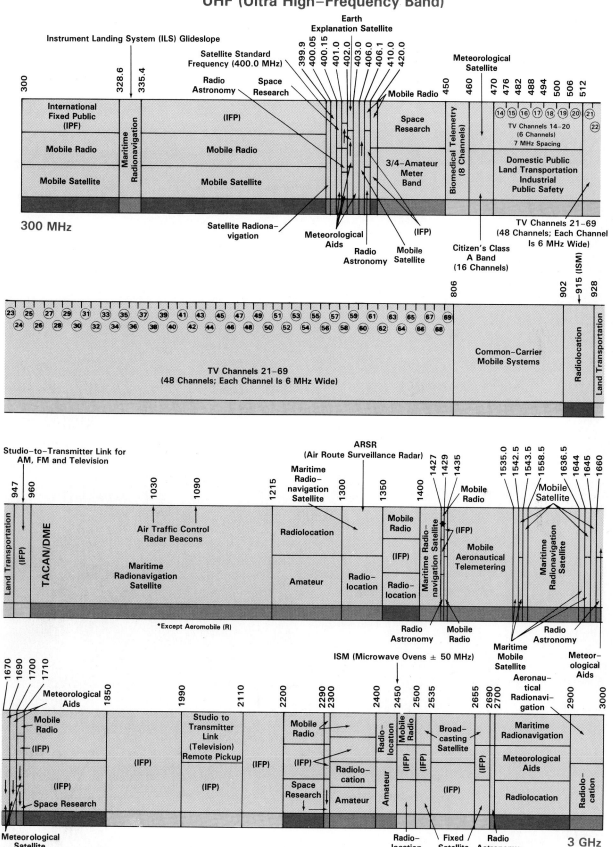

SHF (Super High-Frequency Band)

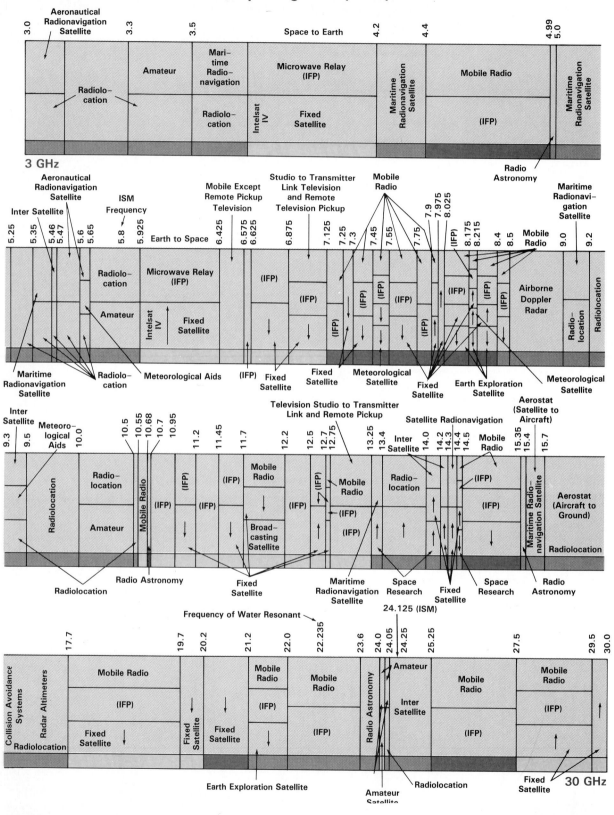

750

EHF (Extremely High–Frequency Band)

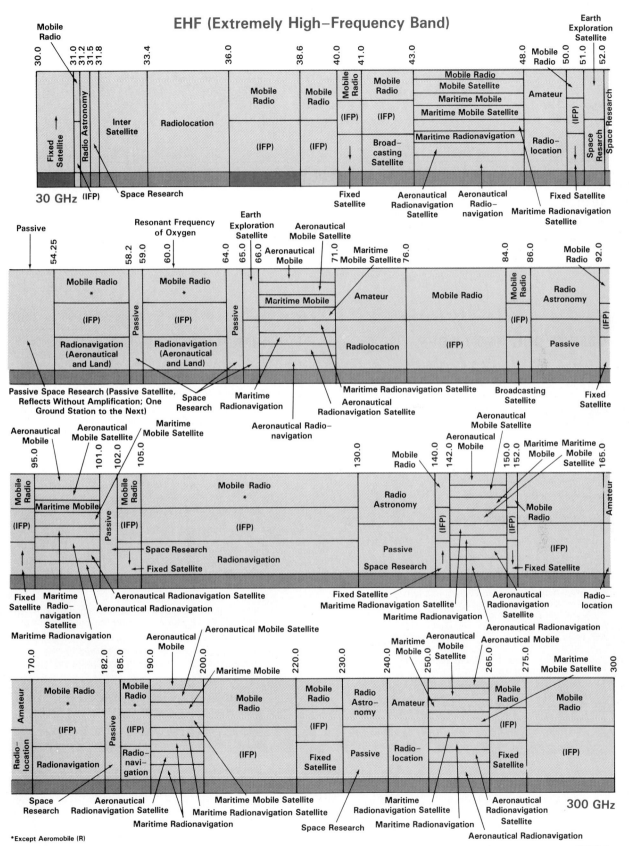

*Except Aeromobile (R)

751

OPTICAL SPECTRUM

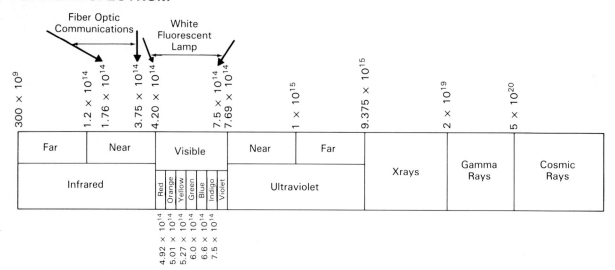

DEFINITIONS AND CODES

USAGE

Exclusively used by government, state, and federal agencies

Shared by government and nongovernment agencies

Nongovernment; publicly used frequencies

TYPES OF EMISSION

Amplitude (AM;Amplitude Modulation)
A0–Steady Unmodulated Pure Carrier
A1–Telegraphy on Pure Continuous Waves
A2–Amplitude Tone–modulated Telegraphy
A3–AM Telephony Including Single and Double Sideband
   with full, Reduced, or Suppressed Carrier
A4–Facsimile (Pictures)
A5–Television

Frequency FM;Frequency Modulation)
F0–Steady Unmodulated Pure Carrier
F1–Carrier–Shift Telegraphy
F2–Audio–frequency–shift Telegraphy
F3–Frequency or Phase–modulated Telephony
F4–Facsimile
F5–Television

Pulse (PM;Pulse Modulation)
P0–Pulsed Radar
P1–Telegraphy: On/Off Keying of Pulsed Carrier
P2–Telegraphy: Pulse Modulation of Pulse Carrier,
   Pulse Width, Phase or Position Tone Modulated
P3–Telephony: Amplitude (PAM), Width (PWM),
   Phase or Position (PPM) Modulated Pulses

ABBREVIATED TERMS

(R), International major air route (air traffic control)
(OR), Off route military (air traffic control)
S Assigned satellite frequency
↑ Earth to space communication (uplink)
↓ Space to earth communication (downlink)

ISM: industrial, scientific, medical
Passive communication equipment:
  generates no electromagnetic radiation (receive only)
Active communication equipment:
  generates electromagnetic radiation (transmits)

APPLICATIONS

- International Fixed: public radiocommunication service. Transmitters designed for one frequency.

- Aeronautical Fixed: a service intended for the transmission of air navigation and preparation for safety of flight.

- Fixed Satellite: satellites that maintain a geostationary orbit (same position above the earth)

- Radiolocation: radio waves used to detect an object's direction, position, or motion.

- Radio Astronomy: radio waves emitted by celestial bodies that are used to obtain data about them.

- Space Operations

- Space Research

APPLICATIONS (continued)

- Mobile: Radio service between a fixed location and one or more mobile stations or between mobile stations.

- Aeronautical Mobile: As above, except mobile stations are aircraft.

- Maritime Mobile: As above, except mobile stations are marine.

- Land Mobile: As above, except mobile stations are automobiles.

- Mobile Satellite

- Aeronautical Mobile Satellite

- Maritime Mobile Satellite

- Radionavigation (Aeronautical and Land): Navigational use of radiolocation equipment such as direction finders, radio compass, radio homing beacons, etc.

- Aeronautical Radionavigation

- Maritime Radionavigation

- Satellite Radionavigation

- Aeronautical Radionavigation Satellite

- Maritime Radionavigation Satellite

- Inter Satellite

APPLICATIONS (continued)

- Broadcasting: Transmission of speech, music, or visual programs for commercial or public service purposes.

- Broadcasting Satellite

- Amateur: a frequency used by persons licensed to operate radio transmitters as a hobby. Person is also called a radio ham.

- Amateur Satellite: Communication by radio hams via satellite.

- Citizen's: a radiocommunications service of fixed, land, and mobile stations intended for short-distance personal or business purposes.

- Standard Frequency: Highly accurate signal broadcasted by the national bureau of standards (NBS) radio station (WWV) to provide frequency, time, solar flare, and other standards.

- Standard Frequency Satellite: NBS broadcast via satellite.

- Meteorological Aids: a radio service in which the emission consists of signals used solely for meteorological use.

- Meteorological Satellite: Meteorological broadcast via satellite.

- Earth Exploration Satellite: Radio frequencies used for earth exploration.

Electronic components such as resistors, capacitors, diodes, transistors, and intergrated circuits are combined to form circuits, which in turn are combined to form electronic systems. These components have their leads interconnected and bonded with solder. Like the expression the "straw that broke the camels back," one sloppy solder connection can cause an entire electronic system to fail.

M.1
SOLDERING TOOLS

Soldering is a skill that can be developed with practice and knowledge. A solder connection is the joining together of two metal parts by applying both heat and solder. The heat provided by the soldering iron is at a high enough temperature to melt the solder, making it a liquid that flows onto and slightly penetrates the two metal surfaces that need to be connected. Once the soldering iron and therefore the heat are removed, the solder cools. In so doing,

FIGURE M-1   **Basic Electronic Tools.**

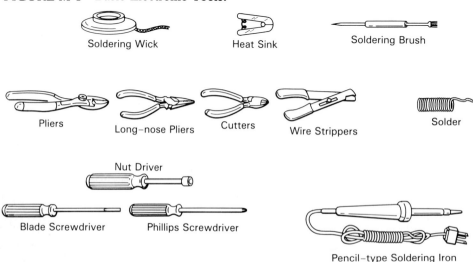

Soldering Wick   Heat Sink   Soldering Brush

Pliers   Long-nose Pliers   Cutters   Wire Strippers   Solder

Nut Driver

Blade Screwdriver   Phillips Screwdriver   Pencil-type Soldering Iron

it solidifies and bonds the two metal surfaces, producing a good *electrical and mechanical connection*. This is the purpose of soldering: to bond two metal surfaces together so that they are both electrically and mechanically connected.

You should at this time have a set of tools and tool box with all the needed components for the experiments listed in this text's associated lab manual. Figure M-1 illustrates the set of basic electronic tools that will be needed for soldering and experimentation. These tools include:

1. Solder wick: Used to remove solder from terminals.
2. Heat sink: During the soldering process, this device is clamped onto the lead of especially heat sensitive components to conduct the heat generated by the soldering iron away from the component.
3. Soldering brush: Used to clean off flux after soldering.
4. Pliers
5. Long-nose pliers: Used for gripping and bending; they are also known as needle-nose pliers.
6. Cutters: Also called dikes; these are available in many different sizes and are used to cut wires or cables.
7. Wire strippers: The strippers are designed to be adjustable or they have a variety of holes for stripping the insulating sleeve off a wire.
8. Solder: 60/40 rosin core solder is most commonly used in electronics.
9. Nut driver
10. Blade screwdriver
11. Phillips screwdriver
12. Soldering iron

M.1.1 WETTING

Everytime you make a good electrical and mechanical connection, the solder will flow, when heated to its melting temperature, over the lead and terminal to be connected, as seen in Figure M-2. The solder, in fact, actually penetrates into the metals, and this embedding of the solder into the metal is called

FIGURE M-2 Wetting.

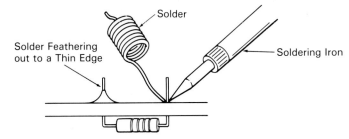

wetting. If the solder feathers out to a thin edge, good wetting is said to have occurred, and it is this that gives the connection its physical strength and electrical connection.

M.1.2 SOLDER AND FLUX

Solder is a mixture of tin and lead, both of which have a low melting temperature with respect to other metals. This is necessary so that the soldering iron melts the solder and not the terminals or leads.

Different proportions of tin and lead are available to produce solder with different characteristics. For example, Figure M-3(a) shows a 60/40 solder, which is a mixture of 60% tin and 40% lead, whereas Figure M-3(b) shows a 40/60 solder that consists of 40% tin and 60% lead. The proportions of tin to lead determine the melting temperature of the solder; for example, 60/40 solder has a lower melting temperature than 40/60 solder. In electronics, 60/40 is most commonly used because of its low melting temperature, which means that a component lead or terminal will not have to be heated to a high temperature in order to make a connection. 63/37 solder has an even lower melting temperature and is therefore even safer than 60/40 solder; however, it is more expensive.

Any lead or terminal is always exposed to the air, which forms an invisible insulating layer on the surface of leads, pins, terminals, and any other surface. A chemical substance is needed to remove this layer; otherwise, the solder would not be able to flow and stick to the metal contacts. Flux removes this invisible insulating oxide layer, and nearly all solder used in electronics contains the flux inside a type of solder tube, as seen in Figure M-4. As the solder is applied to a heated connection, the flux will automatically remove any oxide. The two most common types of flux are rosin and acid. Acid core solder is only used in sheet metal work and should never be used in electronics since it is highly corrosive. In electronics, you should only use rosin core solder, and this is normally indicated as seen in Figure M-3.

A variety of solder diameters is available. The larger-diameter solder is used for terminals and large component leads and connections, whereas the smaller-diameter solder is used for soldering terminals that are very close to one another on a printed circuit board, and so the amount of solder being applied to the connection needs to be carefully controlled.

FIGURE M-3 **Tin/Lead Solder Ratios. (a) 60/40. (b) 40/60.**

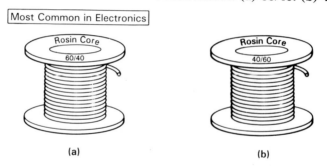

(a)   (b)

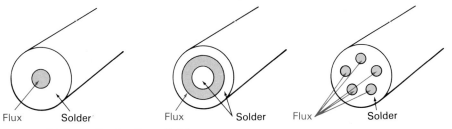

Flux   Solder   Flux   Solder   Flux   Solder

FIGURE M-4   Rosin Core Solder.

M.1.3   SOLDERING IRONS

Figure M-5 illustrates the two basic types of soldering irons. They are rated in terms of wattage, which indicates the amount of power consumed. More importantly, the wattage rating of a soldering iron indicates the amount of heat it produces. When an iron is applied to a connection, heat transfer takes place and heat is drained away from the iron to the metals to be connected; so a larger connection will need a larger-wattage soldering iron. A 25- to 60-W pencil iron is ideal for most electronic work.

For your safety and to protect voltage- and current-sensitive components, you should always use a soldering iron with a three-pin plug and three-wire ac power cord. The third ground wire will ground the iron's exposed metal areas and the tip to prevent electrical shock. This grounded tip will protect delicate MOS integrated circuits from leakage electricity and static charges. On the subject of safety, always turn off the equipment before soldering as the grounded tip may cause a short circuit in the equipment and possible damage.

The tip of a soldering iron can sometimes be changed, and a different temperature tip will allow your iron to be used for different applications. The temperature of the tip selected should be governed by the size of the connection and the temperature sensitivity of the components. In general, a 700°F tip is ideal for most electronic applications; however, for delicate printed circuit boards, use a 600°F tip.

The tip of a soldering iron comes in a variety of shapes and sizes, as seen in Figure M-5(a). The heat produced by the soldering iron heats the connection and melts the solder so that it flows over the connection. A tip shape should be chosen that will conduct the heat to the connection as quickly as possible, so as not to damage the component or PCB terminal. The tip shape that makes the best contact with the connection, and will therefore conduct the most heat, should be used.

As you regularly use your soldering iron, the tip will naturally accumulate dirt and oxide, and this contamination will reduce its effectiveness. The tip should be regularly cleaned by wiping it across a damp sponge, as seen in Figure M-6(a). Soldering iron tips are generally made of copper, plated with either iron or nickel, and so you should never clean a tip by filing the end, as this will probably remove the plating.

The tip of the soldering iron should always be tinned after it has been cleaned. Tinning the tip can be seen in Figure M-6(b) and is achieved by applying a small layer of rosin core solder to the tip. This will protect the

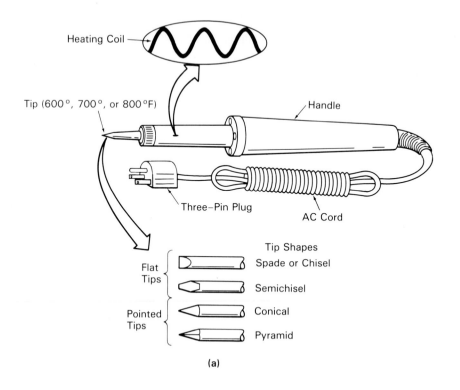

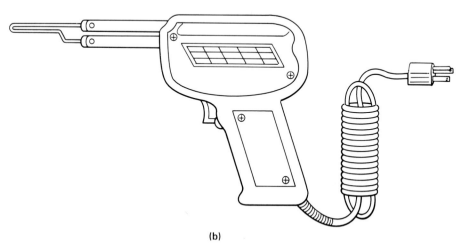

FIGURE M-5 Types of Soldering Irons. (a) Pencil. (b) Gun.

Appendix M   Soldering Tools and Techniques

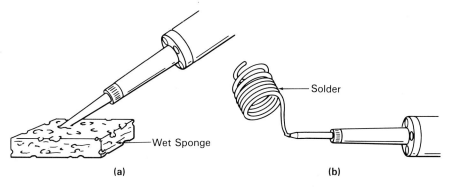

FIGURE M-6 Cleaning a Soldering Iron. (a) Cleaning the tip. (b) Tinning the tip.

tip from oxidation and also increase the amount of heat transfer to the connection.

In summary, it is important to remember that a soldering iron should not be used to melt the solder; its purpose is to heat the connection so that the solder will melt when it makes contact with the connection.

M.2
SOLDERING TECHNIQUES

Before carrying out the five basic soldering steps, wires and components should be prepared. Wires should be wrapped around a terminal to give it a more solid physical support, as seen in Figure M-7(a). Their insulation should be stripped back so as to leave a small gap. Too large a gap will expose the bare wire and possibly cause a short to another terminal, whereas too small a gap will cause the insulation to burn during soldering. Stranded wire should be tinned, as seen in Figure M-7(b), and components should be mounted flat on the board, as seen in Figure M-7(c).

FIGURE M-7 Wire and Component Preparation.

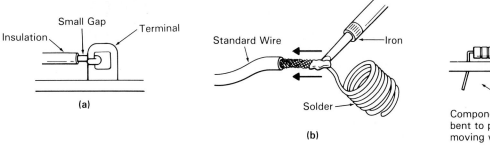

M.2.1 SIX-STEP SOLDERING PROCEDURE

To begin, always remember to wear safety goggles and a protective apron, and then proceed.

STEP 1   Clean the tip on a damp sponge.

STEP 2   Tin the tip if necessary.

STEP 3   Heat the connection.

STEP 4   Apply the solder to the opposite side of the connection.

STEP 5   Leave the tip of the soldering iron on the connection only long enough to melt the solder.

STEP 6   Remove the solder and then the iron; then let the solder cool and solidify undisturbed.

Once this procedure is complete, you should remove the excess leads with the cutters and remove the flux residue, because it collects dust and dirt that could produce electrical leakage paths. When inspecting your work, you should notice the following:

1. The connection is *smooth and shiny*.
2. The solder should *feather out to a thin edge*.

M.2.2 POOR SOLDER CONNECTIONS

1. *Cold solder joint*: This connection has a dull gray appearance; it makes a poor mechanical and electrical connection and is normally caused due to lead movement while the solder was cooling or insufficient heat [Figure M-8(a)]. The cure is to resolder the connection.

2. *Excessive solder joint*: This is caused by too much solder being applied to the connection or by using too large a diameter solder [Figure M-8(b)]. The cure is to solder away the excess with a wick.

3. *Insufficient solder joint*: Not enough solder was applied, and so a poor bond exists between the lead and the terminal [Figure M-8(c)]. The cure is to resolder as if it were a new joint.

FIGURE M-8   Poor Solder Connections.

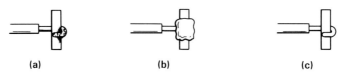

(a)   (b)   (c)

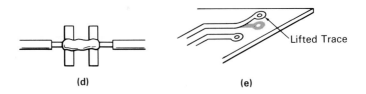

(d)   (e)

4. *Solder bridge*: This occurs when adjacent terminals or traces on printed circuit boards are accidentally connected due to using too much solder [Figure M-8(d)].

5. *Excessive heat*: On printed circuit boards, too much heat can lift the traces or track and therefore ruin the whole board [Figure M-8(e)].

M.3

DESOLDERING

If a faulty component has to be replaced or if a component or wire was soldered in an incorrect place, it will have to be desoldered. Figure M-9 illustrates some of the tools used to desolder. The easiest way to remove a component is to remove all the solder and then disconnect the leads. All three tools seen in Figure M-9 are designed to remove solder.

Use of the braided wick to remove solder can be seen in Figure M-10(a). The braid contains many small wire stands, and when it is placed between the solder and the iron, the melted solder flows into the braid by capillary action. The solder-filled braid is then cut off and discarded.

The desoldering bulb or spring-loaded plunger relies on suction rather than capillary action to remove the solder, as seen in Figure M-10(b). The steps to follow are:

Step 1 Squeeze the desoldering bulb or cock the spring-loaded plunger.

Step 2 Melt the solder with the soldering iron.

Step 3 Remove the iron's tip.

Step 4 Quickly insert the tip of the bulb or plunger into the molten solder and then activate the suction.

When as much of the solder as possible has been removed, use the long-nose pliers to hold the lead and remove the component. It may be necessary to use the iron to slightly heat the lead to loosen up the residual solder so that the component lead can be removed with the pliers. Always be careful not to apply an excessive amount of heat or stress to either the component or board.

FIGURE M-9   **Desoldering Tools.**

Desoldering
Braid

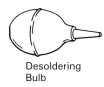

Desoldering
Bulb

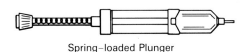

Spring–loaded Plunger

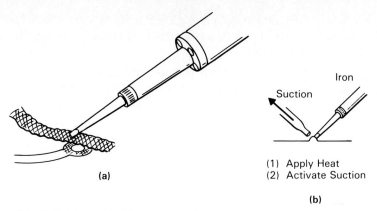

(a)

Suction   Iron

(1) Apply Heat
(2) Activate Suction

(b)

FIGURE M-10 Solder Removal.

M.4
SAFETY

Molten solder, like any other liquid, can splash or spill, and soldering irons are even hotter than the solder. Consequently, you should treat both with respect and always wear protective safety glasses and an apron to protect yourself from burns.

APPENDIX N   Answers to Self-Test Review Questions

STR (§1.1)

1. Current, voltage, resistance, power

2. Resistors, capacitors, inductors

3. Amplifier, oscillator

4. Communications

STR (§1.2)

1. Ohm (Georg S. Ohm)

2. Ampere (André M. Ampère)

3. Volt (Alessandro Volta)

4. Watt (James Watt)

STR (§1.3)

Lasers and robotics

STR (§2.1)

1. Elements are made up of similar atoms; compounds are made up of similar molecules.

2. Protons, neutrons, electrons

3. Copper

4. Like charges repel; unlike charges attract.

STR (§2.2)

1. Amp

2. Current $= Q/t$ (number of coulombs divided by time in seconds).

3. The direction of flow is different: negative to positive is termed electron current flow; positive to negative is known as conventional current flow.

4. (a) Set to highest scale and work down.
   (b) Connect positive lead to positive circuit point and negative lead to negative circuit point.
   (c) Place meter in the path of current flow.

STR (§2.3)

1. Volts

2. 3000 kV.

3. (a) Set to highest scale and work down.
   (b) Connect positive lead to positive circuit point and negative lead to negative circuit point.
   (c) Place meter across component.

4. They are directly proportional.

STR (§2.4)

1. False

2. (a) Electrons are far away from nucleus.
   (b) Many electrons per unit of volume.
   (c) Incomplete valence shell.

3. Copper

4. 28.6 millisiemens

STR (§2.5)

1. True

2. Mica

3. The voltage needed to cause current to flow through a material

4. Small

STR (§§2.6 and 2.7)

1. A material whose characteristics fall between a conductor and an insulator

2. Silicon

3. No

4. A short circuit provides an unintentional path for current to flow. A closed circuit has a complete path for current.

STR (§3.1)

1. It is the opposition to current flow.

2. The difference is in how much or how little current will flow.

3. They are inversely proportional.

4. Small

STR (§3.2)

1. A circuit is said to have a resistance of one ohm when one volt produces a current of one amp.

2. $I = V/R = 24\text{ V}/6\ \Omega = 4$ amps.

3. A memory aid to help remember Ohm's law:

4. Current is proportional to voltage and inversely proportional to resistance.

STR (§3.3)

1. Light, heat, magnetic, chemical, electrical, mechanical

2. When energy is transformed, work is done. Power is the rate at which work is done or energy is transformed.

3. $W = Q \times V, P = V \times I$

4. When 1 kW of power is used in one hour

STR (§3.4)

1. Type of conducting material used, cross-sectional area, length, temperature

2. The resistance of a conductor is proportional to its length and resistivity, and inversely proportional to its cross-sectional area.

3. True

4. True

STR (§4.1)

1. Carbon composition, carbon film, metal film, wirewound, metal oxide, thick film

2. SIPs have one row of connecting pins, while DIPs have two rows of connecting pins.

3. Rheostat has two terminals; potentiometer has three terminals.

4. Linear means that the resistance changes in direct proportion to the amount of change of the input, while a tapered potentiometer varies nonuniformly.

5. True

6. Photoresistor

STR (§4.2)

1. Ohmmeter

2. As $22 \times 1000 = 22,000 = 22 \text{ k}\Omega$

3. As $470 \times 1 = 470 \ \Omega$

4. Short together meter leads and adjust zero ohms control; turn off power and remove part; connect meter leads across component and adjust scale for mid-scale reading and multiply if necessary.

STR (§§4.3 and 4.4)

1. General purpose are $\pm 5\%$ or greater; precision are $\pm 2\%$ or less.

2. $222 \times 0.1, \pm 0.25\% = 22.2 \ \Omega, \pm 0.25\%$

3. Yellow, violet, red, silver

4. False

STR (§4.5)

1. Opens

2. Ohmmeter

STR (§5.1)

1. Load resistance is the device or equipment resistance. Load current is the amount of current drawn from the source by the device or equipment. A dc voltage source is one that will supply a constant polarity output voltage, while a dc current is a unidirectional (one direction) current.

2. False

3. They are inversely proportional.

4. A load that will draw a large amount of current.

STR (§5.2.1)

1. Friction forces electrons out of their orbits to create a difference in potential.

2. Due to the piezoelectric effect, compounds such as quartz will produce a voltage when pressure is applied.

STR (§5.2.2)

1. Thermocouple

2. Used to control a gas shut-off valve as in a water heater.

STR (§5.2.3)

1. Photocell

2. Used as a light meter on a camera to help control exposure

STR (§5.2.4)

1. Lodestone

2. Like poles repel; unlike poles attract.

3. Permanent, temporary, electromagnets

4. DC electric generator

STR (§5.2.5)

1. Negative plate, positive plate, and an electrolyte

2. Carbon–zinc, alkaline–manganese, mercury, silver oxide, lithium

3. Lead-acid, nickel-cadmium

4. False

5. Maximum power transfer is possible when the internal resistance of the battery is equal to the load resistance.

6. No battery is an ideal voltage source and is therefore not 100% efficient.

STR (§5.2.6)	**1.** AC voltage
	2. False
	3. False
	4. More accurate voltage source, easy to vary output, never runs down
STR (§5.3)	**1.** Indicate the maximum current that can flow without blowing the fuse and the maximum voltage value that will not cause an arc.
	2. Fast blow fuses blow almost instantly after an excessive current occurs, whereas a slow blow fuse will have a certain time delay until it blows.
	3. Thermal, magnetic, thermomagnetic
	4. SPST, SPDT, DPST, DPDT, NOPB, NCPB, Rotary, and DIP.

STR (§6.1)	**1.** False
	2. Electron
	3. To determine the direction of the magnetic force
	4. North
STR (§6.2)	**1.** Spiral coil
	2. True
STR (§6.3)	**1.** Number of lines of force (or maxwells) in webers
	2. Number of magnetic lines of flux per square meter
	3. The magnetic pressure that produces the magnetic field
	4. Magnetomotive force divided by length of coil
	5. Opposition or resistance to the establishment of a magnetic field
	6. A measure of how easily a material will allow a magnetic field to be set up within it
STR (§6.4)	**1.** Magnetic-type circuit breaker, relays
	2. An NO relay is open when off; NC is closed when off.
	3. A reed relay has an electromagnet in it, while a reed switch needs an external magnet to operate.
	4. Home security circuits

STR (§7.1)	**1.** A circuit in which current has only one path
	2. 8 A

STR (§7.2)

1. $R_T = R_1 + R_2 + R_3 \ldots$

2. $R_T = R_1 + R_2 + R_3 = 2 \text{ k}\Omega + 3 \text{ k}\Omega + 4700 = 9.7 \text{ k}\Omega$

STR (§7.3)

1. True

2. True

3. $R_T = R_1 + R_2 = 6 + 12 = 18 \ \Omega; I_T = \dfrac{V_S}{R_T} = \dfrac{18}{18} = 1 \text{ A}$

$V_{R1} = 1 \text{ A} \times 6 \ \Omega = 6 \text{ V}$
$V_{R2} = 1 \text{ A} \times 12 \ \Omega = 12 \text{ V}$

4. $V_X = (R_X/R_T) \times V_S$

5. Potentiometer

6. No

STR (§7.4)

1. $P = I \times V$ or $P = V^2/R$ or $P = I^2 \times R$

2. $P = V^2/R = 12^2/12 = 144/12 = 12 \text{ W}$

3. Wirewound, at least 12 W, ideally a 15 W

4. $P_T = P_1 + P_2 = 25 \text{ W} + 3800 \text{ mW} = 28.8 \text{ W}$

STR (§7.5)

1. Component will open, component's value will change, and component will short.

2. No current will flow; source voltage dropped across it.

3. False

4. No voltage drop across component, yet current still flows in circuit; resistance equals zero for component.

STR (§8.1)

1. When two or more components are connected to the same voltage source so that current can branch out over two or more paths

2. False

3. $V_{R1} = V_S = 12 \text{ V}$

4. No

STR (§8.2)

1. The sum of all currents entering a junction is equal to sum of all currents leaving that same junction.

2. $I_2 = I_T - I_1 = 4 \text{ A} - 2.7 \text{ A} = 1.3 \text{ A}$

3. $I_X = (R_T/R_X) \times I_T$

4. $I_T = V_T/R_T = 12/1 \text{ k}\Omega = 12 \text{ mA}; I_1 = 1 \text{ k}\Omega/2 \text{ k}\Omega \times 12 \text{ mA} = 6 \text{ mA}$

STR (§8.3)

1. $R_T = \dfrac{R_1 \times R_2}{R_1 + R_2}$

2. $R_T = \dfrac{1}{(1/R_1) + (1/R_2) + (1/R_3)\ .\ .\ .}$

3. $R_T = \dfrac{\text{common value of resistors}}{\text{number of parallel resistors}}$

4. $R_T = \dfrac{1}{(1/2.7\ \text{k}\Omega) + (1/24\ \text{k}\Omega) + (1/1\text{M}\Omega)} = 2.421\ \text{k}\Omega$

STR (§8.4)

1. True

2. $P_1 = I_1 \times V = 2\ \text{mA} \times 24\ \text{V} = 48\ \text{mW}$

3. $P_T = P_1 + P_2 = 22\ \text{mW} + 6400\ \mu\text{W} = 28.4\ \text{mW}$

4. Yes

STR (§8.5)

1. No current will flow in the open branch; total current will decrease.

2. Maximum current is through shorted branch; total current will increase.

3. Will cause a corresponding opposite change in branch current and total current

4. False

STR (§9.1)

1. By tracing current to see if it has one path or more than one path

2. $R_{1,2} = R_1 + R_2 = 12\ \text{k}\Omega + 12\ \text{k}\Omega = 24\ \text{k}\Omega;\ R_{1,2,3} = \dfrac{R_{1,2} \times R_3}{R_{1,2} + R_3} =$

$\dfrac{24\ \text{k}\Omega \times 6\ \text{k}\Omega}{24\ \text{k}\Omega + 6\ \text{k}\Omega} = \dfrac{144\ \text{k}\Omega}{30\ \text{k}\Omega} = 4.8\ \text{k}\Omega$

3. Find equivalent resistances of series connected resistors; find equivalent resistances of parallel-connected combinations; find equivalent resistances of remaining series-connected resistances.

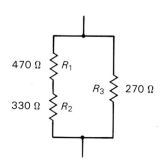

470 Ω R_1

R_3 270 Ω

330 Ω R_2

Figure § 9-1, 4

4. $R_{1,2} = R_1 + R_2 = 470 + 330 = 800\ \Omega$

$R_{1,2,3} = \dfrac{R_{1,2} \times R_3}{R_{1,2} + R_3} = \dfrac{800 \times 270}{800 + 270} = \dfrac{216\ \text{k}\Omega}{1.07\ \text{k}\Omega} = 201.9\ \Omega$

STR (§9.2)

1. Find total resistance; find total current; find voltage drop with $I_T \times R_X$.

2. The voltage drops previously calculated would not change.

STR (§§9.3, 9.4, 9.5)

1. Find total resistance; find total current; find voltage across each series and parallel combination resistors; find current through each branch of parallel resistors; find total and individual power dissipated.

2. This is a do-it-yourself question; each answer will vary.

STR (§9.6)

1. When a load resistance changes the circuit and lowers output voltage

2. Used to check for an unknown resistor's resistance

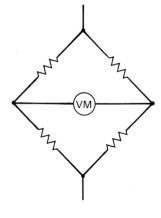

Figure § 9-6, 2

3. *R*

4. Current divider for a digital-to-analog converter

STR (§9.7)

1. Small, large

2. Voltage source

3. Voltage, resistor

4. Current

STR (§9.8)

1. If component is in series with circuit, no current will flow and source voltage will be across bad component. If component is parallel, no current will flow in that branch.

2. Total resistance will decrease and bad component will have 0 volts dropped across it.

3. Will cause the circuit's behavior to vary

STR (§10.1)

1. Ammeter, voltmeter, ohmmeter

2. Analog, digital

STR (§10.1.1)

1. Meter resistance is the resistance of the coil of the armature; sensitivity is the current needed to cause FSD.

2. Shunt resistors

3. False

4. Multiplier resistors

5. Ohms-per-volt rating is the reciprocal of the FSD current of the coil. The larger this value is, the less current it takes to cause FSD, so the more sensitive it is.

6. Infinite (∞)

7. Ammeters and voltmeters get their power from the circuit under test, whereas an ohmmeter has its own battery. They are all used to measure different electrical quantities.

8. Short leads and adjust zero ohm control; ensure no power to component; connect leads across component and read resistance on scale, multiplying by range selected.

STR (§10.1.2)

1. Ease of reading, increased accuracy

2. Lower cost, faster response

3. Display will flash on and off or display OL.

4. 373 Ω

STR (§11.1)

1. (a) Alternating current
   (b) Direct current

2. DC; current only flows in one direction

3. AC

4. DC

5. Power transfer, information transfer

6. DC flows in one direction, whereas AC first flows in one direction and then in the opposite direction.

STR (§11.1.1)

1. AC generators can be larger, less complex, and cheaper to run; transformers can be used with ac to step up/down, so low current power lines can be used; easy to change ac to dc, but hard the other way around.

2. False

3. $P = I^2 \times R$

4. A device that can step up or down ac voltages

5. 120 VAC

6. AC, DC

STR (§11.1.2)

1. The property of a signal or message that conveys something meaningful to the recipient; the transfer of information between two points

2. Sound, electromagnetic, field, electrical

3. (a) Sound wave

    (b) Electromagnetic wave

    (c) Electrical wave

4. 1133 feet per second; 186,000 miles per second

5. Sound, electromagnetic, electrical

6. (a) Microphone

    (b) Speaker

    (c) Antenna

    (d) Human ear

STR (§11.1.3)

1.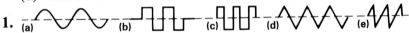

2. Time, frequency

3.

$$\lambda \text{ (mm)} = \frac{344.4 \text{ m/s}}{f \text{ (Hz)}};$$

$$\lambda \text{ (m)} = \frac{3 \times 10^8 \text{ m/s}}{f \text{ (Hz)}};$$

different because sound waves travel at a different speed than do electromagnetic waves.

4. Odd

STR (§11.1.4)

1. Electronic

2. Electrical

STR (§12.1.1)

1. Sensing: equipment that is used to measure some quantity that is in the circuit
Generating: equipment that creates a signal so the response can be observed

2. True; no

3. RF probe is used to measure high-frequency waves and uses a special high frequency rectifier. HV probe is used to measure high voltage and uses multiplier resistors.

4. Pros: easy to read, more accurate. Cons: not as fast as analog in its response.

STR (§12.1.2)

1. To analyze the frequency of a periodic wave at its input jack and display the frequency on its readout

2. Sensing

STR (§12.1.3)

1. CRT

2. Period = 20 μs/div × 4 div = 80 μs; f = 1/period = 1/80 μs = 12.5 kHz

3. Peak = 2 V/cm × 2 div = 4 V; peak to peak = 2 V/cm × 4 cm = 8 V

4. Allows comparisons of phase, amplitude, shape, timing

STR (§12.2)

1. The frequencies they output: AF = 20 Hz to 20 kHz; RF = 3 kHz to 3 GHz

2. AM will vary the amplitude; CW will not.

3. Sine, square, rectangular, triangular, sawtooth

4. No, only electrical waves.

STR (§13.1)

1. False

2. Used to determine the magnetic polarity; placing your left hand fingers in the direction of the current, your thumb points in the direction of the north magnetic pole.

3. AC

4. Number of turns, current, length of coil

STR (§13.2)

1. True

2. The state beyond which an electromagnet is incapable of further magnetic strength

STR (§13.3)

1. The voltage induced or produced in a coil as the magnetic lines of force link with the turns of a coil

2. Faraday's: When the magnetic flux linking a coil is changing, an EMF is induced.
Lenz's: The current induced in a coil due to the change in the magnetic flux is such as to oppose the cause producing it.

3. Sine wave

4. Because it changes sound waves of air pressure (acoustical) to electrical waves (electro)

STR (§14.1)

1. Two plates, dielectric

2. Dielectric

STR (§14.2)

1. True

2. None

STR (§14.3)

1. B

2. Field strength (V/cm) = charge difference (V), volts/distance between plates (d), meters

3. The displacement of an atom's electrons within the dielectric toward the positive plate

4. "Di" because it is between two plates; "electric" because an electric field exists within it

STR (§14.4)

1. The farad

2. Capacitance, C (farads) = charge, Q (coulombs)/voltage, V (volts)

3. $30,000 \times 10^{-6} = 0.03 \times 10^{0} = 0.03$ F

4. $C = Q/V = 17.5/9 = 1.94$ F

STR (§14.5)

1. Plate area, distance between plates, type of dielectric used

2. $C = \dfrac{(8.85 \times 10^{-12}) \times K \times A}{d}$

3. Double

4. Double

STR (§14.6)

1. Glass

2. False

STR (§§14.7 and 14.8)

1. Large, small

2. $C_T = \dfrac{1}{(1/C_1) + (1/C_2) + (1/C_3)} = \dfrac{1}{(1/2\ \mu F) + (1/3\ \mu F) + (1/5\ \mu F)}$
   $= 0.968\ \mu F$

3. $C_T = C_1 + C_2 + C_3 = 7$ pF + 2 pF + 14 pF = 23 pF

4. $V_{CX} = (C_T/C_X) \times V_T$

5. True

STR (§14.9)

1. Mica, ceramic, paper, plastic, electrolytic

2. **(a)** Electrolytic; **(b)** ceramic

3. Air, mica, ceramic, plastic

4. **(a)** Air; **(b)** mica, ceramic, plastic

5. Electrolytic

STR (§14.10)

1. 470 pF, 2% tolerance

2. 0.47 μF or 470 nF, 5% tolerance

STR (§14.11)

1. The time it takes a capacitor to charge to 63.2%

2. 63.2%

3. 36.8%

4. False

STR (§14.12)   **1.** Capacitive reactance is low for ac (short) and high for dc (open)

2. True

STR (§14.13)   **1.** In a pure capacitive circuit, the phase of the voltage and the phase of the current are 90° apart.

2. False

3. False

4. False

STR (§15.1)   **1.** Opposition to current flow without the dissipation of energy

2. $X_C = 1/(2\pi fC)$

3. When frequency or capacitance goes up, there is more charge and discharge current; so X_C is lower.

4. $X_C = 1/(2\pi fC) = 1/(2\pi \times 4 \text{ kHz} \times 4 \text{ }\mu\text{F}) = 9.95 \text{ }\Omega$

STR (§15.2)   **1.** Current leads voltage by some phase angle less than 90%.

2. An arrangement of vectors to illustrate the magnitude and phase relationships between two or more quantities of the same frequency

3. Z = total opposition to current flow; $Z = \sqrt{R^2 + X_C^2}$

4. **(a)** 0°
       (b) 90°
       (c) Between 0° and 90°

STR (§15.3)   **1.** Resistor current is in phase with voltage; capacitor current is 90° out of phase (leading) with voltage.

2. No

3. **(a)** $I_T = \sqrt{I_R^2 + I_C^2}$; **(b)** $Z = V_S/I_T$; also $Z = (R \times X_C)/\sqrt{R^2 + X_C^2}$

4. Lead

STR (§§16.1 and 16.2)   **1.** It's the action of a magnetic field being created from current flowing through a conductor.

2. Electromagnetism; when a magnetic field is caused by current flow. Electromagnetic induction; current flow created from a magnetic field.

STR (§16.3)

1. When the current-carrying coil of a conductor induces a voltage within itself

2. The induced voltage, which opposes the applied emf; $V_{ind} = L \times (\Delta i / \Delta t)$

3. $V_{ind} = L \times (\Delta i / \Delta t) = 2 \text{ mH} \times 4 \text{ kA/s} = 8 \text{ V}$

STR (§16.4)

1. Different applications: An electromagnet is used to generate a magnetic field; an inductor is used to oppose any changes of circuit current.

2. True

STR (§16.5)

1. Number of turns, area of coil, length of coil, core material used

2. $L = \dfrac{N^2 \times A \times \mu}{l}$

STR (§16.6)

1. False

2. (a) $L_T = L_1 + L_2 + L_3 \ldots$
   (b) $L_T = 1/(1/L_1) + (1/L_2) + (1/L_3) \ldots$

3. (a) $L_T = L_1 + L_2 = 4 \text{ mH} + 2 \text{ mH} = 6 \text{ mH}$
   (b) $L_T = \dfrac{L_1 \times L_2}{L_1 + L_2}$ (using product over sum) $= \dfrac{4 \text{ mH} \times 2 \text{ mH}}{4 \text{ mH} + 2 \text{ mH}} = \dfrac{8 \text{ mH}}{6 \text{ mH}} = 1.33 \text{ mH}$

STR (§16.7)

1. Air core, iron core, ferrite core

2. Air core

3. Chemical compound, basically powdered iron oxide and ceramic

4. Toroidal-type inductors have a greater inductance.

5. Ferrite core variable inductor

6. Core permeability

STR (§16.8)

1. Current in an inductive circuit builds up in the same time that voltage does in a capacitive circuit, but the capacitive time constant is proportional to resistance, where the inductive time constant is inversely proportional to resistance.

2. True

3. False

4. It will continuously oppose the alternating current.

STR (§16.9)

1. The opposition to current flow offered by an inductor without the dissipation of energy: $X_L = 2 \times \pi \times f \times L$

2. The larger the frequency (and therefore the faster the change in current) or the larger the inductance of the inductor, then the larger the magnetic field created, the larger the counter emf will be to oppose applied emf; so the inductive reactance (opposition) will be large also.

3. Inductive reactance is an opposition, so it can be used in Ohm's law in place of resistance.

4. False

STR (§16.10)

1. False

2. $V_S = \sqrt{V_R^2 + V_L^2} = \sqrt{4^2 + 2^2} = \sqrt{16 + 4} = \sqrt{20} = 4.47$ V

3. The total opposition to current flow offered by a circuit with both resistance and reactance: $Z = \sqrt{R^2 + X_L^2}$

4. $+ 45°$

5. Power consumption above the zero line, caused by positive current and voltage or negative current and voltage

6. Quality factor of an inductor that is the ratio of the energy stored in the coil by its inductance to the energy dissipated in the coil by the resistance: $Q = X_L/R$

7. True power is energy dissipated and lost by resistance; reactive power is energy consumed and then returned by a reactive device.

8. PF = true power (TP)/apparent power (AP) or PF = R/Z or PF = $\cos \theta$

STR (§16.11)

1. False

2. $I_T = \sqrt{I_R^2 + I_L^2}$

STR (§16.12)

1. **(a)** With an ohmmeter; resistance will be infinite instead of low.
(b) With an inductor analyzer to check its inductance value

2. An open

STR (§16.13)

1. *RL* integrator, *RL* differentiator, *RL* filter

2. The integrator will approximate a dc level; the differentiator will output spikes of current.

3. Can be made to function as either a high- or low-pass filter

STR (§17.1)

1. Mutual inductance is the process by which an inductor induces a voltage in another inductor, whereas self-inductance is the process by which a coil induces a voltage within itself.

2. False

STR (§17.2)	**1.** True
	2. True
	3. Primary and secondary

| STR (§17.3) | **1.** True |
| | **2.** False |

STR (§17.4)

1. It is the ratio of the number of magnetic lines of force that cut the secondary compared to the total number of magnetic flux lines being produced by the primary: k = flux linking secondary coil/total flux produced by primary.

2. True

3. How close together the primary and secondary are to one another and the type of core material used

STR (§17.5)

1. Turns ratio = N_s/N_p = 1608/402 = 4; step up

2. $V_s = N_s/N_p \times V_p$

3. False

4. $I_s = N_p/N_s \times I_p$

5. Turns ratio = $\sqrt{Z_L/Z_S}$ = $\sqrt{75\ \Omega/25\ \Omega}$ = $\sqrt{3}$ = 1.732

6. $V_s = (N_s/N_p) \times V_p$ = (200/112) × 115 = 205.4 V

STR (§17.6)

1. If both primary and secondary are wound in the same direction, output is in phase with input; but if primary and secondary are wound in different directions, the output will be 180° out of phase with the input.

2. When there is a positive on the primary dot, there will be a positive on the secondary dot. If there is a negative on the primary dot, there will be a negative on the secondary dot.

STR (§17.7)

1. Air core, iron or ferrite core

2. Center-tapped secondary, multiple-tapped secondary, multiple winding, single winding

3. False

4. To be able to switch between two primary voltages and obtain the same secondary voltage

5. Automobile ignition system

6. Smaller, cheaper, lighter than normal separated primary/secondary transformer types

Appendix N   Answers to Self-Test Review Questions

STR (§17.8)	**1.** 10 kVA is the apparent power rating, 200 V the maximum primary voltage, 100 V the maximum secondary voltage, at 60 cycles per second.
	2. $R_L = (V_s/I_s = 1$ kV/8 A $= 125\ \Omega$; $125 > 100$, so the transformer will overheat and possibly burn out.
STR (§17.9)	**1.** Copper losses, hysteresis, eddy-current loss, magnetic leakage
	2. Eddy-currents cannot cross the laminated parts of the core, so reduced eddy currents mean reduced opposition to main flux, reducing the loss.

STR (§18.1)

1. Calculate the inductive and capacitive reactance (X_L and X_C), the circuit impedance (Z), the circuit current (I), the component voltage drops (V_R, V_L, and V_C), and the power distribution and power factor (PF)

2. (a) $Z = \sqrt{R^2 + (X_L \sim X_C)^2}$
(b) $I = V_s/Z$
(c) Apparent power $= V_s \times I$ (volt-amps)
(d) $V_S = \sqrt{V_R^2 + (V_L \sim V_C)^2}$
(e) True power $= I^2 \times R$ (watts)
(f) $V_R = I \times R$
(g) $V_L = I \times X_L$
(h) $V_C = I \times X_C$

3. A circuit condition that occurs when the inductive reactance (X_L) and the capacitive reactance (X_C) have been balanced

4. A series *RLC* circuit that at resonance X_L equals X_C, so V_L and V_C will cancel, and $Z = R$

5. Voltage across L and C will measure 0; impedance only equals R; voltage drops across inductor or capacitor can be higher than source voltage

6. Q factor indicates the quality of the series resonant circuit, or is the ratio of the reactance to the resistance

7. Group or band of frequencies that causes the larger current flow

8. BW $= f_0/Q = 12$ kHz/1000 $= 12$ Hz

STR (§18.2)

1. (a) $I_R = V/R$
(b) $I_T = \sqrt{I_R^2 + I_X^2}$

(c) $I_C = V/X_C$

(d) $I_L = V/X_L$

2. In a series resonant RLC circuit, source current is maximum; in a parallel resonant circuit, source current is minimum at resonance.

3. Oscillating effect with continual energy transfer between capacitor and inductor

4. $Q = X_L/R = 50\ \Omega/25\ \Omega = 2$

5. Yes

6. Ability of a tuned circuit to respond to a desired frequency and ignore all others

STR (§18.3)

1. **(a)** High pass
   (b) Low pass
   (c) Band stop
   (d) Band pass

2. Television, radio

STR (§18.4)

1. Real, imaginary

2. Magnitude $= \sqrt{5^2 + 6^2} = 7.81$
   Angle $= \arctan (6/5) = 50.2°$
   Rectangular number $5 + j6 =$ polar number, $7.81 \angle 50.2°$

3. Real number $= 33 \cos 25° = 29.9$
   Imaginary number $= 33 \sin 25° = 13.9$
   Polar number, $33 \angle 25° =$ rectangular number, $29.9 + j13.9$

4. Combination of a real and imaginary number

APPENDIX O   Answers To Odd-Numbered Problems

Chapter 1

1. d	**5.** a	**9.** b	**13.** b
3. d	**7.** c	**11.** c	

(The answers to Essay Questions 14 through 28 can be found in the sections indicated that follow the questions.)

Chapter 2

1. a	**7.** a	**13.** b	**19.** d
3. d	**9.** a	**15.** d	
5. b	**11.** a	**17.** d	

(The answers to Essay Questions 21 through 30 can be found in the sections indicated that follow the questions.)

31. $G = 1/100 = 0.01$ or 10 millisiemens

33. $I = Q/t = 3/4 = 0.75$ amps or 750 milliamps

35. (a) 1.473 V
    (b) 7.143 kV
    (c) 139 V
    (d) 390 kV

37. 24,000 V/70 kV/cm = 0.343 cm

39. (a) 200,000 kV/meter
    (b) 25 kV/mm

Chapter 3

1. a	**5.** d	**9.** a	**13.** c
3. a	**7.** a	**11.** b	**15.** d

Figure Ch. 3, 41(b)

(*The answers to Essay Questions 16 through 40 can be found in the sections indicated that follow the questions.*)

41. (a) $I = V/R = 120\ \text{V}/6\ \Omega = 20$ amps

43. $R = V/I = 120\ \text{V}/10\ \text{A} = 12\ \Omega$

45. $P = I \times V, I = P/V$
   (a) $300/120 = 2.5$ A
   (b) $100/120 = 833$ mA
   (c) $60/120 = 0.5$ A
   (d) $25/120 = 208.3$ mA

47. (a) Volts
   (b) Milliamps
   (c) Kilowatts
   (d) Megaohm

49. 13

51. $I = V/R = 12\ \text{V}/12\ \text{mA} = 1\ \text{k}\Omega$

53. $R = V/I = 120\ \text{V}/500\ \text{mA} = 240\ \Omega$
   $P = I \times V = 500\ \text{mA} \times 120\ \text{V} = 60\ \text{W}$

55. $P = V \times I$
   (a) 120×20 mA $= 2.4$ W
   (b) $12 \times 2 = 24$ W
   (c) $9 \times 100\ \mu\text{A} = 900\ \mu\text{W}$
   (d) 1.5×4 mA $= 6$ mW

57. (a) 1 kW
   (b) 345 mW
   (c) 1.25 MW
   (d) 1250 μW

59. (a) 27,000 kilowatt-hours
   (b) 32,538 kilowatt-hours

Chapter 4

1. b	**7.** c	**13.** c	**19.** d
3. c	**9.** c	**15.** c	
5. a	**11.** b	**17.** b	

(*The answers to Essay Questions 21 through 35 can be found in the sections indicated that follow the questions.*)

37. (a) $P = I^2 \times R = 50 \text{ mA}^2 \times 10 \text{ k}\Omega = 25$ W; (b) no

39. Question 31

(a) $1.2 \text{ M}\Omega \times 0.1 = \pm 120 \text{ k}\Omega = 1.08$ to $1.32 \text{ M}\Omega$

(b) $10 \times 0.05 = \pm 0.5 = 9.5$ to $10.5 \ \Omega$

(c) $27.3 \text{ k}\Omega \times 0.2 = \pm 5.46 \text{ k}\Omega = 21.84$ to $32.76 \text{ k}\Omega$

(d) $273 \text{ k}\Omega \times 0.005 = \pm 1.365 \text{ k}\Omega = 271.635$ to $274.365 \text{ k}\Omega$

Question 32

(a) $33 \times 0.2 = \pm 6.6 = 26.4$ to $39.6 \ \Omega$

(b) $22.5 \text{ k}\Omega \times 0.02 = \pm 450 = 22.05$ to $22.95 \text{ k}\Omega$

(c) $10 \text{ k}\Omega \times 0.05 = \pm 500 = 9.5$ to $10.5 \text{ k}\Omega$

(d) $910 \times 0.1 = \pm 91 = 819 \ \Omega$ to $1.001 \ \Omega$

41. See Section 4-5.

43. In tolerance

45. Calibrate the ohmmeter first, turn off power and remove part, multiply reading by the scale used, remember resistor tolerance, avoid touching conductive parts of probes or component leads.

Chapter 5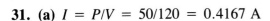

1. b	**7.** b	**13.** a	**19.** a
3. d	**9.** a	**15.** c	
5. c	**11.** d	**17.** b	

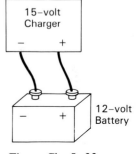

Figure Ch. 5, 33

(The answers to Essay Questions 21 through 30 can be found in the sections indicated that follow the questions.)

31. (a) $I = P/V = 50/120 = 0.4167$ A

33. (a) $150 \text{ Ah} \times 3600 = 54 \times 10^4$ coulombs

(b) $150 \text{ Ah} \div 5 \text{ A} = 30$ hours

35. (a) $3.5 \text{ A} \times 4 \text{ hours} = 14 \text{ Ah}, 14 \text{ Ah} \times 3600 = 50.4 \times 10^3$ As.

(b) $14 \text{ Ah} \div 2 \text{ A} = 7$ hours

Chapter 6

1. a	**7.** b	**13.** d	**19.** a
3. b	**9.** c	**15.** c	
5. c	**11.** b	**17.** a	

(The answers to Essay Questions 21 through 30 can be found in the sections indicated that follow the questions.)

31. $\dfrac{1200 \ \mu Wb}{6.4 \times 10^{-3}} = 0.1875$ teslas

33. $H = (I \times N)/l = (1.2 \times 40)/0.15 = 48/0.15 = 320$ At/m

35. $\mathcal{R} = mmf/\phi$; $mmf = I \times N = 3 \times 36 = 108$
$\mathcal{R} = 108/2.3 \times 10^{-4} = 469.57$K ampere-turns/weber

37. $\mathcal{R} = mmf/\phi = 150/360 \times 10^{-6} = 416.67$ kAt/w

39. $\mu = \mu_r \times \mu_0 = 100{,}000 \times 4\pi \times 10^{-7} = 1.257 \times 10^{-1}$

Chapter 7 _____

1. d	**5.** a	**9.** a	**13.** d
3. c	**7.** d	**11.** c	**15.** d

(The answers to Essay Questions 16 through 25 can be found in the sections indicated that follow the questions.)

27. $I = \dfrac{V_S}{R_T}$, $R_T = R_1 + R_2 = 40 + 35 = 75 \ \Omega$
$I = 24/75 = 320$ mA, $150 \ \Omega$ (double $75 \ \Omega$) needed to halve current

29. $45 \ \Omega$, $15 \ \Omega$, and $30 \ \Omega$

31. $I_{R_1} = I_T = 6.5$ mA

33. $P_T = P_1 + P_2 + P_3 = 120 + 60 + 200 = 380$ W; $I_T = P_T/V_S = 380/120 = 3.17$ A
$V_1 = P_1/I_T = 120 \ W/3.17 \ A = 38$ V.
$V_2 = P_2/I_T = 60 \ W/3.17 \ A = 18.9$ V
$V_3 = P_3/I_T = 200 \ W/3.17 \ A = 63.1$ V

35. **(a)** $I = V/R = 12 \ V/43.7 \ k\Omega = 274.6 \ \mu A$
(b) $R_T = V/I = 12 \ V/10 \ mA = 1.2 \ k\Omega$
$P_T = V \times I = 12 \ V \times 10 \ mA = 120$ mW
(c) $R_T = R_1 + R_2 + R_3 + R_4 = 5 + 10 + 6 + 4 = 25 \ \Omega$;
$V_S = I \times R_T = 100 \ mA \times 25 \ \Omega = 2.5$ V
$V_{R_1} = I \times R_1 = 100 \ mA \times 5 = 500$ mV.
$V_{R_2} = I \times R_2 = 100 \ mA \times 10 = 1$ V.
$V_{R_3} = I \times R_3 = 100 \ mA \times 6 = 600$ mV.
$V_{R_4} = I \times R_4 = 100 \ mA \times 4 = 400$ mV, $P_1 = I \times V_1 = 100$ mA $\times$ 500 mV $= 50$ mW
$P_2 = I \times V_2 = 100 \ mA \times 1 \ V = 100$ mW, $P_3 = I \times V_3 = 100 \ mA \times 600 \ mV = 60$ mW
$P_4 = I \times V_4 = 100 \ mA \times 400 \ mV = 40$ mW

(d) $P_T = P_1 + P_2 + P_3 + P_4 = 12$ mW $+ 7$ mW $+ 16$ mW $+$
3 mW $= 38$ mW
$I = P_T/V_S = 38$ mW/12.5 V $= 3.04$ mA
$R_1 = P_1/I^2 = 12$ mW/(3.04 mA)$^2 = 1.3$ kΩ
$R_2 = P_2/I^2 = 7$ mW/(3.04 mA)$^2 = 757$ Ω
$R_3 = P_3/I^2 = 16$ mW/(3.04 mA)$^2 = 1.7$ kΩ
$R_4 = P_4/I^2 = 3$ mW/(3.04 mA)$^2 = 325$ Ω

37. Zero voltage drop across shorted component, while there is an increase
in voltage across others

39. (a) No current at all (zero)
(b) Go to infinity (∞)
(c) Measure source voltage
(d) No voltage across any other component

Chapter 8

1. b **5.** c **7.** b **9.** a

3. d

*(The answers to Essay Questions 11 through 20 can be found in the
sections indicated that follow the questions.)*

21. $R_T = R/$no. of R's $= 30$ kΩ/4 $= 7.5$ kΩ

23. $R_T = R/$no. of R's $= 25/3 = 8.33$ Ω, $I_T = V_S/R_T = 10/8.33 = 1.2$ A
$I_1 = I_2 = I_3 = R_T/R_X \times I_T = 8.33/25 \times 1.2 = I_T/$no. of R's $= 1.2$
A/3 $= 400$ mA

25. $I_T = V_S/R_T = 14/700 = 20$ mA; $\quad I_X = I_T/$no. of $R = 20$ mA/3 $=$
6.67 mA

27. (a) $R_T = \dfrac{R_1 \times R_2}{R_1 + R_2} = \dfrac{33\,k \times 22\,k}{33\,k + 22\,k} = \dfrac{726\,kΩ}{55\,kΩ} = 13.2$ kΩ
(b) $I_T = V_S/R_T = 20/13.2$ kΩ $= 1.5$ mA
(c) $I_1 = R_T/R_1 \times I_T = (13.2$ kΩ/33 kΩ$) \times 1.5$ mA $= 600$ μA
$I_2 = R_T/R_2 \times I_T = 13.2$ kΩ/22 kΩ $\times 1.5$ mA $= 900$ μA
(d) $P_T = I_T \times V_S = 1.5$ mA $\times 20 = 30$ mW
(e) $P_1 = I_1 \times V_1, V_1 = V_S = 20$ V, 600 μA $\times 20 = 12$ mW
$P_2 = I_2 \times V_2, V_2 = V_S = 20$ V, 900 μA $\times 20 = 18$ mW

29. (a) $R_T = \dfrac{R_1 \times R_2}{R_1 + R_2} = \dfrac{22\,k \times 33\,k}{22\,k + 33\,k} = \dfrac{726\,k}{55\,k} = 13.2$ kΩ, $I_T = \dfrac{V_S}{R_T}$

$= \dfrac{10}{13.2\,k} = 757.6$ μA

$I_1 = \dfrac{R_T}{R_1} \times I_T = \dfrac{13.2\,k}{22\,k} \times 757.6$ μA $= 454.56$ μA

$$I_2 = \frac{R_T}{R_1} \times I_T = \frac{13.2 \text{ k}}{33 \text{ k}} \times 757.6 \text{ μA} = 303.04 \text{ μA}$$

(b) $R_T = \dfrac{1}{(1/R_1) + (1/R_2) + (1/R_3)} = \dfrac{1}{(1/220) + (1/330) + (1/470)} =$ 103 Ω

$I_T = V_S/R_T = 10/103 = 97 \text{ mA}, I_1 = R_T/R_1 \times I_T = 103/220 \times$ 97 mA = 45.4 mA

$I_2 = (R_T/R_2) \times I_T = (103/330) \times 97 \text{ mA} = 30.3 \text{ mA}$

$I_3 = (R_T/R_2) \times I_T = (103/470) \times 97 \text{ mA} = 21.3 \text{ mA}$

31. (a) $G_T = \dfrac{1}{R_1} + \dfrac{1}{R_2} + \dfrac{1}{R_3} = \dfrac{1}{5} + \dfrac{1}{5} + \dfrac{1}{5} = 0.6 \text{ S}, R_T = \dfrac{1}{G} = \dfrac{1}{0.6} =$ 1.67 Ω

(b) $G_T = \dfrac{1}{R_1} + \dfrac{1}{R_2} = \dfrac{1}{200} + \dfrac{1}{200} = 10 \text{ mS}, R_T = \dfrac{1}{G} = \dfrac{1}{10 \text{ mS}} =$ 100 Ω

(c) $G_T = \dfrac{1}{R_1} + \dfrac{1}{R_2} + \dfrac{1}{R_3} = \dfrac{1}{1 \text{ MΩ}} + \dfrac{1}{500 \text{ MΩ}} + \dfrac{1}{3.3 \text{ MΩ}} =$

1.305 μS, $R_T = \dfrac{1}{G} = \dfrac{1}{1.305 \text{ μS}} = 766.3 \text{ kΩ}$

(d) $G_T = \dfrac{1}{R_1} + \dfrac{1}{R_2} + \dfrac{1}{R_3} = \dfrac{1}{5} + \dfrac{1}{3} + \dfrac{1}{2} = 1.033 \text{ S}, R_T = \dfrac{1}{G} =$

$\dfrac{1}{1.033} = 967.7 \text{ mΩ}$

33. (a) $R_T = \dfrac{R_1 \times R_2}{R_1 + R_2} = \dfrac{15 \times 7}{15 + 7} = \dfrac{105}{22} = 4.77 \text{ Ω}$

(b) $R_T = \dfrac{1}{(1/R_1) + (1/R_2) + (1/R_3)} = \dfrac{1}{(1/26) + (1/15) + (1/30)} =$ 7.22 Ω

(c) $R_T = \dfrac{R_1 \times R_2}{R_1 + R_2} = \dfrac{5.6 \text{ kΩ} \times 2.2 \text{ kΩ}}{5.6 \text{ kΩ} + 2.2 \text{ kΩ}} = \dfrac{12.32 \text{ MΩ}}{7.8 \text{ kΩ}} = 1.58 \text{ kΩ}$

(d) $R_T = \dfrac{1}{(1/R_1) + (1/R_2) + (1/R_3) + (1/R_4) + (1/R_5)} =$

$\dfrac{1}{(1/1 \text{ M}) + (1/3 \text{ M}) + (1/4.7 \text{ M}) + (1/10 \text{ M}) + (1/33 \text{ M})} =$ 596.5 kΩ

35. (a) $I_2 = I_T - I_1 - I_3 = 6 \text{ mA} - 2 \text{ mA} - 3.7 \text{ mA} = 300 \text{ μA}$

(b) $I_T = I_1 + I_2 + I_3 = 6 \text{ A} + 4 \text{ A} + 3 \text{ A} = 13 \text{ A}$

(c) $R_T = \dfrac{R_1 \times R_2}{R_1 + R_2} = \dfrac{5.6 \text{ M} \times 3.3 \text{ M}}{5.6 \text{ M} + 3.3 \text{ M}} = \dfrac{18.48 \text{ TΩ}}{8.9 \text{ M}} = 2.08 \text{ MΩ}$

$V_S = I_T \times R_T = 100 \text{ mA} \times 2.08 \text{ M} = 208 \text{ kV}, I_1 = \dfrac{R_T}{R_1} \times I_T$

$$= \frac{2.08 \text{ M}}{5.6 \text{ M}} \times 100 \text{ mA} = 37 \text{ mA}, I_2 = \frac{R_T}{R_2} \times I_T = \frac{2.08 \text{ M}}{3.3 \text{ M}} \times$$
100 mA = 63 mA.

(d) $I_1 = V_{R_1}/R_1 = 2/200 \text{ k}\Omega = 10 \text{ μA}, I_2 = I_T - I_1 = 100 \text{ mA} -$
10 μA = 99.99 mA

$$R_2 = \frac{V_{R_2}}{I_2} = \frac{2}{99.99 \text{ mA}} = 20.002 \text{ }\Omega, P_T = I_T \times V_S = 100 \text{ mA}$$
× 2 V = 200 mW

37. a

39. Total current would increase, and the branch current with the shorted resistor would increase to 20 V/1 Ω = 20 A.

Chapter 9 _____

1. c	**7.** a	**13.** a	**17.** c
3. b	**9.** c	**15.** a	**19.** c
5. c	**11.** b		

(The answers to Essay Questions 21 through 35 can be found in the sections indicated that follow the questions.)

37. $V_A = V_S = 100 \text{ V}, V_B = V_S - V_R = 100 - 11.125 = 88.875 \text{ V}$
$V_C = V_S - V_{R_1} - V_{R_2} = 44.3, V_D = I_{R_4} \times R_4 = 4.43 \text{ mA} \times 2.5$
$\text{k}\Omega = 11.075 \text{ V}, V_e = 0 \text{ V}$

39. (a) $V_{RL} = \frac{R_L}{R_T} \times V_S = \frac{R_L}{R_L + R_{int}} \times V_S = \frac{25}{25 + 15} \times 15 = 9.375$

(b) $V_{RL} = \frac{R_L}{R_T} \times V_S = \frac{R_L}{R_L + R_{int}} \times V_S = \frac{2.5 \text{ k}\Omega}{2.5 \text{ k}\Omega + 15} \times 15 =$
14.91 V

(c) $V_{RL} = \frac{R_L}{R_T} \times V_S = \frac{R_L}{R_L + R_{int}} \times V_S = \frac{2.5 \text{ M}\Omega}{2.5 \text{ M}\Omega + 15} \times 15 =$
14.99991 V

41. (a) For $V_1 : R_T = R_1 + R_{2,3}, R_{2,3} = \frac{2 \text{ k}\Omega \times 6 \text{ k}\Omega}{2 \text{ k}\Omega + 6 \text{ k}\Omega} = \frac{12 \text{ M}\Omega}{8 \text{ k}\Omega} = 1.5$

$\text{k}\Omega, R_T = 8 \text{ k}\Omega + 1.5 \text{ k}\Omega = 9.5 \text{ k}\Omega, I_T = V_S/R_T = \frac{28}{9.5 \text{ k}\Omega} =$

$2.95 \text{ mA}, I_{R_2} = R_{2,3}/R_2 \times I_T = \frac{1.5 \text{ k}\Omega}{2 \text{ k}\Omega} \times 2.95 \text{ mA} = 2.21 \text{ mA}$

For $V_2 : R_T = R_3 + R_{1,2}, R_{1,2} = \frac{8 \text{ k}\Omega \times 2 \text{ k}\Omega}{8 \text{ k}\Omega + 2 \text{ k}\Omega} = \frac{16 \text{ M}\Omega}{10 \text{ k}\Omega} = 1.6$

$\text{k}\Omega, R_T = 6 \text{ k}\Omega + 1.6 \text{ k}\Omega = 7.6 \text{ k}\Omega, I_T = V_S/R_T = 20/7.6 \text{ k}\Omega$
$= 2.63 \text{ mA}, I_{R_2} = R_{1,2}/R_2 \times I_T = 1.6 \text{ k}\Omega/2 \text{ k}\Omega \times 2.63 \text{ mA} =$
$2.1 \text{ mA}, I_{R_2} \text{ total} = 2.21 \text{ mA} + 2.1 \text{ mA} = 4.31 \text{ mA}$

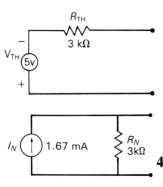

Figure Ch. 9, 43

(b) For V_1 : $R_T = R_1 + R_{2,3} + R_4$, $R_{2,3} = \dfrac{R_2 \times R_3}{R_2 + R_3} = \dfrac{15 \times 75}{15 + 75} = \dfrac{1125}{90} = 12.5\ \Omega$

$R_T = 10 + 12.5 + 5 = 27.5$, $I_T = V_S/R_T = 3.5/27.5 = 127$ mA, $I_{R_3} = (R_{2,3}/R_3) \times I_T = (12.5/75) \times 127$ mA $= 21$ mA

For V_2 : $R_T = R_2 + \dfrac{1}{(1/R_{1,4}) + (1/R_3)}$, $R_{1,4} = R_1 + R_4 = 10\ \Omega$

$+ 5\ \Omega = 15\ \Omega$, $R_T = 15\ \Omega + \dfrac{1}{(1/15\ \Omega) + (1/75\ \Omega)} = 27.5\ \Omega$,

$I_T = V_S/R_T = 1.5$ V/$27.5\ \Omega = 54.5$ mA, $I_{R_3} = (R_{1,3,4}/R_3) \times I_T = (12.5\ \Omega/75\ \Omega) \times 54.5$ mA $= 9$ mA, I_{R_3} total $= 21$ mA $-$ 9 mA $= 12$ mA.

43. $V_{TH} = 5$ V, $R_{TH} = 3$ kΩ

$I_{RL} = V_{TH}/R_T = 5/(3\ \text{k}\Omega + 1\ \text{k}\Omega) = 5/4\ \text{k}\Omega = 1.25$ mA
$I_N = I_{R_2} = V_S/R_2 = 5/3\ \text{k}\Omega = 1.67$ mA $\qquad R_N = 3$ kΩ

45. (a) $V = I \times R = 5$ mA $\times 5$ M$\Omega = 25$ kV
   (b) $V = I \times R = 10$ A $\times 10$ k$\Omega = 100$ kV
   (c) $V = I \times R = 0.0001$ A $\times 2.5$ k$\Omega = 250$ mV

47. This answer will vary with each person.

49. $V_{TH} = \dfrac{R_4}{R_T} \times V_S$, $R_T = R_{1,2} + R_4$, $R_{1,2} = \dfrac{R_1 \times R_2}{R_1 + R_2} = \dfrac{2 \times 3}{2 + 3} = \dfrac{6}{5} = 1.2\ \Omega$

$R_T = 1.2 + 7 = 8.2\ \Omega$, $V_{TH} = \dfrac{7}{8.2} \times 20 = 17.1$ V

$R_{TH} = \dfrac{1}{(1/R_1) + (1/R_2) + (1/R_4)} = \dfrac{1}{(1/2) + (1/3) + (1/7)} = 1.024\ \Omega$

   (a) If R_2 shorts, $V_{TH} = 20$ V, $R_{TH} = 0\ \Omega$

   (b) If R_2 opens, $V_{TH} = \dfrac{R_4}{R_T} \times V_S$, $R_T = R_1 + R_4 = 2 + 7 = 9\ \Omega$,

$V_{TH} = \dfrac{7}{9} \times 20 = 15.56$ V, $R_{TH} = \dfrac{R_1 \times R_4}{R_1 + R_4} = \dfrac{2 \times 7}{2 + 7} = \dfrac{14}{9} = 1.56\ \Omega$.

Chapter 10

1. a	**5.** a	**9.** b	**13.** d
3. c	**7.** a	**11.** d	**15.** c

788

(The answers to Essay Questions 16 through 25 can be found in the sections indicated that follow the questions.)

27. $60,000 \; \Omega/V \times 1000 \; V = 60 \; M\Omega$

29. $I_M = 1/50,000 \; \Omega/V = 20 \; \mu A$
$R = V/I = 10 \; V/20 \; \mu A = 500 \; k\Omega$
$R = V/I = 100 \; V/20 \; \mu A = 5 \; M\Omega$

31. **(a)** $25 \; V \times 20,000 \; \Omega/V = 500 \; k\Omega$
(b) $100 \; V \times 20,000 \; \Omega/V = 2 \; M\Omega$
(c) $1.5 \; V \times 20,000 \; \Omega/V = 30 \; k\Omega$

33. Follow same formulas as Figure 10-17.

35. **(a)** $V_m = I_m \times R_m = 50 \; \mu A \times 1500 \; \Omega = 75 \; mA.$
$I_{RSH_1} = 200 \; \mu A - 50 \; \mu A = 150 \; \mu A$
$R_{SH_1} = 75 \; mV/150 \; \mu A = 500 \; \Omega.$
$I_{RSH_2} = 200 \; mA - 50 \; \mu A = 199.95 \; mA$
$R_{SH_2} = 75 \; mV/199.95 \; mA = 375 \; \Omega.$
$I_{RSH_3} = 2 \; A - 50 \; \mu A = 1.99995 \; A$
$R_{SH_3} = 75 \; mV/1.99995 \; A = 37.5 \; \Omega.$
$I_{RSH_4} = 20 \; A - 50 \; \mu A = 19.99995 \; A$
$R_{SH_4} = 75 \; mV/19.99995 \; A = 3.75 \; \Omega$

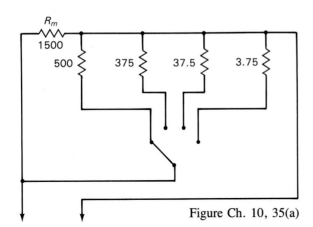

Figure Ch. 10, 35(a)

R_{SH_1} is now 500 Ω, R_{SH_2} is now 375 Ω
R_{SH_3} is now 37.5 Ω, R_{SH_4} is now 3.75 Ω

(b) Decrease the shunt resistor values, and increase the multiplier resistor values appropriately.

Chapter 11

1. c	**7.** b	**13.** b	**19.** d
3. d	**9.** a	**15.** b	
5. b	**11.** d	**17.** c	

(The answers to Essay Questions 21 through 30 can be found in the sections indicated that follow the questions.)

31. Time $= \dfrac{1}{\text{frequency}}$: **(a)** 1/27 kHz $= 37$ μs; **(b)** 1/3.4 MHz $= 294$ ns; **(c)** 1/25 Hz $= 40$ ms; **(d)** 1/365 Hz $= 2.7$ ms; **(e)** 1/60 Hz $= 16.67$ ms; **(f)** $\dfrac{1}{200 \text{ kHz}} = 5$ μs

33. **(a)** RMS $= 0.707 \times$ peak $= 0.707 \times 22$ V $= 15.554$ V
(b) Average $= 0.637 \times$ peak $= 0.637 \times 22$ V $= 14.014$ V
(c) Peak to peak $= 2 \times$ peak $= 2 \times 22$ V $= 44$ V

35. Time $=$ distance/velocity $= 60$ miles/186,000 miles/s $= 322.58$ μs

37. PRT $= 1/\text{PRF} = 1/400$ kHz $= 2.5$ μs

39. 50%

Chapter 12

1. d   **5.** e   **7.** c   **9.** a

3. d

(The answers to Essay Questions 11 through 20 can be found in the sections indicated that follow the questions.)

21. 4.2 kV/100 $= 42$ V

23. $V_{\text{p-p}} = 2 \times V_p$, $V_p = V_{\text{rms}} \times 1.414 = 6 \times 1.414 = 8.484$; $V_{\text{p-p}} = 2 \times 8.484 = 16.968$, 2 V/cm $\times$ 8 would only show 16 V, so 5 V/cm is lowest setting. $t = 1/t = 1/350$ kHz $= 2.857$ μs; 0.2 μs/cm $\times$ 10 would only show 2 μs, so 0.5 μs/cm is lowest setting.

25. $t = 5.5$ cm $\times 1$ μs/cm $= 5.5$ μs; frequency $= 1/t = 1/5.5$ μs $= 181.818$ kHz

Chapter 13

1. c   **5.** d   **9.** b   **13.** d

3. a   **7.** a   **11.** d

(The answers to Essay Questions 15 through 25 can be found in the sections indicated that follow the questions.)

1. b	**7.** d	**13.** d	**19.** c
3. a	**9.** b	**15.** d	
5. c	**11.** a	**17.** b	

(The answers to Essay Questions 21 through 39 can be found in the sections indicated that follow the questions.)

41. Field strength $= V/d = 6$ V/32 μm $= 187.5$ kV/m

43. $C = \dfrac{(8.85 \times 10^{-12}) \times k \times A}{d} =$

$\dfrac{(8.85 \times 10^{-12}) \times 2.5 \times 0.008 \text{ m}^2}{0.00095 \text{ m}} = 186.3$ pF

45. $C_T = \dfrac{1}{(1/C_1) + (1/C_2) + (1/C_3)} =$

$\dfrac{1}{(1/0.025 \ \mu\text{F}) + (1/0.04 \ \mu\text{F}) + (1/0.037 \ \mu\text{F})} = 0.0109 \ \mu\text{F}$

$V_{C1} = \dfrac{C_T}{C_1} \times V_T = \dfrac{0.0109 \ \mu\text{F}}{0.025 \ \mu\text{F}} \times 12 \text{ V} = 5.2 \text{ V}$

$V_{C2} = \dfrac{C_T}{C_2} \times V_T = \dfrac{0.0109 \ \mu\text{F}}{0.04 \ \mu\text{F}} \times 12 \text{ V} = 3.27 \text{ V}$

$V_{C3} = \dfrac{C_T}{C_3} \times V_T = \dfrac{0.0109 \ \mu\text{F}}{0.037 \ \mu\text{F}} \times 12 \text{ V} = 3.53 \text{ V}$

47. **(a)** 47 F
(b) 34×10^{-7} F or 3.4 μF

49. **(a)** $V_{1TC} = 63.2\%$ of $V_S = 0.632 \times 10$ V $= 6.32$ V; 5TC $= 5 \times$ 84 ms $= 420$ ms
(b) $V_{1TC} = 63.2\%$ of $V_S = 0.632 \times 10$ V $= 6.32$ V; 5TC $= 5 \times$ 16.8 ms $= 84$ ms
(c) $V_{1TC} = 63.2\%$ of $V_S = 0.632 \times 10$ V $= 6.32$ V; 5TC $= 5 \times$ 4.08 ms $= 20.4$ ms
(d) $V_{1TC} = 63.2\%$ of $V_S = 0.632 \times 10$ V $= 6.32$ V; 5TC $= 5 \times$ 980 μs $= 4.9$ ms

1. b	**5.** d	**9.** d	**13.** e
3. b	**7.** b	**11.** a	**15.** b

(The answers to Essay Questions 16 through 25 can be found in the sections indicated that follow the questions.)

27. $V_S = \sqrt{V_R^2 + V_C^2} = \sqrt{6^2 + 12^2} = \sqrt{36 + 144} = \sqrt{180} = 13.4\text{V}$

29. (a) $I_R = V/R = 12\text{ V}/4\text{ M}\Omega = 3\ \mu\text{A}$
    (b) $I_C = V/X_C = 12\text{ V}/1.3\text{ k}\Omega = 9.23\text{ mA}$
    (c) $I_T = \sqrt{I_R^2 + I_C^2} = \sqrt{(3\ \mu\text{A})^2 + (9.23\text{ mA})^2} = 9.23\text{ mA}$
    (d) $Z = V/I_T = 12/9.23\text{ mA} = 1.3\text{ k}\Omega$
    (e) $\theta = \arctan(R/X_C) = \arctan(4\text{ M}\Omega/1.3\text{ k}\Omega) = 89.98°$

31. $C = \dfrac{1}{2\pi f X_C} = \dfrac{1}{2\pi\,(20\text{ kHz})\,10\text{ k}\Omega} = 795.8\text{ pF}$

33.

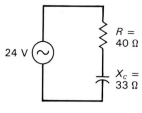

Figure Ch. 15, 33(a)

    (b) $Z = \sqrt{R^2 + X_C^2} = \sqrt{40^2 + 33^2} = 51.9\ \Omega$
    $I = V/Z = 24\text{V}/51.9\ \Omega = 462.4\text{ mA}$
    $V_R = I \times R = 462.4\text{ mA} \times 40\ \Omega = 18.496\text{ V}$
    $V_C = I \times X_C = 462.4\text{ mA} \times 33\ \Omega = 15.2592\text{ V}$
    $I_R = I_C = I_T = 462.4\text{ mA}$
    $\theta = \arctan(X_C/R) = \arctan(33\ \Omega/40\ \Omega)\ 39.5°$

35. (a) $Z = \sqrt{R^2 + X_C^2} = \sqrt{1\text{ M}^2 + 2.5\text{ M}^2} = 2.69\text{ M}\Omega$
    $I = V/Z = 12/2.69\text{ M}\Omega = 4.5\ \mu\text{A}$
    $V_R = I \times R = 4.5\ \mu\text{A} \times 1\text{ M}\Omega = 4.5\text{ V}$
    $V_C = I \times X_C = 4.5\ \mu\text{A} \times 2.5\text{ M}\Omega = 11.25\text{ V}$
    (b) $Z = \sqrt{R^2 + X_C^2} = \sqrt{300^2 + 200^2} = 360.6\ \Omega$
    $I = V/Z = 50\text{ V}/360.6\ \Omega = 138.7\text{ mA}$
    $V_R = I \times R = 138.7\text{ mA} \times 300\ \Omega = 41.61\text{ V}$
    $V_C = I \times R = 138.7\text{ mA} \times 200\ \Omega = 27.74\text{ V}$

37. Lag, 90

39. $X_C = \dfrac{1}{2\pi f c} = \dfrac{1}{2\pi(35\text{ kHz})\,10\ \mu\text{F}} = 455\text{ m}\Omega$
    $Z = \sqrt{R^2 + X_C^2} = \sqrt{100\text{ k}\Omega^2 + 455\text{ m}\Omega^2} = 100\text{ k}\Omega.$
    $I = V/Z = 24\text{V}/100\text{ k}\Omega = 240\ \mu\text{A}$
    True power $= I^2 \times R = 240\ \mu\text{A}^2 \times 100\text{ k}\Omega = 5.76\text{ mVA}$
    Reactive power $= I^2 \times X_C = 240\ \mu\text{A}^2 \times 455\text{ m}\Omega = 26\text{ nVAR}$
    Apparent power $= \sqrt{P_R^2 + P_X^2} = 5.76\text{ mVA}$
    Power factor $= R/Z = 100\text{ k}\Omega/100\text{ k}\Omega = 1$

41. Short: contact from plate to plate
    Open: lead becoming disconnected from its plate
    Breakdown: deterioration of dielectric

43. Capacitor value change, capacitor leakage: leakage current flow between plates.
    Dielectric absorption: will not fully discharge, leaving residual charge.

Equivalent series resistance: resistance of leads, plate connection, electrolyte causing effective circuit capacitance value to change.

45. Shorted, open

Chapter 16

1. c	**9.** a	**17.** a	**25.** a
3. b	**11.** d	**19.** b	
5. a	**13.** e	**21.** b	
7. d	**15.** d	**23.** a	

(The answers to Essay Questions 26 through 35 can be found in the sections indicated that follow the questions.)

37. (a) 22 MΩ
   (b) 78.6 kΩ
   (c) 314.2 kΩ

39. (a) $L_T = L_1 + L_2 + L_3 = 75 \ \mu H + 61 \ \mu H + 50 \ mH$
   $= 50.136 \ mH$
   (b) $L_T = L_1 + L_2 + L_3 = 8 \ mH + 4 \ mH + 22 \ mH = 34 \ mH$

41. (a) 4.36 mH
   (b) 4.8 μH
   (c) 3.82 μH

43. (a) $V_S = \sqrt{V_R^2 + V_L^2} = \sqrt{12^2 + 6^2} = \sqrt{144 + 36} = \sqrt{180}$
   $= 13.4 \ V$
   (b) $I = V_S/2 = 13.4 \ V/14 \ k\Omega = 957.1 \ \mu A$
   (c) $\angle = \arctan V_L/V_R = \arctan 2 = 63.4°.$
   (d) $Q = V_L/V_R = 12/6 = 2.$
   (e) $PF = \cos \theta = 0.448.$

45. $f = X_L/2\pi L = 27 \ k\Omega/2\pi 330 \ \mu H = 13.02 \ MHz$

47. $\tau = \dfrac{L}{R} = \dfrac{400 \ mH}{2 \ k\Omega} = 200 \ \mu s$

   V_L will start at 12 V and then exponentially drop to 0 V

Time	Factor	V_S	V_L
0	1.0	12	12
1 T$_C$	0.365	12	4.416
2 T$_C$	0.135	12	1.62
3 T$_C$	0.05	12	0.6
4 T$_C$	0.018	12	0.216
5 T$_C$	0.007	12	0.084

49. (a) $R_T = R_1 + R_2 = 250 + 700 = 950\ \Omega$

(b) $L_T = L_1 + L_2 = 800\ \mu H + 1200\ \mu H = 2\ mH$

(c) $X_L = 2\pi fL = 2\pi(350\ Hz)\ 2\ mH = 4.4\ \Omega$

(d) $Z = \dfrac{R \times X_L}{\sqrt{R^2 + X_L^2}} = \dfrac{950 \times 4.4}{\sqrt{950^2 + 4.42^2}} = 4.39$

(e) $V_{RT} = V_{LT} = V_S = 20\ V$

(f) $I_{RT} = \dfrac{V_S}{R_T} = \dfrac{20\ V}{950\ \Omega} = 21\ mA,\ I_{LT} = \dfrac{V_{S1}}{L_T} = \dfrac{20\ V}{4.4\ \Omega} = 4.5\ A$

(g) $I_T = \sqrt{I_R^2 + I_L^2} = \sqrt{21\ mA^2 + 4.5\ A^2} = 4.5\ A$

(h) $\theta = \arctan(R/X_L) = \arctan 950/4.4 = 89.7^0$

(i) $TP = I^2 \times R = 21\ mA^2 \times 950 = 418.95\ mW$

$RP = I^2 \times X_L = 4.5\ A^2 \times 4.4 = 89.1$

$AP = \sqrt{TP^2 + RP^2} = \sqrt{418.95\ mW^2 + 89.1\ A^2} = 89.1W$

(j) $PF = IP/AP = 418.95\ mW/89.1\ W = 0.0047$

Chapter 17

1. d   **5.** c   **7.** a   **9.** e

3. c

(The answers to Essay Questions 11 through 25 can be found in the sections indicated that follow the questions.)

27. (a) $V_S = N_S/N_P \times V_P = 24/12 \times 100\ V = 200V$

(b) $V_S = \dfrac{N_S}{N_P} \times V_P = 250/3 \times 100\ V = 8.33\ kV$

(c) $V_S = \dfrac{N_S}{N_P} \times V_P = 5/24 \times 100\ V = 20.83\ V$

(d) $V_S = \dfrac{N_S}{N_P} \times V_P = 120/240 \times 100\ V = 50\ V$

29. Turns ratio $= \sqrt{Z_L/Z_S} = \sqrt{8\ \Omega/24\ \Omega} = \sqrt{1/3} = 0.58$

31. For 16 turns: $V_S = (N_S/N_P) \times V_P = (16/12) \times 24\ V = 32\ V$

For 2 turns: $V_S = (N_S/N_P) \times V_P = (2/12) \times 24\ V = 4\ V$

For 1 turn: $V_S = (N_S/N_P) \times V_P = (1/12) \times 24\ V = 2\ V$

For 4 turns: $V_S = (N_S/N_P) \times V_P = (4/12) \times 24\ V = 8\ V$

33. Follow polarity dots.

35. (a) $I_S = $ apparent power$/V_S = 500\ VA/600\ V = 833.3\ mA$

(b) $R_L = V_S/I_S = 600\ V/833.3\ mA = 720\ \Omega$

1. b	**5.** d	**9.** c	**13.** a
3. b	**7.** c	**11.** a	**15.** d

(The answers to Essay Questions 16 through 30 can be found in the sections indicated that follow the questions.)

31. (a) $X_C = \dfrac{1}{2\pi f C} = \dfrac{1}{2\pi(60)0.02\ \mu F} = 132.6\ k\Omega$

(b) $X_C = \dfrac{1}{2\pi f C} = \dfrac{1}{2\pi(60)18\mu F} = 147.4\ \Omega$

(c) $X_C = \dfrac{1}{2\pi f C} = \dfrac{1}{2\pi(60)360\ pF} = 7.37\ M\Omega$

(d) $X_C = \dfrac{1}{2\pi f C} = \dfrac{1}{2\pi(60)2700\ nF} = 982.4\ \Omega$

(e) $X_L = 2\pi f L = 2\pi(60)4\ mH = 1.5\ \Omega$
(f) $X_L = 2\pi f L = 2\pi(60)8.18\ H = 3.08\ k\Omega$
(g) $X_L = 2\pi f L = 2\pi(60)150\ mH = 56.5\ \Omega$
(h) $X_L = 2\pi f L = 2\pi(60)2\ H = 753.98\ \Omega$

33. (a) $X_L = 2\pi f L = 2\pi(60\ Hz)150\ mH = 56.5\ \Omega$

(b) $X_C = \dfrac{1}{2\pi f c} = \dfrac{1}{2\pi(60\ Hz)20\ \mu F} = 132.6\ \Omega$

(c) $I_R = V/R = 120\ V/270\ \Omega = 444.4\ mA$
(d) $I_L = V/X_L = 120\ V/56.5\ \Omega = 2.12\ A$
(e) $I_C = V/X_C = 120\ V/132.6\ \Omega = 905\ mA$
(f) $I_T = \sqrt{I_R^2 + I_X^2} = \sqrt{(444.4\ mA)^2 + (1.215)^2} = 1.29\ A$
(g) $Z = V/I_T = 120\ V/1.29\ A = 93.02\ \Omega$
(h) Q factor $= X_L/R = 56.5\ \Omega/270\ \Omega = 0.209$

(i) Resonant frequency $= \dfrac{1}{2\pi\sqrt{LC}} = \dfrac{1}{2\pi\sqrt{150\ mH \times 20\ \mu F}} =$
91.89 Hz
(j) Bandwidth $= f_0/Q = 91.89\ Hz/0.209 = 439.7\ Hz$

35. Using a source voltage of 1 volt:
$Z = V/I_T,\ I_T = \sqrt{I_R^2 + I_X^2},\ I_R = V/R = 1/750 = 1.33\ mA$
$I_L = V/X_L = 1/25 = 40\ mA,\ I_C = V/X_C = 1/160 = 6.25\ mA$
$I_X = I_L - I_C = 40\ mA - 6.25\ mA = 33.75\ mA$
$I_T = \sqrt{(1.33\ mA)^2 + (33.75\ mA)^2} = 33.78\ mA$
$Z = 1/33.78\ mA = 29.6\ \Omega$

37. (a) Real number $= 25\cos 37° = 19.97$; imaginary number $= 25$
$\sin 37° = 15,\ 19.97 + j15$
(b) Real number $= 19\cos - 20° = 17.9$; imaginary number $=$
$19\sin -20° = -6.5,\ 17.9 - j6.5$

(c) Real number $= 114 \cos -114° = -46.4$; imaginary number $= 114 \sin -114° = -104.1$, $-46.4 - j104.1$*

(d) Real number $= 59 \cos 99° = +9.2$; imaginary number $= 59 \sin 99° = 58.3$, $+9.2 + j58.3$

39. (a) $(4 + j3) + (3 + j2) = (4 + 3) + (j3 + j2) = 7 + j5$

(b) $(100 - j50) + (12 + j9) = (100 + 12) + (-j50 + j9) = 112 - j41$

41. (a) $Z_T = 73 - j23$, $\sqrt{73^2 + 23^2} = 76.5$, $\angle = \arctan(23/73) = -17.5°$, $76.5 \angle -17.5°$; $Z_T = 76.5\ \Omega$ at -17.5 phase angle

(b) $Z_T = 40 + j15$, $\sqrt{40^2 + 15^2} = 42.7$, $\angle = \arctan(15/40) = 20.6°$, $42.7 \angle 20.6°$; $Z_T = 42.7\ \Omega$ at $20.6°$ phase angle

(c) $Z_T = 8\ \text{k}\Omega - j3\ \text{k}\Omega + j20\ \text{k}\Omega = 8\ \text{k}\Omega + 17\ \text{k}\Omega$, $\sqrt{8\ \text{k}\Omega^2 + 17\ \text{k}\Omega^2} = 18.8\ \text{k}\Omega$, $\angle = \arctan(17\ \text{k}\Omega/8\ \text{k}\Omega) = 64.8°$, $18.8\ \text{k}\Omega \angle 64.8°$; $Z_T = 18.8\ \text{k}\Omega$ at $64.8°$ phase angle

43. (a) $Z_T = 47 - j40 + j30 = 47 - j10$, $\sqrt{47^2 + 10^2} = 48.05$, $\angle = \arctan(-10/47) = -12°$, $Z_T = 48.05 \angle -12°$

(b) $I = \dfrac{V_S}{Z_T} = \dfrac{20 \angle 0°}{48.05 \angle -12} = \dfrac{20}{48.05 \angle 0 - (-12)°} = 416.2\ \text{mA} \angle 12°$

(c) $V_R = I \times R = 416.2\ \text{mA} \angle 12° \times 47 \angle 0 = 416.2\ \text{mA} \times 47 \angle (12 + 0) = 19.56\ \text{V} \angle 12°$

$V_C = I \times X_C = 416.2\ \text{mA} \angle 12° \times 40 \angle -90° = 416.2\ \text{mA} \times 40 \angle (12 - 90) = 16.65 \angle -78°$

$V_L = I \times X_L = 416.2\ \text{mA} \angle 12° \times 30 \angle 90° = 416.2\ \text{mA} \times 30 \angle (12° + 90°) = 12.49\ \text{V} \angle 102°$

(d) V_C lags I by $90°$, V_L leads I by $90°$, V_R is in phase with I.

45. (a) $Z_1 = 0 + j27 - j17 = j10$, $\sqrt{0^2 + 10^2} = 10 \angle = 90°$, $Z_1 = 10 \angle 90°$; $Z_2 = 37 - j20 = 42.06 \angle -28.39°$; Z combined = Product/Sum; Product $= 420.59 \angle 61.61°$; Sum $= 37 - j10 = 38.33 \angle -15.12°$; Z combined $= 10.97 \angle 76.73°$

(b) $I_1 = 3.72 \angle -34.48$ A; $I_2 = 884 \angle 83.91$ mA;

(c) $I_T = 3.39 \angle -21.21°$ A;

(d) $Z_T = 29.52 \angle 21.21°\ \Omega$

47. $V_R = 19.56 \angle 12° = 19.56 \cos 12° + j19.56 \sin 12° =$

$V_C = 16.65 \angle -78° = 16.65 \cos -78° + j16.65 \sin -78° =$

$V_L = 12.49 \angle 102° = 12.49 \cos 102° + j12.49 \sin 102° =$

49. Easier with use of complex numbers

$19.13 + j4.07$

$3.46 - j16.29$

$-2.6 + j12.22$

$\overline{19.99 + j0}$

---

* These examples were for practice purposes only; real numbers are always positive when they represent impedances.

INDEX